# TIMBER
# DESIGNERS'
# MANUAL

# TIMBER DESIGNERS' MANUAL

J. A. BAIRD
E. C. OZELTON

**GRANADA**
London Toronto Sydney New York

Granada Technical Books
Granada Publishing Ltd
8 Grafton Street, London W1X 3LA

First Edition published in Great Britain 1976 by
Crosby Lockwood Staples (original ISBN 0 258 97028 6)
Reprinted 1981, 1982 by Granada Publishing Ltd
Second Edition 1984

*British Library Cataloguing in Publication Data*
Baird, J.A.
Timber designers' manual.–2nd ed.
1. Timber        2. Structural design
3. Building, wooden
I. Title        II. Ozelton, E.C.
624.1'84        TA666

ISBN 0-246-12375-3

Printed and bound in Great Britain by
Mackays of Chatham, Kent

# Contents

# Foreword to 1st Edition (1976)

by J. G. Sunley, MSc, FIStruct. E, FIWSc
Director of the Timber Research and
Development Association
Chairman, Code of Practice on the Structural Use of Timber,
British Standards Institution
Co-ordinator, CIBWI8 Timber Structures
Leader, IUFRO Wood Engineering Group

The first British timber design code, CP 112 'The structural use of timber in buildings', was published in 1952. It is surprising, therefore, that not until now has there been available a comprehensive design manual to assist engineers in their everyday work of designing to CP 112. This manual, which deals in a simple and logical manner with the actual design of timber structures, complements Booth and Reece's *The Structural Use of Timber* which gives background information to the provisions of the current timber design code.

It is essential that the writers of a manual such as this should be professional engineers who have been immersed in the day-to-day problems of a timber design office. Jack Baird and Carl Ozelton have spent a large part of their careers actually designing timber structures with timber engineering and manufacturing companies, during the period in which timber engineering became a recognised and reputable industry. I cannot think of anyone better qualified than these two to have undertaken such a work.

The manual will obviously be of invaluable use to regular timber designers, who will find the advice and the short-cuts which the authors have developed particularly useful in their day-to-day work. It will also be a useful reference work for those designers who spend most of their time on materials other than timber and hence are not fully conversant with the subject. In many design offices only a small proportion of time is spent in designing timber structures and a well laid out reference work is essential if the errors of detail frequently found in timber design are to be avoided.

Others who will find the book most useful will be teachers in timber engineering and officers of approving authorities. With regard to the latter, the manual should help to reduce the considerable amount of time wasted by qualified engineers in argument as to whether or not a particular design accords with the Code of Practice, and hence is acceptable.

The logical layout of chapters with contents covering all aspects of timber design, from beams, columns and glulam to components and frame structures, makes the book very easy to use for everyday reference. In addition, the chapter on timber supplies and grades should help to remove some of the impossible specifications asked for in many contracts.

I have no hesitation in recommending this book to all who have an interest in, or a reason to design, timber structures.                    J. G. S.

# Introduction

When we entered timber engineering we experienced unnecessary difficulties due to the lack of any publication to use for guidance and reference when designing to CP 112 (now re-numbered BS 5268). Quite simply, the object of this manual is to fill that gap. The reception which the 1st Edition received has encouraged us to rewrite and update it now that BS 5268: Part 2 has replaced CP 112:Part 2.

The manual is not an explanation of BS 5268. An engineer designing to BS 5268 will still have to refer to the actual document. However, as one of our objects is to give guidance on practical design, the design examples are in accordance with it. We believe that, if an engineer is designing to another Code or Standard he or she will still find the manual useful and be able to distinguish between those parts which are based on pure theory or experience and those related only to BS 5268.

The manual is written mainly for engineers and structural designers, but much of the content will assist architects and builders in understanding timber components and performance. In this volume we have limited ourselves to the design of the components which constitute the bulk of the work in timber engineering which is mainly beams, columns and trusses.

Our aim has been to cover both overall and detail design. We are not conscious of having omitted any of those points (e.g. shear deflection, partial restraint, stiffener design) which are so often glossed over in design publications. Many of the notes are based on our own experiences in practice, and these should prove particularly valuable to the reader.

We believe parts of this manual to be original: e.g. the sections on lateral stability of partially restrained beams and the deflection coefficients for tapered beams, but even these are presented in a way which the engineer can use directly in calculations, thus saving a considerable amount of design office time. Some of the original work from the 1st Edition of the manual has formed the basis of additions to BS 5268.

Throughout the writing of this manual we have been particularly mindful to provide tables and coefficients which will save the practising engineer many design hours, and we are sure that these aids will prove to be indispensable time-savers.

Codes and Standards are constantly under review, however, we have tried to ensure that all references to them are up to date at the time of going to press.

The system of paragraph numbering is designed for quick reference. We have permitted ourselves some duplication to reduce cross-referencing.

J. A. B.

E. C. O.

# Acknowledgements

The authors wish to express their gratitude to those authorities, associations, companies and individuals who have assisted in the preparation of this manual or who have given permission for information to be extracted from their publications. Particular thanks are due to the Princes Risborough Laboratory of the Building Research Establishment and the British Standards Institution; also to the Council of the Forest Industries of British Columbia, Finnish Plywood International, the Timber Research and Development Association, the American Plywood Association, and the Swedish Finnish Timber Council.

The authors are also grateful for the encouragement given at the time of writing the 1st Edition by Walter Holme and Sons Ltd, Prestoplan Homes Ltd and Träinformation; and to Margaret and Jo Baird for the many hundreds of hours of typing from indecipherable roughs; and to Harry Ashbridge for his great effort on the diagrams.

References to British Standards are made by permission of the British Standards Institution, 2 Park Street, London WIA 2BS, from whom copies of the complete publications can be obtained.

# Notation

## Force and stress

$F$        total force or total load (occasionally $W$ or $P$ has been used)

$p$        UDL

$\sigma$        stress

$\tau$        shear stress

## Subscripts

c        compression

m        bending

t        tension

v        shear

r        rolling stress

p        panel stress

L        long term

M        medium term

S        short term

VS        very short term

## Bending

$\sigma_{m,a,par}$        applied bending stress parallel to grain

$\sigma_{m,g,par}$        grade bending stress parallel to grain

$\sigma_{m,adm,par}$        permissible bending stress parallel to grain

$M$        external bending moment

$\bar{M}$        moment capacity of the section

$M_f$        moment taken by the flanges

$M_w$        moment taken by the web

$M_0$        moment at mid span

Deflection

| | |
|---|---|
| $E$ | modulus of elasticity |
| $E_{mean}$ | mean value of $E$ |
| $E_{min}$ | minimum value of $E$ |
| $E_N$ | statistical minimum value of $E$ for $N$ pieces acting together |
| $N$ | number of pieces or members acting together to support a common load |
| $E_{fN}$ | modulus of elasticity of flanges of $N$ pieces acting together |
| $E_w$ | modulus of elasticity of the web or webs |
| $\delta_t$ | total deflection |
| $\delta_m$ | bending deflection |
| $\delta_v$ | shear deflection |
| $K_m$ | constant related to bending deflection |
| $K_v$ | constant related to shear deflection |
| $G$ | shear modulus (modulus of rigidity) |

Compression

| | |
|---|---|
| $\sigma_{c,a,par}$ | applied compression stress parallel to grain |
| $\sigma_{c,g,par}$ | grade compression stress parallel to grain |
| $\sigma_{c,adm,par}$ | permissible compression stress parallel to grain |
| $\sigma_{c,a,tra}$ | applied compression stress perpendicular to grain |
| $\sigma_{c,g,tra}$ | grade compression stress perpendicular to grain |
| $\sigma_{c,adm,tra}$ | permissible compression stress perpendicular to grain |
| $\sigma_e$ | the Euler critical stress |

Shear

| | |
|---|---|
| $F_v$ | vertical external shear |
| $\overline{F}_v$ | shear capacity |
| $\tau_a$ | applied shear parallel to grain stress |
| $\tau_g$ | grade shear parallel to grain stress |
| $\tau_{adm}$ | permissible shear parallel to grain stress |
| $\tau_r$ | rolling shear stress |
| $\tau_p$ | panel shear stress |
| $G$ | shear modulus (modulus of rigidity) |

Tension

| | |
|---|---|
| $\sigma_{t,a,par}$ | applied tensile stress parallel to grain |
| $\sigma_{t,g,par}$ | grade tensile stress parallel to grain |
| $\sigma_{t,adm,par}$ | permissible tensile stress parallel to grain |

Sectional properties

| | |
|---|---|
| $Z$ | section modulus |
| $Z_X$ | section modulus about $XX$ axis |
| $Z_Y$ | section modulus about $YY$ axis |
| $I$ | second moment of area |
| $I_X$ | second moment of area about $XX$ axis |
| $I_Y$ | second moment of area about $YY$ axis |
| $S_{X,f}$ | first moment of area of flanges about $XX$ axis |
| $S_{X,w}$ | first moment of area of web about $XX$ axis |
| $h$ | depth (width) of beam; greater transverse dimension of a tension or compression member |
| $b$ | breadth of beam ($B$ used for overall breadth of a ply web section) |
| $t$ | thickness; thickness of lamination |
| $A$ | total cross-sectional area |
| $KAR$ | knot area ratio |
| $i$ | radius of gyration |
| $J$ | torsional constant |
| $K_{form}$ | form factor |
| $L_e$ | effective length of column or effective design span of beam |
| $\lambda$ | slenderness ratio (expressed in terms of radius of gyration) |
| $L_e/t$ | slenderness ratio (expressed in terms of thickness of section) |
| $L$ | length; span |
| $L_b$ | laterally unrestrained length of beam |
| $h_e$ | effective depth |
| $d$ | diameter |
| $r$ | radius of curvature |

Other symbols used

| | |
|---|---|
| $\alpha$ | angle between direction of load and angle of grain |
| $\theta$ | angle between the longitudinal axis of a member and a connector axis |
| $\eta$ | eccentricity factor |
| $\omega$ | moisture content |
| $s$ | standard deviation |

Loading coefficients – wind

$S_1$      topography factor
$S_2$      ground roughness, building size and height above ground factor
$S_3$      a statistical factor
$V$      basic wind speed
$V_s$      design wind speed
$q$      dynamic pressure of wind
$C_p$      pressure coefficient
$C_{pe}$      external pressure coefficient
$C_{pi}$      internal pressure coefficient

Greek symbols used

| | | | |
|---|---|---|---|
| $\alpha$ | alpha | $\rho$ | rho |
| $\beta$ | beta | $\Sigma, \sigma$ | sigma |
| $\Delta, \delta$ | delta | $\tau$ | tau |
| $\eta$ | eta | $\phi$ | phi |
| $\theta$ | theta | $\psi$ | psi |
| $\lambda$ | lambda | $\omega$ | omega |
| $\pi$ | pi | | |

Modification factors from BS 5268:Part 2:1984, *The Structural Use of Timber*

$K_1$      modification factor to convert geometrical properties of timber for the dry exposure condition to values for the wet exposure condition

$K_2$      modification factor to convert dry stresses and moduli of timber to stresses and moduli for the wet exposure condition

$K_3$      modification factor for duration of loading
$K_4$      modification factor for bearing stress for length of bearing
$K_5$      modification factor for shear strength at notched end of beams
$K_6$      form factor for circular and square sections loaded on a diagonal
$K_7$      depth factor for bending
$K_8$      modification factor for load sharing including for trimmer joists and lintels

$K_9$      modification factor to obtain $E$ value of composite trimmer joists and lintels

$\left.\begin{array}{l} K_{10} \\ K_{11} \end{array}\right\}$      size factors for certain North American stress grades

$K_{12}$      modification factor for slenderness of compression members
$K_{13}$      modification factor for the effective length of spaced columns
$K_{14}$      width factor for tension

$\left.\begin{array}{l} K_{15} \\ K_{16} \\ K_{17} \\ K_{18} \\ K_{19} \\ K_{20} \end{array}\right\}$  modification factors for single grade glued laminated members and horizontally glued laminated beams

$\left.\begin{array}{l} K_{21} \\ K_{22} \\ K_{23} \\ K_{24} \\ K_{25} \\ K_{26} \end{array}\right\}$  modification factors for combined grade glued laminated members and horizontally glued laminated beams

$\left.\begin{array}{l} K_{27} \\ K_{28} \\ K_{29} \end{array}\right\}$  modification factors for vertically glued laminated beams

$\left.\begin{array}{l} K_{30} \\ K_{31} \\ K_{32} \end{array}\right\}$  modification factors for individually designed glued end joints in horizontally glued laminated members

$\left.\begin{array}{l} K_{33} \\ K_{34} \\ K_{35} \end{array}\right\}$  modification factors for curved glued laminated members

$K_{36}$  modification factor to convert dry stresses and moduli of plywood to stresses and moduli for the wet exposure condition

$K_{37}$  stress concentration factor for certain glue lines between plywood, or tempered hardboard, and timber

$K_{38}$  modification factor for duration of loading on tempered hardboard

$\left.\begin{array}{l} K_{39} \\ K_{40} \\ K_{41} \end{array}\right\}$  modification factors for the effect of magnitude of load on tempered hardboard

$K_{42}$  modification factor for fasteners in certain North American stress grades

$K_{43}$  modification factor for lateral load for nails driven into the end grain of timber

$\left.\begin{array}{l} K_{44} \\ K_{45} \end{array}\right\}$  modification factors for improved nails

$K_{46}$  modification factor for lateral load on nails or coach screws in a steel plate to timber joint

$K_{47}$  modification factor for lateral load on nails in thick plywood

$K_{48}$  modification factor for duration of load in a nailed joint

$K_{49}$  modification factor for moisture content in a nailed joint

$K_{50}$  modification factor for the number of nails in line

$K_{51}$  modification factor for lateral load on screws in thick plywood

$K_{52}$  modification factor for duration of load in a screwed joint

| | |
|---|---|
| $K_{53}$ | modification factor for moisture content in a screwed joint |
| $K_{54}$ | modification factor for the number of screws in line |
| $K_{55}$ | modification factor for duration of load in a bolted joint |
| $K_{56}$ | modification factor for the moisture content in a bolted joint |
| $K_{57}$ | modification factor for the number of bolts in line |
| $K_{58}$ | modification factor for duration of load in a toothed-plate connector joint |
| $K_{59}$ | modification factor for moisture content in a toothed-plate connector joint |
| $K_{60}$ | modification factor for end distance, edge distance and spacing of a toothed-plate connector joint |
| $K_{61}$ | modification factor for the number of toothed-plate connectors in each line |
| $K_{62}$ | modification factor for duration of load in a split-ring connector joint |
| $K_{63}$ | modification factor for moisture content in a split-ring connector joint |
| $K_{64}$ | modification factor for end distance, edge distance and spacing in a split-ring connector joint. |
| $K_{65}$ | modification factor for the number of split-ring connectors in line |
| $K_{66}$ | modification factor for duration of load in a shear-plate connector |
| $K_{67}$ | modification factor for moisture content in a shear-plate connector |
| $K_{68}$ | modification factor for end distance, edge distance and spacing in a shear-plate connector joint |
| $K_{69}$ | modification factor for the number of shear-plate connectors in line |
| $K_{70}$ | modification factor for a glue line assembled with glue/nailing |
| $K_{71}$ | modification factor for the strength test of tempered hardboard structures |
| $K_{72}$ | modification factor for the deflection test of tempered hardboard structures |
| $K_{73}$ | modification factor for acceptance by test of timber or plywood structures |
| $K_S$ $K_C$ $K_D$ | modification factors for end, edge or spacing dimensions of certain connectors |

# Chapter One
# The Materials Used in Timber Engineering

## 1.1   TIMBER

### 1.1.1   General

There is a multiplicity of timber species which can be split into two categories, softwoods and hardwoods. Softwood is the timber of a conifer whereas hardwood is that of a deciduous tree. Some softwoods can in fact be quite hard (e.g. Douglas Fir), some hardwoods soft (e.g. balsa).

This handbook deals almost entirely with design in softwood, because nearly all timber engineering in the UK is carried out ·with softwood. Hardwoods are used however to a certain extent (e.g. in harbour works, farm buildings etc.) and may be used more extensively in future. Chapter 25 deals with aspects which must be considered if hardwood is to be used.

Several countries export softwoods to the UK and a small percentage (about 8% of the import figure) is grown in the UK. About 90% of the import comes from Sweden, Finland, Russia, Canada, Poland, Czechoslovakia and Norway. The import from Sweden, Finland, Russia, Poland, Czechoslovakia and Norway amounts to about 70% of the import figure and is of European Whitewood (*Picea abies*) and European Redwood (*Pinus sylvestris*). Canada supplies:

| | |
|---|---|
| Spruce-Pine-Fir which consists of: | White Spruce (*Picea glauca*) |
| | Engelmann Spruce (*Picea engelmannii*) |
| | Lodgepole Pine (*Pinus contorta*) |
| | Alpine Fir (*Abies lasiocarpa*) |
| (and from the East): | Red Spruce (*Picea rubens*) |
| | Black Spruce (*Picea mariana*) |
| | Jack Pine (*Pinus banksiana*) |
| | Balsam Fir (*Abies balsamea*) |
| | |
| Hemlock (known as Hem-Fir when stress graded) which consists of: | Western Hemlock (*Tsuga heterophylla*) |
| | Amabilis Fir (*Abies amabilis*) |
| | Grand Fir (*Abies grandis*) |

Douglas Fir-Larch which consists of:

Douglas Fir (*Pseudotsuga menziesii*)
Western Larch (*Larix occidentalis*)

Western Red Cedar (*Thuja plicata*)

In addition the UK grows Sitka Spruce (*Picea sitchensis*), which is the most common, Norway Spruce (*Picea abies*), Scots Pine (*Pinus sylvestris*), Douglas Fir (*Pseudotsuga menziesii*) and others. In BS 5268, British-grown Norway Spruce is referred to as European Spruce (British grown).

In recent years the USA has exported Southern Pine surfaced to ALS (American Lumber Standards) (§ 1.3.5). Currently it is not known what quantity will be exported in future to the UK.

For normal structural uses the designer should normally think in terms of using European Whitewood, European Redwood, Canadian Spruce-Pine-Fir or Hem-Fir. Supplies of British-grown Sitka Spruce or Norway Spruce are also available. If a large size is necessary it may be advantageous to use Douglas Fir-Larch. The permissible stresses for these and other species are given in Chapter 2 and, of course, in BS 5268 : Part 2.

European Redwood and Whitewood are allocated the same strength properties, therefore the structural designer can usually consider the two species to be structurally interchangeable, with a bias towards Whitewood for normal structural uses, and Redwood being chosen if a more 'warm' appearance is required. Redwood can be rather more expensive than Whitewood in the higher commercial grades, although the difference is usually less in the stress grades. Redwood is easier to preserve to high levels of preservative retention than Whitewood, but, particularly since organic-solvent treatment processes have become popular, both species can be preserved commercially for many applications. Whitewood and Redwood are readily available up to a size of 75 × 225 mm (see § 1.5.2) and are nearly all dried before being imported to the UK (§ 1.5.5). The majority is kiln dried to around 18–20% moisture content.

Canadian Spruce-Pine-Fir surfaced to CLS (Canadian Lumber Standards) (§ 1.5.7) is usually kiln dried (to 19% m.c.). Sawn sizes of Spruce-Pine-Fir, Hem-Fir and Douglas Fir-Larch, and Hem-Fir surfaced to CLS are usually imported undried ('green'). Canadian timber is available in larger sizes than European timber. The thicker sizes are usually green therefore, particularly if over 75 mm in thickness, and may fissure if left to dry in position. If fissures are unacceptable from an appearance point of view it is better to use a built-up or composite section rather than a solid section of over 75 mm thickness.

## 1.2   COMMERCIAL GRADES AVAILABLE

### 1.2.1   Swedish and Finnish commercial grades

For the Whitewood and Redwood from Sweden and Finland there are six theoretical basic grades numbered I, II, III, IV, V and VI. There are agreed descriptions for these grades, but these are only guiding principles as most well-known mills practise a stricter sorting in their 'usual bracking'. Most have their own mill-grading rules. Figure 1.1 is intended to show typical examples of the different qualities of 100 mm wide sawn boards, and is reproduced by permission of Träinformation, Sweden.

The basic qualities I–IV are usually grouped together for export. Because they are not sorted into separate grades, the grouped grade is sold with the title 'unsorted'. This is traditionally a joinery grade. The V quality is sold separately (or, if a mill has little unsorted it may sell 'fifths and better') and is traditionally a grade for building and construction. The VI quality is sold separately and is traditionally a grade for lower quality uses in building and for packaging.

The mill grading for Whitewood and Redwood from Poland and Norway, and the Whitewood from Czechoslovakia may be considered to be based on similar rules to those of Sweden and Finland.

Further information about Swedish and Finnish Whitewood and Redwood can be obtained from the Swedish Finnish Timber Council.

### 1.2.2   Russian commercial grades

The Russian commercial grades are somewhat similar to those of Sweden and Finland except that they are divided into five basic grades. The basic qualities I–III form the unsorted, and IV and V are similar to the Swedish/Finnish V and VI respectively.

### 1.2.3   Canadian commercial grades

The usual import from Canada of non-stress-graded timber is of:

Spruce-Pine-Fir and Hem-Fir in green sawn sizes up to a maximum quoted green size of 4 inch × 12 inch in a mix of 'No. 1 Merchantable' and 'No. 2 Merchantable' and occasionally in 'No. 3 Common' grade to the 'Export R List Grading and Dressing Rules'.

Douglas Fir-Larch or Hemlock in 'Door Stock' to the NLGA rules in grade mixes known as 'Factory Select', 'No. 1 Shop (Door Stock)', 'No. 2 Shop (Door Stock)' or 'No. 3 Shop (Door Stock)', imported green in sawn sizes up to a maximum quoted green size of 2 inch × 8 inch.

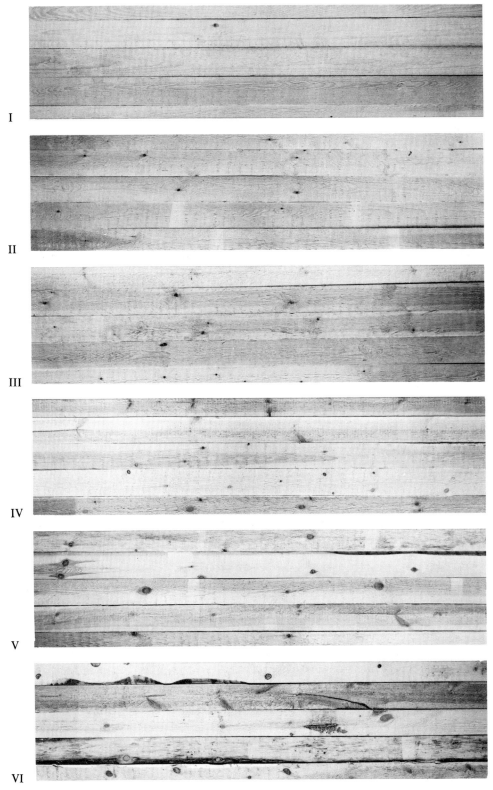

Fig. 1.1: *Typical examples of basic qualities, Swedish boards 100 mm wide.*

Douglas Fir-Larch, Hemlock or Western Red Cedar to the 'Export R List' in 'No. 2 Clear and Better', 'No. 3 Clear' and 'No. 4 Clear', imported green in sawn sizes up to a maximum quoted green size of 4 inch × 12 inch.

Western Red Cedar 'Shop Lumber' in green sawn sizes up to a maximum quoted green size of 2 inch × 12 inch.

Hemlock, Douglas Fir-Larch and Western Red Cedar in green sawn sizes up to a maximum quoted green size of 4 inch × 12 inch.

'Beams and Stringers' to the 'Export R List' up to a maximum quoted green size of 6 inch × 16 inch in Hemlock, Douglas Fir-Larch and Western Red Cedar.

'Posts and Timbers' to the 'Export R List' up to a maximum quoted green size of 16 inch × 16 inch (8 inch × 8 inch in Spruce-Pine-Fir) in Hemlock, Douglas Fir-Larch, Western Red Cedar or Spruce-Pine-Fir.

Further information on Canadian timber can be obtained from the Council of the Forest Industries of British Columbia and the Canadian High Commission.

Canadian timber is often costed on a price per 'Standard'. This refers to a 'Petrograd Standard' which equals 4.671 m³.

### 1.2.4 Commercial rules for British-grown timber

British Standard BS 3819:1964 gave the 'Grading rules for sawn home-grown softwood'. There were four basic grades, HGI, HGII, HGIII and HGIV, with the addition of a grade HGI Clear. These rules applied to British-grown Sitka Spruce, Scots Pine and Douglas Fir. However, British sawmills did not generally grade to BS 3819 and it has been withdrawn. At present the general practice is for mills to produce 'run of the mill' or grade for uses (e.g. structural use, pallets, fencing, furniture).

## 1.3 STRESS GRADES AVAILABLE

### 1.3.1 General

When CP 112:Part 1:1967 was written, stress grades were described in an appendix. The term 'basic stresses' was introduced to define stresses for clear timber with no defects. The permitted defects for various grades were then described and the 'basic stresses' were reduced to 'grade stresses'. The grades were described as percentages of the basic strength. Thus a 75 grade was permitted to have defects which were considered to have strength-reducing characteristics of 25% and the main permissible stresses were 75% of the basic stresses. Grades 75, 65, 50 and 40 plus a composite 50/40 grade were described. Experience has shown that too many grades were described and also that the limitations placed on edge knots were too severe and led to unnecessarily high reject rates.

Later experience and knowledge of the strength of timber has been incorporated into stress grades described in BS 4978:1973, 'Timber grades for structural use', and the description of grading rules has been deleted from the Code of Practice. The stress grades in BS 4978 are described in § 1.3.2. A stress grade for hardwood is described in BS 5756:1980, 'Tropical hardwoods graded for structural use', BS 4978 is being revised.

As well as giving strength values for softwoods stress graded to BS 4978, BS 5268:Part 2 also gives strength values for Canadian timber stress graded to the 'National Grading Rules for Dimension Lumber' (NLGA), and for USA timber stress graded to the 'National Grading Rules for Softwood Dimension Lumber' (NGRDL) (§ 1.3.5).

Strength values are not currently given in BS 5268:Part 2 for the Nordic T-Timber stress grades T18, T24, T30 or Ö-Virke. Many Swedish and Finnish mills are authorised to stress grade at source, visually or by machine, to the BS 4978 rules.

The latest knowledge on the strength of softwood has been incorporated into European Standard, 'ECE recommended standard for stress grading of coniferous sawn timber 1982' (§ 1.3.6).

'Basic stresses' have been dropped in BS 5268:Part 2. Machine stress grades M50 and M75 are retained, although visual versions of 75, 60, 50 etc. have been deleted.

### 1.3.2   BS 4978 stress grades

BS 4978:1973 describes two primary grades for visual stress grading which can also be selected by machine, and also makes provision for grades (e.g. M50 and M75) to be selected by machine only. In addition, three glulam grades are described for visual selection.

The two primary grades are General Structural and Special Structural, which can be described by the abbreviations GS and SS when visually graded and MGS and MSS when machine graded. The glulam grades are LA, LB and LC. (It is so difficult to develop an end joint in LA laminates that LA can usually be disregarded.)

BS 4978 requires that every piece of graded timber shall be marked to identify the grade and the company or machine which graded the piece. The mark has to appear at least once on an edge, face or end. This clause may be tightened in future to call for face marking. Where a mark is eliminated by processing or cross cutting the processing company is permitted to re-mark the piece, but must add a mark to identify the company, and the letter R to denote that the piece has been re-marked.

Providing that any processing does not remove more than the amount tabulated for 'Constructional timber' in Table 3 of BS 4471:Part 1:1978 (see Table 1.9) the grade is deemed not to have been changed. If a graded piece is resawn in cross-section then each piece must be regraded and marked to BS 4978 if it is to be used structurally. If a graded piece is cut in length then the grade of each piece is not reduced. This is because BS 4978 grading is a full-length grading. The stress grading can be carried out in the country of origin or the country of import.

The quality assurance of timber machine stress graded in the UK or abroad to the BS 4978 stress grades is currently administered under the BSI Kitemark scheme and the timber is marked with the Kitemark, the licence number of the machine, BS 4978, as well as the stress grade mark (and usually a mark to identify the grading company). Currently this scheme is operated in Sweden and Finland as well as the UK. BS 4978 does permit quality assurance of machine stress grades by another approved authority, but so far, all timber machine stress graded to BS 4978 has been Kitemarked. Currently third-party certification for visual stress grading to BS 4978 is voluntary. In the UK, visual graders can be trained and certified in the TRADAMARK scheme run by the Timber Research and Development Association. Sweden and Finland have Stress Grading and Management Committees which are responsible for training and certifying companies to grade visually to the GS and SS rules. Further information, lists of approved companies etc. can be obtained from the Swedish Finnish Timber Council.

### 1.3.3   Stress-graded Whitewood and Redwood

European Whitewood and Redwood is available stress graded to the BS 4978 stress grades GS, MGS, M50, SS, MSS and M75. General Structural is rather a low grade for machine stress grading therefore may be satisfied more by visual than machine stress grading. The stress grading can be carried out in the UK or in the country of origin. In future, Whitewood and Redwood stress graded to the ECE rules S6 and S8 (§ 1.3.6) may be available in the UK. BS 5268 : Part 2 states that S6 and S8 may be considered to be inter-changeable with General Structural and Special Structural respectively.

### 1.3.4   Stress-graded Canadian timber

Canadian timber can be stress graded to the BS 4978 stress grades but timber stress graded in Canada to the NLGA rules is imported to the UK. It is convenient to divide this into three categories, which cover the majority of the supply:

(a) Canadian Lumber Standards (CLS) surfaced on four sides available in Spruce-Pine-Fir, Hem-Fir, and occasionally in Douglas Fir-Larch in finished sizes, when dry, of 38 × 140, 38 × 184, 38 × 235 and 38 × 285 mm. These four sizes have been available in a mix of 'No. 1 and No. 2 Structural Joist and Plank'. 'Select Structural Joist and Plank' may be available in future particularly because BS 5268:Part 2 allocates the same stress values to 'No. 1 and No. 2 Structural'. The Spruce-Pine-Fir is imported kiln dried, the other species green. CLS is often referred to by 'nominal' sizes (e.g. 2 in × 6 in for 38 × 140 mm) and costed as such.

(b) CLS surfaced on four sides available in Spruce-Pine-Fir, Hem-Fir, and occasionally in Douglas Fir-Larch in finished sizes when dry of 38 × 38, 38 × 63 and 38 × 89 mm. These three sizes have been available in a mix of 'No. 1 and No. 2 Structural Light Framing'; or a mix of 'Construction Light Framing' and 'Standard Light Framing'. This pattern of supply may alter in future because BS 5268:Part 2 allocates the same stresses to 'No. 1 and No. 2 Structural' and rather low stresses to 'Construction'. The Spruce-Pine-Fir is imported kiln dried, the other species green.

(c) Sawn sizes are available in Spruce-Pine-Fir and Hem-Fir (perhaps occasionally in Douglas Fir-Larch) in a stress-graded mix of 'No. 1 and No. 2 Structural Joist and Plank'. This pattern of supply may alter in future because BS 5268:Part 2 allocates the same stresses to 'No. 1 and No. 2 Structural'. The Spruce-Pine-Fir and Hem-Fir is usually imported green but some SPF is imported kiln dried. The sizes available are usually in quoted green thicknesses of $1\frac{7}{8}''$ in quoted green widths from $4''$ to $12''$. See §1.5.1 on permitted tolerances. Assuming minus tolerances as Paragraph 747 of the NLGA rules and a 2% drying reduction to 20% moisture content, the approximate minimum dry sizes are 44 × 95, 44 × 145, 44 × 193, 44 × 243 and 44 × 292 mm.

SPF is the usual abbreviation for Spruce-Pine-Fir.

Canadian stress-graded timber is usually priced on the 'nominal' or quoted size rather than the actual dry size and is often priced per 'Standard'. This refers to a 'Petrograd Standard' which equals 4.671 m³.

No machine stress-graded timber has been imported from Canada but Canadian mills are said to be interested in shipping this in future. The stress grades will be different to those detailed above. Also, there has been some experimental visual grading to the GS and SS rules in Canada.

### 1.3.5   Stress-graded USA timber

Although the USA is still a net importer of timber (mainly from Canada) some consignments of USA stress-graded timber have been imported to the UK and permissible stresses are tabulated in BS 5268:Part 2. The import is of Southern Pine in American Lumber Standards sizes (ALS which are the same as CLS) stress graded to the NGRDL rules. It is too early to be certain what the pattern and volume of supply will be. Grade descriptions for visual stress grades are identical to the Canadian NLGA rules but permissible stresses are not all identical even when the species are the same. No machine stress-graded timber has been imported from the USA but American mills are said to be interested in shipping this in future. The stress grades will be different to the visual stress grades.

### 1.3.6   European stress grades

During recent years, based on latest knowledge on the strength and grading of softwoods, a serious attempt has been made to agree common European stress grades which could be used by exporting and importing countries. Three stress grades have been agreed and have been given the designations, S6, S8 and S10. They can be graded visually or by machine. The descriptions appear in 'ECE recommended standard for stress grading of coniferous sawn timber 1982'. The data on which the ECE Standard was based is largely the same as that on which BS 4978 was based.

BS 5268:Part 2 states that S6 and S8 can be used interchangeably with General Structural and Special Structural respectively. At present S10 is set slightly too low to be interchangeable with M75 with all European White-wood and Redwood and discussion has been reopened on this point.

### 1.3.7   Strength Classes

In an attempt to deal with the multiplicity of stress grade/species combinations, BS 5268:Part 2 introduces nine Strength Classes having grade bending stresses parallel to grain of 2.8, 4.1, 5.3, 7.5, 10.0, 12.5, 15.0, 17.5 and 20.5 N/mm² for SC1–SC9 respectively. It should be realised, however, that Strength Classes do not have to be used in designs to match BS 5268: Part 2. Designs can still be produced using stress grade/species combinations, although fixings must be designed as for a Strength Class.

Although some softwood designs may be at the SC2 level, most softwood designs will be at the SC3, SC4 and SC5 levels which, for example, align with European Whitewood GS, SS and M75 respectively.

If a Strength Class is being satisfied by visual stress grading it will always be necessary for the timber to be graded to a stress grade even though it may be given double marking (e.g. GS and SC3). However, if the timber is being machine stress graded it is possible to set the machine to a Strength Class level and mark accordingly. When machine stress grading, the company has the choice of whether or not to double mark with the stress grade and Strength Class.

Although it may be advantageous for a structural designer to design to a Strength Class this can bring complications. For example the amenability to preservation differs for many species within one Strength Class; the nailability of species may differ (e.g. in the case of dry Hem-Fir); and even the permissible load on fasteners. Some low stress grades of North American timber, although tabled in BS 5268:Part 2 as satisfying a Strength Class, are not allocated any strength in tension therefore would have to be excluded in specifying for a tension member. Another complication and possible pitfall is the different sizes and tolerances of North American and European timbers which can be used in the same Strength Class (see § 1.5.1).

Therefore, when specifying using a particular Strength Class it is necessary to exclude any species which is not acceptable from the point of view of size of timber or a material property. If the structural designer knows the species which will be used and/or which he or she wishes to be used it may be preferable to base the specification on this.

## 1.4   BS 1186 JOINERY GRADING

There will be cases where a designer will not have to call for a stress grade yet will not be happy to specify a commercial grade or leave the grade unspecified. In these cases he or she may consider specifying one of the grades described in BS 1186:Part 1:1984, 'Quality of timber in manufactured joinery'. Four classes are described: Class 1, Class 2 and Class 3 plus Class CSH (which is intended to cover 'Clear' grades of softwoods and clear hardwoods). (Publication of the 1984 version of this Standard may be delayed.)

Class 1 (and CSH) are the top grades but most ordinary applications other than top-class joinery can be satisfied by Class 2 or Class 3. BS 1186 distinguishes between surfaces which are exposed, concealed or semi-concealed. An example of a semi-concealed surface is one which is only exposed when a window is open. BS 1186:Part 2:1971 covers workmanship and is being revised.

It is not the intention here to attempt to cover the two parts of BS 1186 but Fig. 1.2, 'Limits of knot size for round or oval knots' is based on Fig. 1 of BS 1186:Part 1, to indicate the level of the various classes.

## 1.5   SIZES AVAILABLE

### 1.5.1   General

The sizes tabulated in § 1.5.2 are largely based on those internationally agreed and mainly as presented in BS 4471:Part 1 and elsewhere. They can be assumed to be available in commercial qualities of European timber, but they will certainly not all be available stress graded. For stress-graded sizes reference should be made to §§ 1.5.3 and 1.5.7. Smaller sizes can be obtained by resawing any of the tabulated sizes and in these cases a reduction of at least 2.5 mm should be allowed for the saw thickness. A 100 mm size, if split equally into two, would normally be considered to yield two 48 mm sizes.

The term 'basic size' refers to the sawn size at a moisture content of 20% to which the limited minus (and plus) tolerances of BS 4471 apply.

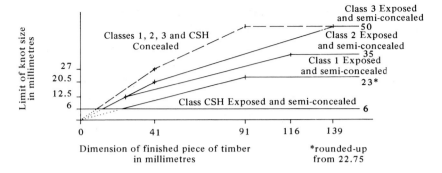

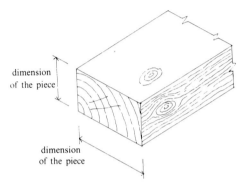

(a) Dimension of the piece; rectangular section

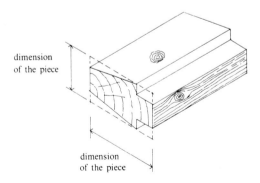

(b) Dimension of the piece; moulded section

Fig. 1.2: *Limits of knot size for round or oval knots. BS 1186 Classes.*

$$\text{Moisture content \%} \ = \ \frac{\text{weight of moisture in the timber}}{\text{weight of oven-dried timber}} \times 100\%$$

If timber is dried to a low moisture content (e.g. 10–12%) it must be realised that the timber will be smaller than the basic size (see § 1.5.4).

Clause 8 of BS 5268:Part 2:1984 makes it clear that when sawn or processed sizes of softwood are in accordance with BS 4471, the geometrical properties given in Appendix D of BS 5268:Part 2 may be used in designs. (The same comment applies to sizes of hardwood in accordance with BS 5450 and Appendix E of BS 5268:Part 2.) Therefore the limited minus tolerances on sawn sizes given in BS 4471:Part 1 (see § 1.5.4) need not be deducted from 'basic' sawn sizes in preparing a structural design to BS 5268: Part 2.

However, BS 5268:Part 2:1984 is not specific in stating what size to use in designs when the minus tolerances are larger than those given in BS 4471 (other than the reference in Clause 5.9 to the use of 'actual' cross-section). Therefore there is uncertainty about knowing what size to take in designs using sawn timber of North American origin, particularly as North American sawn sizes are mostly imported 'green'. (There is no uncertainty about the dry surfaced CLS/ALS sizes which comply with BS 4471 – see § 1.5.7.)

'Occasional' (i.e. 10%) pieces of consignments of sawn timber graded to the NLGA rules are permitted to have the full sawing tolerances, which are given in Para. 747 of the NLGA rules as:

$1''$ nominal $-\frac{1}{16}''$ under or $\frac{1}{8}''$ over
$2''$ nominal $-\frac{1}{8}''$ under or $\frac{1}{4}''$ over
$3''$ to $7''$ $\quad -\frac{3}{16}''$ under or $\frac{3}{8}''$ over
$8''$ or larger $-\frac{1}{4}''$ under or $\frac{1}{2}''$ over

and, in addition, for green timber drying to around 20% m.c. has to be taken into account. Similar tolerances (sometimes tighter, sometimes more generous, depending on grade) appear in Para. 11 of the Export R List grading rules. Tolerances to Para. 156 of the NLGA rules are tighter. See Para. 726 of the NLGA rules for an explanation of 'occasional'. See §§ 1.5.6 and 1.5.7 for the links between tolerances and grades.

Therefore, the designer is faced with the choice of working to sizes based on the largest permitted minimum tolerances, or having a certain amount of selection carried out, and/or using a variation on the concept of regularizing as defined in BS 4471 (see § 1.6.2).

The approximate dry (20% m.c.) minimum sizes of sawn timber given in §§ 1.5.6 and 1.5.7 are based on the largest permitted minus tolerances and a 2% drying reduction (taken from Clause 5.4 of BS 4471:Part 1, assuming a 10% reduction in moisture content from 'green' to 20% m.c.). However, the designer may decide on other assumptions or action regarding sawn North American sizes.

### 1.5.2 Basic sawn sizes of Whitewood and Redwood available in commercial qualities

These sizes are given in Table 1.1 which is largely as presented in BS 4471 : Part 1.

**Table 1.1**

| Width (mm) → Thickness (mm) ↓ | 75 | 100 | 125 | 150 | 175 | 200 | 225 | 250 | 275 | 300 |
|---|---|---|---|---|---|---|---|---|---|---|
| 16 | X | X | X | | | | | | | |
| 19 | X | X | X | | | | | | | |
| 22† | X | X | X | X | X | X | | | | |
| 25‡ | | X | X | X | X | X | | | | |
| 32 | X | X | X | X | X | X | | | | |
| 38 | | X | X | X | X | X | X | X | X | X |
| 44 | | X | X | X | | | | | | |
| 47* | X | X | X | X | X | X | X | | | |
| 50 | X | X | X | X | X | X | X | X | X | X |
| 63 | | X | X | X | X | X | X | | | |
| 75 | | X | X | X | X | X | X | X | X | X |
| 100 | | X | | | X | X | | | | |

(Diagram labels: width, edge, thickness, face. "— rather difficult to obtain" indicates the dashed region of widths 250, 275, 300 etc.)

Thicknesses of 36 mm may be available.
\* This range of widths for 47 mm thickness will usually be found to be available in constructional quality only.
† Usually Whitewood thicknesses.
‡ Usually Redwood thicknesses.

### 1.5.3 Basic sawn sizes of Whitewood and Redwood available stress graded

Basic sawn sizes are given in Table 1.2 but other sizes, including widths of 250, 275 and 300 mm may be made available by special arrangement. The size 38 × 225 mm is available often as a scaffold board grade.

Some Nordic mills will stress grade sizes the same as CLS (§ 1.5.7).

**Table 1.2**

| Width (mm) → Thickness (mm) ↓ | 75 | 100 | 125 | 150 | 175 | 200 | 225 |
|---|---|---|---|---|---|---|---|
| 38 | X | X | X | X | X | X | X |
| 47 | X | X | X | X | X | X | X |
| 50 | X | X | X | X | X | X | X |
| 63 | | | | X | X | X | X |
| 75 | | | | X | X | X | X |

### 1.5.4 Tolerances on Whitewood and Redwood sawn sizes

Basic sawn sizes are measured as at 20% m.c. and may have tolerances according to BS 4471 of minus 1 mm, plus 3 mm on sizes up to 100 mm, or minus 2 mm, plus 6 mm on sizes over 100 mm. The minus tolerances are permitted on only 10% of the pieces in a parcel as imported, and may be disregarded in structural designs.

BS 4471 (Clause 5.4) states that actual sizes with higher than 20% m.c. (up to 30% m.c.) shall be greater by 1% for every 5% of moisture content and may be smaller by 1% for every 5% of moisture content below 20%. The geometrical properties of sawn sections used for structural design are given in BS 5268:Part 2 on the basis of the dry basic sizes which have limited minus tolerances.

### 1.5.5   Moisture content of Whitewood and Redwood

Of the Whitewood and Redwood shipped from Sweden and Finland approximately 90% is kiln dried to an 'average' of 18% m.c. (plus 4% minus 2%). The remainder is air dried to 22% m.c. or less before being shipped.

Of the Whitewood and Redwood imported from Russia, Poland, Czechoslovakia and Norway about a third is kiln dried to a similar level to the Swedish and Finnish production. Much of the remainder is air dried to 22% m.c. or less, although some deliveries can be a few percent higher. The amount which is kilned is expected to increase.

### 1.5.6   Sizes and tolerances of Canadian timber available in commercial qualities

Canadian mills are still sawing to Imperial sizes. When considering the size of Canadian timber available in commercial qualities (i.e. not stress graded) it is necessary to link grade to size and tolerances because both differ with grade. The approximate minimum sizes tabulated in §1.5.6 for 20% m.c. take account of the maximum permitted minus tolerances (see §1.5.1).

A drying reduction of 2% in size is taken below in arriving at sizes at 20% m.c. This is based on assuming the green size to be at about 30% m.c. and allowing the drying estimate from Clause 5.4 of BS 4471 (i.e. 1% in size for every change of 5% m.c.). The sizes given below are limited to the 'Merchantable' grades in the 'R List', the most common sizes of 'Door Stock', 'Clear' grades of Douglas Fir-Larch or Hemlock, and 'Large-sized Timber': 'Beams and Stringers' and 'Posts and Timbers'.

*'Merchantable' grades in the 'R List' (see §1.2.3)*
Available in SPF and Hemlock imported green in the sizes shown in Table 1.3 with tolerances to Para. 11 of the Canadian 'Export R List'. In addition to the tabulated sizes, widths of 5, 7 and 9 inch are sometimes available.

**Table 1.3**

| Quoted green size (inch) | Approximate minimum green size (inch) deducting Para. 11 minus tolerances | Approximate minimum size at 20% m.c. (mm) (see § 1.5.1) |
|---|---|---|
| $1\frac{7}{8}$ X 4 | $1\frac{3}{4}$ X $3\frac{7}{8}$ | 44 X  96 |
| 6 | $5\frac{7}{8}$ | 146 |
| 8 | $7\frac{3}{4}$ | 193 |
| 10 | $9\frac{3}{4}$ | 243 |
| 12 | $11\frac{3}{4}$ | 292 |
| 3 X 4 | $2\frac{7}{8}$ X $3\frac{7}{8}$ | 72 X  96 |
| 6 | $5\frac{7}{8}$ | 146 |
| 8 | $7\frac{3}{4}$ | 193 |
| 10 | $9\frac{3}{4}$ | 243 |
| 12 | $11\frac{3}{4}$ | 292 |
| 4 X 4 | $3\frac{7}{8}$ X $3\frac{7}{8}$ | 96 X  96 |
| 6 | $5\frac{7}{8}$ | 146 |
| 8 | $7\frac{3}{4}$ | 193 |
| 10 | $9\frac{3}{4}$ | 243 |
| 12 | $11\frac{3}{4}$ | 292 |

*'Door Stock'* (*see* § *1.2.3*)
Available in Douglas Fir-Larch and Hemlock imported green, mostly in the sizes shown in Table 1.4, with tolerances to Para. 156 of the NLGA rules.

**Table 1.4**

| Quoted green size (inch) | Approximate minimum size at 20% m.c. (mm) (see § 1.5.1) |
|---|---|
| 2 X 4 | 50 X  96 |
| 2 X 6 | 50 X 146 |
| 2 X 8 | 50 X 193 |

*'Clear' Grades* (*see* § *1.2.3*)
Available in Douglas Fir-Larch and Hemlock (also Western Red Cedar in a larger range of sizes) imported green in the sizes shown in Table 1.5 with tolerances to Para. 11 of the 'Export R List'.

**Table 1.5**

| Quoted green size (inch) | Approximate minimum green size (inch) deducting Para. 11 minus tolerances | Approximate minimum size at 20% m.c. (mm) (see §1.5.1) |
|---|---|---|
| 2 × 4 | $1\frac{15}{16} \times 3\frac{7}{8}$ | 48 × 96 |
| 6 | $5\frac{7}{8}$ | 146 |
| 8 | $7\frac{7}{8}$ | 196 |
| 10 | $9\frac{7}{8}$ | 246 |
| 12 | $11\frac{7}{8}$ | 296 |
| 3 × 4 | $2\frac{7}{8} \times 3\frac{7}{8}$ | 72 × 96 |
| 6 | $5\frac{7}{8}$ | 146 |
| 8 | $7\frac{7}{8}$ | 196 |
| 10 | $9\frac{7}{8}$ | 246 |
| 12 | $11\frac{7}{8}$ | 296 |
| 4 × 4 | $3\frac{7}{8} \times 3\frac{7}{8}$ | 96 × 96 |
| 6 | $5\frac{7}{8}$ | 146 |
| 8 | $7\frac{7}{8}$ | 196 |
| 10 | $9\frac{7}{8}$ | 246 |
| 12 | $11\frac{7}{8}$ | 296 |

*'Large-sized Timber'*

The title 'Large-sized Timber' covers 'Beams and Stringers' and 'Posts and Timbers' graded to the NLGA rules. It is possible to obtain such timber stress graded but import to the UK is usually of non-stress-graded sections. The quality is a mixture of the Merchantable grades. Import is of green sizes in Hemlock, Douglas Fir-Larch, Western Red Cedar and SPF (up to a maximum of 8 inch × 8 inch in the last case).

'Beams and Stringers' refers to sizes which are more than 2 inch 'off-square'. The most common quoted size is 6 inch × 12 inch, with 16 inch being usually the largest imported width. Minus tolerances vary from $\frac{1}{8}$ inch to $\frac{1}{4}$ inch.

'Posts and Timbers' refers to sizes which are not more than 2 inch 'off-square'. The largest quoted size is usually 16 inch × 16 inch (or 8 inch × 8 inch in SPF). The smallest quoted size is 5 inch × 5 inch. Common sizes in Douglas Fir-Larch and Hemlock are 12 × 12, 10 × 12, 10 × 10 and 8 × 10 inch.

**1.5.7   Sizes and tolerances of Canadian timber available stress graded at source**

Canadian timber is available stress graded at source to the NLGA rules both in sawn sizes, and in sizes surfaced on four sides and with 'eased' corners (rounded not more than 3 mm radius). The surfaced sizes are known as CLS (Canadian Lumber Standards). Dry sizes are given in Appendix A to BS 4471 for the CLS sizes. See § 1.3.4 for grades and species available. Although the rounded corners reduce the extreme fibre width to 32 mm, this is usually disregarded in structural designs. The CLS sizes available are shown in Table 1.6.

**Table 1.6**

| Nominal size (inch) | Dry minimum size (mm) |
|---|---|
| 2 X 2 | 38 X 38 |
| 2 X 3 | 38 X 63 |
| 2 X 4 | 38 X 89 |
| 2 X 6 | 38 X 140 |
| 2 X 8 | 38 X 184 |
| 2 X 10 | 38 X 235 |
| 2 X 12* | 38 X 285 |

\* Not normally imported to the UK.

No minus tolerance is permitted by BS 4471 on the dry minimum sizes. Plus tolerance is not limited (but is usually small).

Although some SPF sawn sizes are imported kiln dried, it is more normal for green sawn sizes to be available with tolerances to Para. 747 of the NLGA rules. The minus tolerances are taken into account in arriving at the dry minimum sizes shown in Table 1.7. As in § 1.5.6 a 2% drying reduction is taken into account in arriving at sizes at 20% m.c.

**Table 1.7**

| Quoted green size* (inch) | Approximate minimum green size (inch) deducting Para. 747 minus tolerances | Approximate minimum size at 20% m.c. (mm) (see § 1.5.1) |
|---|---|---|
| $1\frac{7}{8}$ X 4 | $1\frac{3}{4}$ X $3\frac{13}{16}$ | 44 X 95 |
| 6 | $5\frac{13}{16}$ | 145 |
| 8 | $7\frac{3}{4}$ | 193 |
| 10 | $9\frac{3}{4}$ | 243 |
| 12† | $11\frac{3}{4}$ | 292 |

\* Occasionally quoted thicknesses of $1\frac{3}{4}$ inch green may be available in the UK.
† Not normally imported to the UK.

### 1.5.8  Sizes and tolerances of USA timber stress graded at source

At present the pattern of supply to the UK of USA stress-graded timber has not been established but it seems that ALS sizes (same as CLS) are those which are currently being delivered in Southern Pine.

## 1.5.9   Sizes of British-grown softwood

The British sawmilling industry tends to work to the sizes tabulated in BS 4471 but is often willing to cut to customers' specifications rather than hold stocks of British-grown softwood.

## 1.5.10   British-grown softwood stress graded to BS 4978

British-grown softwood (usually the spruces) is being stress graded to BS 4978 generally to the sizes indicated in § 1.5.3. Permissible stresses are given in Table 2.2 from which it can be seen that M75 British-grown Sitka Spruce has approximately the same strength as GS European Whitewood.

## 1.5.11   Tolerances on British-grown sizes

Although sawn British sizes are often sold green, they are usually sawn oversize so that when the section dries to about 20% it is the size as tabulated in BS 4471.

## 1.5.12   Moisture content of British-grown timber

British-grown softwood is often sold green but a few of the mills or merchants sell British-grown timber kiln dried or air dried to around 22% m.c. or less.

## 1.5.13   Lengths of imported Whitewood and Redwood

The agreed European standard lengths as tabulated in BS 4471 and elsewhere are shown in Table 1.8.

**Table 1.8**

| Basic lengths (m) | | | | | |
|---|---|---|---|---|---|
| 1.80 | 2.10 | 3.00 | 4.20 | 5.10 | 6.00 |
| | 2.40 | 3.30 | 4.50 | 5.40 | 6.30 |
| | 2.70 | 3.60 | 4.80 | 5.70 | |
| | | 3.90 | | | |

No minus tolerance is permitted.
Overlength is not limited.

It may be difficult to obtain lengths of 6.30 m, although some mills store longer logs (up to 8.00 or 9.00 m) and can supply long lengths by special agreement. Finger jointing or other jointing methods permit a designer to obtain lengths restricted only by transport or handling limitations. The finger jointing can be carried out in the UK or abroad (usually in Sweden or Finland, which have control organisations).

### 1.5.14    Lengths of Canadian timber

Lengths are 8 ft to 24 ft in 2 ft increments in the commercial and stress grades and, in addition, some of the larger commercially graded pieces are available in lengths up to 40 ft.

### 1.5.15    Lengths of British-grown softwood

These lengths can be assumed to be the same as those for imported Whitewood or Redwood.

## 1.6    PROCESSING

### 1.6.1    General

For many applications in building, the surface and tolerances of timber as originally sawn from the log will be adequate without any further processing. There are however cases in which processed surfaces or closer tolerances are desirable or essential and the types available are described below. When size is important it is preferable to specify the actual size required, preferably working to BS 4471 processing allowances (e.g. 47 × 97 regularized from 47 × 100), rather than state, for example, *ex.* 47 × 100 mm without giving the final actual size. (At the same time it is as well to remember that a millimetre is a fairly small unit.)

### 1.6.2    Regularizing

Regularizing is normally applicable only to structural members such as joists or studs. BS 4471 : Part 1 : 1978 defines regularizing as a 'process by means of which every piece of a batch of constructional timber is sawn and/or machined to a uniform width'. (Note, only on one dimension.)

Regularizing, as described in BS 4471, is intended to make the width more uniform, but there is no reason why, for a special process, the specifier or manufacturer should not make the thickness more uniform as well. Indeed there are cases, such as trussed rafter material or stud material where the thickness is made uniform by skimming-off 1 mm. This is similar to the idea behind 'precision timber' as described in BS 4471 : Part 1 : 1969 which was deleted in the 1978 version.

Regularizing can be carried out by processing on one edge or both edges by machining or sawing. The surface produced may not be suitable for gluing. As covered by BS 4471, regularizing is required to reduce width by 3 mm on widths up to and including 150 mm, and 5 mm for widths over 150 mm.

Regularized timber of a stress grade or commercial quality may have the defects permitted within the grade. This includes wane unless 'no wane' is specified.

BS 4471 does not specify that regularizing must be carried out within certain moisture contents but it is sensible to specify that it should be carried out at or below 20–22% moisture content particularly as the tolerance permitted on the regularized size is only plus or minus 1 mm. If a section is regularized only in the width, the other dimension (i.e. thickness) is still permitted to have the tolerances of a sawn size.

### 1.6.3   Surfaced constructional timber

Whereas regularizing can be carried out by processing on only one edge, planing or surfacing requires processing on two opposed faces or edges. When calling for a section to be planed or surfaced it is normal for it to be processed on both dimensions. BS 4471 does not distinguish between surfacing and planing in the main body of the Standard (as it did in the 1969 version) although, in Appendix A, reference is made to Canadian surfaced constructional timber. Despite this change it is useful to think of the normal processing of constructional timber as being more a surfacing than the type of finish one expects with planed joinery timber. Table 3 of BS 4471 : Part 1 : 1978 gives reductions appropriate to a number of purposes and these are given in Table 1.9. The permissible deviations on finished sizes are plus or minus 0.5 mm.

**Table 1.9  Reductions from basic size to finished size by planing of two opposed faces**

(All dimensions are in millimetres)

| | Reduction from basic size for sawn sizes of width or thickness | | | |
|---|---|---|---|---|
| Purpose | 15 up to and including 35 | Over 35 up to and including 100 | Over 100 up to and including 150 | Over 150 |
| Constructional timber | 3 | 3 | 5 | 6 |
| Matching*; interlocking boards | 4 | 4 | 6 | 6 |
| Wood trim not specified in BS 584 | 5 | 7 | 7 | 9 |
| Joinery and cabinet work | 7 | 9 | 11 | 13 |

* The reduction of width is overall the extreme size and is exclusive of any reduction of the face by the machining of a tongue or lap joint.

### 1.6.4   Planed all round

Certain structural members may be used also in the joinery sense (e.g. curtain walling either withstanding wind only or also carrying vertical loading) and may have to be planed all round or even moulded. The allowances for reductions from basic size are given in Table 1.9.

### 1.6.5   Resawing

When a piece of timber is resawn in cross-section, this will almost certainly change the grade whether it be a commercial or stress grade. It may be adequate to use the resawn commercial pieces without regrading if they are used in a non-structural application, but it will certainly be essential for the pieces to be regarded to BS 4978, 'Timber grades for structural use', if they are to be used in a structural application.

During resawing the saw will remove approximately 2.5 mm. Not more than 2 mm reduction of size of each piece produced by resawing shall be allowed (BS 4471). BS 4471 requires sellers who offer resawn timber to describe it as 'resawn *ex.* larger'. Thus if a 50 × 100 mm section is resawn equally into two sections each piece can be described as 50 × 50 mm resawn *ex.* larger. Alternatively, each piece can be described as 50 × 48 mm (i.e. the new actual size).

### 1.7   MOISTURE CONTENT

Moisture content is given as a percentage:

$$\text{Moisture content} = \frac{\text{weight of moisture in the timber}}{\text{weight of oven-dried timber}} \times 100\%$$

The two most usual methods of measuring the moisture content of solid timber are use of a portable moisture meter, or oven drying of small samples cut from the item being checked. Obviously the second method is destructive therefore is usually used only in the case of test units or dispute. Although sometimes said to be more accurate than a moisture meter it is as well to realise that the result is an average value for the sample (which is why the sample must be small). Providing a moisture meter is calibrated correctly and used correctly it is usually accurate enough for most checking required in a factory or site. If insulated probes are used it is possible to plot the moisture gradient through a section.

Moisture meters measure electrical resistance, therefore readings are affected by whatever is added between the tips of the probes. Normal organic-solvent preservatives are said not to influence readings to any significant extent, but readings are affected by water-borne preservatives and fire-retardants. The manufacturer of the meter may be able to give a correction factor. If the moisture content of a board (e.g. plywood) is required, the manufacturer of the meter should be consulted. (See §24.8 for further reading on moisture content.)

## 1.8   SPECIFYING

### 1.8.1   Specifying for structural uses

One needs to specify:

(a) Stress grade and species; or Strength Class with a limit on the species permitted if this is relevant (§ 1.3.7).
(b) Finished size, processing and tolerance. If no processing is specified a sawn surface will normally be assumed by the supplier. Unless there is a special reason the tolerances of BS 4471:Part 1 will apply. If specifying stress grading to BS 4978 it is not essential to refer to BS 4471 tolerances because compliance with BS 4978 requires compliance with BS 4471.
(c) Length – or refer to drawings.
(d) Moisture content – where important.
(e) Preservation – where necessary. (See Chapter 24 for guidance.)

*Examples*

Preamble. Timber to be European Whitewood or Redwood stress graded and marked to BS 4978, dried to 22% moisture content or less.

   47 X 150 sawn X 5.10 m GS grade
   38 X 97 X 4.80 m regularized from 38 X 100 SS grade

Preamble. Joists to be Canadian SPF stress graded No. 2 Structural Joist and Plank, dried to 22% m.c. or less before installation.

   38 X 184 CLS X 14 ft 0 in
   38 X 235 CLS X 16 ft 0 in

Preamble. Timber to be European Redwood stress graded and marked to BS 4978, treated by double-vacuum organic-solvent, dried to 22% m.c. or less.

   MSS grade   38 X 122 X 5.10 m regularized from 38 X 125
   MSS grade   38 X 195 X 4.80 m regularized from 38 X 200
   GS grade    38 X  97 X 2.70 m regularized from 38 X 100
   GS grade    38 X 100 X 5.10 m sawn

Preamble. Structural timber to be to Strength Class SC4 in any European species of softwood, tolerances to BS 4471.

   38 X 150 X 3.60 m

### 1.8.2 Specifying for uses other than structural

One needs to specify:

(a) Species, if this is important.

(b) Finished size, surface and tolerance. If no surface finish is specified a sawn surface will normally be assumed by the supplier. Unless there is a special reason the tolerances of BS 4471:Part 1 will apply.

(c) Length – or refer to drawings.

(d) Grade – If an appearance grade is important it may be relevant to refer to one of the Classes of BS 1186:Part 1 (see § 1.4). If grade is not particularly important it may be adequate to refer to a commercial grade (see § 1.2) remembering that resawing invalidates the original grading; or indeed simply to state what is important (e.g. straight, no wane, no decay).

(e) Moisture content – where important.

(f) Preservation – where necessary. (See Chapter 24 for guidance.)

*Examples*

Preamble. Timber to be European Whitewood Vth quality, tolerances to BS 4471, dried to 22% moisture content or less.

32 × 200 × 4.20 m
32 × 195 × 4.20 m regularized from 32 × 200
43 × 191 finished planed all round × 1.80 m

Preamble. Timber to be European Redwood to Class 3 of BS 1186, tolerances to BS 4471. Exposed and concealed surfaces as indicated on Drawings . . . Preservation to be by organic-solvent double-vacuum process to Table 2 of BS 5589, desired service life 60 years.

## 1.9  PLYWOOD

### 1.9.1  General

Several countries produce plywood either from softwood or hardwood logs or a mixture of both, however, very few plywoods are considered suitable for structural use. As a general rule only plywood bonded with an exterior-quality resin adhesive should be used for structural design.

As far as the UK is concerned, the principal structural plywoods are:

Canadian Douglas Fir-faced plywood commonly referred to as Canadian Douglas Fir plywood.
Canadian Softwood plywood
Finnish Birch plywood (birch throughout)
Finnish Birch-faced plywood

Finnish Conifer plywood
American Construction and Industrial plywood
Swedish Softwood plywood

§§1.9.2 – 1.9.5 give outline details of the sizes and qualities of these plywoods and the name of the organisations from which further details can be obtained. Permissible stresses are tabulated in BS 5268:Part 2. Before specifying a particular plywood it is as well to check on its availability. The permissive stresses are at the 15% moisture content level as in BS 5268:Part 2.

When designing with plywood, the designer should be careful to distinguish between the 'nominal' thickness and the actual thickness. Designs to BS 5268:Part 2 must use the 'minimum' thickness which is based on the assumption that all veneers have minus tolerance deducted. The designer should also check to see whether sanded or unsanded sheets will be supplied as this can make a difference of several percent to the thickness and permissible stresses.

In reading various design manuals the designer should be aware that there are at least two methods of expressing permissible stresses. One is to express the stresses assuming that the whole area of the plywood has equal strength. This is the 'full area' method. Another is to express the stresses as though the veneers which run perpendicular to the direction of stress have no strength at all. This is the 'parallel ply method'. Yet another method is the 'layered' approach in which the perpendicular veneers are assumed to have some strength but much less than the parallel veneers. If used correctly any method should give similar results, but the designer must be careful to use the correct geometrical properties (either full area or parallel veneers only) with the stated permissible stress. BS 5268:Part 2 is written on the full area method which is also the method used in this manual. The permissible stresses differ with thickness and quality.

If appearance is important it is prudent to check whether the description of the face veneer relates to both outer veneers or only to one face, the other outer veneer being referred to as the back veneer.

When a plywood is referred to as an exterior grade of plywood this reference is to the durability of the adhesive and not to the durability of the species used in the plywood construction.

If plywood is to be glued and used as part of an engineered component (e.g. a ply web beam or stress skin panel) the designer should satisfy himself as to the suitability of the face quality and integrity of the chosen plywood. Some qualities are really intended mainly for wall or floor sheathing rather than glued components.

### 1.9.2 Available sizes and quality of Canadian plywood

Canada has not yet adopted metric measure therefore the size of plywood sheets usually available is 2440 × 1220 mm (8 ft × 4 ft). The face veneer runs parallel to the longer side (Fig. 1.3). 2400 × 1200 mm sheets are available if a sufficiently large order is placed.

Fig. 1.3: *Canadian Fir-faced plywood.*

Two basic types of Canadian exterior grade plywood are available in the UK. These are a plywood with one or both outer veneers of Douglas Fir and inner veneers of other species which is known commercially as Douglas Fir plywood, and an exterior Canadian Softwood plywood which is made from much the same species as Douglas Fir plywood but without a Douglas Fir face veneer. Appendix B of BS 5268:Part 2 lists the species which are used.

The face quality grades available for both plywoods are:

'Sheathing Grade' with Grade C veneers throughout, unsanded.

'Select Grade' with Grade B face veneer and Grade C inner and back veneers, unsanded.

'Select Tight Face Grade' with Grade B 'filled' face veneer and Grade C inner and back veneers, unsanded.

'Good One Side Grade' with Grade A face veneer and Grade C inner and back veneers, sanded. (The Douglas Fir version of this grade is permitted to have only the face veneer of Douglas Fir in thicknesses of 6, 8, 11 and 14 mm.)

'Good Two Sides Grade' with Grade A face and back veneers and Grade C inner veneers, sanded.

For further information refer to BS 5268:Part 2, and contact the Council of the Forest Industries of British Columbia.

### 1.9.3 Available sizes and quality of Finnish construction plywood

The most commonly available size is 1220 × 2440 mm. The face veneer runs parallel to the shorter side (Fig. 1.4). Other sizes which are said to be generally available are 1220 × 2440, 1200 × 2400, 1525 × 3050, 1500 × 3000, 1525 × 3660, 1500 × 3600, 1220 × 3660, 1200 × 3600, 1525 × 2440, 1500 × 2400, 1525 × 2745, 1500 × 2700. The face veneer runs parallel to the first dimension stated.

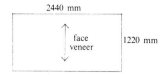

Fig. 1.4: *Finnish Birch, Birch-faced or Conifer plywood.*

Three basic types of Finnish exterior grade construction plywood are available in the UK as well as a flooring plywood. They are a plywood made entirely of birch veneers known as Finnish Birch (or all-birch) plywood; a plywood known as Finnish Birch-faced plywood which has the two outer veneers on both faces made from birch with alternate inner veneers of spruce and birch (except for 4 mm and 6.5 mm thicknesses which have only one birch veneer on both faces); and an all-conifer plywood referred to as Finnish Conifer plywood which is made mainly from spruce with fir permitted.

The face qualities available are grades:

I/I, I/II, I/III, II/II, II/III, III/III, III/IV and IV/IV.

The first figure indicates the quality of the face veneer, the second figure the quality of the back veneer. The combinations most likely to be available are:

I/III, II/III, III/III, II/IV and III/IV

With Birch plywood the inner veneers of birch and spruce are of at least IV Grade and the plywoods are sanded.

Likewise with Conifer plywood the inner veneers are of at least IV Grade and the plywoods are sanded.

For further information refer to BS 5268:Part 2, and contact Finnish Plywood International.

### 1.9.4   Available sizes and quality of American Construction plywood

The volume of American Construction and Industrial plywood exported to the UK has increased in recent years, mainly in the C–D Grade for use as sheathing. The most commonly available sheet size is 2440 × 1220 mm. The face veneer runs parallel to the longer side (as Fig. 1.3).

The type of American exterior grade construction plywood covered in BS 5268:Part 2:1984 is a plywood with one or both outer veneers of 'Group 1' species (usually Douglas Fir for export to the UK) and inner veneers of Group 1 or Group 2 species. Appendix B of BS 5268:Part 2 lists the species which are used in category Groups 1 and 2 as referenced below. In designing to BS 5268 and using C–D Grade plywood it is important to ensure that the quality is as referenced in BS 5268.

The face quality grades are:

'C–D Grade' unsanded which has a face veneer of Grade C of species Group 1, and back and inner veneers of Grade D of species Groups 1 or 2.

'C–C Grade' unsanded which has veneers throughout of Grade C of species Group 1.

'B–C Grade' sanded which has a face veneer of Grade B and back and inner veneers of Grade C, all veneers being of species Group 1.

'A–C Grade' sanded which has a face veneer of Grade A and back and inner veneers of Grade C, all veneers being of species Group 1.

BS 5268:Part 2 points out that C–D Grade is not suitable for prolonged use in damp or wet conditions, and that panels less than 600 mm wide for any of the American plywoods listed above should have certain stresses reduced (to 50% at 200 mm or less). Grade C–D should not be used for gussets for trussed rafters (BS 5268:Part 3).

For further information refer to BS 5268:Part 2, and contact the American Plywood Association.

### 1.9.5   Available sizes and quality of Swedish Construction plywood

Sheet sizes of 2400 × 1200 and 2440 × 1220 mm are available. The face veneer is parallel to the longer side (Fig. 1.5). Spruce (Whitewood) is mainly used in manufacture, but fir (Redwood) is permitted.

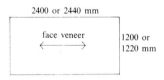

Fig. 1.5: *Swedish softwood plywood.*

There are several face qualities, but the most common for structural uses is C/C. However, the important reference for structural purposes is either P30 or P40 grade. The figures 30 or 40 are a reference in $N/mm^2$ to the minimum permitted ultimate failure stress in bending at the lower 5% exclusion limit. P30 and P40 are the references used in BS 5268:Part 2 and there is no reference to C/C etc. P30 and P40 can be obtained either unsanded or sanded and different permissible stresses are applicable to each.

For further information refer to BS 5268:Part 2 and the Swedish Forest Products Laboratory.

## 1.10   TEMPERED HARDBOARD

For the first time, BS 5268:Part 2 tabulates permissible stresses, moduli and coefficients for tempered hardboard for structural purposes, for dry service conditions. They apply only to TE grade tempered hardboard of nominal thicknesses from 3.2 to 8.0 mm and complying with BS 1142: Part 2, 'Fibre building boards. Medium board and hardboard'. The coefficients are particular for tempered hardboard, but follow the general pattern for design in BS 5268 for plywood.

There are special requirements for 'conditioning' boards before use.

Tempered hardboard has been used for the webs of Box or I beams and for sheathing. Design is not covered in this manual and the designer should refer to BS 5268:Part 2 and the Fibre Building Board Development Organisation (FIDOR) for detailed information on design and supply. The most common size of tempered hardboard is 1220 × 2440 mm.

## 1.11   CHIPBOARD

Although flooring-grade chipboard is used in the UK for flooring no stresses are given in BS 5268:Part 2. For information on strength, grades, sizes and suitability for purpose, contact the individual manufacturers or the Chipboard Development Association, and refer to BS 5669:1979, 'Specification for wood chipboard and methods of test for particle board.'

## 1.12   MECHANICAL FASTENERS

### 1.12.1   General

There are several mechanical fasteners which can be used in timber constructions. Some are multi-purpose fasteners, some specially produced for timber engineering. These are described briefly here and covered in more detail in Chapter 18.

### 1.12.2   Nails

Nails are satisfactory for lightly loaded connections where the nails are in shear. Ordinary nails used structurally in the UK are usually circular as they are cut from wire coil. Nails can be unprotected or treated against corrosion. Nails will slightly indent the timber when loaded in shear. This is not usually serious but must be realised because, for example, in a stress skin panel, a fully rigid joint cannot be claimed between the plywood skin and the timber joists if the joint is made only by nails. Nails can be used to give close contact during curing of a glued–nailed joint. Withdrawal loads are small if driven into side grain and zero in end grain. Nails can be 'fired' from a purpose-made gun.

### 1.12.3 Improved nails

Improved nails are used to a limited extent. Square nails (twisted or untwisted) are permitted to take higher lateral loads than round wire nails. Annular-ringed shank or helical-threaded shank nails are permitted to take higher withdrawal loads, and can be useful in situations where tension or vibration can occur during construction, such as will occur during the construction of the membrane of a shell roof. Improved nails can be unprotected or treated against corrosion. Like ordinary round wire nails they do not give a fully rigid joint.

### 1.12.4 Staples

Staples can be used for lightly loaded connections between a sheet material and solid timber, and to give close contact during curing of a glued joint. It is important that they are not over-fired by a stapling gun.

### 1.12.5 Screws

Screws are very much slower to insert than nails or improved nails and are therefore only used where nails are considered unsuitable (e.g. against withdrawal loads) or perhaps where a joint has to be demountable.

### 1.12.6 Bolts

The bolts usually used in timber engineering are of ordinary mild steel, either 'black' or galvanised. Where only bolts are used, the strength of the joint is achieved by the bearing of the bolt shank on to the timber therefore there is no advantage in using high-strength steel. Likewise there seems little advantage in using friction-grip bolts except where a special joint can be designed. It is usual to use large washers under the head and nut. If a bolt is galvanised the designer should ensure that the nut will still fit. It may be necessary to 'tap' out the nut or 'run down' the thread before galvanising.

### 1.12.7 Toothed-plate connector units

The relevant British Standard is BS 1579. Toothed-plate connectors can be either single or double sided, square or round, and can be placed on one or both sides of the timber. This type of connector is always used in conjunction with a bolt, complete with washers or steel plate under head and nut. The flat part of the connector sits proud of the timber.

The hole in the connector is a tolerance hole, therefore with single-sided connectors there is always bolt slip. Using double-sided connectors it may be possible with careful detailing to eliminate slip in the connection. The British Standard calls for an anti-corrosion treatment, but does not specify which one. This is left to the individual manufacturers.

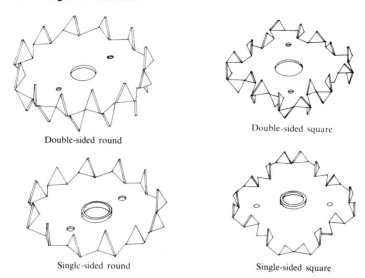

Double-sided round

Double-sided square

Single-sided round

Single-sided square

Fig. 1.6: *Toothed-plate connectors.*

BS 1579:1960 lists connector reference numbers for each connector detailed in the Standard, but this classification system is not used widely enough to enable a specifier to use it and rely on its being understood by all concerned. The Standard has not been metricated but BS 5268:Part 2 uses metric equivalents in referring to connectors.


### 1.12.8   Split-ring connector units

The relevant British Standard is BS 1579. A split ring may have bevelled sides (Fig. 1.7) or parallel sides. It takes shear and necessitates the timbers

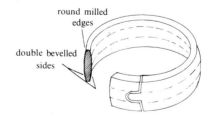

round milled
edges

double bevelled
sides

Fig. 1.7

being held together by another connector – usually a bolt (Fig. 1.8). A small amount of joint slip is likely to take place. The British Standard requires the connectors to have an anti-corrosion treatment. A special tool is required to groove out the timber for the shape of the split-ring and a special 'drawing' tool is required to assemble the unit under pressure. Units can carry a relatively high lateral load.

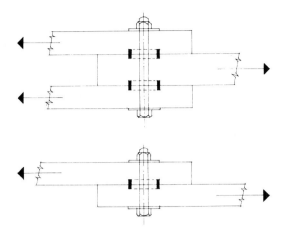

Fig. 1.8

### 1.12.9   Shear-plate connector units

The relevant British Standard is BS 1579. Shear-plate connectors are of pressed steel (Fig. 1.9) or malleable cast iron (Fig. 1.10). A special cutter is required to cut out the timber to take each connector but no tool is required to draw the timber together. A connector surface is flush with the surface of

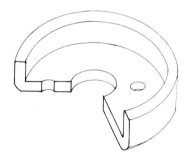

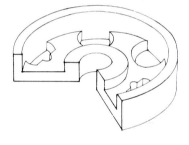

Fig. 1.9: *Pressed-steel shear plate 67 mm outside diameter.*

Fig. 1.10: *Malleable cast-iron shear plate 102 mm outside diameter.*

the timber. The timbers must be held together with a bolt. A small amount of joint slip is likely to take place. Shear-plate connectors are located in place for transit with two locating nails. The connectors must be given an anti-corrosion treatment.

### 1.12.10   Punched metal-plate fasteners

Punched metal-plate fasteners are usually fabricated from 18 or 20 gauge thick plain sheet steel in accordance with BS 2989, Grade Z2. Thicknesses up to 14 gauge are available. For trussed rafters, the Code sets a minimum thickness of 0.91 mm inclusive of the thickness of zinc coating. With the majority of plates, integral teeth are pressed out at right angles to give spikes at close centres which will subsequently be pressed into the timber by specialist machinery. As an alternative, plates with pre-punched holes through which treated nails can be driven can be used. Many manufacturers have been granted Agrément certificates for their plates. The primary use in the UK has been for gusset plates to trussed rafters (Chapter 20). Joint slip will take place. The manufacturers of the plate should be consulted for load/slip characteristics.

Some punched metal-plate fasteners with integral teeth are available in stainless steel but at considerable extra cost.

### 1.12.11   Other fasteners, gussets and hangers

Rag-bolts, rawl-bolts, ballistic nails etc. all have their use with connections for timber. Steel gussets or shoe plates etc. can be used with timber in a similar way to that in which they are used with structural steelwork, except that in this case nails and screws can be used with them as well as bolts.

Proprietary joist hangers or purpose-made steel hangers etc. can also be used. If a tight fit is required between timber and steel (e.g. to take compression or to prevent a moisture trap) an epoxy resin or similar mix can be used.

Obviously the finish of any fastener, hanger etc. must be suitable for the service conditions.

## 1.13   ADHESIVES USED IN TIMBER ENGINEERING

### 1.13.1   General

Glue manufacturers produce several adhesives to their own formulations but from the point of view of a structural designer these fall into three main categories as described in §§ 1.13.2 – 1.13.4. See Chapter 19 for detailed information on glue joints.

### 1.13.2   Weather proof and boil proof glues

When a component or joint is likely to be directly exposed to weather, either during its life or for lengthy periods during erection, it is necessary to use an adhesive which can match the 'weather proof and boil proof' (WBP) requirements of BS 1204:Part 1:1979. These glues are normally gap-filling thermosetting resorcinol–formaldehyde (RF), phenol–formaldehyde (PF) or phenol/resorcinol formaldehyde (PF/RF) resin adhesives. They are required where there is high hazard from weather, temperatures approaching 50°C or higher, high humidity, or chemically polluted atmosphere. They are often used (sometimes with a filler) for structural finger jointing. These glue types require careful quality control during storage, mixing, application and curing, and are therefore normally intended for factory use.

### 1.13.3  Boil resistant or moisture resistant glues

When a component is unlikely to be subjected to any serious atmospheric conditions once in place and is only likely to receive slight wettings during transit and erection it may be possible to use an adhesive which matches the 'boil resistant' (BR) requirements of BS 1204:Part 1:1979. These are normally gap-filling melamine/urea–formaldehyde (MF/UF) or certain other modified UF adhesives which can match the BR requirements. They are usually cheaper than WBP adhesives. For certain low hazard situations it might be possible to use an adhesive which matches only the 'moisture resistant and moderately weather resistant' (MR) requirements of BS 1204: Part 1. These are normally urea–formaldehyde (UF) adhesives, perhaps modified, to match the MR requirements. They require careful quality control during storage, mixing, application and curing, and are therefore normally intended for factory use.

### 1.13.4  Interior glues

A structural component is not normally glued with a glue which matches only the 'interior' (INT) requirements of BS 1204:Part 1. There are exceptions, perhaps where a site joint is required or where control of the gluing can not be guaranteed to a sufficient degree to enable a WBP, BR or MR glue to be used. An 'interior' glue should be used only when the designer can be certain that the glue joint will not be subjected to moisture or high temperature in place. A cold-setting casein adhesive to BS 1444:1970 could be used. (Even though a casein adhesive is not a resin adhesive it is normal to judge it against the requirements for interior adhesives laid down in BS 1204.)

One example where 'interior' casein adhesives have been used is site gluing of the layers of tongued and grooved boards on shell roofs. Tarpaulins or similar covers are used to protect the joints during construction, but even if the moisture content of the timber rises or the temperature drops during construction, the glue will cure eventually, whereas a WBP glue may not cure at all if the initial conditions are not correct.

(It is possible that BS 1444 will be withdrawn by the BSI but objections to this have been lodged on the basis that it is referenced in several British Standards and Codes.)

### 1.13.5  Epoxy resins

Epoxy resins are not normally used in timber engineering, but can be extremely useful in localised situations such as ensuring true bearing in a compression joint between timber and a steel bearing plate or, for example, in sealing the possible gaps between a timber post and a steel shoe. To reduce the cost in such compression joints the resin can be mixed with sand. Some timber engineering companies use their normal resin adhesive in these compression or sealing joints instead of making a special mix of epoxy resin.

## 1.13.6   Gluing

Gluing, including gluing of finger joints, is described in more detail in Chapter 19. Basically the quality-control requirements to obtain a sound glue joint are correct storage, mixing and application of the adhesive, correct surface conditions, moisture content and temperature of the timber, and correct temperature of the air during application of the adhesive and during curing. Draughts must be avoided. The instructions of the adhesive manufacturer must be followed. The joints must be held together during gluing, either by clamps or by nails or staples (glued–nailed, i.e. nails left in although their only duty is to hold the surfaces in close contact during curing, design being based on glue stresses only).

With some of the resin adhesives the adhesive manufacturer states requirements for the initial curing time, but does not make it clear that the edges of the glue joints must not be subject to running water or rain until at least seven days after gluing. This point should be watched in relation to delivery of components.

Several sources point out that the glue must be compatible with any preservative used. This is not usually much of a problem although the designer should check on compatibility the first time he uses a particular preservative, particuarly if it contains a water repellent or other additive. It must be realised that a water-borne preservative will increase the moisture content of the timber quite considerably and may raise grain. If the timber is to be glued after treatment (and re-drying) it is likely that the surfaces will have to be processed, even if lightly, which will remove some of the preservation. If a fire retardant containing ammonia or inorganic salts is to be used, then gluing by resorcinol types (and perhaps others) should not take place after treatment, nor should treatment take place until at least seven days after gluing. (Also see § 19.3.10.)

For certain components or joints, accelerated curing may be necessary, but this is expensive and is usually avoided where possible with structural components.

# Chapter Two
# Stress Levels for Solid Timber

## 2.1 INTRODUCTION

For the designer to achieve the most economical design of several components it is necessary to understand the way in which the stresses and $E$ values for the various stress grades and Strength Classes are derived. This is explained in this chapter in relation to the stresses quoted in BS 5268:Part 2:1984. For further reading the designer is referred to the publications of the Princes Risborough Laboratory. Grade stresses and $E$ values for the most commonly used softwoods are tabulated in Tables 2.2–2.5 and modification factors which apply to them are detailed in the various design chapters in this manual.

Many publications comment on the variable nature of timber, but what must be emphasised is that, for many properties of species, the variability has been established and can be interpreted by statistical methods. The method of dealing with the variability in setting permissible stresses for use in designs is such that almost always the erected components have a higher factor of safety than the required minimum, while deflection can be calculated within acceptable limits. What many designers do not appreciate is that all materials are variable and an understanding of the method of deriving stress and $E$ values for timber can help in an appreciation of the performance in practice of other materials. (For some other materials however, the variability is not as great as for timber and a statistical approach to the derivation of stresses is not thought to be warranted.)

Extensive testing, linked with measurement of actual defects in commercial sizes of timber, and with known statistical relations, has led to the setting of stress grades which can be selected by practical visual methods. Such UK stress grades are described in BS 4978 and are given the titles 'General Structural' and 'Special Structural'. Stress grading by machine has also become well established in grading to BS 4978. Machines check timber for a stress grade in a way different to visual stress grading. Machines are set to accept/reject at particular values of 'machine $E$' and the stresses and stiffnesses tabulated for design in a particular stress grade are arrived at through statistical relations which have been established with the machine $E$ value. Two of the machine stress grades of BS 4978 are set to equate with the visual stress grades General Structural and Special Structural. Initially they were given slightly higher $E$ values than the visual grades but, for design convenience, they are now exactly interchangeable and have identical stresses and '$E$' values with the visual versions.


## 2.2   BASIC STRESS

When CP 112:Part 2:1967 was written the concept of 'basic stress' was used as 'the stress which can safely be permanently sustained by timber containing no strength-reducing characteristics'. Even though 'basic stresses' have not been retained for BS 5268:Part 2 there are some interesting points to be learnt in looking at the history of basic and grade stresses.

Timber as sawn from the log with dimensions suitable for structural use inevitably contains characteristics (defects) which reduce the strength and make the use of basic stresses inappropriate in practical design. Therefore the stresses used for various stress grades of timber (i.e. grade stresses) in practical design to CP 112:Part 2 were based on basic stresses reduced by factors to take account of various characteristics.

The method which had been used extensively to derive stresses was to test many small clear 'green' specimens for the various species, to accumulate results which were then statistically analysed to give basic stresses. These basic stresses were modified by reduction factors which were related to the maximum size of defects permitted by the grade and to the exposure conditions, to arrive at the grade stresses. However, the latest work on strength properties has been carried out on full-size members containing defects representative of the grade and, in general, has shown that some of the earlier assumptions were conservative. In particular, it has been confirmed that if a piece of timber has relatively high density for the species, this can counteract the strength-reducing effect of defects. Therefore, with a grading method which is influenced by density (e.g. machine stress grading) this knowledge is taken into account and is the reason why, with machine stress grading, one may see knots which are larger than one would expect to see in an equivalent grade selected by visual stress grading.

A well-quoted example of tests on small clear specimens which illustrates the method which used to be used to derive basic stresses is that carried out at the Princes Risborough Laboratory to establish the modulus of rupture of 'green' Baltic Redwood. Two thousand, seven hundred and eight specimens, 20 mm square × 300 mm long, were tested in bending under a central point load. (BS 373 details the standard test method.) The failure stresses for all specimens were plotted to form a histogram as shown in Fig. 2.1. The histogram shows the natural variation of the modulus of rupture about the mean value for essentially similar pieces of timber. Superimposed upon the histogram is the normal (Gaussian) distribution curve and this is seen to give a sufficiently accurate fit to the histogram to justify the use of statistical

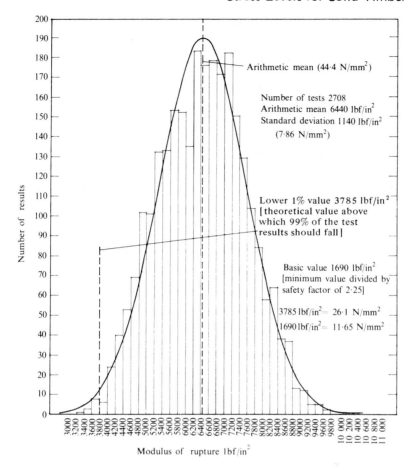

Fig. 2.1: *Variability of modulus of rupture of green Baltic Redwood (courtesy Princes Risborough Laboratory).*

methods related to this distribution to derive basic stresses. In the example shown, the mean modulus of rupture for the green timber is 44.4 N/mm² and the standard deviation is 7.86 N/mm². (When this work was carried out the UK used to work to the lower 1 in 100 exclusion level.)

The standard deviation is the square root of the average of the square of the differences between each test value and the mean. In a mathematical form this may be written as:

$$s = \sqrt{\left[\frac{\Sigma\,(x - \bar{x})^2}{N - 1}\right]}$$

where $s$ = standard deviation
   $x$ = individual test value
   $\bar{x}$ = mean of the test values
   $N$ = number of tests.

The standard deviation may be regarded as being analogous to the radius of gyration in the theory of second moments of area. There may be doubts as to whether $N-1$ or $N$ should be used in this formula for standard deviation. Statisticians advise that it should be $N-1$ for this particular timber example.

The area under the normal distribution curve represents a probability of 1.0, and the area between any two vertical ordinates represents the probability that any test chosen at random will have a value between the values of the ordinates. Normally one selects one low vertical ordinate, being interested in the probability of a small number of pieces falling below this ordinate. The normal distribution curve extends to infinity in either direction, however, for practical purposes it may be regarded as terminating at three times standard deviation on each side of the arithmetic mean.

From the distribution curve it is simple to predict a stress level below which a specimen is not expected to fall, according to a chosen level of probability. What is not so straightforward is to choose which probability level to take for various stress conditions and which factors of safety to assign to the lower exclusion level. Where over-stress can lead to failure, it used to be the practice in the UK to take the 'lower 1% exclusion level'. In such a case the probability is that 99% of the values will fall above this level. This probability level can be described as 1 in 100, and for CP 112: Part 2:1971 this applied to bending, tension, shear and compression parallel to the grain. Where over-stress would not lead to other than local failure (as with compression perpendicular to the grain) the probability level for CP 112:Part 2:1971 was taken as 1 in 40.

In BS 5268:Part 2:1984 it is the 'lower 5% exclusion level' (i.e. 1 in 20) which is taken for bending and compression parallel to the grain. (If the designer sees a reference, for example, to '5% exclusion level' he should check whether the intention is to refer to the 'lower 5% exclusion level' or $2\frac{1}{2}\%$ exclusion at each extreme of the distribution curve. Sometimes the phrase '5 percentile' is used to indicate the lower exclusion level.)

To determine the deviation from the mean appropriate to a selected probability level, reference can be made to Fig. 2.2 which deals with the area of the normal distribution curve. The area increases from 0.135% (of the total area) to 50% as $k_p$ times the standard deviation reduces from three to zero, representing a change in the probability level from 1 in 740 to 1 in 2. Probability levels of 1 in 20 and 1 in 40 have standard deviations of 1.645 and 1.96 respectively.

In BS 5268:Part 2:1984 the concept of 'basic stress' has been discontinued and it is therefore no longer possible to quote grade stresses as a factor of basic stresses. For BS 5268, the procedure explained briefly below has been adopted. Characteristic 5 percentile failure values in bending and compression parallel to the grain, 5 percentile $E$ values and mean $E$ values were derived from in-grade structural-sized members of certain species. Grade

| Coefficient $k_p$ | Area to left of $k_p s$ line | Probability (related to lower exclusion level) |
|---|---|---|
| 0 | 0.500 0 | 1 in 2 |
| 0.5 | 0.308 5 | 1 in 3 |
| 1.0 | 0.158 7 | 1 in 6 |
| 1.5 | 0.066 8 | 1 in 15 |
| 1.645 | 0.050 0 | 1 in 20 |
| 1.96 | 0.025 0 | 1 in 40 |
| 2.0 | 0.022 8 | 1 in 44 |
| 2.33 | 0.010 0 | 1 in 100 |
| 3.00 | 0.001 35 | 1 in 740 |

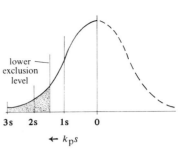

Fig. 2.2

relativity factors established from the same series of tests were applied to the SS grade stresses to provide values for other stress grades of these species. The analysis of these in-grade test results was based on the three-parameter Weibull distribution or a more sophisticated statistical approach.

For species for which test results from commercial sizes were not available, stresses and $E$ values were derived from the ratios of the mean strengths of small clear specimens of the tested and untested species.

Structural-sized tests for shear and compression perpendicular to the grain have not been conducted at PRL. The grade shear stresses were derived from small clear tests by a method essentially the same as that described for CP 112 except that fifth percentiles (rather than one percentiles) are used, and the safety factors have been slightly increased. The compression perpendicular to the grain grade stresses were derived from fifth percentile small clear bending results.

The structural-sized tests, the small clear tests, the methods of analysis and the derivation of the grade stresses were all undertaken by the Princes Risborough Laboratory and presented in technical papers to the Code committee.

To take account of the differences in load duration between testing and load duration in use, and to give a factor of safety, the tabulated long term grade stresses are derived by multiplying the relevant characteristic stress by 0.346 to obtain bending stress values, and 0.408 to obtain values for compression parallel to the grain. Grade bending values are further modified to the 300 mm depth level (except for those cases for which the tabulated values are for one specific width–e.g. 89 mm width for some North American stress grades).

Although, with a slight modification to the yield of machine stress-graded timber, tension grade stresses could probably have been allocated higher values, for BS 4978 stress grades they are quoted as 60% of bending values. The shear modulus is given as one sixteenth of the modulus of elasticity.

BS 5268:Part 2 requires that 'wet stresses' be taken for any condition where the equilibrium moisture content in service is above 18%. Dry stresses apply to moisture contents of 18% or below. Dry stresses are tabulated in BS 5268. For wet conditions, the geometrical properties of sections are obtained from dry values by multiplying by a factor $K_1$, (which is different for size, area etc.), and the stresses and moduli are multiplied by factor $K_2$.

Although BS 5268 gives one moisture content level at which the change-over from wet to dry is said to occur, strength and stiffness varies with variations in moisture content up to the fibre saturation point (around 26–28% moisture content for Redwood, 30–34% for Whitewood but often rounded off at 30% for both species). In a centrally heated house the moisture content of the timber is almost certainly below 18%. Should the designer be able to take advantage of a moisture content of less than 18% (e.g. by designing for a country where a lower design moisture content is accepted), the relationship between stress or $E$ value and moisture content is generally taken as:

$$\log \sigma_{M1} = \log \sigma_{M2} + C(M_2 - M_1)$$

$$\log E_{M1} = \log E_{M2} + C(M_2 - M_1)$$

where   $\sigma_{M1}, E_{M1}$ = the stress and $E$ values at moisture content $M_1$

$\sigma_{M2}, E_{M2}$ = the stress and $E$ values at moisture content $M_2$

$C$ = a constant (see §4.7).

Only dry stresses are dealt with in this manual, because the designer is invariably concerned with design in conditions where the timber will be at less than 18% equilibrium moisture content. The stresses and moduli for wet conditions derived from BS 5268 by multiplying dry stresses and moduli by $K_2$ are those appropriate to the fibre saturation point.


## 2.3   MODULUS OF ELASTICITY

There is no factor of safety on modulus of elasticity. In BS 5268 two values of $E$ are tabulated for each stress grade of each species, or for each Strength Class. These are given the title $E_{mean}$ and $E_{min}$. They are 'true' values of $E$ (i.e. they do not contain an element of shear modulus) therefore, for example, beams must be designed for shear deflection as well as bending deflection. (In CP 112, the $E$ values were 'apparent' $E$ values, and shear deflection of solid timber and glulam was considered to be accounted for in bending deflection calculations.)

$E_{mean}$ is the arithmetic mean from a number of test results. BS 5268: Part 2 permits this value to be used in certain cases where four or more members act together to support a common load. By and large, this usage of

$E_{mean}$ in designs leads to acceptable components, however, it can be shown that statistically a far greater number of pieces would have to act together to justify the use of $E_{mean}$ if one required probability that $E_{mean}$ would be achieved in service. However, for example, span tables for joists and rafters in Schedule 6 of the Building Regulations for England and Wales have been based on $E_{mean}$ for a number of years.

$E_{min}$ is a statistical minimum value which, for the purposes of BS 5268, is set at the lower 5 percentile level. BS 5268 requires the use of $E_{min}$ in designs when one piece of solid timber acts alone.

In the First Edition of this manual the authors described a method for calculating the values of $E_2$, $E_3$ etc. for cases where two, three or more pieces act together and where the use of $E_{mean}$ is not relevant. This method is based on the assumption that if $E_{mean}$ is the arithmetic mean of a normally distributed sample, for the lower 5 percentile level, then:

$$E_N = E_{mean} - \frac{1.645s}{\sqrt{N}}$$

where $E_N$ = the statistical minimum value of $E$ appropriate to the number of pieces ($N$) acting together

$s$ = the standard deviation.

When $N = 1$, then the value of $E_N$ is the tabulated value of $E_{min}$ and, by knowing this and $E_{mean}$ (from tables in BS 5268) it is a simple matter to use the formula to calculate the standard deviation for the stress grade/species combination being considered, and then to calculate the statistical minimum value of the required $E_N$.

Many designers, including the authors, have used this method of calculating $E$ for commercial designs (although there was no mention of it in CP 112:Part 2:1971). A simplified method of calculating values for $E_2$, $E_3$ etc. has now been introduced into BS 5268:Part 2:1984 which permits $E_{min}$ to be increased by a factor $K_9$ for composite trimmers or lintels and by a factor $K_{28}$ for composites such as ply web beams. One set of values is given for $K_9$ and $K_{28}$ for all softwoods and one set for all hardwoods. Because the ratio of $E_{mean}$ to $E_{min}$ for the various softwoods given in BS 5268 varies from about 1.5 to 1.6 it follows that $K_9$ and $K_{28}$ should differ for each stress grade or Strength Class. However the variation was not thought to be significant therefore separate values are not given.

If, however, a designer is calculating a camber for a particularly long-span principal beam which is required to deflect close to the horizontal under permanent loading, it would be as well to consider the actual ratio of $E_{mean}$ and $E_{min}$ for the species and stress grade of the design, as well as the use of $K_{28}$, remembering that both methods give a statistical minimum value, therefore the probability is that the beam will deflect to a lesser extent. Otherwise, significant over-camber may occur.

As can be seen by reference to Fig. 2.3, the values given in BS 5268:Part 2

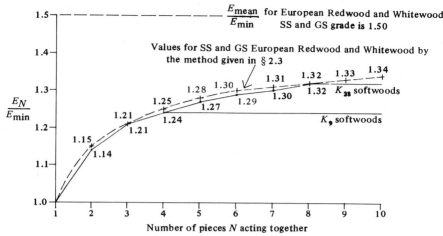

Fig. 2.3

for $K_9$ and $K_{28}$ for softwoods are close to the values which would be derived for European Redwood and Whitewood SS and GS grades by the method described above, except that there are stop-off points for $K_9$ and $K_{28}$ at $N$ equal to 4 and 8 respectively.

## 2.4   GRADE STRESS

### 2.4.1   General

Grade stress is defined as 'the stress which can safely be permanently sustained by material of a specific section size and of a particular Strength Class, or species and (stress) grade'. The reference to section size had to be introduced because BS 5268 has introduced a depth factor for solid timber in bending and a width factor for solid timber in tension.

In 1973, BS 4978 introduced a method of visual stress grading which, although used in North America, was new to Europe. The method of grading gives a more realistic assessment of the effects of knots on the strength of timber, being based on the area and disposition of knots as they affect cross-section rather than the way they affect surfaces, as had been the general case in Europe. Two primary visual grades are described which are General Structural (GS) and Special Structural (SS), which can also be graded out by machine, in which case they are described by the abbreviations MGS and MSS. BS 4978 permits higher machine stress grades (e.g. M75) but no higher visual grades.

In visual stress grading to BS 4978, the main characteristics against which quality is assessed are knots, fissures, slope of grain, wane, rate of growth and distortion. These characteristics influence the determination of grade stresses and the method of allowing for them is best illustrated by considering each of the types of stress in turn. In the notes below mainly grading to BS 4978 is considered although BS 5268 does tabulate stresses for stress grades to other standards (i.e. North American and European standards).

Note that in BS 5268 the term 'grade stress' is applied to Strength Classes as well as stress grades.

### 2.4.2 Grade bending stress

For visual stress grades, bending stress is influenced mainly by the presence of knots and their effective reduction of the modulus of the section, therefore the knot area ratio ($KAR$) and the disposition of the knots in the area is important.

The knot area ratio is defined in BS 4978 as 'the ratio of the sum of projected cross-sectional areas of all knots at a cross-section to the cross-sectional area of the piece'. In making the assessment, knots of less than 5 mm may be disregarded and no distinction need be made between knot holes, dead knots and live knots. Figure 2.4 illustrates some typical knot arrangements and their $KAR$ values.

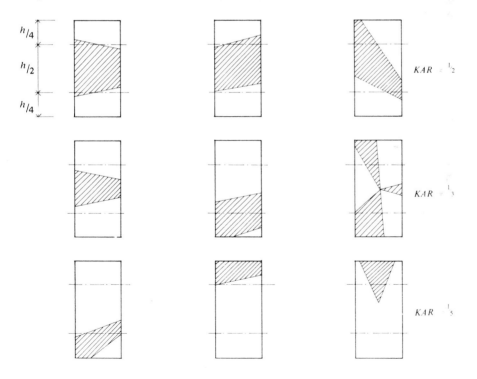

Fig. 2.4

To take account of the fact that a knot near an edge has more effect on the bending strength than a knot near the centre of the piece, the concept of a margin and a margin condition is introduced. For the purposes of BS 4978 a margin is either of the outer quarters of area, and a margin condition is said to exist if more than half the area of either margin is occupied by the projected area of knots.

To qualify as visual SS grade the *KAR* must not exceed $\frac{1}{5}$ if a margin condition exists and $\frac{1}{3}$ if a margin condition does not exist. To qualify as visual GS grade the *KAR* must not exceed $\frac{1}{3}$ if a margin condition exists and $\frac{1}{2}$ if a margin condition does not exist.

The most onerous (theoretical) arrangement of knots corresponding to these limits is illustrated in Fig. 2.5. The ratio $Z_{net}/Z_{gross}$ is shown alongside

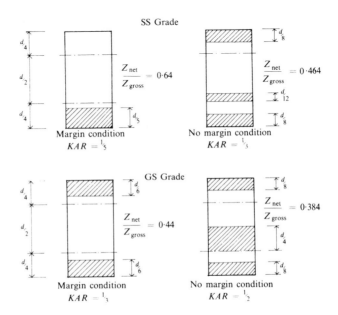

Fig. 2.5

each sketch. From these ratios one could deduce that the ratio of bending stresses between SS and GS grades would be in the order of

$$\frac{0.44}{0.64} = 0.69$$

or

$$\frac{0.384}{0.464} = 0.83$$

depending on the extent to which a margin condition is or is not relevant in commercial sizes. In fact the ratio on which stresses tabulated in BS 5268 are based is 0.7.

As explained in §2.2 grade bending stresses are obtained by multiplying the relevant characteristic stress at a depth of 200 mm by 0.346, and adjusting to the 300 mm depth level (unless the tabulated value is only for one specific depth, such as 89 mm, in which case the adjustment is to that depth).

Where a machine is set to give an equivalent stress grade it is allocated the same stresses and $E$ values as the visual stress grade.

BS 5268:Part 2 introduces a depth factor $K_7$ which, for most species/stress grade combinations, and Strength Classes, permits an increase of up to 17% in bending stress at depths of 72 mm or less. (Note that BS 4978 stress grades apply to sizes only down to 35 × 60 mm, although this point is under review.) At depths of over 300 mm, $K_7$ leads to a reduction in stress.

For the purpose of $K_7$, 'depth' is the dimension in the plane of bending, and applies to solid timber and glulam.

It is important to note that the grade bending stresses tabulated in BS 5268:Part 2 for Strength Classes and timber graded to the BS 4978 stress grading rules and for the North American 'Joist and Plank' grades, are related to a depth of 300 mm. However, the bending stresses tabulated for North American 'Structural Light Framing', 'Light Framing', 'Stud' and 'Utility' grades are related to a depth of 89 mm. Therefore, perhaps the easiest way to avoid confusion is to consider $K_7$ as being 1.00 for North American sizes of 38 × 89 mm, and to apply the factors from Table 12 of BS 5268:Part 2 in those cases where North American sizes of 38 × 38 and 38 × 63 mm are to be used. (Factors for other sizes/grade combinations of North American timber are given in Table 12 but these are for sizes rarely available in the UK.)

Figure 4.4 gives a graph for $K_7$ relevant to solid timber graded to BS 4978, North American 'Joist and Plank' grades, and glulam, and Strength Classes.

### 2.4.3   Grade tension stress

Prior to the evaluation of the results of testing work and research commenced in 1968 as a collaboration between timber research laboratories in the UK, Sweden, Finland and Canada, and even when CP 112:Part 2:1971 was initially published, grade tension stresses were defined in the UK as being the same as bending stresses. In BS 5268:Part 2, tension stress for BS 4978 stress grades are tabulated as being 60% of the bending stress for the same stress grade or Strength Class. This change is a result of the test programme for machine grading carried out at PRL and a rationalisation of the levels of safety attributed to the different loading conditions.

Where a stress grade can be selected by visual methods or by machine, the decision was taken to adopt identical tension values. Had a slightly lower machine yield been accepted, it is likely that higher tension values could have been justified up to a level around 70% or more of the bending values, which is the level used where a grade is tabulated as being selected only by visual methods (e.g. the NLGA grades).

BS 5268:Part 2 introduces a width factor $K_{14}$ for tension members which, for widths between 300 and 72 mm, is numerically equal to $K_7$ (see Fig. 4.4). For the purpose of $K_{14}$, 'width' is the greater dimension (of a rectangular section), and applies to solid timber and glulam.

It is important to note that the tension stresses tabulated in BS 5268: Part 2 for timber graded to the BS 4978 stress grading rules and for the North American 'Joist and Plank' grades are related to a width of 300 mm. However, the tension stresses tabulated for North American 'Structural Light Framing' (Select Structural, No. 1 and No. 2) grades are related to a depth of 89 mm. Therefore, perhaps the easiest way to avoid confusion is to consider $K_{14}$ as being 1.00 for North American sizes of 38 × 89 mm, and to apply the factors from Table 12 of BS 5268:Part 2 in those cases where North American sizes of 38 × 38, 38 × 63 mm (and 63 × 63, 63 × 89 and 89 × 89 mm if available) are to be used.

Note that North American 'Structural Light Framing No. 3', 'Light Framing', 'Utility' and 'Stud' grades are not to be used in tension, hence no $K_{14}$ factor is given for them.

A graph and values of $K_{14}$ are given with $K_7$ in Fig. 4.4. The values are for solid timber graded to BS 4978, North American 'Joist and Plank' grades, and glulam, and Strength Classes.

### 2.4.4   Grade stress for compression parallel to the grain

Grade stresses for compression parallel to the grain are determined from structural-sized tests as explained in § 2.2. Whereas, in BS 5268:Part 2, tension stresses are a smaller percentage of bending stresses than they were in CP 112:Part 2:1971, compression stresses vary from being about 20% higher for lower stress grades (such as GS) to being 20% lower for the highest stress grades (such as M75).

When the slenderness ratio is more than five the grade stresses must be modified by $K_{12}$ (which is based on a different formula to the $K_{18}$ and $K_{19}$ coefficients which applied in CP 112:Part 2:1971).

### 2.4.5   Grade stress for compression perpendicular to the grain

The grade stresses for compression perpendicular to the grain (i.e. bearing) take account of the wane which is permitted up to the grade limit (e.g. one quarter thickness for SS and one third for GS disregarding local permitted deviation). In considering the case where wane is excluded it is no longer possible to refer to the basic stress as was the case with CP 112:Part 2:1971. When wane is excluded one is permitted to work to the SS grade value for the species multiplied by 1.33 (not given a coefficient number). When designing to a Strength Class and wane is excluded, then a separate value is tabulated which, for softwood, represents a variable increase of between 75% for SC1 and 16.6% for SC5. These two methods can lead to significantly different values for the same timber. For example, SS grade European Whitewood with no wane has a grade stress of 1.33 × 2.1 = 2.8 N/mm², whereas the same timber if treated as belonging to SC4 would be considered to have a grade stress of 2.4 N/mm² if wane is excluded.

## 2.4.6   Grade shear stress parallel to grain

The presence of fissures is the main influence on shear stress, the size being measured as shown in Fig. 2.6. For GS and SS stress grades, fissures are limited generally to half the thickness of the piece therefore the grade shear stress is the same for both grades (although is not the same value for all stress grades of a species). Each Strength Class has a different value. GS and SS European Whitewood have a grade stress of 0.82 N/mm² but if designed as SC3 and SC4 would be considered to have a grade stress of 0.67 and 0.71 N/mm² respectively.

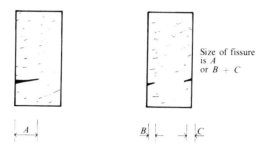

Fig. 2.6

## 2.4.7   Grade values for modulus of elasticity

Modulus of elasticity is related to density which is also related to strength, therefore the modulus of elasticity increases for the higher stress grades. Values of $E_{mean}$ and $E_{min}$ (as explained in § 2.3) are tabulated in BS 5268: Part 2 for stress grade/species combinations and Strength Classes. These are 'true' $E$ values and do not include an allowance for shear modulus, which BS 5268:Part 2 states should be taken as one sixteenth of the corresponding $E$ value. The normal distribution of the modulus of elasticity for SS and GS grades of dry Whitewood and Redwood is illustrated in Fig. 2.7.

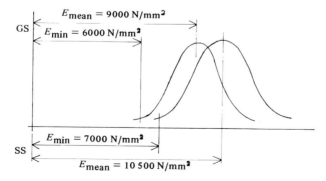

Fig. 2.7: *Normal distribution of the modulus of elasticity for SS and GS grades of dry Whitewood and Redwood.*

## 2.5   LOAD SHARING

### 2.5.1   *E* values

The tabulated values of $E_{min}$ in BS 5268 are for $N = 1$ and are the modulus of elasticity below which only 1 in 20 pieces of timber will fall, therefore $E_1 = E_{mean} - 1.645s$ from which the standard deviation $s$ is determined as:

$$s = \frac{E_{mean} - E_1}{1.645}$$

The type of normal distribution curve considered so far in this chapter is that for individual pieces of timber, and the larger the sample considered, the closer the average value approximates to $E_{mean}$ and the histogram of results approximates to the true Gaussian curve.

For cases where two or more pieces of timber act together in the cross-section of one component, the spread of an associated Gaussian curve diminishes as the number $N$ of pieces acting together increases. Figure 2.8 shows how the distribution of $E_N$ becomes more compact as $N$ increases. The distribution retains its normal character, but its standard deviation decreases as the inverse of the square root of $N$. If $s$ is the standard deviation for the case for members of individual pieces then:

$$E_N = E_{mean} - \frac{1.645s}{\sqrt{N}} = E_{mean} - \frac{E_{mean} - E_1}{\sqrt{N}}$$

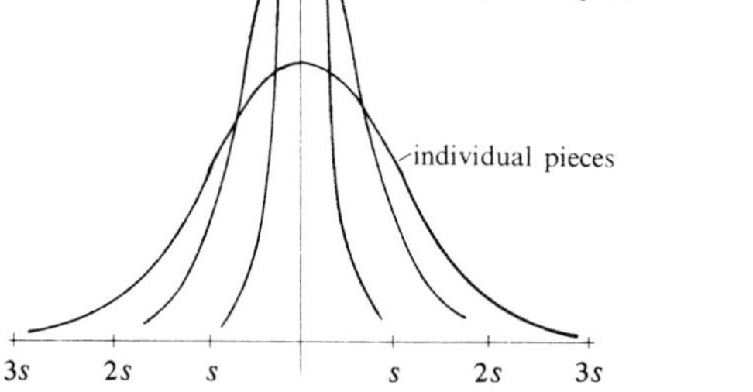

eight pieces in cross-section per component

four pieces in cross-section per component

individual pieces

$3s$   $2s$   $s$     $s$   $2s$   $3s$

Fig. 2.8: *Distribution of E.*

Values of $E_{mean}$ and $E_1$ ($E_{min}$) are given in BS 5268:Part 2 from which $E_N$ can be calculated as explained above and in §2.3. However, the method given in BS 5268 is to multiply $E_{min}$ by a factor $K_9$ for trimmer joists and lintels, or $K_{28}$ for vertically glued laminated beams or composites such as ply web beams. Although $K_9$ and $K_{28}$ vary with values of $N$ they are constant for all softwoods even when the ratio of $E_{mean}$ to $E_{min}$ differs. The effect of this is discussed in § 2.3. For European Redwood or Whitewood SS or GS grade with a ratio of $E_{mean}$ to $E_{min}$ equal to 1.50, the difference in the $E$ values calculated by $K_9$ and $K_{28}$ or the formula above is not large until one reaches the stop-off points for $K_9$ and $K_{28}$, as can be seen from Fig. 2.3 and Table 2.1.

**Table 2.1   Values of $E_N$ ($N/mm^2$) for dry Redwood and White-wood, SS grade**

| $N$ | From BS 5268 by multiplying $E_{min}$ by $K_9$ | From BS 5268 by multiplying $E_{min}$ by $K_{28}$ | From statistics as detailed in § 2.5.1 |
|---|---|---|---|
| 1 | 7000 | 7000 | 7000 |
| 2 | 7980 | 7980 | 8025 |
| 3 | 8470 | 8470 | 8479 |
| 4 | 8680 | 8680 | 8750 |
| 6 | 8680 | 9030 | 9071 |
| 8 | 8680 | 9240 | 9262 |
| 10 | 8680 | 9240 | 9393 |
| 12 | 8680 | 9240 | 9490 |
| ∞ | — | — | 10500 |

From Table 2.1 it can be seen that the value calculated to BS 5268: Part 2 for $E_4$ is only about 83% of the value for $E_{mean}$, yet for four or more members acting in a load-sharing system BS 5268 permits the use of $E_{mean}$ in designs. The use of $E_{mean}$ in these cases is based on the experience of timber in use.

In view of the values of $E_9$ and $E_{28}$ varying with values of $N$ it is a little surprising that, whereas CP 112 : Part 2 tabulated, for glulam beams, stiffness coefficients ($K_2$ and $K_7$) which increased with increasing numbers of laminates, BS 5268 : Part 2 gives coefficients ($K_{20}$ and $K_{26}$ which must be multiplied by the $E_{mean}$ for SS grade) which do not increase with increasing numbers of laminates. For a glulam beam of European Whitewood with LB laminates throughout, factor $K_{20}$ is 0.90 which gives an $E$ value for the beam of $0.9 \times 10\,500 = 9450\ N/mm^2$ no matter how many laminates are used, which compares with an $E$ value for a composite of SS grade (using $E_{min}$ multiplied by $K_{28}$) varying from $8680\ N/mm^2$ (for $N = 4$) to $9240\ N/mm^2$ (for $N = 8$ or more).

The term 'load sharing' is used in BS 5268, however, some designers find the concept of 'load relieving' easier to understand. If, for example, several joists are supporting a common load and one (or more) of the joists has an $E$ value less than the mean of the group, it will tend to deflect more than the average. In doing so, if the construction is able to share load, load will be transferred from this joist to the stiffer joists. Thus load will actually be relieved from the less stiff joist and consequently it will not deflect more than the rest. This is not the way that the design of the less stiff joist is presented (i.e. it is not calculated as carrying less load) but the effect of the BS 5268 design method is the same in calculating deflection.

### 2.5.2   Stress values

In theory, the statistical approach to the determination of $E$ values for several members acting together could be applied similarly to bending stresses. After all, if the grade stress for one member acting alone is set at the lower 5 percentile level (as is $E_{min}$) one could expect the average grade stress for a group of members acting together to increase with increases in $N$ towards the mathematical mean, as one would expect with $E$ values.

However, although the relation of bending stress to stiffness is sufficiently consistent to permit, for example, the development of stress grading machines with which strength can be allocated from measurements of stiffness, it is not a 100% relationship. Therefore, referring to the last paragraph of §2.4, one can not be certain that the least stiff member (or members) from which load is being relieved is also the least strong member and *vice versa*.

In BS 5268:Part 2 increases in permissible stress are permitted with increases in the number of members acting together, but the increases are conservative compared to those one might expect purely from a reference to a normal distribution curve. See Fig. 2.9 for a comparison of $K_{27}$ and $K_8$ as tabulated in BS 5268 with a theoretical curve for European Whitewood or Redwood SS and GS grades. This comparison is for bending stresses.

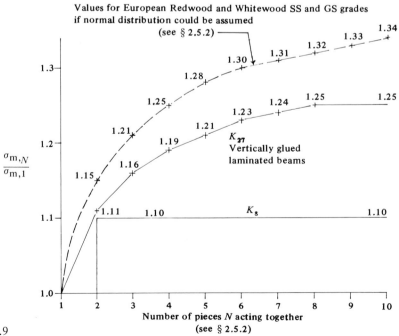

Fig. 2.9

The value given for $K_8$ is 1.1 and is applicable to grade stresses in bending, shear and compression perpendicular to grain of any trimmer of two or more members acting together to support a common load, and for load-sharing systems of four or more members spaced at not more than 610 mm centres (in which compression parallel to grain can also be increased, as in a stud wall).

The values given for $K_{27}$ are applicable in bending, tension and shear to vertically glued laminated beams of structural timber grades (not the laminating grades). It is not possible to plot in Fig. 2.9 a true comparison with the coefficients for bending of horizontally laminated beams because the grading of laminates differs from the grading of structural timber grades.

## 2.6   STRENGTH CLASSES

Strength Classes have been introduced into BS 5268:Part 2:1984 for the first time (see § 1.3.7). A stress grade/species combination is said to qualify for a Strength Class if the grade stresses in bending and tension (at a depth/width of 300 mm) are not less than the class values, and the stresses in compression parallel to the grain and shear, and the moduli of elasticity are not less than 95% of the class values. Thus some advantage may accrue in design if a stress grade is assigned to a Strength Class.

Despite this definition which includes a required level for tension, some of the lower stress grade/species combinations of North American timber (e.g. 'Joist and Plank No. 3 grade') which are not given any value in tension, are allocated to a Strength Class in BS 5268. In the relevant tables in BS 5268 a note is added that such stress grades should not be used in a tension member, however this is a source of potential confusion if a specification given to a timber supplier simply states the Strength Class without stating whether or not the timber will be called upon to take tension. Fortunately of the stress grades involved only 'Construction' and 'Standard' Light Framing are usually imported to the UK.

Several stress grade/species combinations allocated to a Strength Class must have fixings designed as though the timber is of a lower Strength Class (e.g. nearly all softwoods which are allocated to SC5 should have fasteners designed as for SC3/SC4).

Even in the cases where a designer does not design to a Strength Class, where fasteners are involved it is necessary to allocate the chosen stress grade/species combination to a Strength Class to derive permissible fixing values.

## 2.7   TABULATED GRADE STRESSES AND *E* VALUES INCLUDING FOR STRENGTH CLASSES

Grade stresses and *E* values are given in Tables 2.2–2.5 for dry exposure conditions for European Redwood and Whitewood, Canadian Spruce-Pine-Fir, Canadian Hem-Fir, Canadian Douglas Fir-Larch, and British-grown Sitka Spruce; and also for Strength Classes SC1–SC5 for softwoods.

**Table 2.2**  Grade stresses and moduli for certain softwoods stress graded to the BS 4978 rules for the dry exposure condition. All values are in N/mm²

| Softwoods | Stress grade | Bending parallel to grain* | Tension parallel to grain** | Compression parallel to grain | Compression perpendicular to grain with wane | Compression perpendicular to grain without wane | Shear parallel to grain | Modulus of elasticity mean | Modulus of elasticity minimum | Shear modulus mean | Shear modulus minimum | Strength Class allocated in BS 5268 |
|---|---|---|---|---|---|---|---|---|---|---|---|---|
| European Whitewood or Redwood imported | GS/MGS | 5.3 | 3.2 | 6.8 | 1.8 | 2.79 | 0.82 | 9000 | 6000 | 562.5 | 375 | SC3 |
|  | M50 | 6.6 | 4.0 | 7.3 | 2.1 | 2.79 | 0.82 | 9000 | 6000 | 562.5 | 375 | SC3 |
|  | SS/MSS | 7.5 | 4.5 | 7.9 | 2.1 | 2.79 | 0.82 | 10500 | 7000 | 656.3 | 437.5 | SC4 |
|  | M75 | 10.0 | 6.0 | 8.7 | 2.4 | 2.79 | 1.32 | 11000 | 7000 | 687.5 | 437.5 | SC5† |
| Canadian Spruce-Pine-Fir | GS/MGS | 5.3 | 3.2 | 6.8 | 1.6 | 2.39 | 0.68 | 8500 | 5500 | 531.3 | 343.8 | SC3‡ |
|  | M50 | 6.2 | 3.7 | 7.1 | 1.8 | 2.39 | 0.68 | 9000 | 5500 | 562.5 | 343.8 | SC3‡ |
|  | SS/MSS | 7.5 | 4.5 | 7.9 | 1.8 | 2.39 | 0.68 | 10000 | 6500 | 625 | 406.3 | SC4‡ |
|  | M75 | 9.7 | 5.8 | 8.5 | 2.1 | 2.39 | 1.10 | 10500 | 7000 | 656.3 | 437.5 | SC4‡ |
| Canadian Hem-Fir | GS/MGS | 5.3 | 3.2 | 6.8 | 1.7 | 2.53 | 0.68 | 9000 | 6000 | 562.5 | 375 | SC3 |
|  | M50 | 6.6 | 4.0 | 7.7 | 2.1‡‡ | 2.53 | 0.71 | 10500 | 7000 | 656.3 | 437.5 | SC3 |
|  | SS/MSS | 7.5 | 4.5 | 7.9 | 1.9‡‡ | 2.53 | 0.68 | 11000 | 7500 | 687.5 | 468.8 | SC4 |
|  | M75 | 10.0 | 6.0 | 9.3 | 2.4 | 2.53 | 1.13 | 12000 | 8000 | 750 | 500 | SC5† |
| Canadian Douglas Fir-Larch | GS | 5.3 | 3.2 | 6.8 | 2.2 | 3.19 | 0.85 | 9500 | 6000 | 593.8 | 375 | SC3 |
|  | SS | 7.5 | 4.5 | 7.9 | 2.4 | 3.19 | 0.85 | 11000 | 7500 | 687.5 | 468.8 | SC4 |
| British-grown Sitka Spruce | GS/MGS | 4.1 | 2.5 | 5.2 | 1.4 | 2.13 | 0.64 | 6500 | 4500 | 406.3 | 281.3 | SC1 |
|  | M50 | 4.5 | 2.7 | 5.5 | 1.6 | 2.13 | 0.64 | 7500 | 5000 | 468.8 | 312.5 | SC2 |
|  | SS/MSS | 5.7 | 3.4 | 6.1 | 1.6 | 2.13 | 0.64 | 8000 | 5000 | 500 | 312.5 | SC2 |
|  | M75 | 6.6 | 4.0 | 6.4 | 1.8 | 2.13 | 1.02 | 9000 | 6000 | 562.5 | 375 | SC3‡† |

*Note*: Refer to BS 5268:Part 2 for the grade stresses of other softwoods stress graded to BS 4978.

\* The depth factors $K_7$ given in Fig. 4.4 are relevant to all these bending stresses.

\*\* The width factors $K_{14}$ given in Fig. 4.4 are relevant to all these tension stresses.

† Design fasteners as though in SC3/SC4.

†† Design fasteners as though in SC1/SC2.

**Table 2.3** Grade stresses and moduli for certain Canadian softwoods stress graded to the NLGA rules in cross-sectional dimensions greater than or equal to 38 × 114 mm. All values are in N/mm² and are for the dry exposure condition

| Canadian softwoods | 'Joist and Plank' stress grades | Bending parallel to grain* | Tension parallel to grain** | Compression | | | Shear parallel to grain | Modulus of elasticity | | Shear modulus | | Strength Class allocated in BS 5268 |
| | | | | parallel to grain | perpendicular to grain | | | | | | | |
| | | | | | with wane | without wane | | mean | minimum | mean | minimum | |
| Spruce-Pine-Fir | Select | 8.0 | 6.2 | 8.8 | 1.8 | 2.39 | 0.68 | 10 500 | 7000 | 656.3 | 437.5 | SC4† |
| | No. 1 | 5.6 | 4.3 | 7.9 | 1.8 | 2.39 | 0.68 | 9 000 | 6000 | 562.5 | 375 | SC3† |
| | No. 2 | 5.6 | 4.3 | 6.9 | 1.6 | 2.39 | 0.68 | 9 000 | 6000 | 562.5 | 375 | SC3† |
| | No. 3 | 4.1 | – | 5.3 | 1.2 | 2.39 | 0.45 | 9 000 | 5500 | 562.5 | 343.8 | SC1 |
| Hem-Fir | Select | 8.0 | 6.2 | 8.8 | 2.1 | 2.53 | 0.71 | 12 000 | 8500 | 750 | 531.3 | SC4 |
| | No. 1 | 5.6 | 4.3 | 7.9 | 2.1 | 2.53 | 0.71 | 10 500 | 7000 | 656.3 | 437.5 | SC3 |
| | No. 2 | 5.6 | 4.3 | 6.9 | 1.8 | 2.53 | 0.71 | 10 500 | 7000 | 656.3 | 437.5 | SC3 |
| | No. 3 | 4.1 | – | 5.3 | 1.4 | 2.53 | 0.47 | 10 500 | 6500 | 656.3 | 406.3 | SC1 |
| Douglas Fir-Larch | Select | 8.0 | 6.2 | 8.8 | 2.7 | 3.19 | 0.93 | 12 500 | 8500 | 781.3 | 531.3 | SC4 |
| | No. 1 | 5.6 | 4.3 | 7.9 | 2.7 | 3.19 | 0.93 | 11 000 | 7500 | 687.5 | 468.8 | SC3 |
| | No. 2 | 5.6 | 4.3 | 6.9 | 2.4 | 3.19 | 0.93 | 11 000 | 7500 | 687.5 | 468.8 | SC3 |
| | No. 3 | 4.1 | – | 5.3 | 1.8 | 3.19 | 0.61 | 10 500 | 6500 | 656.3 | 406.3 | SC1 |

\* The depth factors $K_7$ given in Fig. 4.4 are relevant to all these bending stresses.
\*\* The width factors $K_{14}$ given in Fig. 4.4 are relevant to all these tension stresses.
† Lateral load perpendicular to the grain for bolts and timber connectors should be multiplied by $K_{42} = 0.9$.
– Where, in the column for tension stresses, a dash is given, this signifies that no tension value is given in BS 5268 for this stress grade.

**Table 2.4**  Grade stresses and moduli for certain Canadian softwoods stress graded to the NLGA rules in cross-sectional dimensions of 38 × 89 mm. All values are in N/mm² and are for the dry exposure condition

| Canadian softwoods | Stress grade | Bending parallel to grain* | Tension parallel to grain** | Compression | | | | Modulus of elasticity | | Shear modulus | | Strength Class allocated in BS 5268 |
| | | | | parallel to grain | perpendicular to grain | | Shear parallel to grain | mean | minimum | mean | minimum | |
| | | | | | with wane | without wane | | | | | | |
| Spruce-Pine-Fir | *'Structural Light Framing'* | | | | | | | | | | | |
| | Select | 9.1 | 7.1 | 8.8 | 1.8 | 2.39 | 0.68 | 10 500 | 7000 | 656.3 | 437.5 | SC4‡ |
| | No. 1 | 6.4 | 4.9 | 7.9 | 1.8 | 2.39 | 0.68 | 9 000 | 6000 | 562.5 | 375 | SC3‡ |
| | No. 2 | 6.4 | 4.9 | 6.9 | 1.6 | 2.39 | 0.68 | 9 000 | 6000 | 562.5 | 375 | SC3‡ |
| | No. 3 | 4.7 | — | 5.3 | 1.2 | 2.39 | 0.45 | 9 000 | 5500 | 562.5 | 343.8 | SC1 |
| | *'Light Framing'* | | | | | | | | | | | |
| | Construction | 5.4 | — | 6.1 | 1.8 | 2.39 | 0.68 | 9 000 | 6000 | 562.5 | 375 | SC2 |
| | Standard | 4.0 | — | 4.6 | 1.6 | 2.39 | 0.68 | 9 000 | 6000 | 562.5 | 375 | SC1 |
| Hem-Fir | *'Structural Light Framing'* | | | | | | | | | | | |
| | Select | 9.1 | 7.1 | 8.8 | 2.1 | 2.53 | 0.71 | 12 000 | 8500 | 750 | 531.3 | SC4 |
| | No. 1 | 6.4 | 4.9 | 7.9 | 2.1 | 2.53 | 0.71 | 10 500 | 7000 | 656.3 | 437.5 | SC3 |
| | No. 2 | 6.4 | 4.9 | 6.9 | 1.8 | 2.53 | 0.71 | 10 500 | 7000 | 656.3 | 437.5 | SC3 |
| | No. 3 | 4.7 | — | 5.3 | 1.4 | 2.53 | 0.47 | 10 500 | 6500 | 656.3 | 406.3 | SC1 |
| | *'Light Framing'* | | | | | | | | | | | |
| | Construction | 5.4 | — | 6.1 | 2.1 | 2.53 | 0.71 | 10 500 | 7000 | 656.3 | 437.5 | SC2 |
| | Standard | 4.0 | — | 4.6 | 1.9 | 2.53 | 0.71 | 10 500 | 7000 | 656.3 | 437.5 | SC1 |

*Note:* For sizes of 38 × 63 and 38 × 38 refer to factors in Table 12 of BS 5268:Part 2. Refer to BS 5268:Part 2 for the grade stresses for other Canadian stress grade/species combinations.

 * In using these values, $K_7$ should be taken as 1.00.
 ** In using these values, $K_{14}$ should be taken as 1.00.
 ‡ Lateral load perpendicular to the grain for bolts and timber connectors should be multiplied by $K_{42} = 0.9$.
 − Where, in the column for tension stresses, a dash is given, this signifies that no tension value is given in BS 5268 for this stress grade.

**Table 2.5  Grade stresses and moduli for Strength Classes SC1 – SC5 for the dry exposure condition. All values are in N/mm²**

| Strength Class | Bending parallel to grain* | Tension parallel to grain**‡ | Compression | | | Shear parallel to grain | Modulus of elasticity | | Shear modulus | |
|---|---|---|---|---|---|---|---|---|---|---|
| | | | parallel to grain | perpendicular to grain | | | | | | |
| | | | | with wane | without wane | | mean | minimum | mean | minimum |
| SC1 | 2.8 | 2.2 | 3.5 | 1.2 | 2.1 | 0.46 | 6 800 | 4500 | 425 | 281.3 |
| SC2 | 4.1 | 2.5 | 5.3 | 1.6 | 2.1 | 0.66 | 8 000 | 5000 | 500 | 312.5 |
| SC3 | 5.3 | 3.2 | 6.8 | 1.7 | 2.2 | 0.67 | 8 800 | 5800 | 550 | 362.5 |
| SC4 | 7.5 | 4.5 | 7.9 | 1.9 | 2.4 | 0.71 | 9 900 | 6600 | 618.8 | 412.5 |
| SC5 | 10.0 | 6.0 | 8.7 | 2.4 | 2.8 | 1.00 | 10 700 | 7100 | 668.8 | 443.8 |

*Note*: Refer to BS 5268 : Part 2 for the grade stresses for SC6 – SC9.
*   The depth factors $K_7$ as given in Fig. 4.4 are relevant to all these bending stresses, even to 89 mm depths of North American stress grades.
**  The width factors $K_{14}$ as given in Fig. 4.4 are relevant to all these tension stresses, even to 89 mm widths of North American stress grades.
‡  The North American stress grades 'Light Framing', 'Stud', 'Structural Light Framing No. 3', and 'Joist and Plank No. 3' should not be used for members taking tension.

# Chapter Three
# Loading

## 3.1 TYPES OF LOADING

The types of loading normally taken by a timber component or structure are the same as those taken by components or structures of other materials. These are:

self-weight of the components
other dead loading such as finishes, walls, tanks and contents, partitions etc.
imposed loading such as stored materials
imposed loading caused by people either as part of an overall uniformly distributed loading or as an individual concentrated load representing one person
imposed loading such as snow on roofs
wind loading, either vertical, horizontal, inclined, external, internal, pressure or suction, or drag.

## 3.2 LOAD DURATION

Timber has the property of being able to withstand higher stresses for short periods of time than those it can withstand for longer periods or permanently. It is therefore necessary to know whether a particular type of loading is 'long term', 'medium term', 'short term' or 'very short term'. BS 5268: Part 2 gives values for a load-duration factor $K_3$ which varies from 1.00 to 1.75.

Dead loading due to self-weight, finishes, partitions etc. is obviously in place all the time and is long term loading with a $K_3$ value of 1.00. Snow occurs in the UK for periods of a few weeks, either once a year or less frequently, and falls into the category of medium term loading having a $K_3$ value of 1.25. In some countries snow may have to be treated as a long term load. Wind loading in the UK is based on the 3, 5 or 15 second gust loading from CP 3 : Chapter V : Part 2. A loading combination which contains wind at the 15 second gust level can be taken as short term loading in designs to BS 5268 and has a $K_3$ value of 1.50. A loading condition which contains wind at the 3 or 5 second gust level can be taken as very short term loading with a $K_3$ value of 1.75.

For cases other than domestic floor loading, the possible concentrated load caused by an individual person can usually be taken as a short term load. However, for domestic floors, BS 5268:Part 2 states that the concentrated load should be added to the dead load and both treated as of medium term duration.

Dead and imposed loads are covered by CP 3:Chapter V:Part 1 and the Building Regulations. These documents are under review and the designer should work to the latest document which may vary some of the comments in this chapter. A proposed Part 7 to BS 5268 may define a 'slab load' for short-span domestic joists.

When CP 3:Chapter V:Part 1 is replaced by BS6399:Part 1, it is likely that the concept of a 125 mm or 300 mm wide area on which a concentrated load has been considered to act, will disappear. Also, because this is unlikely to be defined as a loading from a man, the concept of a minimum height in which the load can act will also disappear.

## 3.3   DEAD LOADING

The dead load of materials in a construction can be established with a fair degree of accuracy. However, in some circumstances (e.g. investigating unexplained deflection on site) it is important to realise that quoted dead loadings have a tolerance and the density of materials can vary. The anticipated thickness can vary, thus affecting weight. Even changes in moisture content of a material can affect weight. The designer has to base his design on the best information available at the time of preparing the design. With regard to stress, the factor of safety takes account of any normal variation or tolerance in the weight of known materials, but with regard to deflection, any increase or decrease in the weight of materials causes a corresponding increase or decrease in the deflection. This variation normally falls within the degree of tolerance one obtains with normal building practice; however, if the designer is not informed of any major change (e.g. changing a plywood membrane for wood wool slabs) and given a chance to change the design, trouble may occur.

Section 27.1 lists the generally accepted weight of the more common building materials. When the manufacturer of an individual item is known, the manufacturer's quoted weights should be used in preference to these guide weights. Further reference can be made to BS 648:1964.

## 3.4   IMPOSED LOADINGS FOR FLOORS

CP 3:Chapter V:Part 1:1967, 'Dead and imposed loads', gives the imposed loadings to be used on floors. For beams the nature of the imposed loading is twofold, so that two design cases must be considered. In addition to dead load, it is necessary to take into account either (a) a uniformly distributed load or (b) a concentrated load applied, unless otherwise stated, over any square with a 300 mm side, whichever produces the greater stress.

Except with respect to concentrated loads on ceilings, CP 3:Chapter V: Part 1:1967 gives no guidance on the degree of permanence of an imposed load, the implications with regard to the design of timber not having been recognised. Even for domestic flooring BS 5268:Part 2 calls for the imposed uniformly distributed loading to be treated as long term even though it contains an element of medium or short term loading caused by people. If it is necessary to design for the concentrated load on domestic floors (1.4 kN in CP 3:Chapter V:Part 1) BS 5268 permits it to be added to the dead loading and both to be treated as medium term. Clause 4.1 of CP 3: Chapter V:Part 1 permits this concentrated load to be disregarded where there is effective lateral distribution of load (as one would expect with a timber joist floor).

In CP 112:Part 2 there was a limitation that, if the bending stress in a floor joist exceeded 60% of the permissible stress, $E_{min}$ had to be used in deflection calculations. This requirement (modified for domestic floors in BRE Information Sheet 26 of 1975) has been dropped in BS 5268 which permits $E_{mean}$ to be used in the design of any floor joist unless supporting an area intended for mechanical plant and equipment, or for storage, or for floors subject to considerable vibrations (such as a gymnasium or ballroom) in which case $E_{min}$ (or modified $E_{min}$) should be used.

For the special case of a house having not more than three storeys and intended for occupation by one family, The Building Regulations 1976 for England and Wales calls for different imposed loading to CP 3:Chapter V: Part 1:1967. This includes a 'slab loading' or increased uniformly distributed imposed loading which affects the design of beams with a span of less than about 2.4 m. The current (1984) belief is that future editions of the Building Regulations will refer only to CP 3:Chapter V for loading, but there is discussion that a proposed Part 7 of BS 5268 will call for a 'slab' loading for short-span timber joists.

The loading used in design examples in this manual is to CP 3:Chapter V: Part 1:1967. (To become BS 6399:Part 1:1984.)

The concentrated load on a ceiling is defined as short term in CP 3: Chapter V:Part 1.

## 3.5    IMPOSED LOADINGS FOR FLAT ROOFS

CP 3:Chapter V:Part 1:1967 gives the imposed loadings to be used on flat or sloping roofs.

On flat roofs and sloping roofs up to and including 10° for which no access is provided to the roof (other than that necessary for cleaning and repair), the imposed uniformly distributed loading including snow is taken as 0.75 kN/m² (which is usually a medium term loading condition) measured on true plan. The design must be checked for a short term load of 0.9 kN concentrated on a square with a 300 mm side, added to the dead loading but not to the uniformly distributed imposed loading. This latter case is critical only on short spans. The concentrated load should be placed at the position which will give the worst loading condition. Lateral distribution of the concentrated load may reduce the effect on any one member.

On flat roofs and sloping roofs up to and including 10°, for which access (in addition to that necessary for cleaning and repair) is provided, the imposed uniformly distributed loading including snow is taken as 1.5 kN/m² (which is usually a medium term loading condition) measured on true plan. The design must be checked for a short term load of 1.8 kN concentrated on a square with a 300 mm side, added to the dead loading but not to the uniformly distributed loading. This latter case may be critical only on short spans. Lateral distribution of the concentrated load may reduce the effect on any one member.

When the purpose of a roof changes from one without permanent access to one with permanent access, the increase in the imposed loading must be considered as well as any addition to the dead loading.

## 3.6   SNOW LOADING

Freshly fallen snow weighs approximately 0.8 kN/m³ but compacted snow can weigh 3.2 kN/m³. It may be necessary to take note of this latter figure in the design of certain site works. In designs for extreme weather conditions (usually abroad) the designer should also consider the possibility of snow turning into frozen ice, which increases the weight more than threefold.

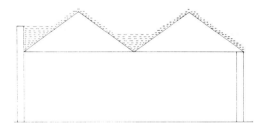

Fig. 3.1: *Build-up of snow behind deep fascia and in a valley.*

The uniformly distributed loading including snow is given for roofs in CP 3:Chapter V:Part 1. In addition, the designer should examine the roof shape to see whether fascias, valleys etc. could lead to a build-up of snow with consequent increase in load (see Fig. 3.1). This is particularly important on slopes approaching 75° because on slopes of 75° or more CP 3:Chapter V: Part 1 requires no allowance to be made for any imposed loading. The inter-polation between 75° and 30° is linear (see Fig. 3.2). It is likely that the revised version of CP 3:Chapter V:Part 1 will amend snow loading.

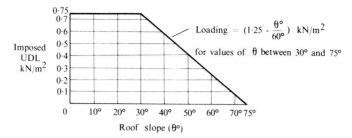

Fig. 3.2: *Imposed uniformly distributed loading, including snow, on sloping roofs without permanent access to the roof.*

The designer is advised to read the clauses on snow loading in BS 5502: Part 1 : Section 1.2, 'Code of practice for design of buildings and structures for agriculture'.

### 3.7   IMPOSED LOADINGS FOR SLOPING ROOFS

CP 3 : Chapter V : Part 1 : 1967 gives the imposed loading to be used on flat or sloping roofs.

On roofs with a slope greater than 10° and with no access provided to the roof (other than that necessary for cleaning and repair) the imposed uniformly distributed loading including snow is taken as $0.75\,kN/m^2$ measured on true plan for slopes up to 30°, and as zero at 75°, with linear interpolation between (see Fig. 3.2). Designs should be checked for a short term load of 0.9 kN concentrated on a square with a 300 mm side, added to the dead loading but not to the uniformly distributed imposed loading. This latter case may be critical only on short spans. Lateral distribution of the concentrated load may reduce the effect on any one member.

### 3.8   IMPOSED LOADINGS FOR CURVED ROOFS

The imposed uniformly distributed loading on a curved roof is calculated by dividing the roof into at least five equal segments and then taking the loading on each, appropriate to its mean slope.

### 3.9   IMPOSED LOADINGS FOR CEILINGS

The supports for ceilings, other than those for which the ceiling space is inaccessible or the ceiling is demountable for access, or the ceiling space is provided with catwalks supported independently, have to be designed for certain imposed loadings. CP 3 : Chapter V : Part 1 : 1967 gives this loading as $0.25\,kN/m^2$ uniformly distributed and 0.9 kN concentrated on a length or square of 125 mm so placed as to provide the maximum stress or deflection in the affected members. The UDL is long term. The concentrated load is short term.

Whereas with floors and roofs the uniformly distributed loading and the concentrated load are alternative imposed loadings, with ceilings they are both additive to the dead loading. However, the design case including the concentrated load is a short term condition, therefore it may not be critical in relation to bending or shear stress. With deflection, however, it will always be more critical than the case of dead load plus the uniformly distributed imposed loading.

The concentrated load is intended to represent the weight from a man. Where, for example, with a ceiling under a pitched roof the vertical height reduces to below 1.0 m, it is usually acceptable to consider that a man would not go into that part of the roof space and therefore that the concentrated load need not be considered to act in that particular area. (Specific regulations may vary this.)

A ceiling construction is not normally a load-sharing system, therefore the full concentrated load usually has to be considered in the design of each member (although the new trussed rafter code may permit some load sharing to be considered).

## 3.10 IMPOSED LOADING FOR ONE-FAMILY HOUSES NOT EXCEEDING THREE STOREYS

Part D2 of The Building Regulations 1976 for England and Wales permits 'the imposed loading on a floor, ceiling or roof of a house having not more than three storeys and intended for occupation by one family only' to be taken as differing somewhat from CP 3 : Chapter V : Part 1 : 1967, the ceiling loading showing the biggest difference. The loading is detailed in Part D2(3). The choice of which loading requirement to take is usually left to the discretion of the designer when designing one-family houses. Schedule 6 of The Building Regulations gives span tables which are calculated on the D2(3) loadings for the floor, roof and ceiling members usually encountered in one-family houses, and no further design is required for these specific cases. The timber sections used to satisfy these tables are based primarily but not exclusively on grades GS and SS of BS 4978 : 1973. It is expected that future editions of The Building Regulations for England and Wales will specify loading to CP 3 : Chapter V : Part 1 as is currently the case with the Regulations for Scotland.

Schedule 6 does not make provision in the span tables for partitions (unless the designer can account for this in the uniform dead loading) and this point must be recognised in calculating design loading.

## 3.11   WIND LOADING

The wind loading for which structures must be designed in the UK is calculated from CP 3:Chapter V:Part 2:1972. Timber is frequently associated with buildings of lightweight construction, and the general lack of mass usually requires attention to be paid to wind loading, as it affects not only the stress in individual members and the overall stability but connections and anchorage. (CP 3:Chapter V:Part 2 is under review.)

The 'basic wind speed' is based on the maximum 3 second gust speed likely to be exceeded on average only once in 50 years at 10 m above the ground in open level country. CP 3:Chapter V:Part 2:1972 gives a map of the UK showing which basic wind speeds are applicable to which areas. The map necessarily incorporates the effects of the general level of the ground above mean sea level. A table is also given of wind speeds relevant to several towns and cities.

The 'design wind speed' $V_s$ is obtained by modifying the basic wind speed $V$ by three factors $S_1$, $S_2$ and $S_3$. The topography factor $S_1$ takes account of large local variations in the ground surface, hills, valleys etc. Three values (1.0, 1.1 and 0.9) are given but, except for very exposed or very sheltered locations, it is usual to take $S_1$ as being 1.0. The factor $S_2$ takes account of 'ground roughness, building size and height above ground'. Table 3 of CP 3:Chapter V:Part 2:1972 lists values of $S_2$ varying from classes of 'open country with no obstructions' to 'city centres' related to cladding or building size (Classes A, B or C) where each class is described as follows:

Class A   All units of cladding, glazing and roofing and their immediate fixings, and individual members of unclad structures.

Class B   All buildings and structures for which neither the greatest horizontal dimension nor the greatest vertical dimension exceeds 50 m.

Class C   All buildings and structures whose greatest horizontal dimension or greatest vertical dimension exceeds 50 m.

Class A is based on a 3 second gust (appropriate to cladding design and fixings), whereas Class B is based on a 5 second gust and Class C on a 15 second gust to take account of the natural period of oscillations of buildings of this size. Most timber design is concerned with Class B.

CP 3:Chapter V:Part 2 states that, for the design of cladding or its fixings to a structural member, Class A should be used. For the design of a structural member carrying the cladding, Class B or C should be used with the pressure coefficient applicable to the area in which the member lies. In considering the design against high local pressures of the structural member carrying the cladding, the secondary effects such as distribution due to the stiffness of the cladding should be taken into account. For main structural members the design should be to Class B or C using the normal coefficients for the whole area.

Factor $S_2$ also deals with height above the surrounding ground. For a low level building in a sheltered place the value of $S_2$ can be as low as 0.52 for Class B and 0.47 for Class C.

Designers in timber are usually concerned with Class B buildings and heights between 3 and 10 m. Table 3.1 gives values of $S_2$ for increments of 0.5 m for $H$ for Class B buildings interpolated from the values in Table 3 of CP 3 : Chapter V : Part 2 : 1972.

**Table 3.1  Factor $S_2$ (ground roughness, and height above ground) for Class B buildings**

| $H$ (m) | (1) Open country with no obstructions | (2) Open country with scattered windbreaks | (3) Country with many windbreaks; small towns; outskirts of large cities | (4) Surface with large and frequent obstructions, e.g. city centres |
|---|---|---|---|---|
| 3 | 0.780 | 0.670 | 0.600 | 0.520 |
| 3.5 | 0.7925 | 0.6875 | 0.6125 | 0.5275 |
| 4 | 0.805 | 0.705 | 0.625 | 0.5350 |
| 4.5 | 0.8175 | 0.7225 | 0.6375 | 0.5425 |
| 5 | 0.830 | 0.740 | 0.650 | 0.550 |
| 5.5 | 0.842 | 0.754 | 0.659 | 0.557 |
| 6 | 0.854 | 0.768 | 0.668 | 0.564 |
| 6.5 | 0.866 | 0.782 | 0.677 | 0.571 |
| 7 | 0.878 | 0.796 | 0.686 | 0.578 |
| 7.5 | 0.890 | 0.810 | 0.695 | 0.585 |
| 8 | 0.902 | 0.824 | 0.704 | 0.592 |
| 8.5 | 0.914 | 0.838 | 0.713 | 0.599 |
| 9 | 0.926 | 0.852 | 0.722 | 0.606 |
| 9.5 | 0.938 | 0.866 | 0.731 | 0.613 |
| 10 | 0.95 | 0.880 | 0.740 | 0.620 |

$S_3$ is a factor, based on statistical concepts, which takes account of the degree of security required and the period of time in years during which there will be exposure to wind. Excluding temporary structures or certain exceptional cases it is usual to take $S_3$ as 1.0. Particularly when a designer is unfamiliar with the area in which a building is to be designed, he should consider asking the local authority for guidance on which wind speed, $S_1$ and $S_2$ values to use and the relevant local authority should be able to supply this information.

The 'dynamic pressure of the wind' $q$ is derived from the design wind speed $V_s$. The value $q = 0.613\ V_s^2$ with $q$ in N/m² and $V$ in m/s. CP 3 : Chapter V : Part 2 : 1972 tabulates values of $q$ for $V_s$.

For the frequently occurring case of a Class B building with $S_1 = S_3 = 1.0$, the dynamic pressure of the wind ($q$) is given by the formula:

$$q = 0.613\ (S_2 V)^2$$

where $q$ = dynamic pressure of the wind in N/m$^2$
$\quad$ $V$ = basic wind speed in m/s.

The actual pressure or suction on walls or roofs is obtained by multiplying $q$ by pressure coefficients $C_{pe}$ to obtain the external pressure or suction and by $C_{pi}$ to obtain the internal pressure or suction. Coefficients $C_{pe}$ are obtained by reference to tables in CP 3:Chapter V:Part 2:1972 which show buildings of many differing proportions. Coefficients $C_{pi}$ are obtained by reference to Appendix E of CP 3. The net pressure on any surface is the algebraic addition of $C_{pe}$ and $C_{pi}$ and is multiplied by the effective area to give the loading on the structure.

On certain buildings it may be necessary to consider wind drag. The designer is referred to the code for full data, but should realise that wind is a complex consideration and the loadings arrived at from CP 3:Chapter V:Part 2:1972 are at best an approximation of an extremely complicated loading condition.

The *Wind Loading Handbook* by BRE gives confirmation that 'local coefficients' which are relevant to areas near the edges, ends or ridges apply to cladding and fixings, not to any structural member.

Where there is a net wind uplift or a net overturning moment due to wind, it is usual practice to increase the overturning effect or uplift value by a factor of 1.33 when designing anchorage. Currently this is not documented in Codes (although the proposal is to use a factor of 1.4 in BS 5268:Part 3).

## 3.12   UNBALANCED LOADING

The critical condition of loading in the overall design of a component is usually the case with the total loading in position. However, particularly with lattice constructions, the maximum loading in some of the internal members occurs with only part of the imposed loading in position.

The vertical shear in diagonal X in Fig. 3.3 is zero with the balanced loading. The maximum force in diagonal X occurs with the unbalanced loading shown in Fig. 3.4. The vertical shear and hence the force in diagonal Y is less with the balanced loading shown in Fig. 3.5 than with the unbalanced loading shown in Fig. 3.6.

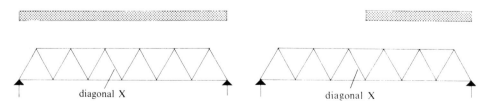

diagonal X                                          diagonal X

Fig. 3.3                          Fig. 3.4

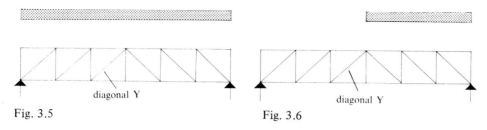

Fig. 3.5 Fig. 3.6

Obviously the positioning of the dead loading cannot vary, but the designer should be alert to the effect on certain members (particularly internal members in frameworks) caused by unbalanced imposed loadings. It is most important to consider this when the balanced loading leads to zero force in any particular member. Unbalanced loading can occur during erection.

## 3.13 COMBINATIONS OF LOADING

The designer must consider all combinations of the loadings of which he is made aware. Common design cases are dead plus imposed, dead plus wind, dead plus imposed plus wind. On normal roofs for example dead plus wind is usually a more critical design case than dead plus imposed plus wind.

An interesting series of loading combinations is the one as detailed in the latest draft of BS 5268:Part 3 for trussed rafter design. These combinations are dead loading (including water tanks) plus the following imposed loading:

a   imposed UDL on ceiling tie     (long term)
b   imposed UDL on rafter and ceiling tie     (medium term)
c   imposed UDL on rafter and ceiling tie, and concentrated load on the ceiling tie     (short term)
d   imposed UDL on ceiling tie, and concentrated load on the rafter     (short term)
e   wind loading (Class B structure)     (very short term)
f   wind loading (Class B structure), and imposed UDL on ceiling tie, and concentrated load on ceiling tie     (very short term)
g   as Case f plus imposed UDL on rafter.

Occasionally Class C wind loading may be relevant.

A small amount of load sharing of the concentrated load on the ceiling may be permitted.

Often a designer will be sufficiently experienced to be able to identify the two or three critical combinations and may not have to work through all possible combinations.

## 3.14    SPECIAL LOADINGS

The designer should be alert to the possible load and deflection implications in buildings from various items not all of which are within his control. A few are discussed below.

### 3.14.1    Sliding doors

A sliding door can be bottom run, in which case the designer has to ensure that the deflection of any component above the door will not prevent the door from opening or closing. Often the extent of vertical adjustment or tolerance built into door runners is unrealistically small.

A sliding door can be top hung, in which case one can either provide a separate support beam or beams (Fig. 3.7) so that deflection of the main support beams will not cause the door to jam, or provide short beams between the main beams (Fig. 3.8), so that if it is found that too little adjustment has been provided in the door mechanism, adjustment to the secondary beams can be carried out without major structural alterations. It is not advisable to hang a door directly under a main beam (also see §12.6.5).

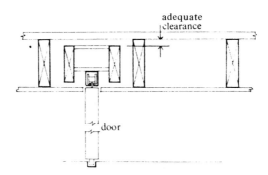

Fig. 3.7

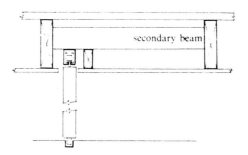

Fig. 3.8

### 3.14.2   Water tanks

Normally the critical design condition occurs with a tank full of water (a long term loading condition), but occasionally an unacceptable gap can open up when the tank is emptied. If the tank is placed to give unbalanced loading on a component (see § 3.12), the act of emptying it can lead to reversal of stress. The case with the tank empty is likely to be a medium term loading condition, but could be long term if a building ceases to be used for a time (or even during construction). If no overflow is provided the tank may become more full than is intended.

### 3.14.3   Roofs under high towers or masts

Roofs under high towers or tall masts have been penetrated by icicles which form then break off. This potential problem should be discussed in the design of such roofs.

### 3.14.4   Expansion/shrinkage

The coefficient of linear thermal expansion of timber is small – around 0.000 005 4 per °C for Redwood and Whitewood – therefore expansion joints are required only very occasionally unless made necessary by the cladding. Expansion due to increase in moisture content in service is not normally considered in structural design because in normal building this will not happen if the detailing and site control is correct. However, consideration must be given to the type of differential settlement which can occur during the initial drying-out of a structure, for example, with a timber-framed house having an outer skin of brickwork. Even when the timber in the stud walls has been dried to 18–20% moisture content before installation, the total settlement within a two-storey height can be in the order of 5–10 mm mainly due to shrinkage across the grain of the joists and horizontal members as the moisture content reduces to about 14–15%.

### 3.14.5   Accidental loading

In certain areas the designer may find himself called upon to take account of possible loading (e.g. 34 kN/m$^2$) from explosion of a gas appliance. In such a case the 'Accidental damage' clause in BS 5268 : Part 2 permits an increase (usually of 100%) in permissible stresses and permissible loads in fasteners. See BS 5268 for further details on the requirements to take account of possible accidental damage and, for a building having five or more storeys, Regulations D17 and D18 of The Building Regulations 1976 for England and Wales and C3 of The Scottish Regulations 1981.

Chapter Four
# The Design of Beams.
# General Notes

## 4.1 RELATED CHAPTERS

This chapter deals in detail with the considerations necessary for the design of beams. As such it is a reference chapter. Several other chapters deal with the actual design of beams, each one being devoted to one main type:

Chapter 5     Beams of Solid Timber plus Timber and Plywood Decking. Principal members, load-sharing systems, tongued and grooved solid decking and floor boards, plywood decking.

Chapter 6     Beams of Simple Composites. Composites of two or three sections, rectangular, Tee or I shapes, glued or nailed composites.

Chapter 7     Glulam Beams. Vertical and horizontal glulam, parallel or tapered beams.

Chapter 8     Ply Web Beams. I beams and Box beams.

Chapter 9     Lateral Stability of Beams. Full or partial lateral restraint.

Chapter 10     Stress Skin Panels. Single and double skin.

Chapter 11     Parallel-chord Lattice Beams.

Chapter 12     Deflection. Practical and Special Considerations.

## 4.2 DESIGN CONSIDERATIONS

The principal considerations in the design of beams are:

bending stress
prevention of lateral buckling
shear stress
deflection: both bending deflection and shear deflection
bearing at supports and under any point loads
prevention of web buckling (with ply web beams)

The size of timber beams may be governed by the requirement of the section modulus to limit the bending stress without lateral buckling of the compression edge or fracture of the tension edge, or of the section inertia to limit deflection, or the requirement to limit the shear stress.

Generally, bending is critical for medium-span beams, deflection for long-span beams and shear only for heavily loaded short-span beams. In most cases it is advisable however to check for all three conditions.

The relevant permissible stresses are computed by modifying the grade stresses by the factors from BS 5268:Part 2 which are discussed in this chapter.

## 4.3   EFFECTIVE DESIGN SPAN

In simple beam design it is normal to assume that the bearing pressure at the end of a simply supported beam is uniformly distributed over the bearing area and no account need be taken of the difference in intensity due to rotation of the ends of the beam (Clause 14.2 of BS 5268:Part 2). If, therefore, only sufficient bearing area is provided to limit the actual bearing pressure to the permissible value, the design span should be taken as the distance between centres of bearing. In many cases, however, the bearing area actually provided is much larger than that required to satisfy the bearing pressure. This obviously does not weaken the beam therefore, in the calculations, the effective design span need be taken only as the distance between the bearing areas which are required for bearing stress limitations, rather than the centres of the bearing areas actually provided (Fig. 4.1).

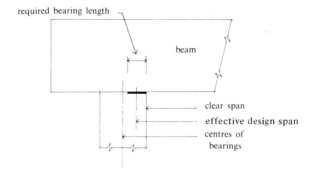

Fig. 4.1

In determining the effective design span from the clear span, as an approximation it is usually acceptable to assume an addition of around 0.05 m for solid timber joists and 0.1 m for built-up beams on spans up to around 12 m, but more care is needed with beams of longer span or with heavily loaded beams. The effective design span can be given the symbol $L_e$ although $L$ is used in most of the design examples in this manual. Even if a beam is built-in and the loading occurs, therefore, only over the clear span it is usual to assume in designs that the loading occurs over the effective design span.

## 4.4   LOAD-SHARING SYSTEMS

### 4.4.1   Lateral distribution of load

When a beam system has adequate provision for lateral distribution of loads it can be considered to be a load-sharing system. In such a system a member less stiff than the rest will tend to deflect more than the other members. As it does, the load on it will be transferred laterally to adjoining stiffer members. Rather than give load-relieving factors, this load sharing is dealt with in BS 5268: Part 2 by permitting higher values of $E$ than $E_{min}$, and increases in grade stresses in the design of certain components.

In the special case of four or more members occurring in a load-sharing system and being spaced not further apart than 610 mm (as with a domestic floor joist system) the appropriate grade stresses may be increased by 10% (factor $K_8$ of 1.1), and $E_{mean}$ (and $G_{mean}$) may be used in calculations of deflection (although BS 5268 calls for the use of $E_{min}$ if a flooring system is supporting an area intended for mechanical plant and equipment, or for storage, or for floors subject to considerable vibrations such as gymnasia and ballrooms).

Factor $K_8$ applies to bearing stresses as well as bending, shear and, where appropriate, to compression or tension stresses parallel to grain.

In the case of two or more members fastened together securely to form a trimmer joist or lintel, the grade stresses in bending and shear parallel to the grain and compression perpendicular to the grain (i.e. bearing) may be increased by 10% (factor $K_8$ of 1.1). Also, $E_{min}$ may be increased by factor $K_9$. See § 2.3 and Fig. 2.3 in this manual

Where members are glued together but (unlike a glulam beam) have their strong axis in the vertical plane of bending, and are of members stress graded GS, SS etc. rather than LB or LC, the bending, tension and shear parallel to grain stresses may be increased by factor $K_{27}$ (see § 2.5 and Fig. 2.9), and $E_{min}$ and compression parallel to grain may be increased by factor $K_{28}$ (see §2.3 and Fig. 2.3). Compression perpendicular to grain stresses may be increased by 10% (i.e. $K_9 = 1.10$) and permissible bearing stresses can be further increased if wane is excluded (to $1.33 \times 1.10 \times K_3 \times$ the grade bearing stress for SS grade for the species with wane present).

If one compares $K_8$ with $K_{27}$ and $K_{28}$ (for compression parallel to the grain) it can be seen that there are some inconsistencies. For example, if a trimmer consists of two members nailed together, the grade stresses for bending and shear parallel to the grain may be multiplied by $K_8 = 1.10$, whereas, if the members are glued together, the relevant factor is $K_{27}$ at a value of 1.11. (The modifications to $E$ value do line through.) Also, if a compression member consists of two members nailed together, the grade stress for compression parallel to the grain can not be increased (although $K_8$ in Clause 13 of BS 5268: Part 2 permits a 10% increase in four or more members acting at 600 mm centres in a stud wall, and $K_8$ in Clause 14.11 of

BS 5268:Part 2 permits a 10% increase to compression perpendicular to the grain for two or more pieces connected together). However, if a compression member consists of two or more members glued together, a $K_{28}$ value of 1.14 may be taken.

This should not be a serious matter for a designer, although, if designing a two-member column which also takes a bending moment, which is supported at one end on an inclined bearing surface, it may seem odd to increase the permissible bending stress parallel to grain for load sharing but not the permissible compression stress parallel to grain. Also, when considering end bearing, the permissible bearing stress perpendicular to grain may be increased for load sharing, but not the permissible bearing stress parallel to grain, even though the bearing stresses being considered are at the same position.

It is not normal to design glulam members (horizontally laminated) as part of a load-sharing system, however, BS 5268:Part 2 recognises the load-sharing effect within each member of four or more laminates by giving various coefficients $K_{15}-K_{20}$ (see Table 7.1) for members made from one grade of laminate, and $K_{21}-K_{26}$ (see Table 7.2) for members made from two grades of laminate.

In the case of a built-up beam such as a ply web beam BS 5268:Part 2 recognises the load-sharing effect by permitting the use in designs of factors $K_{27}$, $K_{28}$ and $K_{29}$ as for vertically glued members. In applying $K_{27}$ to the grade stresses it has been agreed that the value of $N$ is appropriate to the number of members in each flange (see § 2.5.2 and Fig. 2.9). In applying $K_{28}$ to $E_{min}$ it has been agreed that the value of $N$ is appropriate to the total number of solid timber members in cross-section (see Table 2.1). Because $K_{29}$ is the factor which relates to bearing, the same comment given above for vertically glued members applies also to ply web beams.

Very often, ply web beams are placed at 1.2 m centres with secondary noggings fixed at regular intervals between them at the level of the top and bottom flanges. As such, many practising engineers have considered this to be a load-sharing system. Some engineers have used $E_{mean}$ in calculations of deflection whilst some have used $E_N$ (based on guidance in the First Edition of this manual). $E_N$ from BS 5268:Part 2 would now be derived as $E_{min} \times K_{28}$. The extent to which a ply web beam system can be taken as a load-sharing system is discussed in Chapter 8 in relation to $E$ value (and $G$ value) and in § 19.7.3 in relation to finger jointing.

### 4.4.2   Concentrated load. Load-sharing system

When a concentrated load, of the magnitude of that specified in CP 3: Chapter V:Part 1':1967, acts on a floor supported by a load-sharing system, Clause 4.1 of that Code permits it to be disregarded in the design of the members. When the concentrated load acts on a flat roof or sloping roof (§§ 3.5 and 3.7) it must be considered, particularly for short-span beams.

In a load-sharing system lateral distribution of the load will reduce the effect on any one member, particularly if the decking is quite thick and the effect can 'spread' (sideways and along the span) through the thickness of the decking. It is usually acceptable to assume that not more than half the quoted concentrated load will occur on any one member in a load-sharing system.

For domestic floors BS 5268:Part 2 states that the concentrated load defined in CP 3:Chapter V:Part 1 should be taken as medium term. BS 5268:Part 3 gives confirmation that the concentrated load acting at ceiling or rafter level should be taken as short term. This is usually taken as a load from a person.

Other concentrated loads (as from partitions) must be given special consideration. One case which is open to discussion is illustrated in Fig. 4.2. The usual questions are whether or not the Joists B supporting the partition should be designed using $E_{mean}$ when they are part of a load-sharing system, and whether or not the other Joists A will carry a share of the load from the partition. Providing it will not lead to significant differential deflection between Joists A and B it is probably better to assume that Joists B support the total weight of the partition and a share of the floor loading and have an $E$ value of $E_N$. It is prudent to carry out an additional deflection check to ensure that, under dead loading only, the deflection of Joists A and B is similar and that any difference is unlikely to be noticeable.

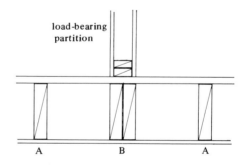

Fig. 4.2

## 4.5   LOAD-DURATION FACTOR

Timber (and plywood) can sustain a very much greater load for a short period of time than it can for a longer period of time or permanently. BS 5268 takes account of this by quoting duration of loading factors $K_3$ for medium term, short term and very short term loadings. These are shown in Table 4.1. These modification values are applicable to all strength properties including bearing, but are not to be applied to moduli of elasticity or shear moduli. No load-duration factors are given in BS 5268 for $E$ or $G$ values.

**Table 4.1**    Modification factor $K_3$ for duration of loading

| Duration of loading | Value of $K_3$ |
|---|---|
| Long term (e.g. dead + permanent imposed)<br>(domestic UD loading is to be considered in this category) | 1.0 |
| Medium term (e.g. dead + snow<br>dead + temporary imposed)<br>(for domestic floors the concentrated-loading condition given in CP 3:Chapter V:Part 1 – which can sometimes be disregarded – is to be considered as in this category) | 1.25 |
| Short term (e.g. dead + weight of a man<br>dead + imposed + weight of a man<br>dead + imposed + wind of Class C)<br>dead + imposed + snow + wind of Class C | 1.50 |
| Very short term (e.g. any combination of loading containing wind of Classes A or B) | 1.75 |

It has been found, however, that, at design load, a timber beam does not deflect by the full calculated amount until several hours after the application of a load. The testing clause of BS 5268 requires that only 80% of the calculated deflection should have taken place after 24 hours. There is therefore a case for disregarding or modifying deflection calculations in some cases when considering wind gust loading. At least it should be safe to assume that only 80% of the deflection calculated by assuming full gust loading will take place during the few seconds the load is assumed to act. This is a point for the designer to discuss with the approving authority.

The design must be checked to ensure that the permissible stresses are not exceeded for any relevant condition of loading. Because of the duration of loading factors it is possible, for example, for the design case of dead loading only, to be more critical than the case of dead plus imposed roof loading.

Different values are given in BS 5268 for load-duration factors on tempered hardboard ($K_{38}$–$K_{41}$ including load-duration factors for $E$ and $G$ values), and for fasteners ($K_{48}$, $K_{52}$, $K_{55}$, $K_{58}$, $K_{62}$, $K_{66}$).

## 4.6   LATERAL STABILITY

There are several methods of linking the degree of lateral stability of a timber beam to the permissible stress in bending. Those from BS 5268: Part 2 are described below and special methods for designing for conditions of partial restraint are covered in Chapter 9.

### 4.6.1   Maximum depth-to-breadth ratios (solid and laminated members)

BS 5268 gives limiting values of depth-to-breadth ratios which, if complied with for solid or laminated members, justify use of the full grade bending stress (increased by the load-duration or load-sharing factor if applicable) in calculations. These ratios are given in Table 17 of BS 5268:Part 2 and are presented in Table 4.2. When the depth-to-breadth ratio exceeds the value corresponding to the appropriate degree of lateral support, BS 5268 permits the designer to check to ensure that there is no risk of lateral buckling under the design loading. This usually requires in designs the use of an appropriately reduced bending stress. One method determined by the authors is discussed in § 9.4.

**Table 4.2**

| Degree of lateral support | Maximum depth-to-breadth ratio |
|---|---|
| No lateral support | 2 |
| Ends held in position | 3 |
| Ends held in position and members held in line, as by purlins or tie rods | 4 |
| Ends held in position and compression edge held in line, as by direct connection of sheathing, deck or joists | 5 |
| Ends held in position and compression edge held in line, as by direct connection of sheathing, deck or joists, together with adequate bridging or blocking spaced at intervals not exceeding six times the depth | 6 |
| Ends held in position and both edges firmly held in line | 7 |

### 4.6.2   Maximum second moment of area ratios (built-up beams)

BS 5268 requires built-up beams such as ply web beams to be checked to ensure that there is no risk of buckling (of the compression flange) under design load. This checking can be by special calculation, or by considering the compression flange as a column, or by providing lateral or end restraint as detailed in Clause 14.10 of BS 5268:Part 2 which is summarised in Table 4.3 of this manual related to the ratio of second moments of area.

Designers often find that it is easier and more acceptable to match a restraint condition appropriate to a higher $I_X/I_Y$ ratio than the one actually occurring, and it must be obvious that this should be acceptable to any approving authority. For example, with beams having an $I_X/I_Y$ ratio between 31 and 40, it is more likely that the designer will provide full restraint to the compression flanges rather than a brace at 2.4 m centres, particularly if the beams are exposed. Although this does not satisfy a pedantic interpretation of BS 5268, it provides more restraint and more than satisfies the intention of the clause.

**Table 4.3    Maximum second moment of area ratios related to degree of restraint**

| $\dfrac{I_X}{I_Y}$ | Degree of restraint |
|---|---|
| Up to 5 | No lateral support required |
| 5–10 | The ends of beams should be held in position at the bottom flange at the supports |
| 10–20 | Beams should be held in line at the ends |
| 20–30 | One edge should be held in line (N.B. not necessarily the compression edge) |
| 30–40 | The beam should be restrained by bridging or bracing at intervals of not more than 2.4 m |
| More than 40 | Compression flange should be fully restrained |

As stated above, BS 5268 permits the compression flange of a ply web beam to be considered as a horizontal column which will tend to buckle laterally between points of restraint. Such an approach is conservative, makes no recognition of the stabilising influence of the web and tension flange and the height of application of the load in the beam depth, but is set out in §9.7 as it can be a useful design method.

In Chapter 9 the authors set out two further methods of considering lateral restraint of built-up beams.

## 4.7   MOISTURE CONTENT

It is considered that solid timber of a thickness of about 100 mm or less (and glulam of any thickness) in a covered building in the UK will have an equilibrium moisture content of 18% or less. BS 5268:Part 2 gives stresses and $E$ values for timber for two service conditions which are:

(i) dry exposure conditions where the moisture content of solid timber does not exceed 18% for any significant period and
(ii) wet exposure conditions where the moisture content of solid timber exceeds 18% for significant periods.

Note that, although BS 5268 still refers to 'green' timber as timber having a moisture content in excess of 18% due to the environmental conditions, the term 'green' which was used in CP 112 in relation to stresses and service conditions is replaced in BS 5268 by 'wet'.

For solid sections (not glulam) exceeding 100 mm in thickness, BS 5268: Part 2 states that generally 'wet' stresses should be used in the structural design because it is considered that full loading could be applied before drying is complete. (Obviously if the sections are carefully dried before installation this restriction does not apply.) Also, it is as well to be aware that sections of 100 mm thickness or thicker are liable to contain shakes after drying and it is advisable not to use such a section if it is to be exposed where appearance is important. It is better to use a glulam or composite section.

BS 5268:Part 2 does not tabulate stresses for wet exposure conditions but gives factors $K_2$ by which dry values should be multiplied, and factors $K_1$ by which section properties should be multiplied, if the design is for a wet exposure condition. The same basic idea is used for plywood with $K_{36}$ factors being given by which stresses should be multiplied, but no factor is included for section properties.

Although it is preferable for timber to be dried before installation to a moisture content within a few percent of the level it will attain in service, it is essential, if dry stresses have been used in the design, for the timber to have dried to around 18% moisture content before a high percentage of the total design load is applied. If it is found that beams are still well over 18% moisture content when the time comes to apply, say, more than about one half of the total dead load then, either loading should be delayed or the beams should be propped until the moisture content is brought down to 18% or less, otherwise 'creep deflection' may take place (§4.16). If beams are erected at a moisture content of 22% or less in a situation where drying can continue freely and only part of the dead load is applied, there is very little likelihood of serious trouble from creep deflection.

BS 5268 tabulates 18% moisture content values as dry values and gives factors from which wet values may be calculâted which are the values appropriate to the fibre saturation point.

A designer may wish to prepare a design for a country or condition where a different moisture content is applicable for dry conditions. The relationship of stress and $E$ values to moisture content for solid timber is given by the formulae:

$$\log \sigma_{M1} = \log \sigma_{Ms} + C\,(M_s - M_1)$$
$$\log E_{M1} = \log E_{Ms} + C\,(M_s - M_1)$$

where

$$\sigma_{M1}, E_{M1} = \text{the stress and } E \text{ values at moisture content } M_1$$
$$\sigma_{Ms}, E_{Ms} = \text{the stress and } E \text{ values at moisture content } M_s$$
$$\text{(moisture saturation point)}$$
$$C = \text{a constant.}$$

The fibre saturation point for European Redwood or Whitewood can generally be taken as 30% moisture content. The value for other timbers can be obtained from timber technology books. The constant $C$ is obtained by substituting into the formula the known values of $\sigma_{M1}$, $\sigma_{Ms}$, $M_s$, $M_1$, $E_{M1}$ and $E_{Ms}$ at the 18% moisture content and fibre saturation points (i.e. the dry and wet values from BS 5268).

## 4.8   BENDING STRESSES

At the stress levels permitted by BS 5268 it is acceptable to assume a straight-line distribution of bending stress in solid beams between the maximum value in the outer compression fibres and the outer tension fibres (Fig. 4.3). See Chapters 8 and 10 for the variations applicable with ply web beams and stress skin panels.

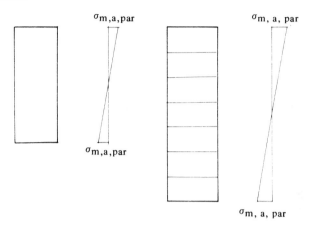

Fig. 4.3

The symbols used for bending parallel to the grain in BS 5268 are:

$\sigma_{m,adm,par}$   permissible bending stress
$\sigma_{m,a,par}$   applied (or actual) bending stress
$\sigma_{m,g,par}$   grade bending stress.

BS 5268 permits some of the subscripts to be omitted if there is no chance of misunderstanding.

Bending stresses are influenced by depth factor (see § 4.9). Load-sharing and load-duration factors are applicable to bending stresses.

## 4.9   DEPTH AND FORM FACTORS

### 4.9.1   Depth factor for flexural members

Whereas CP 112 : Part 2 : 1971 gave a depth factor which applied only to solid and laminated beams having a depth greater than 300 mm, BS 5268 : Part 2 retains the same value for this depth factor (given the number $K_7$) for depths in excess of 300 mm. It also introduces another formula (still given the number $K_7$) for depths between 72 and 300 mm, which is applicable to the grade bending stresses tabulated in BS 5268 for Strength Classes, timber graded to BS 4978, glulam, and North American 'Joist and Plank' grades. The values for $K_7$ for these combinations are:

1.17  for depths of 72 mm or less

$$\left(\frac{300}{h}\right)^{0.11}$$ for solid and glued laminated members having a depth between 72 and 300 mm

$$0.81 \times \left(\frac{h^2 + 92\,300}{h^2 + 56\,800}\right)$$ for solid and glued laminated members having a depth greater than 300 mm

These values are presented in Table 4.4 and in graph form in Fig. 4.4. Depth $h$ is the dimension in the line of bending.

Table 4.4  $K_7$ and $K_{14}$ values for Strength Classes, glulam, solid timber graded to BS 4978, and North American 'Joist and Plank' grades†

| $K_7$ mm | $K_7$ | $K_7$ mm | $K_7$ | $K_{14}$ mm | $K_{14}$ | $K_{14}$ mm | $K_{14}$ |
|---|---|---|---|---|---|---|---|
| 72 | 1.17* | 315 | 0.994 | 72 | 1.17* | 315 | 0.995 |
| 75 | 1.165* | 350 | 0.970 | 75 | 1.165* | 350 | 0.983 |
| 89 | 1.143* | 360 | 0.964 | 89 | 1.143* | 360 | 0.980 |
| 97 | 1.132* | 400 | 0.943 | 97 | 1.132* | 400 | 0.969 |
| 100 | 1.128* | 405 | 0.940 | 100 | 1.128* | 405 | 0.968 |
| 122 | 1.104 | 450 | 0.921 | 122 | 1.104 | 450 | 0.956 |
| 125 | 1.101 | 495 | 0.905 | 125 | 1.101 | 495 | 0.946 |
| 140 | 1.087 | 500 | 0.904 | 140 | 1.087 | 500 | 0.945 |
| 145 | 1.083 | 540 | 0.893 | 145 | 1.083 | 540 | 0.937 |
| 147 | 1.082 | 550 | 0.890 | 147 | 1.082 | 550 | 0.936 |
| 150 | 1.079 | 585 | 0.882 | 150 | 1.079 | 585 | 0.929 |
| 169 | 1.065 | 600 | 0.879 | 169 | 1.065 | 600 | 0.927 |
| 170 | 1.064 | 630 | 0.873 | 170 | 1.064 | 630 | 0.922 |
| 175 | 1.061 | 675 | 0.866 | 175 | 1.061 | 675 | 0.915 |
| 180 | 1.058 | 700 | 0.863 | 180 | 1.058 | 700 | 0.911 |
| 184 | 1.055 | 720 | 0.860 | 184 | 1.055 | 720 | 0.908 |
| 194 | 1.049 | 765 | 0.855 | 194 | 1.049 | 765 | 0.902 |
| 195 | 1.049 | 800 | 0.851 | 195 | 1.049 | 800 | 0.898 |
| 200 | 1.046 | 810 | 0.850 | 200 | 1.046 | 810 | 0.896 |
| 219 | 1.035 | 855 | 0.846 | 219 | 1.035 | 855 | 0.891 |
| 220 | 1.035 | 900 | 0.843 | 220 | 1.035 | 900 | 0.886 |
| 225 | 1.032 | | | 225 | 1.032 | | |
| 235 | 1.027 | | | 235 | 1.027 | | |
| 250 | 1.020 | | | 250 | 1.020 | | |
| 270 | 1.012 | | | 270 | 1.012 | | |
| 285 | 1.006 | | | 285 | 1.006 | | |
| 300 | 1.000 | | | 300 | 1.000 | | |

* Not applicable to North American stress grades. For these, in sizes less than 38 × 114 mm, see text for values of $K_7$ and $K_{14}$ (and $K_{10}$ and $K_{11}$).
† See Fig. 4.4. Also see text for grades for which these values of $K_7$ and $K_{14}$ are applicable (and not applicable).

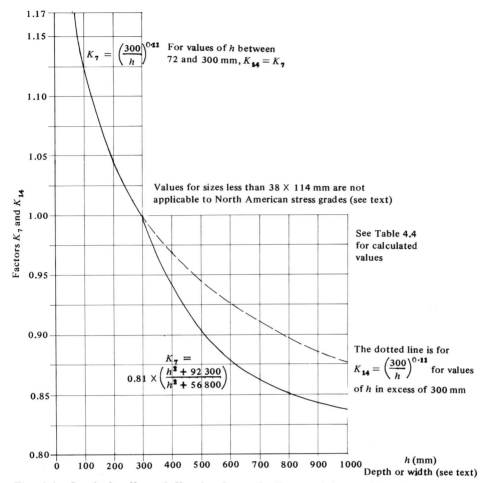

Fig. 4.4: *Graph for $K_7$ and $K_{14}$ for Strength Classes, glulam, solid timber graded to BS 4978, and North American 'Joist and Plank' grades (except for No. 3 which must not be used in tension).*

It is important to realise that these values for $K_7$ do not apply to North American 'Structural Light Framing', 'Light Framing', 'Stud' and 'Utility' grades. To minimise confusion when designing to these grades it is perhaps easiest to take $K_7$ as being 1.00 for sizes of 38 × 89 mm, and to refer to Table 12 of BS 5268:Part 2 when using other sizes in these grades (also see § 2.4.2).

Although the intention of this chapter is to cover beams, it is convenient to discuss the width factor for tension members whilst discussing the depth factor for beams. 'Width' is the greater dimension of a rectangular section in tension and the width factor is given the number $K_{14}$.

For widths between 72 and 300 mm of solid members and glulam the value of $K_{14}$ applied to the grade tension stresses tabulated in BS 5268 for Strength Classes, timber graded to BS 4978, and North American 'Joist and

Plank' grades (except No. 3 which must not be used in tension) is numerically equal to $K_7$ and it is convenient to present values in Table 4.4 and Fig. 4.4. For widths greater than 300 mm the value for $K_{14}$ in these grades is $(300/h)^{0.11}$ where $h$ is the greater dimension (of a rectangular section).

For solid sections graded to the North American 'Structural Light Framing' grades (except No. 3 which must not be used in tension) the value of $K_{14}$ for sizes of 38 × 89 mm can be taken as 1.00. For values of $K_{14}$ for other sizes in 'Structural Light Framing' see Table 12 of BS 5268 : Part 2. 'Light framing', 'Stud' and 'Utility' grades should not be used in tension (see § 2.4.3).

### 4.9.2   Form factor for flexural members

The grade stresses in BS 5268 for flexural members apply to solid and laminated timber of a rectangular cross-section (including square sections with the load parallel to one of the principal axes).

For solid circular sections and solid square sections with the load in the direction of a diagonal, the grade stresses should be multiplied by the factor $K_6$ as indicated in Fig. 4.5.

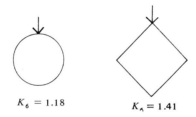

$K_6 = 1.18$                $K_A = 1.41$

Fig. 4.5

## 4.10   BEARING

### 4.10.1   Timber. Bearing across and along the grain

The grade stress for bearing across the grain of timber is the grade stress for compression perpendicular to the grain. The grade stress for bearing on end grain is the grade stress for compression parallel to the grain (at a slenderness ratio of zero). Load-sharing and load-duration factors apply (but see § 4.4.1). The symbols used are:

$\sigma_{c,adm,par}$   permissible bearing stress parallel to grain
$\sigma_{c,a,par}$   applied bearing stress parallel to grain
$\sigma_{c,g,par}$   grade bearing stress parallel to grain
(Note that the symbols are the same as for compression parallel to grain.)

$\sigma_{c,adm,tra}$    permissible bearing stress perpendicular to grain
$\sigma_{c,a,tra}$    applied bearing stress perpendicular to grain
$\sigma_{c,g,tra}$    grade bearing stress perpendicular to grain.

The grade stresses tabulated in BS 5268:Part 2 for compression (or bearing) perpendicular to the grain have been reduced to take account of the amount of wane which is permitted within each stress grade (see Fig. 4.6). If, however, wane is excluded the grade stress may be taken as 1.33 times the SS value for compression perpendicular to grain (see Tables 2.2–2.5) for the relevant species.

Fig. 4.6

The grade stresses tabulated in BS 5268:Part 2 for compression (or bearing) parallel to the grain have *not* been reduced to take account of wane. If the end in question is bearing onto timber of the same species this need not be considered because the critical bearing stress will be perpendicular to the

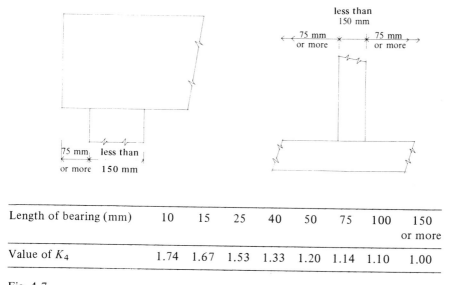

| Length of bearing (mm) | 10 | 15 | 25 | 40 | 50 | 75 | 100 | 150 or more |
|---|---|---|---|---|---|---|---|---|
| Value of $K_4$ | | 1.74 | 1.67 | 1.53 | 1.33 | 1.20 | 1.14 | 1.10 | 1.00 |

Fig. 4.7

grain of the other piece of timber. If, however, the piece being considered is bearing directly onto steel, for example, and wane is permitted, a reduction in the permissible stress (or the area of bearing) should strictly speaking be taken. BS 5268:Part 2 does not give guidance on this point but it would make sense to calculate the reduced end grain area which would remain if wane occurred up to the grade limit. By reference to BS 4978 it can be calculated that the reduced areas for GS, M50, SS and M75 rectangular sections are 94%, 97%, 97% and 98% respectively. As can be seen, even with GS, the reduction in area is not dramatic and can usually be disregarded.

When the bearing length is less than 150 mm and the bearing occurs 75 mm or more from the end of a member (as sketched in Fig. 4.7) the permissible bearing stress may be increased by the factor $K_4$ tabulated in Fig. 4.7. Interpolation is permitted.

### 4.10.2   Timber. Bearing on end stiffeners

The permissible bearing stress along the grain is several times larger than the permissible bearing stress across the grain, and in ply web beams for example, it can be advantageous to continue the end stiffener past the lower flange to obtain end grain bearing (Fig. 4.8), but the shear on line X must be checked, as there is a gap between the lower flange and the end stiffener. The detail may have to be amended. One way would be to glue extra plywood webs (as gussets) over the end stiffener and continue them along the beam for the full depth of the beam, up to and over the first stiffener from the end.

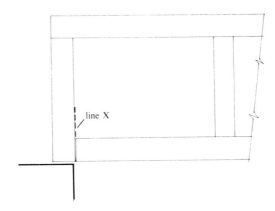

Fig. 4.8

### 4.10.3   Plywood. Bearing on the face

BS 5268:Part 2 gives values in tables for bearing on the face. The load-duration ($K_3$) factor applies.

### 4.10.4   Plywood. Bearing on edge

BS 5268 does not give values for edge bearing of plywood. It is not usual to require plywood to take edge bearing, and if such cases arise it will be necessary to test for the strength of the actual case.

## 4.11   SHEAR

### 4.11.1   Solid sections

BS 5268 gives permissible shear parallel to grain stresses which relate to maximum shear stresses and not to average values (as can occur in BS 449 in the design of steel joists). Load-duration and load-sharing factors apply.
   The formula for calculating shear stress at any level of a built-up section is:

$$\tau = \frac{F_v A_u \bar{y}}{I_X b}$$

where $\tau$ = the shear parallel to grain stress at the level being considered
   $F_v$ = the vertical external shear
   $A_u$ = the area of the beam above the level at which $\tau$ is being calculated
   $\bar{y}$ = the distance from the neutral axis of the beam to the centre of the area $A_u$
   $I_X$ = the complete second moment of area of the beam at the cross-section being considered
   $b$ = the breadth of the beam at the level at which $\tau$ is being calculated. If there are two breadths at any level in the cross-section there will be two values of $\tau$ at this level.

If $\dfrac{F_v A_u \bar{y}}{I_X}$ is evaluated instead of $\dfrac{F_v A_u \bar{y}}{I_X b}$ this gives the total shear force parallel to grain above the level being considered per unit length of beam.

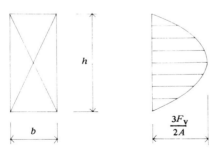

Fig. 4.9

   In the special case of a rectangular beam (Fig. 4.9) the maximum shear stress parallel to grain occurring at the neutral axis becomes:

$$\tau = \frac{3F_v}{2A}$$

where $A$ = total cross-sectional area of the beam. (For the derivation of this simplified formula, see §4.11.3.)

### 4.11.2   Ply web beams. Panel shear stress

Panel shear stress is the term used in timber engineering for the horizontal shear stress in a plywood web. For a symmetrical ply web beam the maximum panel shear stress occurs at the $XX$ axis. The general formula for horizontal shear stress must be modified to allow for the effect of the material in the flanges and web having differing $E$ values thus:

$$\tau_p = \frac{F_v (E_{fN} S_{Xf} + E_w S_{Xw})}{tEI_X}$$

where $\tau_p$ = the panel shear stress
 $F_v$ = the vertical external shear
 $E_{fN}$ = the $E$ value of the flanges taking account of the number of pieces $N$ acting together (note: in the whole beam, not just one flange)
 $E_w$ = the $E$ value of the web
 $S_{Xf}$ = the first moment of area of the flange elements above (or below) the $XX$ axis
 $S_{Xw}$ = the first moment of area of the web above (or below) the $XX$ axis
 $t$ = the thickness of the web.
 $EI_X$ = the sum of the $EI$ values of the separate parts.

Panel shear stress is discussed further in Chapter 8 on ply web beams as is 'rolling shear stress' on the glue lines between flanges and web.

### 4.11.3   Glulam

The horizontal shear stress is a maximum on the neutral axis and is also the maximum stress on the glue line if a glue line occurs at the neutral axis.

The horizontal shear stress at the neutral axis (Fig. 4.10) is as for solid beams:

$$\tau = \frac{F_v \dfrac{A}{2} \bar{y}}{I_X b} = \frac{F_v \times h \times b \times h \times 12}{bh^3 \times 2 \times 4 \times b} = \frac{3 F_v}{2A}$$

where $A$ = total cross-sectional area of the beam.

Shear stress is not usually critical in solid or glulam rectangular sections unless the span is small and the loading is heavy.

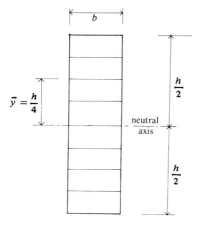

Fig. 4.10

## 4.12 THE EFFECT OF NOTCHES AND HOLES

BS 5268 requires that, in calculating the strength at a notched or drilled section, allowance shall be made for the notches or holes with the effective depth $h_e$ being taken as the minimum depth of the net section.

Notwithstanding this, BS 5268 states that the effect of notches and holes need not be calculated in floor or roof joists not more than 250 mm deep where notches not exceeding 0.125 of the joist depth are located between 0.07 and 0.25 of the span from the support, and holes drilled at the neutral axis with a diameter not exceeding 0.25 of the joist depth (and not less than 3 diameters apart centre-to-centre) are located between 0.25 and 0.4 of the span from the support. These limits are for simply supported joists. In other circumstances, where it is necessary to drill or notch a member, allowance for the hole or notch should be made in the design.

In considering the effect of round holes on the bending stress in solid joists it is usually not necessary to consider any effects of stress concentrations. Deflection will hardly be affected, as it is a function of $EI$ over the full joist span. Holes in ply web beams are discussed in Chapter 8.

Shear stress for an un-notched solid rectangular section is:

$$\tau = \frac{3F_v}{2bh} \leqslant \tau_{adm} \qquad \text{(modified as appropriate by } K_3, K_8 \text{ or } K_{27}, K_{19} \text{ and } K_{25})$$

where $h$ = total depth of section.

Shear stress for a notched solid rectangular section is:

$$\tau = \frac{3F_v}{2bh_e} \leqslant \tau_{adm} \times K_5 \qquad \text{(modified as appropriate by } K_3, K_8 \text{ or } K_{27}, K_{19} \text{ and } K_{25})$$

where $h_e$ = effective depth (see Fig. 4.11).

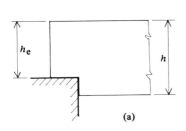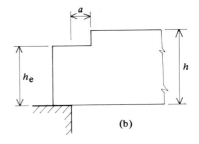

Fig. 4.11

For a notch on the lower edge (see Fig. 4.11(a)):

$$K_5 = \frac{h_e}{h}$$

For a notch on the top edge (see Fig. 4.11(b)):

$$K_5 = \frac{h(h_e - a) + ah}{h_e^2} \quad \text{for } a \leqslant h_e$$

and

$$K_5 = 1.0 \quad \text{for } a > h_e$$

### 4.12.1 Shear stress concentrations

Where beams or joists are notched at the end to suit the support conditions, there is a stress concentration resulting from the rapid change of section.

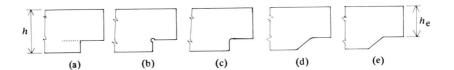

Fig. 4.12

Square cut notches should be avoided wherever possible. Poor workmanship can lead to over-sawing the notch with a tendency to propagate an early shear failure along the dotted line shown in Fig. 4.12(a). One method of avoiding this problem is to specify all notches to be cut to a pre-drilled hole of say 8 mm diameter as shown in Fig. 4.12(b), or to form a generous fillet or taper to the notch as shown in Figs 4.12(c), (d) or (e).

The shear capacity of an un-notched rectangular section is:

$$\bar{F}_v = \frac{2A\tau}{3}$$

### 4.12.2   Notches in the lower edge

Square cut notches formed in the lower edge cause the greatest shear stress concentrations and consequently give the greatest reduction in shear capacity. The net section of timber is subjected to a reduced permissible shear stress proportional to the depth of the notch.

If the full section depth is $h$ and the depth of timber at the notch position is $h_e = \alpha h$, then the permissible shear stress at the notch is $\alpha \tau$ which, when applied to the net area $\alpha A$, gives a shear capacity:

$$\overline{F}_v = \tfrac{2}{3}\alpha^2 A\tau$$

It can be seen from the formula above that the shear capacity of the notched section is equal to the shear capacity of the un-notched section multiplied by $\alpha^2$ and therefore reduces rapidly with increased depth of notch (e.g. with half the full depth remaining the shear capacity is a quarter that of the full section).

### 4.12.3   Notches in the top edge

Notches may be required in the top edge (Figs. 4.13–4.15) to suit fixing details of gutters, reduced fascias etc. Such notches should not exceed 40% of the full beam depth. The projection of the notch $\beta h$ beyond the inside edge of the bearing line influences the shear capacity.

Figure 4.15 is a special condition where $\beta h = 0$ and the end of the notch coincides with the inside edge of the bearing. Providing that $\alpha$ is at least 0.6 the full shear capacity of the un-notched section is realised.

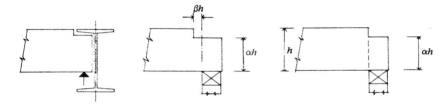

Fig. 4.13                Fig. 4.14                        Fig. 4.15

When the notch extends by an amount $\beta h$ (Fig. 4.14) the reduced shear capacity is given as:

$$\overline{F}_v = \tfrac{2}{3}\left(1 + \beta - \frac{\beta}{\alpha}\right)A\tau$$

If $\beta$ exceeds $\alpha$ this formula is not suitable and the shear capacity is then evaluated on the net depth $\alpha h$ and the shear capacity is given as:

$$\overline{F}_v = \tfrac{2}{3}\alpha A\tau$$

When the top edge is bevelled (Fig. 4.16) the two formulae given above apply with $\alpha h$ measured to the intersection with the inner edge of the support and $\beta h$ measured from the inner edge to the commencement of the bevel on the top edge.

A summary of these formulae is presented in Table 4.5.

### 4.12.4   Summary of formulae for shear capacity of various end details of rectangular solid sections

**Table 4.5**  (see §§ 4.12–4.12.3)

| Detail | Shear capacity |
|---|---|
| | $\overline{F}_{vu}$ |
| | $\alpha^2 \overline{F}_{vu}$ |
| | $\overline{F}_{vu}$, providing that $\alpha$ is not less than 0.6 |
| | $\left(1 + \beta + \dfrac{\beta}{\alpha}\right) \overline{F}_{vu}$ |
| | $\alpha \overline{F}_{vu}$ |

$\overline{F}_{vu}$ is the full shear capacity for the un-notched section.

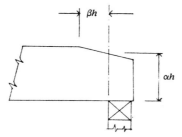

Fig. 4.16

## 4.13 SHEAR IN BEAMS SUPPORTED BY FASTENINGS AND IN ECCENTRIC JOINTS

When a beam is supported along its length as indicated, for example, in Figs 4.17–4.20, the force $F$ being transmitted locally into the beam should be resisted by the net effective area as indicated in the figures. In addition, of course the connectors and supporting members must be able to resist the force $F$, and the beam must be checked for overall shear.

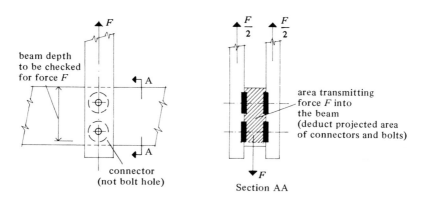

Fig. 4.17

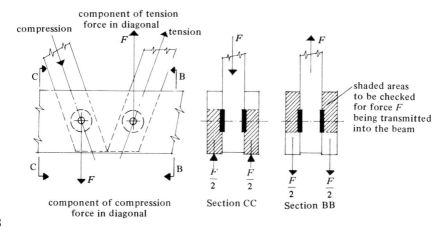

Fig. 4.18

In the case of the bolted joint shown in Fig. 4.19 the area to be considered as locally resisting force $F$ is the net area of section between the edge towards which the load is acting and the bolt furthest from that edge.

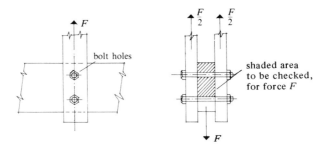

Fig. 4.19

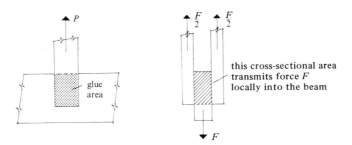

Fig. 4.20

In the case of a glued joint as shown in Fig. 4.20 the cross-sectional area of the beam transmitting force $F$ locally into the beam is the shaded area. However, under normal circumstances, the critical factor at this position would be the rolling shear at the glue interface.

## 4.14   GLUE-LINE STRESSES

Providing that one of the recommended glue types is used correctly (see Chapter 19 on glue joints) the glue is always stronger than the surrounding timber. The strength of glue lines is therefore calculated as the permissible shear on the face of the timbers or plywood being glued. Load-duration and load-sharing factors apply because the stresses are timber stresses.

Glue has little strength in tension, therefore connections should be designed to prevent any tension effect on the glue line. The aim should be to achieve pure shear. The type of overlap joint shown in Figs 4.21(a) and (b) should be symmetrical if glued.

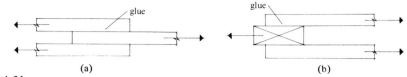

Fig. 4.21

When the forces are stressing the glue along the grain of timber (Fig. 4.21(a)) the permissible stress is the permissible stress for the timber grade for shear parallel to the grain (reduced by a factor $K_{70}$ of 0.9 if the glue-line pressure during curing is obtained by nails or staples rather than clamps – Clause 47.4.2 of BS 5268 : Part 2).

When the forces are stressing the glue across the grain of timber (Fig. 4.21(b)) the permissible stress is one third of the permissible stress for the timber grade for shear parallel to the grain. This shear across the grain is referred to as 'rolling shear' as the tendency is for the outer fibres to roll. If there are likely to be frequent changes of several percent in the moisture content, it is not good practice to glue two or more pieces of solid timber together with the angle between the fibre directions approaching 90°, due to the difference in expansion and contraction along and across the grain.

For forces acting at other than angles of 0° and 90° to the grain (e.g. Area *A* in Fig. 4.22 but not Area *B*) BS 5268 : Part 2 calls for the following formula to be used (instead of Hankinson's formula as in CP 112 : Part 2) in calculating the permissible stress for the glue line:

| $\alpha^{\circ}$ | $K_T$ |
|---|---|
| 0 | 1.00 |
| 5 | 0.94 |
| 10 | 0.88 |
| 15 | 0.83 |
| 20 | 0.77 |
| 25 | 0.72 |
| 30 | 0.67 |
| 35 | 0.62 |
| 40 | 0.57 |
| 45 | 0.53 |
| 50 | 0.49 |
| 55 | 0.45 |
| 60 | 0.42 |
| 65 | 0.40 |
| 70 | 0.37 |
| 75 | 0.36 |
| 80 | 0.34 |
| 85 | 0.34 |
| 90 | 0.33 |

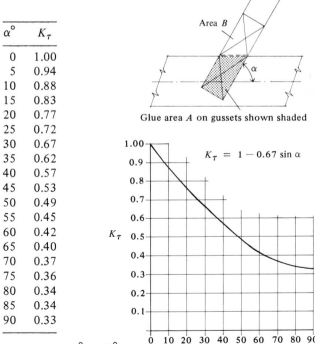

Glue area *A* on gussets shown shaded

$$K_T = 1 - 0.67 \sin \alpha$$

Fig. 4.22: *Graph of $K_T$ for $\alpha$ from 0° to 90°.*

$$\tau_\alpha = \tau_{\text{adm, par}} (1 - 0.67 \sin \alpha) = \tau_{\text{adm, par}} \times K_\tau$$

where $\tau_{\text{adm,par}}$ = the permissible shear parallel to the grain stress for the timber

$\alpha$ = the angle between the direction of the load and the longitudinal axis of the piece of timber (see Fig. 4.22).

Note that when $\alpha = 90°$, $\sin \alpha = 1.0$ and $\tau_\alpha$ ($= \tau_{\text{adm, tra}}$) is one third of $\tau_{\text{adm,par}}$ and is the value for rolling shear. (0.67 in the formula is really $\frac{2}{3}$.)

Values for $1 - 0.67 \sin \alpha$ are tabulated and shown in graph form in Fig. 4.22.

When the forces are stressing the face of plywood either across or along the face veneer (Fig. 4.23) the permissible stress is the permissible 'rolling shear' stress for the plywood grade. Even if the stress is parallel to the face

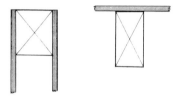

Fig. 4.23

grain it is considered that the next veneer face is so close to the surface that 'rolling' of the fibres in this veneer could occur before the full shear strength of the face veneer is generated. (Take particular note of this point if different species are used within one plywood.)

In the special case of the flange-to-web connection of a ply web beam and the connection of the plywood (or other board) to the outermost joist of a glued stress skin panel (Fig. 4.24), Clause 30 of BS 5268 requires that the permissible shear stress at the glue line be multiplied by 0.5 ($K_{37}$). This is an arbitrary factor to take account of likely stress concentrations, and it is pertinent to add that some manufacturers of proprietary ply web beams who have tested their beams to destruction have found this type of joint to have

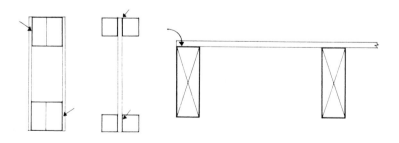

arrows denote glue lines with stress concentrations

Fig. 4.24

factors of safety above ten. Normally, however, a designer would be expected to work to the clause which is in addition to the reduction factor of 0.9 ($K_{70}$) from Clause 47.4.2 of BS 5268:Part 2 which applies if the bonding pressure is obtained by nails or staples.

With the type of glued joint shown in Fig. 4.22, and in the ply web beams shown in Fig. 4.24, and the type of glulam section shown in Fig. 4.10, BS 5268:Part 2 (Clause 47.1) puts a limit of 50 mm on the individual pieces of solid timber being glued. (Note that the flange width of the box beam shown at the left of Fig. 4.24 could be twice 50 mm and still comply with this clause.) It is prudent to take note of this clause, because if thick pieces are distorted in any way it is extremely difficult to hold them in line during assembly and, if they are forced into place before gluing, subsequent release of the built-in fabrication stresses could damage glue lines. By inspection of Fig. 4.22 it is easy to see that a wide incoming member could also cause trouble in assembling the (plywood) gussets if the incoming member is distorted, particularly if it is also thick. If large pieces of solid timber are to be assembled in a frame it may well be better to use bolted steel gussets, or to substitute laminated members for the solid timber.

Similarly, Clause 47.1 of BS 5268:Part 2 puts a limit of 24 mm on the thickness of plywood (or tempered hardboard) on glued joints of the type as shown in Figs 4.22, 4.23 and 4.24.

### 4.14.1  Glued/nailed joints

BS 5268 gives certain clauses relating to glued/nailed joints for fastening plywood to timber. Similar clauses apply to tempered hardboard. A new British Standard (BS 6446) is being produced on the manufacture of glued structural components of timber and wood based panels and, when this is printed (probably in 1984) reference should be made to it for nail centres etc. Also see § 19.4.3.

When gluing plywood to timber the permissible glue-line stresses should be multiplied by a $K_{70}$ factor of 0.9 if assembly is by nailing. If assembly is by clamping this factor does not apply. The 0.5 factor ($K_{37}$) described in § 4.14 may also apply.

The length of round wire nails should be not less than four times the thickness of the plywood in contact with the nail head.

The maximum spacing of nails along the grain should not normally exceed 100 mm for plywood not more than 12.5 mm thick and 150 mm for plywood more than 12.5 mm thick. Nail centres and edge distances should be in accordance with Clause 41.3 of BS 5268:Part 2 on which the values given in Chapter 18 of this manual are based. The side spacing between lines of nails should normally not exceed 100 mm. The nails are not considered to add to the strength of the glue line.

Screws, improved wire nails and power-driven fastenings such as staples may be used if proved to be capable of applying pressure to the glue line at least equal to the nailing requirements described in BS 5268:Part 2.

## 4.15   DEFLECTION

### 4.15.1   Deflection limits and $E$ values

Unlike a stress there is no stage at which deflection leads to failure in the component which is deflecting. There is however a stage at which a deflection would become visually unacceptable to the majority of people or lead to distortion, cracking or failure of, for example, a ceiling or partition under the beam, or a waterproof membrane being supported by the beam.

Any general limit which is placed on deflection is purely arbitrary, usual practice or good practice and may be relaxed by the specifying authority in certain cases if this will lead to acceptable economy; or tightened, perhaps for appearance or if a particularly brittle component is being supported. For example, it used to be common practice to limit deflection to $L/500$ when large cast-iron water tanks were being supported, whereas on storage racks where men can walk, $L/240$ is often used, or $L/180$ where there is storage with no access for personnel.

The usual limit for horizontal beams in buildings is around $L/300$. BS 5268:Part 2 (Clause 14.7) actually recommends a limit of $0.003L$ for general purposes. For longer-span (i.e. over 4.67 m span) domestic floor joists an arbitrary level of 14 mm is added to avoid undue vibration resulting from moving or impact loading. The stiffness of the floor joists is only one of several factors which affect the 'feel' of a floor but, looking at the comparison $E$ value ratios shown in Fig. 2.3, and considering the committee decision to introduce an arbitrary limit of 14 mm for domestic floor joists, one wonders if $E_{mean}$ should really be used in designs. Perhaps a value of about the $E_3$–$E_5$ values would be more representative of the load sharing encountered with some of the boards used for modern flooring. As it stands, some authorities do not permit a stress grade higher than Special Structural to be used for domestic floor joists because of the risk of 'bounce'. However, an architect or specifier should not normally expect the structural designer to work to tighter limits than $0.003L$, or to use other than $E_{mean}$ in calculations unless other instructions are clearly specified. If, on the other hand, a designer feels justified in working to looser limits this should be made absolutely clear in submitting the scheme.

. 'True' $E$ values are tabulated in BS 5268:Part 2 therefore the designer should also include a calculation of shear deflection in determining the stiffness of a beam, even for simple floor joists. (This point will be further clarified in a proposed Part 7 to BS 5268 on the method of arriving at span tables to match BS 5268.)

Deflection is proportional to loading.

All beams carry a dead loading which is in place throughout the life of the member. Most beams also carry an imposed loading which, in the case of a roof beam carrying snow, will only be in place for a small percentage of the life of the beam and may never reach the amount specified in CP 3: Chapter V. In the case of a floor beam, part of the imposed loading will be in place for most of the time (e.g. furniture) whereas the rest of the design imposed load (e.g. people) will only be in place for part of the time and may never reach the amount specified in CP 3:Chapter V. In the case of a ceiling joist in a roof truss, the imposed loading could be in place for years at a time (e.g. suitcases or spare items).

Although BS 5268 (as with similar documents for other materials) only puts a limit on the deflection position below the horizontal with the total load in place, this total load may never occur or may occur only for brief occasional periods, and often the more important consideration is the deflection under the permanent or dead loading. When a beam is built up from several pieces (e.g. a ply web beam or glulam) it is possible to build-in a camber ('subject to consideration being given to the effect of excessive deformation') so that under the action of the dead loading the beam settles approximately to a horizontal position. In these cases, the deflection limit of $0.003L$ applies to deflection under the imposed loading only. If a beam is over-cambered, the deflection limit of $0.003L$ still applies to the total imposed loading, no reduction being allowed to this because of the over-cambering.

In the case of beams of solid sawn timber, there is no chance to camber, however, the span of such beams is limited to around 6 m, at which span the usual deflection limit under total loading is 18 mm (or 14 mm if one is working to the arbitrary limit in BS 5268 for domestic floor joists). There-fore, by proportion, the deflection under permanent loading in most cases is unlikely to be more than around 8–10 mm, which is hardly likely to be unduly noticed unless a beam lines up with a horizontal feature, or the joists are supported by a long trimmer which itself deflects 8–10 mm under permanent loading (see Fig. 12.12).

The decision on which $E$ value (and $G$ value) to use in calculating the deflection of components in various situations has already been covered in §§ 2.3, 3.4 and 4.4.1 of this manual. Also see Fig. 2.3. Therefore the notes given below are a summary. In addition, practical and special considerations of deflection are discussed in Chapter 12. Where a value for $E$ is stated below (e.g. $E_{min}$) it should be understood that the appropriate value of shear modulus (e.g. $G_{min}$) is also intended.

*Single member principal beam.* A beam having one piece of timber in cross-section, acting in isolation, must be regarded as a principal member with deflection calculated using $E_{min}$ (and $G_{min}$).

*Trimmer joist or lintel.* If a trimmer joist or lintel is constructed from two or more pieces solidly fixed together by mechanical fasteners to be able to share the loading, the $E_{min}$ and $G_{min}$ values may be increased by $K_9$ having a stop-off point at four or more pieces.

*Vertically glued laminated beam.* Where a vertically glued laminated beam is constructed from two or more pieces the $E_{min}$ and $G_{min}$ values may be increased by $K_{28}$ having a stop-off point of eight or more pieces (unless of course used as part of a load-sharing system with members at centres not exceeding 610 mm, in which case mean values may be used).

*Built-up beams* (such as ply web beams). The $E$ and $G$ values to use in calculating deflection are the minimum values modified by factor $K_{28}$ for the number of pieces of timber in cross-section. (Clause 14.10 of BS 5268: Part 2.)

*Load-sharing system at close centres.* Where a load-sharing system of four or more members (such as joists or studs) spaced at centres not in excess of 610 mm has adequate provision for lateral distribution of loads, the mean values of moduli may be used in calculations of deflection (unless the area is to support mechanical plant and equipment, or is used for storage, or is subject to vibrations as with a gymnasium or ballroom, in which case minimum values of moduli are to be used).

*Horizontally laminated glulam beams.* Deflection is calculated by multiplying the mean value of the moduli for SS grade of the same species as the laminates by either factor $K_{20}$ or factor $K_{26}$ as appropriate.

### 4.15.2   Camber

When a beam is built up and is to span considerably more than ordinary joists, it is usually good practice to camber so that under the action of dead loading the beam deflects to an approximately horizontal position. The $E$ value of timber is variable, and thought should be given to deciding which $E$ value to use in calculating camber under dead loading and the subsequent limiting deflection of $0.003L$ under imposed loading. The value for $E$ (and $G$) used in designs need not necessarily be the same in each case. Consider the case of built-up ply web beams with several timber members in cross-section spaced at 1.2 m centres, with effective lateral distribution of load, and dead and imposed loading roughly equal. Once the beams and cladding are in place, particularly if there are continuous noggings at the level of the bottom flange, it is quite likely that the beams will perform under subsequent imposed loading with an $E$ and $G$ value approaching the mean values (even though BS 5268 is clear in requiring that $E_N$ should be used in calculations for imposed loading of a cambered built-up beam). However, before the cladding and ceiling and their supports are in place, each beam is likely to be acting mostly on its own (even though the loading may not be a high percentage of the total design load), and certainly $E_N$ is relevant to calculations (using factor $K_{28}$ of BS 5268).

When deciding upon the camber for a beam, particularly a long-span beam it is as well to remember that $E_N$ values are statistical minimum values and that values closer to $E_{\text{mean}}$ could be encountered in practice. If there are more than eight members in cross-section it is as well to study the comparison given in Fig. 2.3. In deciding upon what camber to use when constructing a long-span beam, one can always calculate deflection values under permanent loading using $E_{\text{mean}}$ and $E_N$ values and decide upon the camber from a judgement of both values, which are not likely to be greatly different in real terms, particularly when four or more pieces occur at a cross-section.

In setting a camber, it is as well to remember that a beam, if perfectly horizontal, tends to look as though it has a slight sag purely due to optical illusion (this is even more the case with a lattice beam or truss than a solid or ply web beam). Also, a builder usually finds it easier to adjust the soffit if there is a slight residual camber than if there is a slight sag, and there will be manufacturing tolerances. Take note of any horizontal feature in the building particularly if the feature runs close to a beam position.

A few moments thought and calculation on camber in the design office can save hours of work on site.

### 4.15.3   Variations in dead loading

An engineer may find himself called upon to investigate unexplained deflection or residual camber on site (not only with timber beams). In doing so he should not ignore the possibility that the dead loading may be considerably more (or less) than the specified dead loading. It is fashionable to talk about the variability in strength and $E$ value of timber, but hardly any mention is made of how variable certain building materials can be. A builder can change the design dead load completely, for example, by using the wrong type of wood wool slab or laying screed on such a slab without first putting down a scrim or felt barrier.

### 4.15.4   Short spans

BS 5268 gives the same deflection limitation ratio for all spans plus an arbitrary limit for certain longer spans. There is, however, a case for allowing a larger deflection ratio on short spans. On a span of 1 m a deflection of 4 mm would be $33\frac{1}{3}\%$ too great to comply with BS 5268 yet the deflection is only 1 mm more than the permissible value and to reject the design of a beam, nogging or cladding for non-compliance in such a case would be taking limits to the extreme. It is far more important on short spans that the 'feel' or 'bounce' is acceptable to personnel if they have access to the area. A simple test is often the best way of determining whether or not a particular short-span case is acceptable.

### 4.15.5   Bending deflection and shear deflection

The deflection of any beam is a combination of bending deflection and shear deflection. In addition, if a beam is installed at a high moisture content and is permitted to dry out with a high percentage of the load in place, 'creep' deflection will take place during the drying-out process. That is one reason why it is important to install beams at a moisture content reasonably close to the moisture content they will attain in service. Shear deflection is usually a fairly small percentage of the total deflection of solid sections, but BS 5268:Part 2 requires it to be taken into account even in the design of solid timber joists and glulam (see § 4.15.7 and 4.15.8). Shear deflection is likely to be very significant in the design of ply web beams (see § 8.6).

### 4.15.6   Bending deflection

The formulae for bending deflection of simply supported beams for various conditions of loading are well established. Several are given in Chapter 27 of this handbook for central deflection, and for maximum deflection when this is different to central deflection. It can be shown with a simply supported beam that no matter what the loading system is, if the central bending deflection is calculated instead of the maximum deflection, the error is never more than 2.57%. Also the maximum bending deflection always occurs within $0.0774L$ of the centre of span. For complicated loading conditions for which it can take a considerable time to calculate the position of the maximum deflection even before calculating its magnitude, it is invariably quite satisfactory to calculate the central deflection.

In all the formulae below, $L$ should strictly be taken as $L_e$ (see § 4.3).

With a complicated system of loading, one can calculate deflection for the exact loading, but this can be time consuming, and an easier way is to calculate the deflection from the equivalent uniformly distributed loading which would give the same deflection as the actual loading.

The deflection at mid span $\delta_m$ is calculated as:

$$\delta_m = \frac{5 F_e L^3}{384EI}$$

The equivalent uniform load $F_e$ may be determined for most loading conditions as:

$$F_e = FK_m$$

where $F$ = actual load
$K_m$ = a coefficient taken from Tables 4.8–4.15 according to the nature of the actual load.

Where more than one type of load occurs on a span, $F_e$ is the summation of the individual $FK_m$ values.

Tables 4.8–4.15 give values of $K_m$ for several loading conditions. The following example shows how the tables are used.

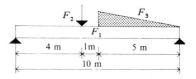

Fig. 4.25

*Example.* Calculate the bending deflection at mid span for the beam shown in Fig. 4.25. Assume an *EI* capacity of 10 100 kN.m².

Calculate $F_e$ in tabular form (Table 4.6).

**Table 4.6**

| Load | Load (kN) | Table for $K_m$ value | $n$ | $K_m$ | $FK_m$ |
|------|-----------|-----------------------|-----|-------|--------|
| $F_1$ | 2 | 4.9 | 1.0 | 1.0 | 2.0 |
| $F_2$ | 6 | 4.15 | 0.4 | 1.51 | 9.06 |
| $F_3$ | 8 | 4.10 | 0.5 | 1.28 | 10.24 |
| | | | | $F_e = $ | 21.30 |

Coefficient $n$ is a ratio of span (see Tables 4.8–4.15).

$$\text{Bending deflection at mid span} = \frac{5 F_e L^3}{384EI} = \frac{5 \times 21.3 \times 10^3}{384 \times 10\,100}$$

$$= 0.0275 \text{ m} = 27.5 \text{ mm}$$

### 4.15.7 Shear deflection

In addition to deflection caused by bending stresses, there is a further deflection caused by shear stresses (except in the special case of a beam subject to pure bending which is free from shear stresses). Shear deflection is often disregarded in most structural materials, although it should certainly be calculated, for example, in the case of a deep heavily loaded steel plate-girder.

Timber beams are frequently deep in relation to their span and have a low $G/E$ value (where $G$ is the modulus or rigidity) taken in BS 5268 as $\frac{1}{16}$ (0.0625) compared to 0.4 for mild steel.

By the method of unit loads:

$$\text{Shear deflection} = K_{\text{form}} \int_0^L \frac{F_v F_i \, dx}{AG}$$

where $K_{form}$ = a form factor dependent on the cross-sectional shape of the beam (equal to 1.2 for a solid rectangle)

$F_v$ = the external shear due to the actual loading

$F_i$ = the shear due to a unit load at the point where the deflection is being calculated

$A$ = the area of the section

$G$ = the modulus of rigidity (taken as $E/16$ in BS 5268).

The shear deflection is normally added to the centre-span bending deflection, therefore it is the centre-span shear deflection in which one is interested. With the unit load placed at centre span, $F_i = 0.5$, and it can be shown that:

Shear deflection at mid span

$$= \frac{K_{form} \times \text{area of shear force diagram to mid span}}{AG}$$

$$= \frac{K_{form} \times M_0}{AG}$$

where $M_0$ = the bending moment at mid span.

To reduce design work $M_0$ for a simple span may be calculated as $F_0 L/8$, where $F_0$ is the equivalent total uniform load to produce the moment $M_0$.

$$M_0 = FK_v$$

where $F$ = actual load

$K_v$ = a coefficient taken from Tables 4.8–4.15 according to the nature of the actual load.

Where more than one type of load occurs on a span, $F_0$ is the summation of the individual $FK_v$ values.

**Table 4.7**

| Load | Load (kN) | Table for $K_v$ value | $n$ | $K_v$ | $FK_v$ |
|------|-----------|-----------------------|-----|-------|--------|
| $F_1$ | 2 | 4.9 | 1.0 | 1.0 | 2.0 |
| $F_2$ | 6 | 4.15 | 0.4 | 1.6 | 9.6 |
| $F_3$ | 8 | 4.10 | 0.5 | 1.333 | 10.67 |
| | | | | $F_0 =$ | 22.27 |

Coefficient $n$ is a ratio of span (see Tables 4.8–4.15).

*Example.* Calculate the shear deflection at mid span for the beam shown in Fig. 4.25. Assume the $AG$ capacity to be 34 660 kN and $K_{form} = 1.2$ (for a solid rectangular beam).

Calculate $F_0$ in tabular form (Table 4.7).

$$M_0 = \frac{F_0 L}{8} = \frac{22.27 \times 10}{8} = 27.8 \,\text{kN.m}$$

$$\text{Shear deflection at mid span} = \frac{K_{\text{form}} \times M_0}{AG}$$

$$= \frac{1.2 \times 27.8}{34\,660}$$

$$= 0.000\,96\,\text{m} = 0.96\,\text{mm}$$

This represents not only a very small actual value but only 3.5% of the bending deflection. It will be seen later in this handbook that in the design of ply web beams shear deflection can be in excess of 15% of the bending deflection and is very significant, but it is usually only a small deflection in real terms with beams of solid timber in the span range where deflection is critical.

### 4.15.8  Magnitude of shear deflection of solid rectangular sections

To give an indication of the proportion of beams where shear deflection is important, the effect of shear deflection on solid uniform rectangular sections subject to uniformly distributed loading is illustrated in Fig. 4.26.

$$\text{Total deflection} = \delta_t = \delta_m + \delta_v$$

where $\delta_m$ = bending deflection
$\delta_v$ = shear deflection.

| $\dfrac{L}{h}$ | $\dfrac{\delta_t}{\delta_m}$ |
|---|---|
| 12 | 1.106 |
| 13 | 1.091 |
| 14 | 1.078 |
| 15 | 1.068 |
| 16 | 1.060 |
| 17 | 1.053 |
| 18 | 1.047 |
| 19 | 1.042 |
| 20 | 1.038 |
| 21 | 1.035 |
| 22 | 1.032 |
| 23 | 1.029 |
| 24 | 1.027 |
| 25 | 1.025 |

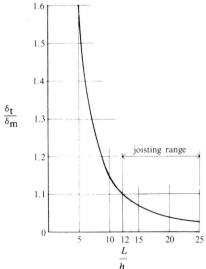

Fig. 4.26

Therefore (with $p$ = load per unit length):

$$\delta_t = \frac{5pL^4}{384EI} + \frac{K_{form} M_0}{GA}$$

but

$$GA = \frac{Ebh}{16} \text{ and } EI = \frac{Ebh^3}{12}$$

therefore

$$GA = \frac{3EI}{4h^2}$$

so that, with $K_{form} = 1.2$:

$$\delta_t = \frac{5pL^4}{384EI} + \frac{1.2pL^2}{8} \times \frac{4h^2}{3EI}$$

The ratio $\delta_t/\delta_m$ indicates the importance of shear deflection:

$$\frac{\delta_t}{\delta_m} = 1 + \frac{\delta_v}{\delta_m} = 1 + 15.36 \left(\frac{h}{L}\right)^2$$

The ratio $\delta_t/\delta_m$ for values of $L/h$ is plotted in Fig. 4.26.

In the normal range of joisting, shear deflection can be expected to add 2.5–10% to the value of bending deflection which, for the spans where deflection is critical, represents a small amount in real terms.

The designer can judge quickly from the graph in Fig. 4.26 whether or not shear deflection is likely to be significant (but note that BS 5268 requires a calculation of shear deflection even for floor or roof joists of solid timber).

Taking the case of a 250 mm deep joist over a 4.6 m span with a total permissible deflection $(0.003L)$ of 13.8 mm for an uncambered beam, shear deflection would be around 0.8 mm $(L/h = 18.4)$.


## 4.16  CREEP DEFLECTION

Creep deflection of timber beams is related to the bending stress level, to the extent of the change in moisture content, and to the time for the change to take place. At the normal stress levels encountered with permanent loading in designs to BS 5268 for beams carrying permanent and live loading, providing that beams are installed within a few per cent of the equilibrium dry exposure moisture content, there is little likelihood of any noticeable creep deflection taking place. However, if a timber beam is erected at a high moisture content and is then allowed to dry out with a considerable amount of the design load in place, creep deflection will take place during the drying-out process (in addition to bending and shear deflection) and can amount to several times the calculated deflection. Once the beam has reached the equilibrium moisture content, the creep deflection will cease.

One cannot design for creep deflection, therefore the engineer must prevent it from happening to any significant extent by ensuring that beams are dried before installation to within a few per cent (say 3–4% or closer) of the moisture content they will attain in service. If the engineer finds that beams have been installed at a high moisture content, he or she should either ensure that they are not loaded until they have dried to the equilibrium moisture content or, preferably, have them propped until this moisture content has been reached. Obviously, more care has to be taken with long- rather than short-span beams, and also when the dead loading is a high proportion of the total loading.

Under normal loading and site conditions there is no evidence to suggest that creep deflection, if it does take place, leads to any permanent damage to the beam. Therefore, providing the deflected form is acceptable or can be adjusted, the beam can usually be left in place and any subsequent deflection under further loading will be that as calculated from the formulae for bending and shear deflection.

In normal dry interior situations (offices, classrooms, halls etc.) where wet processes are absent and where good design practice has been followed, the risk of creep deflection is small and is unlikely to occur to any significant extent. When the designer is aware that intermittent condensation may occur, producing repeated changes in moisture content over a prolonged period, a possible gradual increase in deflection should be anticipated and allowed for in the design.

## 4.17  TABLES OF COEFFICIENTS FOR CALCULATING BENDING AND SHEAR DEFLECTION

Tables 4.8–4.15 give coefficients $K_m$ and $K_v$ which can be used to expedite the calculations of bending and shear deflections in the centre of simply supported beams. The notes and examples in §§ 4.15.6 and 4.15.7 explain how they should be used. Coefficient $n$ in these tables is a ratio of the span.

## 4.18  *E* VALUES AND GRADE STRESSES OF PLYWOOD

When plywood is bent as a plate (Fig. 4.27) the grade stress to use is the one for 'extreme fibre in bending', either the value for face grain parallel to span or face grain perpendicular to span depending on the way the plywood is laid. The *E* value is the one 'in bending'.

$E_{mean}$ and $E_{min}$ values are not given for plywood in BS 5268. One value only is given which should be used in all 'dry' conditions whether load sharing or not. The quoted *E* values are statistical mean values.

When designing plywood in 'wet' exposure conditions the dry stresses and moduli should be multiplied by factor $K_{36}$ (except American C–D grade should not be used. See Table 37 of BS 5268.).

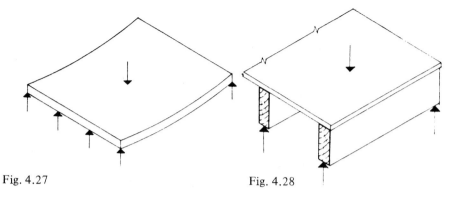

Fig. 4.27                                      Fig. 4.28

When a stress skin panel with one plywood sheet in the compression zone is bent in the main direction of span (Fig. 4.28), the plywood grade stress to use is that for 'compression', either the value for face grain parallel or perpendicular to span (although the actual stress in the plywood is calculated as a bending stress (see Chapter 10)). The $E$ value is that for 'compression'. When a stress skin panel has a plywood sheet both top and bottom (Fig. 4.29) basically the same notes apply as for single-sheet panels with the addition that the stress on the tension side should also be checked. See Chapter 10 for the design of stress skin panels.

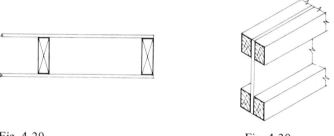

Fig. 4.29                                      Fig. 4.30

With a ply web beam (Fig. 4.30) the grade stress to use in checking the horizontal shear stress is the 'panel shear' value. For the full design of ply web beams see Chapter 8. In the design of ply web beams it is probably incorrect to use the $E$ value for bending on the flat (Fig. 4.27) and it is probably more correct to use the $E$ value for 'tension' or 'compression', whichever is the lesser. (In BS 5268:Part 2 they are given the same values.) If, however, values of $E$ more appropriate to plywood, laterally restrained and bent in a vertical plane, become available they should be used in the design of ply web beams. In the meantime, in Chapter 8 it is the 'tension or compression' values which are used in designs of ply web beams.

Tables 5.28–5.36 inclusive give properties and capacities per metre width for plywood bending 'on the flat' (see Fig. 4.27).

## 4.19   MODULAR RATIO

The term 'modular ratio' is not common or popular in timber engineering but engineers should recognise that ply web beams and stress skin panels, for example, are constructed from two materials with differing $E$ values and strength properties (as are reinforced concrete members). It is feasible to introduce a modular ratio into the design of a ply web beam, usually by increasing the ply thickness in proportion to the modular ratio and dealing with 'softwood equivalent area' properties, but there are certain pitfalls in comparing the applied stresses and the permissible stresses and, in the chapters on ply web beams and stress skin panels in this handbook, the chosen method is to work with the actual $E$ values at each stage of design.

*F is total load in all cases

### Table 4.8

| n | $K_m$ | $K_v$ |
|---|---|---|
| 0 | – | – |
| 0.05 | 0.120 | 0.1 |
| 0.1 | 0.238 | 0.2 |
| 0.15 | 0.355 | 0.3 |
| 0.2 | 0.467 | 0.4 |
| 0.25 | 0.575 | 0.5 |
| 0.3 | 0.677 | 0.6 |
| 0.333 | 0.740 | 0.667 |
| 0.35 | 0.771 | 0.7 |
| 0.4 | 0.858 | 0.8 |
| 0.45 | 0.934 | 0.9 |
| 0.5 | 1.000 | 1.000 |
| 0.55 | 1.054 | 1.082 |
| 0.60 | 1.095 | 1.133 |
| 0.65 | 1.123 | 1.161 |
| 0.667 | 1.130 | 1.167 |
| 0.7 | 1.138 | 1.171 |
| 0.75 | 1.141 | 1.166 |
| 0.8 | 1.133 | 1.150 |
| 0.85 | 1.114 | 1.123 |
| 0.9 | 1.080 | 1.089 |
| 0.95 | 1.046 | 1.047 |

### Table 4.9

| n | $K_m$ | $K_v$ |
|---|---|---|
| 0 | | 2.0 |
| 0.05 | 1.600 | 1.95 |
| 0.1 | 1.598 | 1.9 |
| 0.15 | 1.592 | 1.85 |
| 0.2 | 1.583 | 1.8 |
| 0.25 | 1.570 | 1.75 |
| 0.3 | 1.553 | 1.7 |
| | 1.533 | |
| 0.333 | 1.518 | 1.667 |
| 0.35 | 1.510 | 1.65 |
| 0.4 | 1.485 | 1.6 |
| 0.45 | 1.456 | 1.55 |
| 0.5 | 1.425 | 1.5 |
| 0.55 | 1.391 | 1.45 |
| 0.6 | 1.355 | 1.4 |
| 0.65 | 1.325 | 1.35 |
| 0.667 | 1.311 | 1.333 |
| 0.7 | 1.277 | 1.3 |
| 0.75 | 1.234 | 1.25 |
| 0.8 | 1.190 | 1.2 |
| 0.85 | 1.145 | 1.15 |
| 0.9 | 1.098 | 1.1 |
| 0.95 | 1.049 | 1.05 |

### Table 4.10

| n | $K_m$ | $K_v$ |
|---|---|---|
| 0 | – | – |
| 0.05 | 0.160 | 0.133 |
| 0.1 | 0.317 | 0.266 |
| 0.15 | 0.471 | 0.4 |
| 0.2 | 0.620 | 0.533 |
| 0.25 | 0.760 | 0.567 |
| 0.3 | 0.891 | 0.8 |
| 0.333 | 0.988 | 0.889 |
| 0.35 | 1.010 | 0.933 |
| 0.4 | 1.116 | 1.067 |
| 0.45 | 1.207 | 1.2 |
| 0.5 | 1.280 | 1.333 |
| 0.55 | 1.334 | 1.431 |
| 0.60 | 1.368 | 1.474 |
| 0.65 | 1.382 | 1.477 |
| 0.667 | 1.375 | 1.472 |
| 0.7 | 1.376 | 1.453 |
| 0.75 | 1.351 | 1.407 |
| 0.8 | 1.309 | 1.346 |
| 0.85 | 1.251 | 1.272 |
| 0.9 | 1.180 | 1.188 |
| 0.95 | 1.095 | 1.097 |

**Table 4.11**

| $n$ | $K_m$ | $K_v$ |
|---|---|---|
| 0 | — | — |
| 0.05 | 0.080 | 0.067 |
| 0.1 | 0.159 | 0.133 |
| 0.15 | 0.238 | 0.2 |
| 0.2 | 0.315 | 0.267 |
| 0.25 | 0.370 | 0.333 |
| 0.3 | 0.463 | 0.4 |
| 0.333 | 0.510 | 0.444 |
| 0.35 | 0.532 | 0.467 |
| 0.4 | 0.600 | 0.533 |
| 0.45 | 0.662 | 0.6 |
| 0.5 | 0.720 | 0.667 |
| 0.55 | 0.774 | 0.732 |
| 0.6 | 0.822 | 0.793 |
| 0.65 | 0.864 | 0.845 |
| 0.667 | 0.877 | 0.861 |
| 0.7 | 0.901 | 0.890 |
| 0.75 | 0.932 | 0.925 |
| 0.8 | 0.957 | 0.954 |
| 0.85 | 0.976 | 0.975 |
| 0.9 | 0.990 | 0.989 |
| 0.95 | 0.997 | 0.997 |
| 1.000 | 1.000 | 1.000 |

**Table 4.12**

| $n$ | $K_m$ | $K_v$ |
|---|---|---|
| 0 | 1.600 | 2.0 |
| 0.05 | 1.599 | 1.967 |
| 0.1 | 1.596 | 1.933 |
| 0.15 | 1.591 | 1.9 |
| 0.2 | 1.585 | 1.867 |
| 0.25 | 1.576 | 1.833 |
| 0.3 | 1.566 | 1.8 |
| 0.333 | 1.559 | 1.778 |
| 0.35 | 1.554 | 1.767 |
| 0.4 | 1.541 | 1.733 |
| 0.45 | 1.526 | 1.7 |
| 0.5 | 1.510 | 1.667 |
| 0.55 | 1.492 | 1.633 |
| 0.6 | 1.473 | 1.6 |
| 0.65 | 1.453 | 1.567 |
| 0.667 | 1.446 | 1.556 |
| 0.7 | 1.431 | 1.533 |
| 0.75 | 1.409 | 1.5 |
| 0.8 | 1.385 | 1.467 |
| 0.85 | 1.360 | 1.433 |
| 0.9 | 1.334 | 1.4 |
| 0.95 | 1.307 | 1.367 |
| 1.000 | 1.280 | 1.334 |

**Table 4.13**

| $n$ | $K_m$ | $K_v$ |
|---|---|---|
| 0 | 1.600 | 2.0 |
| 0.05 | 1.597 | 1.933 |
| 0.1 | 1.588 | 1.867 |
| 0.15 | 1.574 | 1.8 |
| 0.2 | 1.555 | 1.733 |
| 0.25 | 1.530 | 1.667 |
| 0.3 | 1.500 | 1.6 |
| 0.333 | 1.479 | 1.556 |
| 0.35 | 1.467 | 1.533 |
| 0.4 | 1.428 | 1.467 |
| 0.45 | 1.386 | 1.4 |
| 0.5 | 1.340 | 1.333 |
| 0.55 | 1.290 | 1.267 |
| 0.6 | 1.237 | 1.2 |
| 0.65 | 1.181 | 1.133 |
| 0.667 | 1.161 | 1.111 |
| 0.7 | 1.122 | 1.067 |
| 0.75 | 1.060 | 1.0 |
| 0.8 | 0.996 | 0.933 |
| 0.85 | 0.929 | 0.867 |
| 0.9 | 0.861 | 0.8 |
| 0.95 | 0.791 | 0.733 |
| 1.000 | 0.72 | 0.667 |

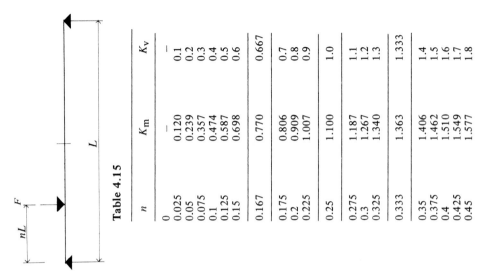

**Table 4.15**

| $n$ | $K_m$ | $K_v$ |
|---|---|---|
| 0 | — | — |
| 0.025 | 0.120 | 0.1 |
| 0.05 | 0.239 | 0.2 |
| 0.075 | 0.357 | 0.3 |
| 0.1 | 0.474 | 0.4 |
| 0.125 | 0.587 | 0.5 |
| 0.15 | 0.698 | 0.6 |
| 0.167 | 0.770 | 0.667 |
| 0.175 | 0.806 | 0.7 |
| 0.2 | 0.909 | 0.8 |
| 0.225 | 1.007 | 0.9 |
| 0.25 | 1.100 | 1.0 |
| 0.275 | 1.187 | 1.1 |
| 0.3 | 1.267 | 1.2 |
| 0.325 | 1.340 | 1.3 |
| 0.333 | 1.363 | 1.333 |
| 0.35 | 1.406 | 1.4 |
| 0.375 | 1.462 | 1.5 |
| 0.4 | 1.510 | 1.6 |
| 0.425 | 1.549 | 1.7 |
| 0.45 | 1.577 | 1.8 |

**Table 4.14**

| $n$ | $K_m$ | $K_v$ |
|---|---|---|
| 0 | 1.28 | 1.333 |
| 0.05 | 1.279 | 1.332 |
| 0.1 | 1.277 | 1.327 |
| 0.15 | 1.273 | 1.320 |
| 0.2 | 1.267 | 1.311 |
| 0.25 | 1.260 | 1.300 |
| 0.3 | 1.252 | 1.287 |
| 0.333 | 1.245 | 1.277 |
| 0.35 | 1.242 | 1.273 |
| 0.4 | 1.230 | 1.257 |
| 0.45 | 1.211 | 1.240 |
| 0.5 | 1.203 | 1.222 |
| 0.55 | 1.188 | 1.203 |
| 0.6 | 1.171 | 1.183 |
| 0.65 | 1.153 | 1.162 |
| 0.667 | 1.147 | 1.155 |
| 0.7 | 1.134 | 1.141 |
| 0.75 | 1.114 | 1.119 |
| 0.8 | 1.093 | 1.090 |
| 0.85 | 1.071 | 1.072 |
| 0.9 | 1.048 | 1.049 |

**Chapter Five**

# Beams of Solid Timber Plus Timber and Plywood Decking

## 5.1   INTRODUCTION

Basically the design of a timber joist is extremely easy. Usually it is the first design carried out at college by a trainee engineer and it will be completed in a few lines. Such a design hardly seems to justify the attention given in this chapter; however, the designer must take account of the various factors discussed in detail in Chapter 4, including shear deflection as now required by BS 5268.

For the majority of uses of solid timber as floor or roof joists, the designer can refer to load/span tables in which the designs have already been carried out by others. Such tables are produced by:

The Swedish Finnish Timber Council for Swedish Redwood and White-wood stress graded to BS 4978.

The Council of the Forest Industries of British Columbia for Canadian timber.

The Timber Research and Development Association for various timbers.

The designer can also refer to Schedule 6 of The Building Regulations 1976 for England and Wales. (Although the tables in Schedule 6 are largely based on General Structural and Special Structural stress graded to BS 4978, the stresses used are in accordance with CP 112:Part 2:1971, not BS 5268: Part 2 and there is no clear indication that they will be brought into line with BS 5268. In fact, in future, they may be withdrawn from the Regulations and be presented elsewhere.) There is also discussion about a possible Part 7 to BS 5268 which could give a guide on how span tables are to be produced, and may eventually contain span tables. Load/span tables are not presented in this manual because they are so readily available else-where to a designer.

## 5.2   DESIGN. GENERAL

The method of explaining the actual design of beams of solid timber (and the other design chapters in this handbook) is to give examples of typical designs.

Chapter 4 details the various factors and aspects which must be taken into account in the design of beams. They are not repeated at length in this chapter, but when a factor is used in the calculations and it is thought that the designer might wish to check back, reference is made to the explanatory item of Chapter 4.

A designer is in one of two different situations in designing beams. One is involved with checking a section and grade which has already been decided, perhaps by checking a design which has already been prepared by another designer. This is a straightforward situation, in which the section properties and the permissible stresses for the grade are known and the designer can immediately calculate the actual stresses and compare them with the permissible. In the other case, the designer knows the loading and span and nothing else. The process of assuming a section and grade and checking for the various stresses can be time-consuming, unless the designer has enough experience or good fortune to choose the correct section and grade at the first attempt.

Both situations are covered by typical examples in this chapter. To eliminate the time-consuming guesswork of the second situation, tables of beam capacities are introduced, and the required capacity values can be compared easily with the actual capacity values tabulated for standard beams in § 5.9, Tables 5.16–5.27.

## 5.3   PRINCIPAL MEMBERS OF SOLID TIMBER

With principal members composed of one piece of timber there is obviously no load sharing and the minimum value of $E$ must be used. The load-duration factor applies to bending, shear and bearing stresses.

### 5.3.1   Example of checking a previously selected floor trimmer beam which is a principal member

For installations within a dormitory floor, check the suitability of a 75 × 225 mm sawn section of SS grade European Whitewood on a clear span of 2.80 m loaded by incoming beams spaced at 0.6 m centres. The dead loading is 0.35 kN/m² inclusive of beam self-weight. The imposed loading is 1.5 kN/m² or a concentrated load of 1.8 kN which can occur anywhere on the span. The incoming beams have a span of 4.20 m, occur on one side of the trimmer and laterally restrain the trimmer. CP 3 : Chapter V : Part 1 is not specific on stating if the concentrated load of 1.8 kN is medium or long term. However Clause 12 of BS 5268 : Part 2 calls for the 1.4 kN load on a domestic floor to be treated as medium term and it seems reasonable to treat the 1.8 kN load also as medium term.

The incoming beams are at close enough centres to justify taking the load on the trimmer as being uniformly distributed. Assuming 0.05 m bearing at each end gives an effective design span of 2.85 m.

Properties and capacities for this section may be taken from Table 5.14. First consider uniformly distributed loading:

$$\text{Dead UDL} = 2.85 \times \frac{4.2}{2} \times 0.35 = 2.09 \,\text{kN}$$

$$\text{Imposed UDL} = 2.85 \times \frac{4.2}{2} \times 1.50 = 8.98 \,\text{kN}$$

$$\text{Total UDL} = 2.09 + 8.98 = 11.1 \,\text{kN long term}$$

$$\text{End reaction} = \text{maximum end shear} = \frac{11.1}{2} = 5.55 \,\text{kN}$$

Bearing stress $\sigma_{c,a,tra}$ (with 50 mm end bearing length)

$$= \frac{5550}{75 \times 50} = 1.48 \,\text{N/mm}^2 \text{ long term}$$

$$\sigma_{c,adm,tra} \text{ (long term)} = 2.10 \,\text{N/mm}^2 \text{ if wane is permitted}$$
$$= 2.79 \,\text{N/mm}^2 \text{ if wane is not permitted}$$

∴ section and bearing is adequate even if wane is permitted to the grade limit. Note that factor $K_4$ does not apply. With a maximum long term shear of 5.55 kN:

$$\text{Applied shear stress } \tau_a = \frac{3F_v}{2bh} = \frac{3 \times 5550}{2 \times 75 \times 225} = 0.49 \,\text{N/mm}^2$$

Permissible shear stress (long term) $= \tau_{adm} = \tau_g = 0.82 \,\text{N/mm}^2$

∴ section is adequate for shear.

From Table 5.16 it can be seen that the long term shear capacity $= \bar{F}_{vL} = 9.22 \,\text{kN}$.

$$\text{Long term bending moment} = \frac{11.1 \times 2.85}{8} = 3.95 \,\text{kN.m}$$

$$\text{Section modulus } Z_X = 0.633 \times 10^6 \,\text{mm}^3 \text{ (Table 5.16)}$$

$$\text{Applied bending stress} = \sigma_{m,a} = \frac{3.95 \times 10^6}{0.633 \times 10^6} = 6.24 \,\text{N/mm}^2$$

$$\text{Permissible bending stress} = \sigma_{m,adm} = \sigma_{m,g} \times K_7$$
$$= 7.50 \times 1.032 = 7.74 \,\text{N/mm}^2$$
$$\text{(see Table 4.4 for } K_7)$$

∴ section is adequate for bending stress.

From Table 5.16 it can be seen that the long term moment capacity = $\overline{M}_L$ = 4.89 kN.m. Since a single principal member is being considered:

$E_{min}$ = 7000 N/mm² is used in calculations of deflection
$G_{min}$ = 437.5 N/mm²
$I_X$ = 71.2 × 10⁶ mm⁴

$$\text{Bending deflection} = \frac{5FL_e^3}{384EI} = \frac{5 \times 11.1 \times 10^3 \times 2.85^3 \times 10^9}{384 \times 7000 \times 71.2 \times 10^6} = 6.71 \text{ mm}$$

$$\text{Shear deflection} = \frac{K_{form} \times M}{AG} = \frac{1.2 \times 3.95 \times 10^6}{16.9 \times 10^3 \times 437.5} = 0.64 \text{ mm}$$
(see § 4.15.7)

$$\frac{L}{h} = 12.67 \qquad\qquad \text{Total deflection} = 7.35 \text{ mm}$$

From Fig 4.26 it could have been estimated that shear deflection is 9.6% of the bending deflection = 0.64 mm (or total deflection is estimated as 1.096 times bending deflection = 7.35 mm).

From Table 5.16, the calculation of bending deflection and shear deflection could be simplified by reading off values of *EI* and *AG*.

$$\text{Permissible deflection} = 0.003 \times 2.85 \times 10^3 = 8.55 \text{ mm}.$$

The second loading condition to be checked is a medium term condition. By inspection, the 1.8 kN point load can be seen to be less critical than the uniform loading condition and no further design check is needed.

### 5.3.2 Example where the section, species and grade are to be determined

Find a section, species and grade of timber to meet the loading conditions of § 5.3.1.

From § 5.3.1 it can be seen that any section to meet the design requirements must have the following capacities:

$$\text{Long term shear capacity} = \overline{F}_{vL} = 5.55 \text{ kN}$$

$$\text{Long term moment capacity} = \overline{M}_L = \frac{11.1 \times 2.85}{8} = 3.95 \text{ kN.m}$$

Because shear deflection must be taken into account it is not possible to calculate an accurate stiffness capacity but a useful method of estimating the required *EI* capacity involves a transposition of the conventional formula. With a total UDL of *F* on span *L*, at a maximum permissible deflection of 0.003*L* and taking account of the ratio $\delta_t/\delta_m$ in Fig. 4.26:

$$\delta_{total} = \frac{5FL^3}{384EI} \times \frac{\delta_t}{\delta_m} = 0.003L$$

from which

$$EI = 4.34FL^3 \times \frac{\delta_t}{\delta_m}$$

(An alternative first approximation would be to increase the constant to say 4.6, rather than estimate $L/h$ and hence $\delta_t/\delta_m$, see §7.10.1.)

For the general case of any loading condition this formula becomes:

$$EI = 4.34F_eL^2 \times \frac{\delta_t}{\delta_m}$$

where $F_e$ is the 'equivalent UDL' which will give the same bending deflection at mid span as the actual loading (see §4.15.6).

In the example being considered with a UDL of 11.1 kN, $F_e$ obviously also equals 11.1 kN and the required $EI$ capacity

$$= 4.34 \times 11.1 \times 2.85^2 \times \frac{\delta_t}{\delta_m}$$

$$= 391 \frac{\delta_t}{\delta_m} \text{ kN.m}^2$$

For the case with dead UDL of 2.09 kN plus a concentrated central load of 1.8 kN:

$$F_e = 2.09 + (1.8 \times 1.6) = 4.97 \text{ kN} \qquad \text{(1.6 is } K_m \text{ from Table 4.15}$$
$$\text{with } n = 0.5\text{)}$$

and

$$EI = 4.34 \times 2.85^2 \times 4.97 \frac{\delta_t}{\delta_m} = 175 \frac{\delta_t}{\delta_m} \text{ kN.m}^2$$

The UDL condition governs the $EI$ requirement.

Without knowing the depth of the joist which will be used the $L/h$ ratio can not be calculated and hence the ratio of $\delta_t/\delta_m$ is not known. However, it can be estimated by experience that $L/h$ is likely to be about 12–15. If 12 is taken to give the first estimate, $\delta_t/\delta_m$ is 1.106 from Fig. 4.26 and (for the case of UDL of 11.1 kN) an $EI$ capacity of 391 × 1.106 = 432 kN.m² is required.

From Table 5.16 it can be seen that a 75 × 225 mm section is satisfactory for shear and bending. The $EI$ capacity of 498 kN.m² indicates that the deflection is satisfactory, subject perhaps to a final design check now that $L/h$ is established.

### 5.3.3   Principal members bending about both the *XX* and *YY* axes

When the direction of load does not coincide with one of the principal axes of a section there is bending about both axes and the design procedure is usually one of trial and error to find a section which will resist the combined bending and deflection of the section.

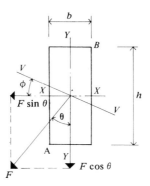

Fig. 5.1

If a load $F$ acts at an angle $\theta$ to the $YY$ axis (Fig. 5.1), then the components of the load acting about the $XX$ and $YY$ axes are $F \cos \theta$ and $F \sin \theta$ respectively. The axis $VV$ is at an inclination $\phi$ to the $XX$ axis, where:

$$\tan \phi = \frac{I_X}{I_Y} \tan \theta = \left(\frac{h}{b}\right)^2 \tan \theta$$

The critical points for bending stress are at A and B which are the fibres furthest from the $VV$ axis and at these points the bending stress is:

$$\sigma_{m,a,par} = \left[\frac{M_X}{Z_X} + \frac{M_Y}{Z_Y}\right] \qquad \text{tension or compression}$$

The direction of the deflection is normal to the $VV$ axis. The total deflection $\delta_t$ is the geometrical sum of the deflections $\delta_X$ and $\delta_Y$.

$$\therefore \ \delta_t = \sqrt{(\delta_X^2 + \delta_Y^2)}$$

where $\delta_X$ = deflection about $XX$ axis.
$\delta_Y$ = deflection about $YY$ axis.

Bending about both the $XX$ and $YY$ axes greatly increases the dimensions of the required rectangular section, consequently the designer should endeavour where possible to place the section so that the load acts directly in the plane of maximum stiffness. For example, a purlin between trusses should preferably be placed in a vertical plane either by employing tapered end blocking pieces or by being supported vertically in metal hangers. Frequently, however, this is not possible. In traditional house construction, a purlin is often placed normal to the roof slope to simplify the fixing of secondary rafters, and the force parallel to the roof slope can be resisted by the secondary system with its roof cladding, sag rods etc. Where such advantageous assumptions are justified, the purlin need only be designed for the force $F \cos \theta$ acting about the $XX$ axis, but attention must be paid to the adequacy of end fixings, horizontal cross ties etc., to ensure that the force $F \sin \theta$ is also adequately resisted or balanced.

Where the permissible bending stress about both axes is the same, the equation for combining bending stresses can be conveniently modified as follows to avoid the need to determine sectional properties about the $YY$ axis. (When the permissible bending stresses differ the method of combining stress ratios as shown in § 7.3.3 for glulam beams must be adopted.)

$$\sigma_{m,a,par} = \frac{M_X}{Z_X} + \frac{M_Y}{Z_Y}$$

$$= \frac{F\cos\theta\,L^2}{8Z_X} + \frac{F\sin\theta\,L^2}{8Z_Y}$$

$$= \frac{FL^2}{8Z_X}\left(\cos\theta + \frac{Z_X}{Z_Y}\sin\theta\right)$$

$$= \frac{M}{Z_X}\left(\cos\theta + \frac{h}{b}\sin\theta\right)$$

where $M$ is the full bending moment applied to the section at $\theta$ to the $YY$ axis.

It can be seen that $\sigma_{m,a,par}$ is a function of $h/b$ and that slender sections with a high $h/b$ ratio have a large bending stress, as would be expected.

If $h/b$ is known, one can calculate the value required for $M_X$ so that $\sigma_{m,a,par}$ does not exceed the permissible stress.

$$\text{Required} \quad \overline{M}_X = M\left[\cos\theta + \frac{h}{b}\sin\theta\right]$$

Similarly the equation for total deflection can be simplified to avoid the need to determine sectional properties about the $YY$ axis.

$$\delta_t = \sqrt{(\delta_X^2 + \delta_Y^2)}$$

which for a uniform load (and disregarding shear deflection) is:

$$\delta_{tm} = \sqrt{\left[\left(\frac{5F\cos\theta\,L^3}{384EI_X}\right)^2 + \left(\frac{5F\sin\theta\,L^3}{384EI_Y}\right)^2\right]}$$

$$= \frac{5FL^3}{384EI_X}\sqrt{\left[\cos^2\theta + \left(\frac{I_X\sin\theta}{I_Y}\right)^2\right]}$$

$$= \frac{5FL^3}{384EI_X}\sqrt{\left\{1 + \left[\left(\frac{h}{b}\right)^4 - 1\right]\sin^2\theta\right\}}$$

If $h/b$ is known, one can calculate the value required for $EI_X$ to limit the deflection $\delta_{tm}$ to a given amount: i.e. for $\delta_{tm}$ not to exceed 0.003 of span, by transposing with $\delta_{tm} = 0.003L$ the formula (disregarding shear deflection) becomes:

$$\text{Required} \quad EI_X = 4.34FL^2\sqrt{\left\{1 + \left[\left(\frac{h}{b}\right)^4 - 1\right]\sin^2\theta\right\}}$$

In considering shear deflection, rather than calculating about both axes, it is almost certainly adequate to increase the required $EI_X$ capacity given above by the ratio shown in Fig. 4.26.

The section with the smallest cross-section for a given loading will be realised if the deflection about both axes can be made the same.

To give equal deflection about the $XX$ and $YY$ axes one can equate $\delta_X = \delta_Y$ and show that $h/b$ must be equal to $\sqrt{(\cot\theta)}$. Values of $\sqrt{(\cot\theta)}$ for values of $\theta$ from $2\frac{1}{2}°$ to $45°$ in $2\frac{1}{2}°$ increments are given in Table 5.1.

**Table 5.1**

| $\theta$ degrees | $\sqrt{\cot\theta}$ |
|---|---|
| $2\frac{1}{2}$ | 4.79 |
| 5 | 3.38 |
| $7\frac{1}{2}$ | 2.76 |
| 10 | 2.38 |
| $12\frac{1}{2}$ | 2.12 |
| 15 | 1.93 |
| $17\frac{1}{2}$ | 1.78 |
| 20 | 1.65 |
| $22\frac{1}{2}$ | 1.55 |
| 25 | 1.46 |
| $27\frac{1}{2}$ | 1.38 |
| 30 | 1.32 |
| $32\frac{1}{2}$ | 1.25 |
| 35 | 1.19 |
| $37\frac{1}{2}$ | 1.14 |
| 40 | 1.09 |
| $42\frac{1}{2}$ | 1.04 |
| 45 | 1.00 |

With the resultant preferred value for $h/b$ the required value of $EI_X$ becomes:

$$\text{Required } EI_X = 4.34FL^2\sqrt{[1 + (\cot^2\theta - 1)\sin^2\theta]}$$
$$= 4.34FL^2\sqrt{(2\cos^2\theta)}$$
$$\text{Required } EI_X = 6.14FL^2\cos\theta \quad \text{with} \quad h/b \leqslant \sqrt{(\cot\theta)}$$

Then an allowance for shear deflection should be made.

*Example of design of beam bending about XX and YY axes.* Determine an economic purlin size (Fig. 5.2) to suit a span of 2.4 m. The medium term vertical load is 3.8 kN/m run and the angle $\theta = 10°$.

$$\text{Total vertical load} = F = 9.12\,\text{kN}$$

$$\text{Total vertical bending moment} = M = \frac{9.12 \times 2.4}{8} = 2.74\,\text{kN.m}$$

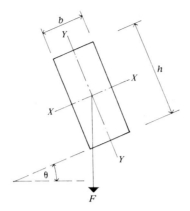

Fig. 5.2

For bending deflection, required $EI_X$ for economical section

$$= 6.14 \times 9.12 \times 2.4^2 \times \cos 10° = 318\,\text{kN.m}^2$$

to which an allowance for shear deflection must be added.

The required $EI$ capacity should be produced by a section having $h/b \leqslant \sqrt{\cot 10°}$ (i.e. $h/b \leqslant 2.38$).

Try a Canadian Spruce-Pine-Fir section of quoted green size 4 $\times$ 8 inch which, with minus tolerances, could have a minimum dry size (§ 1.5.6) of 96 $\times$ 193 mm. Regrade from 'Merchantable' to SS or MSS grade.

$$E_{min} = 6500\,\text{N/mm}^2 \qquad EI_X = 374\,\text{kN.m}^2$$

$$\frac{h}{b} = 2.0 \qquad \frac{L}{h} = \frac{2400}{193} = 12.4$$

Therefore from Fig. 4.26 $\delta_t/\delta_m = 1.100$.

$$\therefore \text{ Required } EI_X = 318 \times 1.100 = 349.8\,\text{kN.m}^2$$

$$\text{Required } \overline{M}_{XM} = 2.74\,(\cos 10° + 2 \sin 10°) = 3.65\,\text{kN.m}$$

$$\text{For the 96} \times \text{193 section } \overline{M}_{XM} = \sigma_{m,g} \times K_3 \times K_7 \times Z_X$$
$$= 7.50 \times 1.25 \times 1.05 \times 0.596 \times 10^6\,\text{N.mm}$$
$$= 5.86\,\text{kN.m}$$

The section meets the deflection and bending criteria. Check this simplified method by the calculations below.

For 96 $\times$ 193 SPF of SS grade, permissible medium term bending stress is 7.50 $\times$ 1.25 $\times$ 1.05 = 9.84 N/mm$^2$ and $E_{min}$ = 6500 N/mm$^2$.

$$Z_X = 0.596 \times 10^6\,\text{mm}^3 \qquad EI_X = 374\,\text{kN.m}^2$$
$$Z_Y = 0.296 \times 10^6\,\text{mm}^3 \qquad EI_Y = 92.5\,\text{kN.m}^2$$
$$M_X = M \cos\theta = 2.74 \cos 10° = 2.70\,\text{kN.m}$$
$$M_Y = M \sin\theta = 2.74 \sin 10° = 0.48\,\text{kN.m}$$

$$\sigma_{m,a} = \frac{2.70 \times 10^6}{0.596 \times 10^6} + \frac{0.48 \times 10^6}{0.296 \times 10^6} = 4.53 + 1.62 = 6.15 \, \text{N/mm}^2$$

Bending deflection (mm):

$$\delta_X = \frac{5 \times 9.12 \times \cos 10° \times 2.4^3 \times 10^3}{384 \times 374} = 4.32 \, \text{mm}$$

$$\delta_Y = \frac{5 \times 9.12 \times \sin 10° \times 2.4^3 \times 10^3}{384 \times 92.5} = 3.08 \, \text{mm}$$

$$\delta_{tm} = \sqrt{(4.32^2 + 3.08^2)} = 5.30 \, \text{mm}$$

Shear deflection (mm) $(G = 6500/16 = 406.3 \, \text{N/mm}^2)$:

$$\delta_X = \frac{K_{form} \times M_X}{AG_{min}} = \frac{1.2 \times 2.70 \times 10^6}{18\,528 \times 406.3} = 0.43 \, \text{mm}$$

$$\delta_Y = \frac{1.2 \times 0.48 \times 10^6}{18\,528 \times 406.3} = 0.076 \, \text{mm}$$

$$\delta_{tv} = \sqrt{(0.43^2 + 0.076^2)} = 0.44 \, \text{mm}$$

Total bending + shear deflection

$$= 5.30 + 0.44 = 5.74 \, \text{mm}$$
$$= (0.0024 \text{ of span})$$

Alternatively:

$$\sigma_{m,a} = \frac{M}{Z_X}\left(\cos\theta + \frac{h}{b}\sin\theta\right) = \frac{2.74 \times 10^6}{0.596 \times 10^6}[0.985 + (2.01 \times 0.174)]$$

$$= 6.14 \, \text{N/mm}^2 \text{ (which is close to 6.15 N/mm}^2 \text{ above)}$$

and bending deflection:

$$= \frac{5FL^3}{384EI_X}\sqrt{\left\{1 + \left[\left(\frac{h}{b}\right)^4 - 1\right]\sin^2\theta\right\}}$$

$$= \frac{5 \times 9.12 \times 2.4^3}{384 \times 374}\sqrt{(1 + 15.32 \times 0.0301)} = 0.0053 \, \text{m}$$

which agrees with the figure of 5.30 mm above, to which shear deflection must be added.

## 5.4  LOAD-SHARING SYSTEMS OF SOLID TIMBER

With a load-sharing system of four or more beams of solid timber spaced at not more than 610 mm centres, the load-sharing factor $(K_8 = 1.1)$ discussed in §4.4 applies to all stresses. $E_{mean}$ is usually applicable subject to the limitation explained in §4.15.1.

## 5.4.1   Example of checking a previously selected load-sharing floor joist system

For installation within a dormitory floor, check the suitability of 47 X 195 mm regularized (width only) European Whitewood of GS grade joists spaced at 0.6 m centres on a clear span of 3.3 m. The uniformly distributed dead loading (including the self-weight of joists) is 0.35 kN/m², the imposed loading is 1.5 kN/m² (all of which must be considered as being permanently in place); the concentrated load may be neglected (see § 3.4). Properties for this section may be taken from Table 5.25. The effective design span may be taken as clear span plus 0.05 m (i.e. 3.35 m).

First consider the uniformly distributed loading:

$$\text{Dead UDL} = 3.35 \times 0.6 \times 0.35 = 0.70$$
$$\text{Imposed UDL} = 3.35 \times 0.6 \times 1.50 = \underline{3.01}$$
$$\text{Total UDL} = 3.71 \text{ kN}$$

$$F_v = \text{maximum end shear (long term)} = \frac{3.71}{2} = 1.85 \text{ kN}$$

$$\sigma_{c,a,tra} \text{ (with 50 mm end bearing length)} = \frac{1850}{47 \times 50} = 0.79 \text{ N/mm}^2$$

$$\sigma_{c,adm,tra} = \sigma_{c,g,tra} \times K_8 = 1.80 \times 1.1 = 1.98 \text{ N/mm}^2$$
$$\text{if wane is permitted}$$
$$= 2.79 \times 1.1 = 3.07 \text{ N/mm}^2$$
$$\text{if wane is excluded}$$

$$\tau_a = \frac{3 \times 1850}{2 \times 47 \times 195} = 0.30 \text{ N/mm}^2$$

$$\tau_{adm} = 0.82 \times 1.1 = 0.90 \text{ N/mm}^2$$

$$\text{Long term bending moment} = M_L = \frac{3.71 \times 3.35}{8} = 1.55 \text{ kN.m}$$

$$Z_X = 0.297 \times 10^6 \text{ mm}^3$$

$$\sigma_{m,a} = \frac{1.55 \times 10^6}{0.297 \times 10^6} = 5.22 \text{ N/mm}^2$$

$$\sigma_{m,adm} = 5.3 \times K_7 \times K_8$$
$$= 5.3 \times 1.048 \times 1.1$$
$$= 6.10 \text{ N/mm}^2$$

$E_{mean} = 9000$ N/mm² can be used in calculations of bending deflection.

The $EI_X$ value of 47 × 195 is 261 kN.m² (Table 5.25).

$$\text{Bending deflection} = \frac{5 \times 3.71 \times 3.35^3 \times 10^3}{384 \times 261} = 6.96 \text{ mm}$$

$$\frac{L}{h} = 17.18 \qquad \therefore \frac{\delta_t}{\delta_m} = 1.052 \text{ (from Fig. 4.26)}$$

∴ Total deflection = 6.96 × 1.052 = 7.3 mm.
Permissible deflection = 0.003 × 3.35 × 10³ = 10.1 mm.

### 5.4.2  Example where the section, species and grade are to be determined

Establish a range of sections, species and grades which will be adequate to support the floor loading described in § 5.4.1.

From the example in § 5.4.1 it can be seen that any section to meet the design requirements must have the following capacities:

$$\overline{F}_{vL} = 1.85 \text{ kN} \quad \overline{M}_L = 1.55 \text{ kN.m}$$

To calculate the required $EI_X$ account must be taken of shear deflection. With $L/h = 17.18$, $\delta_t/\delta_m = 1.052$ (Fig. 4. 26).

$$EI \text{ required} = 4.34FL^2 \frac{\delta_t}{\delta_m}$$

$$= 4.34 \times 3.71 \times 3.35^2 \times 1.052$$
$$= 190 \text{ kN.m}^2$$

The shear, moment and *EI* capacities for various species, sections and grades for load-sharing systems are given in § 5.9. Some relevant values are reproduced in Table 5.2. The required capacities are shown in brackets.

**Table 5.2**

| From Table | Grade | Species | Size | $\overline{F}_{vL}$ (kN) (1.85) | $\overline{M}_L$ (kN.m) (1.55) | $EI$ (kN.m²) (190) |
|---|---|---|---|---|---|---|
| 5.23 | SS/MSS | Whitewood or Redwood | 38 × 195 47 × 170 | 4.45 4.80 | 2.08 1.98 | 246 202 |
| 5.25 | GS/MGS | Whitewood or Redwood | 38 × 220 47 × 195 | 5.02 5.51 | 1.84 1.82 | 303 261 |
| 5.26 | No. 1 and No. 2 | Hem-Fir | 38 × 235 | 4.64 | 2.21 | 431 |
| 5.27 | No. 1 and No. 2 | SPF | 38 × 235 | 4.45 | 2.21 | 369 |

Consideration of the list of possible sections illustrates that shear is seldom critical and the choice depends in most cases on bending and deflection considerations. Deflection tends to be the limit in load-sharing systems and bending tends to be the limit in principal members.

## 5.5 SOLID TIMBER ROOF DECKING

### 5.5.1 Introduction

Solid timber decking is mainly used in roof constructions where the soffit of the decking is exposed to view and the decking spans between glulam beams to give a solid, permanent roof deck which also serves as a ceiling with excellent appearance. Decking is frequently used in swimming pools where the general moisture conditions are a disadvantage for the use of many other structural materials and where ceiling cavities are considered by many to present condensation problems.

Traditionally decking in the UK was undertaken in Western Red Cedar, usually in what is known as a 'random layup', in which end joints of individual boards can occur in the span rather than only at points of support. With random layups a high degree of site attention is required to ensure that boards are laid less randomly than the term suggests. With the introduction of finger jointing machinery as used by most timber engineering companies some simplification of layup has been made possible without high length wastage. This, together with the economic trend towards European timbers, has led most companies to finger joint and profile boards in lengths equal to one, two or three bay modules instead of purchasing random-length boards of Canadian origin already profiled at source, and to use European Whitewood or Redwood.

Basic thicknesses of 38, 50, 63 and 75 mm are used by the larger producers. All decking is tongued and grooved, the thinner boards being tongued

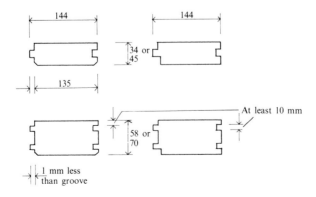

Fig. 5.3: *Typical profiles.*

only once and the thicker boards being double-tongued. Typical profiles are shown in Fig. 5.3. Decking is usually machined from 150 mm wide sections, although sections less than 150 mm wide can be used at a higher cost (both for machining and laying a given area).

The deflection limit usually considered appropriate with solid roof decking is $L/240$.

Some companies prefer to use a finished thickness 2 mm less than those shown in Fig. 5.3 particularly if they finger joint (which can lead to a slight off-set between adjacent pieces (see Fig. 19.4)).

### 5.5.2    Span and end joint arrangements

*Layup 1. Single span*
All boards bear on and are discontinuous at each support (Fig. 5.4). The planks in this arrangement will deflect more than with any other possible arrangement. Depending upon the deck thickness this arrangement may

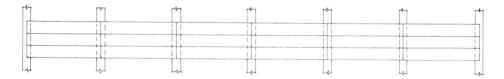

Fig. 5.4: *Layup 1.*

require large widths for the supporting beams to accommodate nail fixings. End fixing is particularly difficult to resolve when the thicker decks are specified, because the longer nails, available only in the larger gauges, require large end distances and spacings. (The requirements for nail fixings are discussed fully in Chapter 18 and particular fixing details for decking are illustrated in § 5.5.3.)

*Layup 2. Double span*
All boards bear on each support and are continuous over two bays (Fig. 5.5). This arrangement has the least deflection of any arrangement, but the loading is not transferred uniformly to supports and beams supporting the decking at

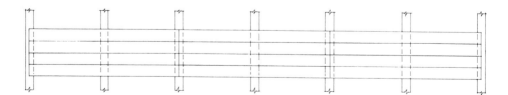

Fig. 5.5: *Layup 2.*

mid length carry more load. For architectural reasons it is usual to require all support beams to be of the same section which, with this particular arrangement, leads to alternate beam sections being over-designed. Similar problems for end fixing as encountered in Layup 1 also apply with this arrangement.

*Layup 3. Treble span*
All boards bear on supports and are continuous over three spans, breaking regularly at every third beam (Fig. 5.6). This arrangement is usually applied with caution, since there are practical difficulties in manufacturing long

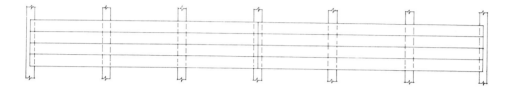

Fig. 5.6: *Layup 3.*

straight lengths of boarding and in handling individual boards. It is recommended from practical experience that the length of individual boards should not exceed 9 m wherever possible.

*Layup 4. Single and double span*
All boards bear on supports. Alternate boards in end bays are single span with the adjacent board being two span continuous. End joints occur over beam supports, being staggered in adjacent courses (Fig. 5.7). Deflections in

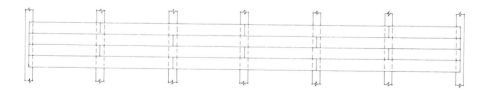

Fig. 5.7: *Layup 4.*

the end bay are intermediate between those of Layups 1 and 2, whilst internal bays deflect approximately the same as for a two span continuous arrangement. This arrangement is probably the one most usually adopted because it avoids the high deflection characteristic of Layup 1 without overloading alternate supporting beams, and the staggering of end joints permits easier end nailing of boards (Fig. 5.14).

*Layup 5. Double and treble span*

All boards bear on supports (Fig. 5.8). Alternate boards commence at one end and span three bays continuously with adjacent boards spanning two bays continuously. Internal spans revert to the two span arrangement of

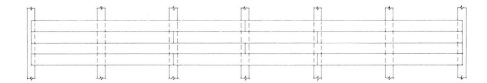

Fig. 5.8: *Layup 5.*

Layup 4. This layup is used only rarely to minimise deflection in the end bays if Layup 4 is not suitable. Care should be taken to avoid board lengths exceeding 9 m, or great care should be taken in selecting the longer boards.

*Layup 6. Cantilevered pieces intermixed*

With this arrangement every third course of boarding is simple span (as Layup 1) and, intermixed between these courses, is a series of cantilever pieces with butt end joints at between the $\frac{1}{3}$ and $\frac{1}{4}$ points of span (Fig. 5.9).

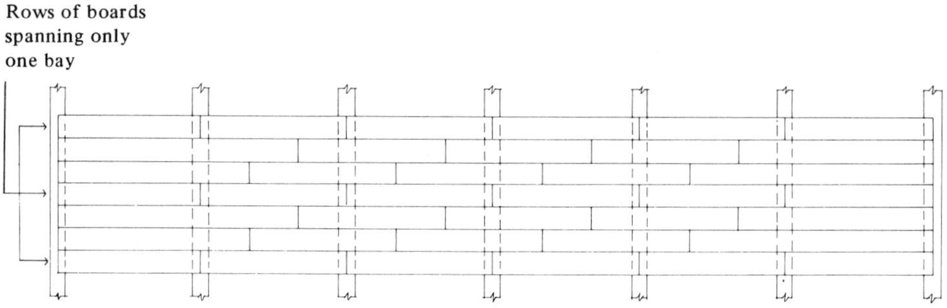

Rows of boards
spanning only
one bay

Fig. 5.9: *Layup 6.*

Each piece of the cantilever system rests on at least one support. This arrangement has some advantageous features. For example, glued end jointing is avoided (although end joints will be visible on the soffit) and short convenient lengths of boards can be used, there being no board length of more than 1.75 times the bay span, and the lengths are predominantly of single bay length. There must be at least three bays for this arrangement and great care is required in the laying of the decking to ensure that the end joints are correctly positioned.

### 5.5.3 Nailing of decking

Decking is laid with the wide face (usually about 135 mm cover width) bearing on supports. Wherever possible each piece of decking should be nailed down to the support beams, which is relatively easy to do where a board is continuous over a support, but requires attention where two boards butt joint over a support beam.

In Layups 1, 2 and 3 end joints occur as shown in Fig. 5.10, from which it can be seen that without pre-drilling the decking (and it is preferable to avoid site drilling) the minimum width of solid timber support beams is $42d$ where $d$ is the nail diameter. This width may be reduced to $26d$ if the decking is pre-drilled.

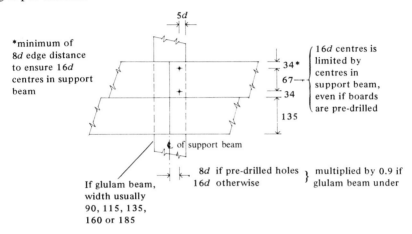

Fig. 5.10: *Nail fixing for 135 mm width decking.*

It is recommended that two nails per end are used for a width of 135 mm.

The centres of nails is obtained from Table 56 of BS 5268:Part 2 multiplied by 0.8 which is a factor which can be applied to all softwoods except Douglas Fir. (Hemlock is prone to split if nailed dry therefore should usually be pre-drilled if used.)

If the support beam is glulam the spacings can be multiplied by a further factor of 0.9.

Table 5.3 indicates the maximum nail sizes which can be taken from normal standard lists (see Table 18.4) to achieve the end distances and spacings shown in Fig. 5.10 to suit 135 mm face width decking of typical thicknesses and support beam widths. Those conditions which require the holes to be pre-drilled are noted. In establishing this table the priorities were:

(a) to try to avoid pre-drilling
(b) to use the biggest diameter nail in the longest length commercially available.

**Table 5.3  Nails for fixing 135 mm face width softwood decking (excluding Douglas Fir) to timber beams**

| Decking thickness (mm) | Maximum nail sizes to be used for typical support beam widths (mm) $d$ = diameter (mm) | | | | |
|---|---|---|---|---|---|
| | 90 | 115 | 135 | 160 | 185 |
| 34 | Pre-drill 3.35$d$ X 75 | 2.65$d$ X 65 | 3$d$ X 65 | 3.75$d$ X 75 | 4$d$ X 100 |
| 45 | Pre-drill 3.35$d$ X 75 | 2.65$d$ X 65 | 3$d$ X 65 | 3.75$d$ X 75 | 4$d$ X 100 |
| 58 | Not recommended with standard nails | Pre-drill 4$d$ X 100 | Pre-drill 4$d$ X 100 | Pre-drill 4$d$ X 100 | 4$d$ X 100 |
| 70 | | Pre-drill 4$d$ X 100 | Pre-drill 4$d$ X 100 | Pre-drill 4$d$ X 100 | 4$d$ X 100 |

Where decking is continuous over a support there is no criterion for end distance spacing in the board. In practice the fixing in these cases would be as shown in Fig. 5.11, using the same nail sizes as the cases with end joints, there being no pre-drilling requirement. Although 5$d$ is the minimum recommended edge distance, this should be increased wherever possible to avoid possible breaking through of the nail point on the vertical face of the support beam due to possible site inaccuracies.

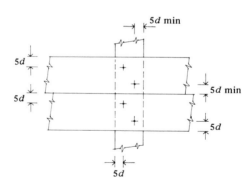

Fig. 5.11

In addition to nailing through the wide face, it is usual to nail horizontally when thicker boards (e.g. 58 and 70 mm) are used. The purpose of this lateral nailing is to draw the boards tightly together during installation and, on occasions, to transfer shear from board to board to provide a diaphragm action. For ease of driving the lateral nails, holes are pre-drilled in pairs at approximately 750 mm centres as shown in Fig. 5.12. Only one nail is driven per pair of holes.

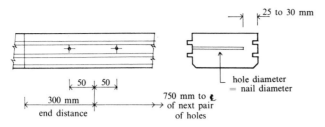

Fig. 5.12

Pairing of holes for the application of a single nail is usually necessary to avoid nail holes coinciding on site, because holes are usually factory drilled on a jig system. If holes are made on site, only a single hole need be drilled to avoid the known positions of existing nails. These lateral nails have to be specially ordered, being of non-standard sizes (i.e. for 70 mm decking use 6.7 × 200 mm and for 58 mm decking use 5.6 × 200 mm).

It is not practical to nail laterally through 34 mm and 45 mm thick decking. Instead, it is necessary to slant nails at 0.75 m centres as illustrated in Fig. 5.13. It is practical for these nails to be driven at an angle of 30° to

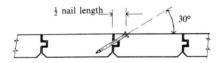

Fig. 5.13

the horizontal, with an initial edge distance of half the nail length. To meet this recommendation, 2.65 × 50 and 3.35 × 75 mm nails can be used for 34 and 45 mm decking respectively.

If the preceding guidelines for lateral nailing are implemented, it can usually be assumed that sufficient lateral shear transfer will occur between boards to give full lateral restraint to any supporting beam. In the particular cases of Layups 4, 5 and 6 the end nailing schedule of Table 5.3 can be relaxed and reliance placed almost exclusively upon the fixings in adjacent boards (Fig. 5.14).

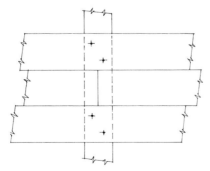

Fig. 5.14

### 5.5.4   Design procedure

All boards are tongued and grooved and act together as a unit to resist vertical shears, moments and deflections. However, an exception for which the designer must make allowance is where moments are not always resisted by the full width of the decking system because butt end joints occur over supports. The effect of concentrated loadings may be ignored because of the lateral distribution achieved by the tongues and the fixing of lateral or slant nails.

Shear stresses need not be considered in the design of decking as they can be shown to be of a very low magnitude. It is therefore necessary to check only deflection and bending stresses. With Layup 4 deflection will be a maximum in the end bay (where continuity is a minimum) and bending stresses will be greatest at the first beam support from the end, where the net resisting section of deck is least due to end joints. For Layups 2 and 3 the full section resists bending stresses over the first support.

Table 5.4 summarises the formulae for end bay deflection, maximum moment under balanced loading and the effective width of decking which

**Table 5.4**

| Layup | Description | Formulae for bending deflection in end bay | Maximum moment under balanced load | Ratio of effective design width as laid at position of $M_{\max}$ | Required $EI$ ($kN.m^2$) with deflection $= L/240$ |
|---|---|---|---|---|---|
| 1 | Single span | $\dfrac{5pL^4}{384\,EI}$ | $\dfrac{pL^2}{8}$ | 1.0 | $3.12pL^3\dfrac{\delta_t}{\delta_m}$ |
| 2 | Double span | $\dfrac{pL^4}{185\,EI}$ | $\dfrac{pL^2}{8}$ | 1.0 | $1.30pL^3\dfrac{\delta_t}{\delta_m}$ |
| 3 | Treble span | $\dfrac{pL^4}{145\,EI}$ | $\dfrac{pL^2}{10}$ | 1.0 | $1.66pL^3\dfrac{\delta_t}{\delta_m}$ |
| 4 | Single and double span | $\dfrac{pL^4}{109\,EI}$ | $\dfrac{pL^2}{8}$ | 0.5 | $2.20pL^3\dfrac{\delta_t}{\delta_m}$ |
| 5 | Double and treble span | $\dfrac{pL^4}{162\,EI}$ | $\dfrac{pL^2}{8}$ | 0.5 | $1.48pL^3\dfrac{\delta_t}{\delta_m}$ |
| 6 | Cantilever pieces intermixed | $\dfrac{pL^4}{105\,EI}$ | $\dfrac{pL^2}{10}$ | 0.667 | $2.29pL^3\dfrac{\delta_t}{\delta_m}$ |

where $p =$ the load per metre run of 1 metre width of decking as laid

$\dfrac{\delta_t}{\delta_m} =$ the allowance for shear deflection as obtained from $L/h$ in Fig. 4.26 ($h$ is the thickness of the decking). See note in § 5.5.6 about shear deflection.

resists this moment. Shear deflection is taken into account by use of the ratio of total deflection to bending deflection which can be calculated, or taken from Fig. 4.26. However, from the example given in § 5.5.6 it can be seen that it is usually safe to disregard shear deflection.

## 5.5.5   Species of decking, grades and capacities

Currently the species most frequently used for decking are Whitewood and Redwood. The choice is usually primarily one of cost, balanced by acceptable architectural appearance. With Whitewood and Redwood, the grading required for appearance usually means that the decking can be considered to match the visual stress grade 'Special Structural'. The Engineer can call for spot checks if thought necessary. Therefore, where load sharing occurs an $E_{mean}$ value of 10 500 N/mm$^2$ and a long term bending stress of 7.50 $\times$ $K_8$ $\times$ $K_7$ N/mm$^2$ can be taken. This applies to Layups 1, 2 and 3. Although, for Layups 4, 5 and 6 the use of $E_{mean}$ in deflection calculations seems justified, at the position of critical stress, load sharing is less positive due to many members being discontinuous along this line, and $K_8$ is hardly justified. In the calculations in Table 5.5, $K_8$ (of 1.1) is used for Layups 1, 2 and 3 but

**Table 5.5   Stiffness ($EI$) and moment capacity ($\overline{M}$) for 1 m width as laid of Whitewood and Redwood decking of SS grade**

| Deck thickness (mm) | $I$ (mm$^4$ $\times$ 10$^6$) | $EI$ (kN.m$^2$) | Layup | $Z_{net}$ (mm$^3$ $\times$ 10$^6$) | $\sigma_{m,adm,par}$ long term (N/mm$^2$) | Moment capacity (kN.m) | | | |
|---|---|---|---|---|---|---|---|---|---|
| | | | | | | $\overline{M}_L$ | $\overline{M}_M$ | $\overline{M}_S$ | $\overline{M}_{VS}$ |
| 34 | 3.27 | 32.7 | 1, 2 & 3 | 0.193 | 9.65 | 1.86 | 2.33 | 2.79 | 3.26 |
| | | | 4 & 5 | 0.0963 | 8.54 | 0.82 | 1.02 | 1.23 | 1.43 |
| | | | 6 | 0.128 | 8.54 | 1.09 | 1.36 | 1.63 | 1.91 |
| 45 | 7.59 | 75.9 | 1, 2 & 3 | 0.337 | 9.65 | 3.25 | 4.07 | 4.88 | 5.69 |
| | | | 4 & 5 | 0.169 | 8.54 | 1.44 | 1.80 | 2.16 | 2.52 |
| | | | 6 | 0.225 | 8.54 | 1.92 | 2.40 | 2.88 | 3.36 |
| 58 | 16.3 | 163 | 1, 2 & 3 | 0.561 | 9.65 | 5.41 | 6.76 | 8.12 | 9.47 |
| | | | 4 & 5 | 0.280 | 8.54 | 2.39 | 2.98 | 3.58 | 4.18 |
| | | | 6 | 0.374 | 8.54 | 3.19 | 3.99 | 4.79 | 5.58 |
| 70 | 28.6 | 286 | 1, 2 & 3 | 0.817 | 9.65 | 7.88 | 9.85 | 11.80 | 13.8 |
| | | | 4 & 5 | 0.408 | 8.54 | 3.48 | 4.35 | 5.22 | 6.09 |
| | | | 6 | 0.544 | 8.54 | 4.64 | 5.80 | 6.96 | 8.13 |

$\overline{M}_L$ = moment capacity, long term
$\overline{M}_M$ = moment capacity, medium term
$\overline{M}_S$ = moment capacity, short term
$\overline{M}_{VS}$ = moment capacity, very short term.

not for Layups 4, 5 and 6. With decking of less than 72 mm thickness $K_7$ is 1.17. A pedantic interpretation of BS 4978 could claim that, because the smallest size covered by BS 4978 is 35 $\times$ 60 mm, the rules and stresses for SS should not be applied to thicknesses of less than 60 mm. However, the authors feel that that would be over-restrictive. The minimum size covered by BS 4978 is to be reconsidered.

### 5.5.6   Example of design of decking

Calculate the thickness of roof decking of Layup 4, single and double span (Fig. 5.7) required on a span of 3.4 m using Redwood of SS grade and permitting a deflection of $L/240$. The dead loading including self-weight is $0.40 \, \text{kN/m}^2$, and the imposed load is $0.75 \, \text{kN/m}^2$ (medium term).

Consider 1.0 m width as laid of decking.

$$\text{Long term loading} = 0.40 \, \text{kN/m}^2$$
$$\text{Medium term loading} = 1.15 \, \text{kN/m}^2$$

From Table 5.4, $M = pL^2/8$.

$$\text{Long term moment (maximum)} = \frac{0.4 \times 3.4^2}{8} = 0.58 \, \text{kN.m}$$

$$\text{Medium term moment (maximum)} = \frac{1.15 \times 3.4^2}{8} = 1.66 \, \text{kN.m}$$

Assume 58 mm thick decking. $L/h = 3400/58 = 58.6$. From Fig. 4.26 it can be seen that the ratio is less than 1.01 which can probably be disregarded. From the formula given in § 4.15.8 the ratio is calculated as 1.0056.
    From Table 5.4:

$$EI \text{ required} = 2.20pL^2 \text{ ($p$ being total loading per metre as there is}$$
$$\text{no pre-camber)}$$
$$= 2.20 \times 1.15 \times 3.4^3 = 99.4 \, \text{kN.m}^2$$

From Table 5.5, 58 mm decking is selected as being suitable.

## 5.6   TONGUED AND GROOVED FLOOR BOARDS

The Building Regulations 1976 for England and Wales (Schedule 6) give the maximum span of softwood tongued and grooved floor boards as in Table 5.6.

Table 5.6

| Finished thickness of board (mm) | Maximum span of board (mm) |
|---|---|
| 16 | 505 |
| 19 | 600 |
| 21 | 635 |
| 28 | 790 |

'Span' in this case means 'the distance between the centres' of adjacent supports.

## 5.7  PLYWOOD DECKING

### 5.7.1.  General

Plywood is often used as a sheathing over timber joists on a flat or low pitch roof as a base for felt and chippings or for asphalt. Plywood is also often used to span over floor joists. In these cases the fixing is usually by simple nailing (but see also Chapter 10 on stress skin panels).

With simple sheathing, as against a stress skin panel, if thin plywood is being used, and particularly if there is access for personnel, it is important to span the face grain at right angles to the supports. The edges of the plywood which are not supported by joists should either be supported by secondary members (noggings) or be tongued and grooved. Small aluminium 'H' clips have been used on the unsupported edges of plywood roof sheathing, but if the plywood is a base for felt or asphalt, there is a risk of tearing or cracking due to differential movement of the plywood between the clips. Also if joints occur parallel to the roof slope, the movement of erectors before felting can cause the clips to slip down the slope.

The required thickness of plywood for roof sheathing or flooring can be calculated, but with the usual short spans (610 mm centres or less) it is more important for the 'feel' of the roof and floor to be correct than for the deflection to be limited to any specific amount.

When asphalt is to be the roof cover, experience has shown that it is unwise to use 8 mm or thinner plywood due to the point loads encountered during construction.

### 5.7.2  Plywood roof sheathing

Table 5.7 gives minimum requirements for Canadian Douglas Fir-faced and Canadian Softwood sheathing plywood where the roof design loading on the

**Table 5.7  Canadian unsanded exterior plywood for roof sheathing**

| Spacing of supports (mm) | Nominal plywood thickness (mm) | Nail length and diameter | Maximum nail spacing (mm) |
|---|---|---|---|
| 300 | 7.5 | 3 mm × 45 mm | 150 along edge |
| 400 | 7.5 | annularly | 300 along |
| 480 | 9.5 | grooved* | intermediate |
| 600 | 9.5 | | supports |

\* If round wire nails are used they should be 3 × 50 mm. 50 × 1.25 mm divergent, corrosion-resistant steel wire staples may be used, in lieu of nails, at half the tabulated nail spacings.

(a) The grade is usually 'Select Sheathing' or 'Sheathing'.
(b) Plywood must be laid with the face grain perpendicular to the supports.
(c) Edge support must be provided by noggings (or the edges must be tongued and grooved).
(d) All end joints must occur over solid bearings not less than 38 mm wide, unless tongued and grooved, and a 2 mm gap should be left between all joints to allow for expansion which can be caused by moisture pick-up during construction.

plywood does not exceed 1.25 kN/m² where there is no access except for maintenance. The information in Table 5.7 is as presented by the Council of the Forest Industries of British Columbia in 'Canadian COFI Exterior Plywood'. For loadings in excess of 1.25 kN/m² or for deflection criteria, the suitability of the plywood thickness can be determined by design using the capacity figures given in Tables 5.28–5.30.

Information in this form is not published in the UK about Finnish Conifer plywood, however, the figures given in Table 5.8 have been provided by Finnish Plywood International for flat roofs with imposed UDL not exceeding 0.75 kN/m², and a concentrated load of 0.9 kN on 300 mm square. Face grain perpendicular to supports, supports to all edges. The apparent discrepancies between the thicknesses given in Table 5.8 and Table 5.7 (and Table 5.9) can be explained by different design assumptions. The designer is referred to Tables 5.28–5.36 for a comparison of strength properties.

**Table 5.8  Finnish sanded Conifer plywood for roof sheathing**

| Spacing of supports (mm) | Nominal plywood thickness (mm) | Nail length and diameter | Maximum nail spacing (mm) |
|---|---|---|---|
| 300 | 12 | 3.35 mm plain wire nails, penetration about 40 mm | 150 along edge |
| 400 | 15 | | |
| 480 | 18 | 4.0 mm ringed shank nails, penetration about 32 mm | 300 along intermediate supports |
| 600 | 18 | | |

Information in this form is not published in the UK about Swedish P30 plywood but the moment and stiffness capacity values given in Table 5.36 can be used for comparison with other plywoods.

Table 5.9 gives guidance for American Construction and Industrial plywood for roof sheathing. This is extracted from 'American plywood for floors, walls and roofs' published by the American Plywood Association. In view of UK experience on the 'feel' of plywood on flat roofs, the figures given in Table 5.9 are actually more conservative than those given by APA

**Table 5.9  American unsanded Construction and Industrial plywood for roof sheathing**

| Spacing of supports (mm) | Nominal plywood thickness (mm) | Nail size and type | Maximum nail spacing (mm) |
|---|---|---|---|
| 300 | 8 | 3 mm X 50 mm round wire | 150 along edge |
| 400 | 8 | | |
| 500 | 9.5 | | 300 along intermediate supports |
| 600 | 9.5 | | |

which are claimed to be able to support $1.675 \, kN/m^2$ live and $0.48 \, kN/m^2$ dead loading. The figures are based on edge support being provided at all joints and the plywood being laid with face grain perpendicular to supports.

### 5.7.3   Plywood floor decking or sheathing

Table 5.10 tabulates minimum requirements for Canadian sheathing plywood when normal domestic floor loads apply. This information is as presented by the Council of the Forest Industries of British Columbia.

**Table 5.10   Canadian exterior plywood for floor decking or sheathing**
Domestic loading where the imposed load will not exceed $1.5 \, kN/m^2$.

| Joist spacing (mm) | Douglas Fir plywood | Canadian Softwood plywood | Nail length and diameter | Maximum nail spacing (mm) |
|---|---|---|---|---|
| | Nominal thickness (mm) | | | |
| 300 | 12.5 | 12.5 | 3 mm X 45 mm annularly grooved* | 150 along edge |
| 400 | 12.5 | 12.5 | | |
| 480 | 15.5 | 15.5 | | 300 along intermediate supports |
| 600 | 15.5 | 18.5 | | |

\* If round wire nails are used they should be 50 mm long.

(a) The grade is usually Select Sheathing.
(b) Plywood must be laid with the face grain perpendicular to the supports.
(c) Edge support must be provided by noggings (or the edges must be tongued and grooved).
(d) All end joints must occur over joists not less than 38 mm wide (unless tongued and grooved).
(e) A 2 mm gap should be left between all joints to allow for expansion which can be caused by moisture pick-up during construction.

Information in this form is not published in the UK about Finnish Conifer plywood, however, the figures given in Table 5.11 have been provided by Finnish Plywood International for floor sheathing when normal domestic floor loads apply. The plywood must be supported at edges and be laid with face grain perpendicular to supports.

**Table 5.11   Finnish sanded Conifer plywood for floor sheathing**
Domestic loading where the imposed load will not exceed $1.5 \, kN/m^2$.

| Joist spacing (mm) | Nominal plywood thickness (mm) | Nail length and diameter | Maximum nail spacing (mm) |
|---|---|---|---|
| 300 | 15 | 3.35 mm plain wire nails, penetration about 40 mm | 150 along edge |
| 400 | 18 | | |
| 450 | 19.5 | 4.0 mm ringed shank nails, penetration about 32 mm | 300 along intermediate supports |
| 600 | 23 | | |

Information in this form is not published in the UK about Swedish P30 plywood but the moment and stiffness capacity values given in Table 5.36 can be used for comparison with other plywoods.

Table 5.12 gives guidance for American Construction and Industrial plywood for floor sheathing for normal domestic loading. This is extracted from 'American plywood for floors, walls and roofs' published by the American Plywood Association. The figures given in Table 5.12 are slightly more conservative than some given in the APA publication. In certain cases it may be possible to use 12.5 mm thick plywood and 50 mm long nails even though 16 and 65 m respectively are the smallest shown in the table.

Table 5.12    **American unsanded Construction and Industrial plywood for floor sheathing**
Domestic loading where the imposed load will not exceed 1.5 kN/m².

| Joist spacing (mm) | Nominal plywood thickness (mm) | Nail length and diameter | Maximum nail spacing (mm) |
|---|---|---|---|
| 300 | 16* | | 150 along edge |
| 400 | 16* | 3.35 mm × 65 mm* | |
| 450 | 16 | round wire | 300 along |
| 600 | 19 | | intermediate supports |

* See text.

## 5.8   GEOMETRICAL PROPERTIES OF SOLID TIMBER JOISTS. DRY

Geometrical properties are given in Tables 5.13, 5.14 and 5.15 for those sizes most likely to be available stress graded in:

dry European sawn sizes
dry European sizes regularized
dry Canadian surfaced sizes (CLS).

**Table 5.13  Geometrical properties of solid timber sawn sizes (for the sizes most readily available stress graded)**

European sizes. Dry.

| Basic size (mm) | Area (mm² × 10³) | Section modulus | | Second moment of area | | Radius of gyration | |
|---|---|---|---|---|---|---|---|
| | | $Z_X$ (mm³ × 10⁶) | $Z_Y$ (mm³ × 10⁶) | $I_X$ (mm⁴ × 10⁶) | $I_Y$ (mm⁴ × 10⁶) | XX axis (mm) | YY axis (mm) |
| 38 × 75 | 2.85 | 0.0356 | 0.0181 | 1.34 | 0.343 | 21.7 | 11.0 |
| 38 × 100 | 3.80 | 0.0633 | 0.0241 | 3.17 | 0.457 | 28.9 | 11.0 |
| 38 × 125 | 4.75 | 0.099 | 0.0301 | 6.18 | 0.572 | 36.1 | 11.0 |
| 38 × 150 | 5.70 | 0.142 | 0.0361 | 10.7 | 0.686 | 43.3 | 11.0 |
| 38 × 175 | 6.65 | 0.194 | 0.0421 | 17.0 | 0.800 | 50.5 | 11.0 |
| 38 × 200 | 7.60 | 0.253 | 0.0481 | 25.3 | 0.915 | 57.7 | 11.0 |
| 38 × 225 | 8.55 | 0.321 | 0.0542 | 36.1 | 1.03 | 65.0 | 11.0 |
| 47 × 75 | 3.53 | 0.0441 | 0.0276 | 1.65 | 0.649 | 21.7 | 13.6 |
| 47 × 100 | 4.70 | 0.0783 | 0.0368 | 3.92 | 0.865 | 28.9 | 13.6 |
| 47 × 125 | 5.88 | 0.122 | 0.0460 | 7.65 | 1.08 | 36.1 | 13.6 |
| 47 × 150 | 7.05 | 0.176 | 0.0552 | 13.2 | 1.30 | 43.3 | 13.6 |
| 47 × 175 | 8.23 | 0.240 | 0.0644 | 21.0 | 1.51 | 50.5 | 13.6 |
| 47 × 200 | 9.40 | 0.313 | 0.0736 | 31.3 | 1.73 | 57.7 | 13.6 |
| 47 × 225 | 10.6 | 0.397 | 0.0828 | 44.6 | 1.95 | 65.0 | 13.6 |
| 50 × 75 | 3.75 | 0.0469 | 0.0313 | 1.76 | 0.781 | 21.7 | 14.4 |
| 50 × 100 | 5.00 | 0.0833 | 0.0417 | 4.17 | 1.04 | 28.9 | 14.4 |
| 50 × 125 | 6.25 | 0.130 | 0.0521 | 8.14 | 1.30 | 36.1 | 14.4 |
| 50 × 150 | 7.50 | 0.188 | 0.0625 | 14.1 | 1.56 | 43.3 | 14.4 |
| 50 × 175 | 8.75 | 0.255 | 0.0729 | 22.3 | 1.82 | 50.5 | 14.4 |
| 50 × 200 | 10.0 | 0.333 | 0.0833 | 33.3 | 2.08 | 57.7 | 14.4 |
| 50 × 225 | 11.3 | 0.422 | 0.0938 | 47.5 | 2.34 | 65.0 | 14.4 |
| 63 × 150 | 9.45 | 0.236 | 0.0992 | 17.7 | 3.13 | 43.3 | 18.2 |
| 63 × 175 | 11.0 | 0.322 | 0.116 | 28.1 | 3.65 | 50.5 | 18.2 |
| 63 × 200 | 12.6 | 0.420 | 0.132 | 42.0 | 4.17 | 57.7 | 18.2 |
| 63 × 225 | 14.2 | 0.532 | 0.149 | 59.8 | 4.69 | 65.0 | 18.2 |
| 75 × 150 | 11.2 | 0.281 | 0.141 | 21.2 | 5.27 | 43.3 | 21.7 |
| 75 × 175 | 13.1 | 0.383 | 0.164 | 33.5 | 6.15 | 50.5 | 21.7 |
| 75 × 200 | 15.0 | 0.500 | 0.188 | 50.0 | 7.03 | 57.7 | 21.7 |
| 75 × 225 | 16.9 | 0.633 | 0.211 | 71.2 | 7.91 | 65.0 | 21.7 |

**Table 5.14   European sizes regularized on width (depth) (from sizes in Table 5.13)**

| Size (mm) | Area (mm² × 10³) | Section modulus | | Second moment of area | | Radius of gyration | |
|---|---|---|---|---|---|---|---|
| | | $Z_X$ (mm³ × 10⁶) | $Z_Y$ (mm³ × 10⁶) | $I_X$ (mm⁴ × 10⁶) | $I_Y$ (mm⁴ × 10⁶) | $XX$ axis (mm) | $YY$ axis (mm) |
| 38 × 72 | 2.74 | 0.0328 | 0.0173 | 1.18 | 0.329 | 20.8 | 11.0 |
| 38 × 97 | 3.69 | 0.0596 | 0.0233 | 2.89 | 0.443 | 28.0 | 11.0 |
| 38 × 122 | 4.64 | 0.0943 | 0.0294 | 5.74 | 0.558 | 35.2 | 11.0 |
| 38 × 147 | 5.59 | 0.137 | 0.0354 | 10.1 | 0.672 | 42.4 | 11.0 |
| 38 × 170 | 6.46 | 0.183 | 0.0409 | 15.6 | 0.777 | 49.1 | 11.0 |
| 38 × 195 | 7.41 | 0.241 | 0.0469 | 23.5 | 0.892 | 56.3 | 11.0 |
| 38 × 220 | 8.36 | 0.307 | 0.0529 | 33.7 | 1.01 | 63.5 | 11.0 |
| 47 × 72 | 3.38 | 0.0406 | 0.0265 | 1.46 | 0.623 | 20.8 | 13.6 |
| 47 × 97 | 4.56 | 0.0737 | 0.0357 | 3.57 | 0.839 | 28.0 | 13.6 |
| 47 × 122 | 5.73 | 0.117 | 0.0449 | 7.11 | 1.06 | 35.2 | 13.6 |
| 47 × 147 | 6.91 | 0.169 | 0.0541 | 12.4 | 1.27 | 42.4 | 13.6 |
| 47 × 170 | 7.99 | 0.226 | 0.0626 | 19.2 | 1.47 | 49.1 | 13.6 |
| 47 × 195 | 9.71 | 0.298 | 0.0718 | 29.0 | 1.69 | 56.3 | 13.6 |
| 47 × 220 | 10.3 | 0.379 | 0.0810 | 41.7 | 1.90 | 63.5 | 13.6 |
| 50 × 72 | 3.50 | 0.0408 | 0.0292 | 1.43 | 0.729 | 20.3 | 14.4 |
| 50 × 97 | 4.85 | 0.0784 | 0.0404 | 3.80 | 1.01 | 28.0 | 14.4 |
| 50 × 122 | 6.10 | 0.124 | 0.0508 | 7.57 | 1.27 | 35.2 | 14.4 |
| 50 × 147 | 7.35 | 0.180 | 0.0613 | 13.2 | 1.53 | 42.4 | 14.4 |
| 50 × 170 | 8.50 | 0.241 | 0.0708 | 20.5 | 1.77 | 49.1 | 14.4 |
| 50 × 195 | 9.75 | 0.317 | 0.0813 | 30.9 | 2.03 | 56.3 | 14.4 |
| 50 × 220 | 11.0 | 0.403 | 0.0917 | 44.4 | 2.29 | 63.5 | 14.4 |
| 63 × 147 | 9.26 | 0.0227 | 0.0972 | 16.7 | 3.06 | 42.4 | 18.2 |
| 63 × 170 | 10.7 | 0.0303 | 0.112 | 25.8 | 3.54 | 49.1 | 18.2 |
| 63 × 195 | 12.3 | 0.0399 | 0.129 | 38.9 | 4.06 | 56.3 | 18.2 |
| 63 × 220 | 13.9 | 0.0508 | 0.146 | 55.9 | 4.58 | 63.5 | 18.2 |
| 75 × 147 | 11.0 | 0.0270 | 0.138 | 19.9 | 5.17 | 42.4 | 21.7 |
| 75 × 170 | 12.8 | 0.0361 | 0.159 | 30.7 | 5.98 | 49.1 | 21.7 |
| 75 × 195 | 14.6 | 0.0475 | 0.183 | 46.3 | 6.86 | 56.3 | 21.7 |
| 75 × 220 | 16.5 | 0.0605 | 0.206 | 66.6 | 7.73 | 63.5 | 21.7 |

**Table 5.15   Geometrical properties of solid timber sizes**
Canadian surfaced sizes (CLS). Dry (for the sizes most readily available), stress graded.

| Size (mm) | Area (mm² × 10³) | Section modulus | | Second moment of area | | Radius of gyration | |
|---|---|---|---|---|---|---|---|
| | | $Z_X$ (mm³ × 10⁶) | $Z_Y$ (mm³ × 10⁶) | $I_X$ (mm⁴ × 10⁶) | $I_Y$ (mm⁴ × 10⁶) | $XX$ axis (mm) | $YY$ axis (mm) |
| 38 × 63 | 2.39 | 0.0251 | 0.0152 | 0.792 | 0.288 | 18.2 | 11.0 |
| 38 × 89 | 3.38 | 0.0502 | 0.0214 | 2.23 | 0.407 | 25.7 | 11.0 |
| 38 × 140 | 5.32 | 0.124 | 0.0337 | 8.69 | 0.640 | 40.4 | 11.0 |
| 38 × 184 | 6.99 | 0.214 | 0.0443 | 19.7 | 0.841 | 53.1 | 11.0 |
| 38 × 235 | 8.93 | 0.350 | 0.0566 | 41.1 | 1.07 | 67.8 | 11.0 |
| 38 × 285 | 10.8 | 0.514 | 0.0686 | 73.3 | 1.30 | 82.3 | 11.0 |

## 5.9   TABLES OF STRENGTH CAPACITIES OF EUROPEAN AND CANADIAN SIZES

### 5.9.1   Introduction to tables

The tables in this section are classified as follows:

(a) Tables 5.16–5.21 are for principal members, Tables 5.22–5.27 for load-sharing members.
(b) Tables 5.16–5.19 and 5.22–5.25 are for dry Whitewood or Redwood, sawn or regularized.
(c) Tables 5.20 and 5.26 are for CLS Hem-Fir.
(d) Tables 5.21 and 5.27 are for CLS Spruce-Pine-Fir.

The abbreviations used are:

$\bar{F}_{vL}$ = long term shear capacity
$\bar{F}_{vM}$ = medium term shear capacity
$\bar{F}_{vS}$ = short term shear capacity
$\bar{F}_{vVS}$ = very short term shear capacity
$\bar{M}_{L}$ = long term moment capacity
$\bar{M}_{M}$ = medium term moment capacity
$\bar{M}_{S}$ = short term moment capacity
$\bar{M}_{VS}$ = very short term moment capacity

**Table 5.16**   Principal member: European Whitewood or Redwood. Basic sawn sizes. Grade SS or MSS

| b (mm) | h (mm) | Self-wt* (kN/m) | Area (mm² × 10³) | Section modulus (mm³ × 10⁶) | Second moment of area (mm⁴ × 10⁶) | $\frac{h}{b}$ | $K_7$ | $\bar{F}_{vL}$ | $\bar{F}_{vM}$ | $\bar{F}_{vS}$ | $\bar{F}_{vVS}$ | $\bar{M}_L$ | $\bar{M}_M$ | $\bar{M}_S$ | $\bar{M}_{VS}$ | EI (kN.m²) | AG (kN) |
|---|---|---|---|---|---|---|---|---|---|---|---|---|---|---|---|---|---|
| | | | | | | | | Shear (kN) | | | | Bending (kN.m) | | | | Deflection | |
| 38 | 75  | 0.015 | 2.85 | 0.0356 | 1.34 | 1.9 | 1.164 | 1.55 | 1.94 | 2.33 | 2.72 | 0.31 | 0.38 | 0.46 | 0.54 | 9 | 1246 |
| 38 | 100 | 0.020 | 3.80 | 0.0633 | 3.17 | 2.6 | 1.128 | 2.07 | 2.59 | 3.11 | 3.63 | 0.53 | 0.67 | 0.80 | 0.93 | 22 | 1662 |
| 38 | 125 | 0.025 | 4.75 | 0.099  | 6.18 | 3.2 | 1.101 | 2.59 | 3.24 | 3.89 | 4.54 | 0.81 | 1.02 | 1.22 | 1.43 | 43 | 2078 |
| 38 | 150 | 0.030 | 5.70 | 0.143  | 10.7 | 3.9 | 1.079 | 3.11 | 3.89 | 4.67 | 5.45 | 1.15 | 1.44 | 1.73 | 2.01 | 74 | 2493 |
| 38 | 175 | 0.035 | 6.65 | 0.194  | 17.0 | 4.6 | 1.061 | 3.63 | 4.54 | 5.45 | 6.36 | 1.54 | 1.92 | 2.31 | 2.70 | 118 | 2909 |
| 38 | 200 | 0.041 | 7.60 | 0.253  | 25.3 | 5.2 | 1.045 | 4.15 | 5.19 | 6.23 | 7.27 | 1.98 | 2.48 | 2.97 | 3.47 | 177 | 3325 |
| 38 | 225 | 0.046 | 8.55 | 0.321  | 36.1 | 5.9 | 1.032 | 4.67 | 5.84 | 7.01 | 8.17 | 2.48 | 3.10 | 3.72 | 4.34 | 252 | 3740 |
| 47 | 75  | 0.019 | 3.53 | 0.0441 | 1.65 | 1.5 | 1.164 | 1.92 | 2.40 | 2.89 | 3.37 | 0.38 | 0.48 | 0.57 | 0.67 | 11 | 1542 |
| 47 | 100 | 0.025 | 4.70 | 0.0783 | 3.92 | 2.1 | 1.128 | 2.56 | 3.21 | 3.85 | 4.49 | 0.66 | 0.82 | 0.99 | 1.16 | 27 | 2056 |
| 47 | 125 | 0.031 | 5.88 | 0.122  | 7.65 | 2.6 | 1.101 | 3.21 | 4.01 | 4.81 | 5.62 | 1.01 | 1.26 | 1.51 | 1.76 | 53 | 2570 |
| 47 | 150 | 0.038 | 7.05 | 0.176  | 13.2 | 3.1 | 1.079 | 3.85 | 4.81 | 5.78 | 6.74 | 1.42 | 1.78 | 2.13 | 2.49 | 92 | 3084 |
| 47 | 175 | 0.044 | 8.23 | 0.240  | 21.0 | 3.7 | 1.061 | 4.49 | 5.62 | 6.74 | 7.86 | 1.90 | 2.38 | 2.86 | 3.34 | 146 | 3598 |
| 47 | 200 | 0.050 | 9.40 | 0.313  | 31.3 | 4.2 | 1.045 | 5.13 | 6.42 | 7.70 | 8.99 | 2.45 | 3.07 | 3.68 | 4.30 | 219 | 4112 |
| 47 | 225 | 0.057 | 10.6 | 0.397  | 44.6 | 4.7 | 1.032 | 5.78 | 7.22 | 8.67 | 10.11 | 3.06 | 3.83 | 4.60 | 5.37 | 312 | 4626 |
| 50 | 75  | 0.020 | 3.75 | 0.0469 | 1.76 | 1.5 | 1.164 | 2.05 | 2.56 | 3.07 | 3.58 | 0.40 | 0.51 | 0.61 | 0.71 | 12 | 1640 |
| 50 | 100 | 0.027 | 5.00 | 0.0833 | 4.17 | 2.0 | 1.128 | 2.73 | 3.41 | 4.10 | 4.78 | 0.70 | 0.88 | 1.05 | 1.23 | 29 | 2187 |
| 50 | 125 | 0.033 | 6.25 | 0.130  | 8.14 | 2.5 | 1.101 | 3.41 | 4.27 | 5.12 | 5.97 | 1.07 | 1.34 | 1.61 | 1.88 | 56 | 2734 |
| 50 | 150 | 0.040 | 7.50 | 0.188  | 14.1 | 3.0 | 1.079 | 4.10 | 5.12 | 6.15 | 7.17 | 1.51 | 1.89 | 2.27 | 2.65 | 98 | 3281 |
| 50 | 175 | 0.047 | 8.75 | 0.255  | 22.3 | 3.5 | 1.061 | 4.78 | 5.97 | 7.17 | 8.37 | 2.03 | 2.53 | 3.04 | 3.55 | 156 | 3828 |
| 50 | 200 | 0.054 | 10.0 | 0.333  | 33.3 | 4.0 | 1.045 | 5.46 | 6.83 | 8.20 | 9.56 | 2.61 | 3.26 | 3.92 | 4.57 | 233 | 4375 |
| 50 | 225 | 0.060 | 11.3 | 0.422  | 47.5 | 4.5 | 1.032 | 6.15 | 7.68 | 9.22 | 10.76 | 3.26 | 4.08 | 4.89 | 5.71 | 332 | 4921 |
| 63 | 150 | 0.051 | 9.45 | 0.236  | 17.7 | 2.3 | 1.079 | 5.16 | 6.45 | 7.74 | 9.04 | 1.91 | 2.39 | 2.86 | 3.34 | 124 | 4134 |
| 63 | 175 | 0.059 | 11.0 | 0.322  | 28.1 | 2.7 | 1.061 | 6.02 | 7.53 | 9.04 | 10.54 | 2.55 | 3.19 | 3.83 | 4.47 | 196 | 4823 |
| 63 | 200 | 0.068 | 12.6 | 0.420  | 42.0 | 3.1 | 1.045 | 6.88 | 8.61 | 10.33 | 12.05 | 3.29 | 4.11 | 4.94 | 5.76 | 294 | 5512 |
| 63 | 225 | 0.076 | 14.2 | 0.532  | 59.8 | 3.5 | 1.032 | 7.74 | 9.68 | 11.62 | 13.56 | 4.11 | 5.14 | 6.17 | 7.20 | 418 | 6201 |
| 75 | 150 | 0.060 | 11.3 | 0.281  | 21.1 | 2.0 | 1.079 | 6.15 | 7.68 | 9.22 | 10.76 | 2.27 | 2.84 | 3.41 | 3.98 | 147 | 4921 |
| 75 | 175 | 0.070 | 13.1 | 0.383  | 33.5 | 2.3 | 1.061 | 7.17 | 8.96 | 10.76 | 12.55 | 3.04 | 3.80 | 4.56 | 5.33 | 234 | 5742 |
| 75 | 200 | 0.081 | 15.0 | 0.500  | 50.0 | 2.6 | 1.045 | 8.20 | 10.25 | 12.30 | 14.35 | 3.92 | 4.90 | 5.88 | 6.86 | 350 | 6562 |
| 75 | 225 | 0.091 | 16.9 | 0.633  | 71.2 | 3.0 | 1.032 | 9.22 | 11.53 | 13.83 | 16.14 | 4.89 | 6.12 | 7.34 | 8.57 | 498 | 7382 |

*Self-weight based on 5.4 kN/m³.

$E$ ... 7000 N/mm²

**Table 5.17 Principal member: European Whitewood or Redwood. Regularized only on width. Grade SS or MSS**

| b (mm) | h (mm) | Self-wt* (kN/m) | Area (mm² × 10³) | Section modulus (mm³ × 10⁶) | Second moment of area (mm⁴ × 10⁶) | $h/b$ | $K_7$ | $\bar{F}_{vL}$ | $\bar{F}_{vM}$ | $\bar{F}_{vS}$ | $\bar{F}_{vVS}$ | $\bar{M}_L$ | $\bar{M}_M$ | $\bar{M}_S$ | $\bar{M}_{VS}$ | $EI$ (kN.m²) | $AG$ (kN) |
|---|---|---|---|---|---|---|---|---|---|---|---|---|---|---|---|---|---|
| 38 | 72 | 0.014 | 2.74 | 0.0328 | 1.18 | 1.8 | 1.169 | 1.49 | 1.86 | 2.24 | 2.61 | 0.28 | 0.36 | 0.43 | 0.50 | 8 | 1197 |
| 38 | 97 | 0.019 | 3.69 | 0.0596 | 2.89 | 2.5 | 1.132 | 2.01 | 2.51 | 3.02 | 3.52 | 0.50 | 0.63 | 0.75 | 0.88 | 20 | 1612 |
| 38 | 122 | 0.025 | 4.64 | 0.0943 | 5.75 | 3.2 | 1.104 | 2.53 | 3.16 | 3.80 | 4.43 | 0.78 | 0.97 | 1.17 | 1.36 | 40 | 2028 |
| 38 | 147 | 0.030 | 5.59 | 0.137 | 10.1 | 3.8 | 1.081 | 3.05 | 3.81 | 4.58 | 5.34 | 1.11 | 1.38 | 1.66 | 1.94 | 70 | 2443 |
| 38 | 170 | 0.034 | 6.46 | 0.183 | 15.6 | 4.4 | 1.064 | 3.53 | 4.41 | 5.29 | 6.18 | 1.46 | 1.82 | 2.19 | 2.55 | 108 | 2826 |
| 38 | 195 | 0.040 | 7.41 | 0.241 | 23.5 | 5.1 | 1.048 | 4.05 | 5.06 | 6.07 | 7.08 | 1.89 | 2.36 | 2.84 | 3.31 | 164 | 3241 |
| 38 | 220 | 0.045 | 8.36 | 0.307 | 33.7 | 5.7 | 1.034 | 4.57 | 5.71 | 6.85 | 7.99 | 2.37 | 2.97 | 3.56 | 4.16 | 236 | 3657 |
| 47 | 72 | 0.018 | 3.38 | 0.0406 | 1.46 | 1.5 | 1.169 | 1.84 | 2.31 | 2.77 | 3.23 | 0.35 | 0.44 | 0.53 | 0.62 | 10 | 1480 |
| 47 | 97 | 0.024 | 4.56 | 0.0737 | 3.57 | 2.0 | 1.132 | 2.49 | 3.11 | 3.73 | 4.36 | 0.62 | 0.78 | 0.93 | 1.09 | 25 | 1994 |
| 47 | 122 | 0.030 | 5.73 | 0.117 | 7.11 | 2.5 | 1.104 | 3.13 | 3.91 | 4.70 | 5.48 | 0.96 | 1.20 | 1.44 | 1.68 | 49 | 2508 |
| 47 | 147 | 0.037 | 6.91 | 0.169 | 12.4 | 3.1 | 1.081 | 3.77 | 4.72 | 5.66 | 6.60 | 1.37 | 1.71 | 2.05 | 2.40 | 87 | 3022 |
| 47 | 170 | 0.043 | 7.99 | 0.226 | 19.2 | 3.6 | 1.064 | 4.36 | 5.45 | 6.55 | 7.64 | 1.80 | 2.25 | 2.71 | 3.16 | 134 | 3495 |
| 47 | 195 | 0.049 | 9.17 | 0.298 | 29.0 | 4.1 | 1.048 | 5.01 | 6.26 | 7.51 | 8.76 | 2.34 | 2.92 | 3.51 | 4.09 | 203 | 4009 |
| 47 | 220 | 0.055 | 10.3 | 0.379 | 41.7 | 4.6 | 1.034 | 5.65 | 7.06 | 8.47 | 9.89 | 2.94 | 3.67 | 4.41 | 5.14 | 291 | 4523 |
| 50 | 72 | 0.019 | 3.60 | 0.0432 | 1.56 | 1.4 | 1.169 | 1.96 | 2.46 | 2.95 | 3.44 | 0.37 | 0.47 | 0.56 | 0.66 | 10 | 1575 |
| 50 | 97 | 0.026 | 4.85 | 0.0784 | 3.80 | 1.9 | 1.132 | 2.65 | 3.31 | 3.97 | 4.63 | 0.66 | 0.83 | 0.99 | 1.16 | 26 | 2121 |
| 50 | 122 | 0.032 | 6.10 | 0.124 | 7.57 | 2.4 | 1.104 | 3.33 | 4.16 | 5.00 | 5.83 | 1.02 | 1.28 | 1.54 | 1.79 | 52 | 2668 |
| 50 | 147 | 0.039 | 7.35 | 0.180 | 13.2 | 2.9 | 1.081 | 4.01 | 5.02 | 6.02 | 7.03 | 1.46 | 1.82 | 2.19 | 2.55 | 92 | 3215 |
| 50 | 170 | 0.045 | 8.50 | 0.241 | 20.5 | 3.4 | 1.064 | 4.64 | 5.80 | 6.97 | 8.13 | 1.92 | 2.40 | 2.88 | 3.36 | 143 | 3718 |
| 50 | 195 | 0.052 | 9.75 | 0.317 | 30.9 | 3.9 | 1.048 | 5.33 | 6.66 | 7.99 | 9.32 | 2.49 | 3.11 | 3.73 | 4.36 | 216 | 4265 |
| 50 | 220 | 0.059 | 11.0 | 0.403 | 44.4 | 4.4 | 1.034 | 6.01 | 7.51 | 9.02 | 10.52 | 3.12 | 3.91 | 4.69 | 5.47 | 310 | 4812 |
| 63 | 147 | 0.050 | 9.26 | 0.227 | 16.7 | 2.3 | 1.081 | 5.06 | 6.32 | 7.59 | 8.85 | 1.84 | 2.30 | 2.76 | 3.22 | 116 | 4051 |
| 63 | 170 | 0.057 | 10.7 | 0.303 | 25.8 | 2.6 | 1.064 | 5.85 | 7.31 | 8.78 | 10.24 | 2.42 | 3.02 | 3.63 | 4.23 | 180 | 4685 |
| 63 | 195 | 0.066 | 12.3 | 0.399 | 38.9 | 3.0 | 1.048 | 6.71 | 8.39 | 10.07 | 11.75 | 3.13 | 3.92 | 4.70 | 5.49 | 272 | 5374 |
| 63 | 220 | 0.074 | 13.9 | 0.508 | 55.9 | 3.4 | 1.034 | 7.57 | 9.47 | 11.36 | 13.25 | 3.94 | 4.92 | 5.91 | 6.90 | 391 | 6063 |
| 75 | 147 | 0.059 | 11.0 | 0.270 | 19.9 | 1.9 | 1.081 | 6.02 | 7.53 | 9.04 | 10.54 | 2.19 | 2.73 | 3.28 | 3.83 | 138 | 4823 |
| 75 | 170 | 0.068 | 12.8 | 0.361 | 30.7 | 2.2 | 1.064 | 6.97 | 8.71 | 10.45 | 12.19 | 2.88 | 3.60 | 4.32 | 5.04 | 214 | 5578 |
| 75 | 195 | 0.078 | 14.6 | 0.475 | 46.3 | 2.6 | 1.048 | 7.99 | 9.99 | 11.99 | 13.99 | 3.73 | 4.67 | 5.60 | 6.54 | 324 | 6398 |
| 75 | 220 | 0.089 | 16.5 | 0.605 | 66.6 | 2.9 | 1.034 | 9.02 | 11.27 | 13.53 | 15.78 | 4.69 | 5.86 | 7.04 | 8.21 | 465 | 7218 |

* Self-weight based on 5.4 kN/m³.
Permissible long term shear stress = 0.82 N/mm².
Permissible long term bending stress = 7.5 N/mm² × $K_7$.

$E_{min} = 7000$ N/mm².
$G = E_{min}/16 = 437.5$ N/mm².

**Table 5.18   Principal member: European Whitewood or Redwood. Basic sawn sizes. Grade GS or MGS**

| $b$ (mm) | $h$ (mm) | Self-wt* (kN/m) | Area (mm² × 10³) | Section modulus (mm³ × 10⁶) | Second moment of area (mm⁴ × 10⁶) | $\frac{h}{b}$ | $K_7$ | Shear (kN) $\bar{F}_{vL}$ | $\bar{F}_{vM}$ | $\bar{F}_{vS}$ | $\bar{F}_{vVS}$ | Bending (kN.m) $\bar{M}_L$ | $\bar{M}_M$ | $\bar{M}_S$ | $\bar{M}_{VS}$ | Deflection $EI$ (kN.m²) | $AG$ (kN) |
|---|---|---|---|---|---|---|---|---|---|---|---|---|---|---|---|---|---|
| 38 | 75  | 0.015 | 2.85 | 0.0356 | 1.34 | 1.9 | 1.164 | 1.55 | 1.94 | 2.33 | 2.72 | 0.21 | 0.27 | 0.32 | 0.38 | 8   | 1068 |
| 38 | 100 | 0.020 | 3.80 | 0.0633 | 3.17 | 2.6 | 1.128 | 2.07 | 2.59 | 3.11 | 3.63 | 0.37 | 0.47 | 0.56 | 0.66 | 19  | 1425 |
| 38 | 125 | 0.025 | 4.75 | 0.099  | 6.18 | 3.2 | 1.101 | 2.59 | 3.24 | 3.89 | 4.54 | 0.57 | 0.72 | 0.86 | 1.01 | 37  | 1781 |
| 38 | 150 | 0.030 | 5.70 | 0.143  | 10.7 | 3.9 | 1.079 | 3.11 | 3.89 | 4.67 | 5.45 | 0.81 | 1.01 | 1.22 | 1.42 | 64  | 2137 |
| 38 | 175 | 0.035 | 6.65 | 0.194  | 17.0 | 4.6 | 1.061 | 3.63 | 4.54 | 5.45 | 6.36 | 1.09 | 1.36 | 1.63 | 1.90 | 101 | 2493 |
| 38 | 200 | 0.041 | 7.60 | 0.253  | 25.3 | 5.2 | 1.045 | 4.15 | 5.19 | 6.23 | 7.27 | 1.40 | 1.75 | 2.10 | 2.45 | 152 | 2850 |
| 38 | 225 | 0.046 | 8.55 | 0.321  | 36.1 | 5.9 | 1.032 | 4.67 | 5.84 | 7.01 | 8.17 | 1.75 | 2.19 | 2.63 | 3.06 | 216 | 3206 |
| 47 | 75  | 0.019 | 3.53 | 0.0441 | 1.65 | 1.5 | 1.164 | 1.92 | 2.40 | 2.89 | 3.37 | 0.27 | 0.34 | 0.40 | 0.47 | 9   | 1321 |
| 47 | 100 | 0.025 | 4.70 | 0.0783 | 3.92 | 2.1 | 1.128 | 2.56 | 3.21 | 3.85 | 4.49 | 0.46 | 0.58 | 0.70 | 0.81 | 23  | 1762 |
| 47 | 125 | 0.031 | 5.88 | 0.122  | 7.65 | 2.6 | 1.101 | 3.21 | 4.01 | 4.81 | 5.62 | 0.71 | 0.89 | 1.07 | 1.24 | 45  | 2203 |
| 47 | 150 | 0.038 | 7.05 | 0.176  | 13.2 | 3.1 | 1.079 | 3.85 | 4.81 | 5.78 | 6.74 | 1.00 | 1.26 | 1.51 | 1.76 | 79  | 2643 |
| 47 | 175 | 0.044 | 8.23 | 0.240  | 21.0 | 3.7 | 1.061 | 4.49 | 5.62 | 6.74 | 7.86 | 1.34 | 1.68 | 2.02 | 2.36 | 125 | 3084 |
| 47 | 200 | 0.050 | 9.40 | 0.313  | 31.3 | 4.2 | 1.045 | 5.13 | 6.42 | 7.70 | 8.99 | 1.73 | 2.17 | 2.60 | 3.03 | 188 | 3525 |
| 47 | 225 | 0.057 | 10.6 | 0.397  | 44.6 | 4.7 | 1.032 | 5.78 | 7.22 | 8.67 | 10.11 | 2.16 | 2.71 | 3.25 | 3.79 | 267 | 3965 |
| 50 | 75  | 0.020 | 3.75 | 0.0469 | 1.76 | 1.5 | 1.164 | 2.05 | 2.56 | 3.07 | 3.58 | 0.28 | 0.36 | 0.43 | 0.50 | 10  | 1406 |
| 50 | 100 | 0.027 | 5.00 | 0.0833 | 4.17 | 2.0 | 1.128 | 2.73 | 3.41 | 4.10 | 4.78 | 0.49 | 0.62 | 0.74 | 0.87 | 25  | 1875 |
| 50 | 125 | 0.033 | 6.25 | 0.130  | 8.14 | 2.5 | 1.101 | 3.41 | 4.27 | 5.12 | 5.97 | 0.75 | 0.94 | 1.13 | 1.32 | 48  | 2343 |
| 50 | 150 | 0.040 | 7.50 | 0.188  | 14.1 | 3.0 | 1.079 | 4.10 | 5.12 | 6.15 | 7.17 | 1.07 | 1.34 | 1.60 | 1.87 | 84  | 2812 |
| 50 | 175 | 0.047 | 8.75 | 0.255  | 22.3 | 3.5 | 1.061 | 4.78 | 5.97 | 7.17 | 8.37 | 1.43 | 1.79 | 2.15 | 2.51 | 133 | 3281 |
| 50 | 200 | 0.054 | 10.0 | 0.333  | 33.3 | 4.0 | 1.045 | 5.46 | 6.83 | 8.20 | 9.56 | 1.84 | 2.30 | 2.77 | 3.23 | 200 | 3750 |
| 50 | 225 | 0.060 | 11.3 | 0.422  | 47.5 | 4.5 | 1.032 | 6.15 | 7.68 | 9.22 | 10.76 | 2.30 | 2.88 | 3.46 | 4.03 | 284 | 4218 |
| 63 | 150 | 0.051 | 9.45 | 0.236  | 17.7 | 2.3 | 1.079 | 5.16 | 6.45 | 7.74 | 9.04 | 1.35 | 1.68 | 2.02 | 2.36 | 106 | 3543 |
| 63 | 175 | 0.059 | 11.0 | 0.322  | 28.1 | 2.7 | 1.061 | 6.02 | 7.53 | 9.04 | 10.54 | 1.80 | 2.26 | 2.71 | 3.16 | 168 | 4134 |
| 63 | 200 | 0.068 | 12.6 | 0.420  | 42.0 | 3.1 | 1.045 | 6.88 | 8.61 | 10.33 | 12.05 | 2.32 | 2.90 | 3.49 | 4.07 | 252 | 4725 |
| 63 | 225 | 0.076 | 14.2 | 0.532  | 59.8 | 3.5 | 1.032 | 7.74 | 9.68 | 11.62 | 13.56 | 2.90 | 3.63 | 4.36 | 5.08 | 358 | 5315 |
| 75 | 150 | 0.060 | 11.3 | 0.281  | 21.1 | 2.0 | 1.079 | 6.15 | 7.68 | 9.22 | 10.76 | 1.60 | 2.01 | 2.41 | 2.81 | 126 | 4218 |
| 75 | 175 | 0.070 | 13.1 | 0.383  | 33.5 | 2.3 | 1.061 | 7.17 | 8.96 | 10.76 | 12.55 | 2.15 | 2.69 | 3.22 | 3.76 | 200 | 4921 |
| 75 | 200 | 0.081 | 15.0 | 0.500  | 50.0 | 2.6 | 1.045 | 8.20 | 10.25 | 12.30 | 14.35 | 2.77 | 3.46 | 4.15 | 4.84 | 300 | 5625 |
| 75 | 225 | 0.091 | 16.9 | 0.633  | 71.2 | 3.0 | 1.032 | 9.22 | 11.53 | 13.83 | 16.14 | 3.46 | 4.32 | 5.19 | 6.05 | 427 | 6328 |

\* Self-weight based on 5.4 kN/m³.

Permissible long term shear stress = 0.82 N/mm²

$E_{mean} = 6000$ N/mm²

**Table 5.19** Principal member: European Whitewood or Redwood. Regularized only on width. Grade GS or MGS

| b (mm) | h (mm) | Self-wt* (kN/m) | Area (mm² × 10³) | Section modulus (mm³ × 10⁶) | Second moment of area (mm⁴ × 10⁶) | h/b | $K_7$ | $\bar{F}_{vL}$ | $\bar{F}_{vM}$ | $\bar{F}_{vS}$ | $\bar{F}_{vVS}$ | $\bar{M}_L$ | $\bar{M}_M$ | $\bar{M}_S$ | $\bar{M}_{VS}$ | EI (kN.m²) | AG (kN) |
|---|---|---|---|---|---|---|---|---|---|---|---|---|---|---|---|---|---|
| | | | | | | | | Shear (kN) | | | | Bending (kN.m) | | | | Deflection | |
| 38 | 72 | 0.014 | 2.74 | 0.0328 | 1.18 | 1.8 | 1.169 | 1.49 | 1.86 | 2.24 | 2.61 | 0.20 | 0.25 | 0.30 | 0.35 | 7 | 1026 |
| 38 | 97 | 0.019 | 3.69 | 0.0596 | 2.89 | 2.5 | 1.132 | 2.01 | 2.51 | 3.02 | 3.52 | 0.35 | 0.44 | 0.53 | 0.62 | 17 | 1382 |
| 38 | 122 | 0.025 | 4.64 | 0.0943 | 5.75 | 3.2 | 1.104 | 2.53 | 3.16 | 3.80 | 4.43 | 0.55 | 0.68 | 0.82 | 0.96 | 34 | 1738 |
| 38 | 147 | 0.030 | 5.59 | 0.137 | 10.1 | 3.8 | 1.081 | 3.05 | 3.81 | 4.58 | 5.34 | 0.78 | 0.98 | 1.17 | 1.37 | 60 | 2094 |
| 38 | 170 | 0.034 | 6.46 | 0.183 | 15.6 | 4.4 | 1.064 | 3.53 | 4.41 | 5.29 | 6.18 | 1.03 | 1.29 | 1.54 | 1.80 | 93 | 2422 |
| 38 | 195 | 0.040 | 7.41 | 0.241 | 23.5 | 5.1 | 1.048 | 4.05 | 5.06 | 6.07 | 7.08 | 1.33 | 1.67 | 2.00 | 2.34 | 140 | 2778 |
| 38 | 220 | 0.045 | 8.36 | 0.307 | 33.7 | 5.7 | 1.034 | 4.57 | 5.71 | 6.85 | 7.99 | 1.68 | 2.10 | 2.52 | 2.94 | 202 | 3135 |
| 47 | 72 | 0.018 | 3.38 | 0.0406 | 1.46 | 1.5 | 1.169 | 1.84 | 2.31 | 2.77 | 3.23 | 0.25 | 0.31 | 0.37 | 0.44 | 8 | 1269 |
| 47 | 97 | 0.024 | 4.56 | 0.0737 | 3.57 | 2.0 | 1.132 | 2.49 | 3.11 | 3.73 | 4.36 | 0.44 | 0.55 | 0.66 | 0.77 | 21 | 1709 |
| 47 | 122 | 0.030 | 5.73 | 0.117 | 7.11 | 2.5 | 1.104 | 3.13 | 3.91 | 4.70 | 5.48 | 0.68 | 0.85 | 1.02 | 1.19 | 42 | 2150 |
| 47 | 147 | 0.037 | 6.91 | 0.169 | 12.4 | 3.1 | 1.081 | 3.77 | 4.72 | 5.66 | 6.60 | 0.97 | 1.21 | 1.45 | 1.69 | 74 | 2590 |
| 47 | 170 | 0.043 | 7.99 | 0.226 | 19.2 | 3.6 | 1.064 | 4.36 | 5.45 | 6.55 | 7.64 | 1.27 | 1.59 | 1.91 | 2.23 | 115 | 2996 |
| 47 | 195 | 0.049 | 9.17 | 0.298 | 29.0 | 4.1 | 1.048 | 5.01 | 6.26 | 7.51 | 8.76 | 1.65 | 2.06 | 2.48 | 2.89 | 174 | 3436 |
| 47 | 220 | 0.055 | 10.3 | 0.379 | 41.7 | 4.6 | 1.034 | 5.65 | 7.06 | 8.47 | 9.89 | 2.07 | 2.59 | 3.11 | 3.63 | 250 | 3877 |
| 50 | 72 | 0.019 | 3.60 | 0.0432 | 1.56 | 1.4 | 1.169 | 1.96 | 2.46 | 2.95 | 3.44 | 0.26 | 0.33 | 0.40 | 0.46 | 9 | 1350 |
| 50 | 97 | 0.026 | 4.85 | 0.0784 | 3.80 | 1.9 | 1.132 | 2.65 | 3.31 | 3.97 | 4.63 | 0.47 | 0.58 | 0.70 | 0.82 | 22 | 1818 |
| 50 | 122 | 0.032 | 6.10 | 0.124 | 7.57 | 2.4 | 1.104 | 3.33 | 4.16 | 5.00 | 5.83 | 0.72 | 0.90 | 1.08 | 1.27 | 45 | 2287 |
| 50 | 147 | 0.039 | 7.35 | 0.180 | 13.2 | 2.9 | 1.081 | 4.01 | 5.02 | 6.02 | 7.03 | 1.03 | 1.29 | 1.54 | 1.80 | 79 | 2756 |
| 50 | 170 | 0.045 | 8.50 | 0.241 | 20.5 | 3.4 | 1.064 | 4.64 | 5.80 | 6.97 | 8.13 | 1.35 | 1.69 | 2.03 | 2.37 | 122 | 3187 |
| 50 | 195 | 0.052 | 9.75 | 0.317 | 30.9 | 3.9 | 1.048 | 5.33 | 6.66 | 7.99 | 9.32 | 1.76 | 2.20 | 2.64 | 3.08 | 185 | 3656 |
| 50 | 220 | 0.059 | 11.0 | 0.403 | 44.4 | 4.4 | 1.034 | 6.01 | 7.51 | 9.02 | 10.52 | 2.21 | 2.76 | 3.31 | 3.87 | 266 | 4125 |
| 63 | 147 | 0.050 | 9.26 | 0.227 | 16.7 | 2.3 | 1.081 | 5.06 | 6.32 | 7.59 | 8.85 | 1.30 | 1.62 | 1.95 | 2.27 | 100 | 3472 |
| 63 | 170 | 0.057 | 10.7 | 0.303 | 25.8 | 2.6 | 1.064 | 5.85 | 7.31 | 8.78 | 10.24 | 1.71 | 2.13 | 2.56 | 2.99 | 154 | 4016 |
| 63 | 195 | 0.066 | 12.3 | 0.399 | 38.9 | 3.0 | 1.048 | 6.71 | 8.39 | 10.07 | 11.75 | 2.21 | 2.77 | 3.32 | 3.88 | 233 | 4606 |
| 63 | 220 | 0.074 | 13.9 | 0.508 | 55.9 | 3.4 | 1.034 | 7.57 | 9.47 | 11.36 | 13.25 | 2.78 | 3.48 | 4.18 | 4.87 | 335 | 5197 |
| 75 | 147 | 0.059 | 11.0 | 0.270 | 19.9 | 1.9 | 1.081 | 6.02 | 7.53 | 9.04 | 10.54 | 1.54 | 1.93 | 2.32 | 2.70 | 119 | 4134 |
| 75 | 170 | 0.068 | 12.8 | 0.361 | 30.7 | 2.2 | 1.064 | 6.97 | 8.71 | 10.45 | 12.19 | 2.03 | 2.54 | 3.05 | 3.56 | 184 | 4781 |
| 75 | 195 | 0.078 | 14.6 | 0.475 | 46.3 | 2.6 | 1.048 | 7.99 | 9.99 | 11.99 | 13.99 | 2.64 | 3.30 | 3.96 | 4.62 | 278 | 5484 |
| 75 | 220 | 0.089 | 16.5 | 0.605 | 66.6 | 2.9 | 1.034 | 9.02 | 11.27 | 13.53 | 15.78 | 3.31 | 4.14 | 4.97 | 5.80 | 399 | 6187 |

* Self-weight based on 5.4 kN/m³.
Permissible long term shear stress = 0.82 N/mm².
Permissible long term bending stress = 5.3 N/mm² × $K_7$.

$E_{min} = 6000$ N/mm².
$G = E_{min}/16 = 375$ N/mm².

**Table 5.20   Principal member: Hem-Fir CLS. No. 1 or No. 2 Joist and Plank Grade**

| $b$ (mm) | $h$ (mm) | Self-wt* (kN/m) | Area (mm² × 10³) | Section modulus (mm³ × 10⁶) | Second moment of area (mm⁴ × 10⁶) | $\frac{h}{b}$ | $K_7$ | Shear (kN) | | | | Bending (kN.m) | | | | Deflection | |
|---|---|---|---|---|---|---|---|---|---|---|---|---|---|---|---|---|---|
| | | | | | | | | $\bar{F}_{vL}$ | $\bar{F}_{vM}$ | $\bar{F}_{vS}$ | $\bar{F}_{vVS}$ | $\bar{M}_L$ | $\bar{M}_M$ | $\bar{M}_S$ | $\bar{M}_{VS}$ | $EI$ (kN.m²) | $AG$ (kN) |
| 38 | 140 | 0.028 | 5.32 | 0.124 | 8.68 | 3.6 | 1.087 | 2.51 | 3.14 | 3.77 | 4.40 | 0.75 | 0.94 | 1.13 | 1.32 | 60 | 2327 |
| 38 | 184 | 0.037 | 6.99 | 0.214 | 19.72 | 4.8 | 1.055 | 3.30 | 4.13 | 4.96 | 5.79 | 1.26 | 1.58 | 1.90 | 2.21 | 138 | 3059 |
| 38 | 235 | 0.048 | 8.93 | 0.350 | 41.09 | 6.1 | 1.027 | 4.22 | 5.28 | 6.34 | 7.39 | 2.01 | 2.51 | 3.01 | 3.52 | 287 | 3906 |

* Self-weight based on 5.4 kN/m³.
Permissible long term shear stress = 0.71 N/mm².
Permissible long term bending stress = 5.6 N/mm² × $K_7$.

$E_{min} = 7000$ N/mm².
$G = E_{min}/16 = 437.5$ N/mm².

**Table 5.21   Principal member: Spruce-Pine-Fir CLS. No. 1 or No. 2 Joist and Plank Grade**

| $b$ (mm) | $h$ (mm) | Self-wt* (kN/m) | Area (mm² × 10³) | Section modulus (mm³ × 10⁶) | Second moment of area (mm⁴ × 10⁶) | $\frac{h}{b}$ | $K_7$ | Shear (kN) | | | | Bending (kN.m) | | | | Deflection | |
|---|---|---|---|---|---|---|---|---|---|---|---|---|---|---|---|---|---|
| | | | | | | | | $\bar{F}_{vL}$ | $\bar{F}_{vM}$ | $\bar{F}_{vS}$ | $\bar{F}_{vVS}$ | $\bar{M}_L$ | $\bar{M}_M$ | $\bar{M}_S$ | $\bar{M}_{VS}$ | $EI$ (kN.m²) | $AG$ (kN) |
| 38 | 140 | 0.028 | 5.32 | 0.124 | 8.68 | 3.6 | 1.087 | 2.41 | 3.01 | 3.61 | 4.22 | 0.75 | 0.94 | 1.13 | 1.32 | 52 | 1995 |
| 38 | 184 | 0.037 | 6.99 | 0.214 | 19.72 | 4.8 | 1.055 | 3.16 | 3.96 | 4.75 | 5.54 | 1.26 | 1.58 | 1.90 | 2.21 | 118 | 2622 |
| 38 | 235 | 0.048 | 8.93 | 0.350 | 41.09 | 6.1 | 1.027 | 4.04 | 5.06 | 6.07 | 7.08 | 2.01 | 2.51 | 3.01 | 3.52 | 246 | 3348 |

* Self-weight based on 5.4 kN/m³.
Permissible long term shear stress = 0.68 N/mm².
Permissible long term bending stress = 5.6 N/mm² × $K_7$.

$E_{min} = 6000$ N/mm².
$G = E_{min}/16 = 375$ N/mm².

**Table 5.22  Load-sharing members: European Whitewood or Redwood. Basic sawn sizes. Grade SS or MSS**

| $b$ (mm) | $h$ (mm) | Self-wt* (kN/m) | Area (mm² × 10³) | Section modulus (mm³ × 10⁶) | Second moment of area (mm⁴ × 10⁶) | $\frac{h}{b}$ | $K_7$ | $\bar{F}_{vL}$ | $\bar{F}_{vM}$ | $\bar{F}_{vS}$ | $\bar{F}_{vVS}$ | $\bar{M}_L$ | $\bar{M}_M$ | $\bar{M}_S$ | $\bar{M}_{VS}$ | $EI$ (kN.m²) | $AG$ (kN) |
|---|---|---|---|---|---|---|---|---|---|---|---|---|---|---|---|---|---|
| | | | | | | | | Shear (kN) | | | | Bending (kN.m) | | | | Deflection | |
| 38 | 75 | 0.015 | 2.85 | 0.0356 | 1.34 | 1.9 | 1.164 | 1.71 | 2.14 | 2.57 | 2.99 | 0.34 | 0.42 | 0.51 | 0.59 | 14 | 1870 |
| 38 | 100 | 0.020 | 3.80 | 0.0633 | 3.17 | 2.6 | 1.128 | 2.28 | 2.85 | 3.42 | 3.99 | 0.58 | 0.73 | 0.88 | 1.03 | 33 | 2493 |
| 38 | 125 | 0.025 | 4.75 | 0.099 | 6.18 | 3.2 | 1.101 | 2.85 | 3.57 | 4.28 | 4.99 | 0.89 | 1.12 | 1.34 | 1.57 | 64 | 3117 |
| 38 | 150 | 0.030 | 5.70 | 0.143 | 10.7 | 3.9 | 1.079 | 3.42 | 4.28 | 5.14 | 5.99 | 1.26 | 1.58 | 1.90 | 2.22 | 112 | 3740 |
| 38 | 175 | 0.035 | 6.65 | 0.194 | 17.0 | 4.6 | 1.061 | 3.99 | 4.99 | 5.99 | 6.99 | 1.69 | 2.12 | 2.54 | 2.97 | 178 | 4364 |
| 38 | 200 | 0.041 | 7.60 | 0.253 | 25.3 | 5.2 | 1.045 | 4.57 | 5.71 | 6.85 | 7.99 | 2.18 | 2.73 | 3.27 | 3.82 | 266 | 4987 |
| 38 | 225 | 0.046 | 8.55 | 0.321 | 36.1 | 5.9 | 1.032 | 5.14 | 6.42 | 7.71 | 8.99 | 2.73 | 3.41 | 4.09 | 4.77 | 378 | 5610 |
| 47 | 75 | 0.019 | 3.53 | 0.0441 | 1.65 | 1.5 | 1.164 | 2.11 | 2.64 | 3.17 | 3.70 | 0.42 | 0.52 | 0.63 | 0.74 | 17 | 2313 |
| 47 | 100 | 0.025 | 4.70 | 0.0783 | 3.92 | 2.1 | 1.128 | 2.82 | 3.53 | 4.23 | 4.94 | 0.72 | 0.91 | 1.09 | 1.27 | 41 | 3084 |
| 47 | 125 | 0.031 | 5.88 | 0.122 | 7.65 | 2.6 | 1.101 | 3.53 | 4.41 | 5.29 | 6.18 | 1.11 | 1.38 | 1.66 | 1.94 | 80 | 3855 |
| 47 | 150 | 0.038 | 7.05 | 0.176 | 13.2 | 3.1 | 1.079 | 4.23 | 5.29 | 6.35 | 7.41 | 1.56 | 1.96 | 2.35 | 2.74 | 138 | 4626 |
| 47 | 175 | 0.044 | 8.23 | 0.240 | 21.0 | 3.7 | 1.061 | 4.94 | 6.18 | 7.41 | 8.65 | 2.10 | 2.62 | 3.15 | 3.67 | 220 | 5397 |
| 47 | 200 | 0.050 | 9.40 | 0.313 | 31.3 | 4.2 | 1.045 | 5.65 | 7.06 | 8.47 | 9.89 | 2.70 | 3.37 | 4.05 | 4.73 | 329 | 6168 |
| 47 | 225 | 0.057 | 10.6 | 0.397 | 44.6 | 4.7 | 1.032 | 6.35 | 7.94 | 9.53 | 11.12 | 3.37 | 4.22 | 5.06 | 5.90 | 468 | 6939 |
| 50 | 75 | 0.020 | 3.75 | 0.0469 | 1.76 | 1.5 | 1.164 | 2.25 | 2.81 | 3.38 | 3.94 | 0.45 | 0.56 | 0.67 | 0.78 | 18 | 2460 |
| 50 | 100 | 0.027 | 5.00 | 0.0833 | 4.17 | 2.0 | 1.128 | 3.00 | 3.75 | 4.51 | 5.26 | 0.77 | 0.96 | 1.16 | 1.35 | 43 | 3281 |
| 50 | 125 | 0.033 | 6.25 | 0.130 | 8.14 | 2.5 | 1.101 | 3.75 | 4.69 | 5.63 | 6.57 | 1.18 | 1.47 | 1.77 | 2.06 | 85 | 4101 |
| 50 | 150 | 0.040 | 7.50 | 0.188 | 14.1 | 3.0 | 1.079 | 4.51 | 5.63 | 6.76 | 7.89 | 1.66 | 2.08 | 2.50 | 2.92 | 147 | 4921 |
| 50 | 175 | 0.047 | 8.75 | 0.255 | 22.3 | 3.5 | 1.061 | 5.26 | 6.57 | 7.89 | 9.20 | 2.23 | 2.79 | 3.35 | 3.90 | 234 | 5742 |
| 50 | 200 | 0.054 | 10.0 | 0.333 | 33.3 | 4.0 | 1.045 | 6.01 | 7.51 | 9.02 | 10.52 | 2.87 | 3.59 | 4.31 | 5.03 | 350 | 6562 |
| 50 | 225 | 0.060 | 11.3 | 0.422 | 47.5 | 4.5 | 1.032 | 6.76 | 8.45 | 10.14 | 11.83 | 3.59 | 4.49 | 5.38 | 6.28 | 498 | 7382 |
| 63 | 150 | 0.051 | 9.45 | 0.236 | 17.7 | 2.3 | 1.079 | 5.68 | 7.10 | 8.52 | 9.94 | 2.10 | 2.62 | 3.15 | 3.68 | 186 | 6201 |
| 63 | 175 | 0.059 | 11.0 | 0.322 | 28.1 | 2.7 | 1.061 | 6.62 | 8.28 | 9.94 | 11.60 | 2.81 | 3.51 | 4.22 | 4.92 | 295 | 7235 |
| 63 | 200 | 0.068 | 12.6 | 0.420 | 42.0 | 3.1 | 1.045 | 7.57 | 9.47 | 11.36 | 13.25 | 3.62 | 4.52 | 5.43 | 6.34 | 441 | 8268 |
| 63 | 225 | 0.076 | 14.2 | 0.532 | 59.8 | 3.5 | 1.032 | 8.52 | 10.65 | 12.78 | 14.91 | 4.52 | 5.65 | 6.78 | 7.92 | 627 | 9302 |
| 75 | 150 | 0.060 | 11.3 | 0.281 | 21.1 | 2.0 | 1.079 | 6.76 | 8.45 | 10.14 | 11.83 | 2.50 | 3.13 | 3.75 | 4.38 | 221 | 7382 |
| 75 | 175 | 0.070 | 13.1 | 0.383 | 33.5 | 2.3 | 1.061 | 7.89 | 9.86 | 11.83 | 13.81 | 3.35 | 4.18 | 5.02 | 5.86 | 351 | 8613 |
| 75 | 200 | 0.081 | 15.0 | 0.500 | 50.0 | 2.6 | 1.045 | 9.02 | 11.27 | 13.53 | 15.78 | 4.31 | 5.39 | 6.46 | 7.54 | 525 | 9843 |
| 75 | 225 | 0.091 | 16.9 | 0.633 | 71.2 | 3.0 | 1.032 | 10.14 | 12.68 | 15.22 | 17.75 | 5.38 | 6.73 | 8.08 | 9.42 | 747 | 11074 |

\* Self-weight based on 5.4 kN/m³.

Permissible long term shear stress = 0.82 N/mm² × $K_8$.

Permissible long term bending stress = 7.5 N/mm² × $K_7$ × $K_8$.

$E_{mean} = 10\,500$ N/mm².

$G = E_{mean}/16 = 656.3$ N/mm².

**Table 5.23   Load-sharing members: European Whitewood or Redwood. Regularized only on width. Grade SS or MSS.**

| b (mm) | h (mm) | Self-wt* (kN/m) | Area (mm² × 10³) | Section modulus (mm³ × 10⁶) | Second moment of area (mm⁴ × 10⁶) | h/b | $K_7$ | $\bar{F}_{vL}$ | $\bar{F}_{vM}$ | $\bar{F}_{vS}$ | $\bar{F}_{vVS}$ | $\bar{M}_L$ | $\bar{M}_M$ | $\bar{M}_S$ | $\bar{M}_{VS}$ | EI (kN.m²) | AG (kN) |
|---|---|---|---|---|---|---|---|---|---|---|---|---|---|---|---|---|---|
| 38 | 72  | 0.014 | 2.74 | 0.0328 | 1.18 | 1.8 | 1.169 | 1.64 | 2.05 | 2.46 | 2.87 | 0.31 | 0.39 | 0.47 | 0.55 | 12  | 1795 |
| 38 | 97  | 0.019 | 3.69 | 0.0596 | 2.89 | 2.5 | 1.132 | 2.21 | 2.77 | 3.32 | 3.87 | 0.55 | 0.69 | 0.83 | 0.97 | 30  | 2418 |
| 38 | 122 | 0.025 | 4.64 | 0.0943 | 5.75 | 3.2 | 1.104 | 2.78 | 3.48 | 4.18 | 4.87 | 0.85 | 1.07 | 1.28 | 1.50 | 60  | 3042 |
| 38 | 147 | 0.030 | 5.59 | 0.137  | 10.1 | 3.8 | 1.081 | 3.35 | 4.19 | 5.03 | 5.87 | 1.22 | 1.52 | 1.83 | 2.13 | 105 | 3665 |
| 38 | 170 | 0.034 | 6.46 | 0.183  | 15.6 | 4.4 | 1.064 | 3.88 | 4.85 | 5.82 | 6.79 | 1.60 | 2.00 | 2.41 | 2.81 | 163 | 4239 |
| 38 | 195 | 0.040 | 7.41 | 0.241  | 23.5 | 5.1 | 1.048 | 4.45 | 5.56 | 6.68 | 7.79 | 2.08 | 2.60 | 3.12 | 3.64 | 246 | 4862 |
| 38 | 220 | 0.045 | 8.36 | 0.307  | 33.7 | 5.7 | 1.034 | 5.02 | 6.28 | 7.54 | 8.79 | 2.61 | 3.27 | 3.92 | 4.57 | 354 | 5486 |
| 47 | 72  | 0.018 | 3.38 | 0.0406 | 1.46 | 1.5 | 1.169 | 2.03 | 2.54 | 3.05 | 3.56 | 0.39 | 0.48 | 0.58 | 0.68 | 15  | 2220 |
| 47 | 97  | 0.024 | 4.56 | 0.0737 | 3.57 | 2.0 | 1.132 | 2.74 | 3.42 | 4.11 | 4.79 | 0.68 | 0.86 | 1.03 | 1.20 | 37  | 2991 |
| 47 | 122 | 0.030 | 5.73 | 0.117  | 7.11 | 2.5 | 1.104 | 3.44 | 4.31 | 5.17 | 6.03 | 1.06 | 1.32 | 1.59 | 1.85 | 74  | 3762 |
| 47 | 147 | 0.037 | 6.91 | 0.169  | 12.4 | 3.1 | 1.081 | 4.15 | 5.19 | 6.23 | 7.27 | 1.51 | 1.88 | 2.26 | 2.64 | 130 | 4534 |
| 47 | 170 | 0.043 | 7.99 | 0.226  | 19.2 | 3.6 | 1.064 | 4.80 | 6.00 | 7.20 | 8.40 | 1.98 | 2.48 | 2.98 | 3.47 | 202 | 5243 |
| 47 | 195 | 0.049 | 9.17 | 0.298  | 29.0 | 4.1 | 1.048 | 5.51 | 6.88 | 8.26 | 9.64 | 2.57 | 3.22 | 3.86 | 4.50 | 304 | 6014 |
| 47 | 220 | 0.055 | 10.3 | 0.379  | 41.7 | 4.6 | 1.034 | 6.21 | 7.77 | 9.32 | 10.88 | 3.23 | 4.04 | 4.85 | 5.66 | 437 | 6785 |
| 50 | 72  | 0.019 | 3.60 | 0.0432 | 1.56 | 1.4 | 1.169 | 2.16 | 2.70 | 3.24 | 3.78 | 0.41 | 0.52 | 0.62 | 0.72 | 16  | 2362 |
| 50 | 97  | 0.026 | 4.85 | 0.0784 | 3.80 | 1.9 | 1.132 | 2.91 | 3.64 | 4.37 | 5.10 | 0.73 | 0.91 | 1.09 | 1.28 | 39  | 3182 |
| 50 | 122 | 0.032 | 6.10 | 0.124  | 7.57 | 2.4 | 1.104 | 3.66 | 4.58 | 5.50 | 6.41 | 1.12 | 1.41 | 1.69 | 1.97 | 79  | 4003 |
| 50 | 147 | 0.039 | 7.35 | 0.180  | 13.2 | 2.9 | 1.081 | 4.41 | 5.52 | 6.62 | 7.73 | 1.60 | 2.00 | 2.41 | 2.81 | 138 | 4823 |
| 50 | 170 | 0.045 | 8.50 | 0.241  | 20.5 | 3.4 | 1.064 | 5.11 | 6.38 | 7.66 | 8.94 | 2.11 | 2.64 | 3.17 | 3.70 | 214 | 5578 |
| 50 | 195 | 0.052 | 9.75 | 0.317  | 30.9 | 3.9 | 1.048 | 5.86 | 7.32 | 8.79 | 10.26 | 2.74 | 3.42 | 4.11 | 4.79 | 324 | 6398 |
| 50 | 220 | 0.059 | 11.0 | 0.403  | 44.4 | 4.4 | 1.034 | 6.61 | 8.26 | 9.92 | 11.57 | 3.44 | 4.30 | 5.16 | 6.02 | 465 | 7218 |
| 63 | 147 | 0.050 | 9.26 | 0.227  | 16.7 | 2.3 | 1.081 | 5.56 | 6.96 | 8.35 | 9.74 | 2.02 | 2.53 | 3.03 | 3.54 | 175 | 6077 |
| 63 | 170 | 0.057 | 10.7 | 0.303  | 25.8 | 2.6 | 1.064 | 6.44 | 8.05 | 9.66 | 11.27 | 2.66 | 3.33 | 3.99 | 4.66 | 270 | 7028 |
| 63 | 195 | 0.066 | 12.3 | 0.399  | 38.9 | 3.0 | 1.048 | 7.38 | 9.23 | 11.08 | 12.92 | 3.45 | 4.31 | 5.18 | 6.04 | 408 | 8062 |
| 63 | 220 | 0.074 | 13.9 | 0.508  | 55.9 | 3.4 | 1.034 | 8.33 | 10.41 | 12.50 | 14.58 | 4.33 | 5.42 | 6.50 | 7.59 | 586 | 9095 |
| 75 | 147 | 0.059 | 11.0 | 0.270  | 19.9 | 1.9 | 1.081 | 6.62 | 8.28 | 9.94 | 11.60 | 2.41 | 3.01 | 3.61 | 4.21 | 208 | 7235 |
| 75 | 170 | 0.068 | 12.8 | 0.361  | 30.7 | 2.2 | 1.064 | 7.66 | 9.58 | 11.50 | 13.41 | 3.17 | 3.96 | 4.75 | 5.55 | 322 | 8367 |
| 75 | 195 | 0.078 | 14.6 | 0.475  | 46.3 | 2.6 | 1.048 | 8.79 | 10.99 | 13.19 | 15.39 | 4.11 | 5.13 | 6.16 | 7.19 | 486 | 9597 |
| 75 | 220 | 0.089 | 16.5 | 0.605  | 66.6 | 2.9 | 1.034 | 9.92 | 12.40 | 14.88 | 17.36 | 5.16 | 6.45 | 7.74 | 9.03 | 698 | 10828 |

* Self-weight based on 5.4 kN/m³.

Permissible long term shear stress = 0.82 N/mm² × $K_7$

$E_{mean} = 10\,500$ N/mm²

**Table 5.24** Load-sharing members: European Whitewood or Redwood. Basic sawn sizes. Grade GS or MGS

| $b$ (mm) | $h$ (mm) | Self-wt* (kN/m) | Area (mm² ×10³) | Section modulus (mm³ ×10⁶) | Second moment of area (mm⁴ ×10⁶) | $h/b$ | $K_7$ | $\bar{F}_{vL}$ | $\bar{F}_{vM}$ | $\bar{F}_{vS}$ | $\bar{F}_{vVS}$ | $\bar{M}_L$ | $\bar{M}_M$ | $\bar{M}_S$ | $\bar{M}_{VS}$ | $EI$ (kN.m²) | $AG$ (kN) |
|---|---|---|---|---|---|---|---|---|---|---|---|---|---|---|---|---|---|
| | | | | | | | | | Shear (kN) | | | | Bending (kN.m) | | | Deflection | |
| 38 | 75  | 0.015 | 2.85 | 0.0356 | 1.34 | 1.9 | 1.164 | 1.71 | 2.14 | 2.57 | 2.99 | 0.24 | 0.30 | 0.36 | 0.42 | 12  | 1603 |
| 38 | 100 | 0.020 | 3.80 | 0.0633 | 3.17 | 2.6 | 1.128 | 2.28 | 2.85 | 3.42 | 3.99 | 0.41 | 0.52 | 0.62 | 0.72 | 28  | 2137 |
| 38 | 125 | 0.025 | 4.75 | 0.099  | 6.18 | 3.2 | 1.101 | 2.85 | 3.57 | 4.28 | 4.99 | 0.63 | 0.79 | 0.95 | 1.11 | 55  | 2671 |
| 38 | 150 | 0.030 | 5.70 | 0.143  | 10.7 | 3.9 | 1.079 | 3.42 | 4.28 | 5.14 | 5.99 | 0.89 | 1.12 | 1.34 | 1.56 | 96  | 3206 |
| 38 | 175 | 0.035 | 6.65 | 0.194  | 17.0 | 4.6 | 1.061 | 3.99 | 4.99 | 5.99 | 6.99 | 1.19 | 1.49 | 1.79 | 2.09 | 152 | 3740 |
| 38 | 200 | 0.041 | 7.60 | 0.253  | 25.3 | 5.2 | 1.045 | 4.57 | 5.71 | 6.85 | 7.99 | 1.54 | 1.93 | 2.31 | 2.70 | 228 | 4275 |
| 38 | 225 | 0.046 | 8.55 | 0.321  | 36.1 | 5.9 | 1.032 | 5.14 | 6.42 | 7.71 | 8.99 | 1.92 | 2.41 | 2.89 | 3.37 | 324 | 4809 |
| 47 | 75  | 0.019 | 3.53 | 0.0441 | 1.65 | 1.5 | 1.164 | 2.11 | 2.64 | 3.17 | 3.70 | 0.29 | 0.37 | 0.44 | 0.52 | 14  | 1982 |
| 47 | 100 | 0.025 | 4.70 | 0.0783 | 3.92 | 2.1 | 1.128 | 2.82 | 3.53 | 4.23 | 4.94 | 0.51 | 0.64 | 0.77 | 0.90 | 35  | 2643 |
| 47 | 125 | 0.031 | 5.88 | 0.122  | 7.65 | 2.6 | 1.101 | 3.53 | 4.41 | 5.29 | 6.18 | 0.78 | 0.98 | 1.17 | 1.37 | 68  | 3304 |
| 47 | 150 | 0.038 | 7.05 | 0.176  | 13.2 | 3.1 | 1.079 | 4.23 | 5.29 | 6.35 | 7.41 | 1.10 | 1.38 | 1.66 | 1.94 | 118 | 3965 |
| 47 | 175 | 0.044 | 8.23 | 0.240  | 21.0 | 3.7 | 1.061 | 4.94 | 6.18 | 7.41 | 8.65 | 1.48 | 1.85 | 2.22 | 2.59 | 188 | 4626 |
| 47 | 200 | 0.050 | 9.40 | 0.313  | 31.3 | 4.2 | 1.045 | 5.65 | 7.06 | 8.47 | 9.89 | 1.91 | 2.38 | 2.86 | 3.34 | 282 | 5287 |
| 47 | 225 | 0.057 | 10.6 | 0.397  | 44.6 | 4.7 | 1.032 | 6.35 | 7.94 | 9.53 | 11.12 | 2.38 | 2.98 | 3.57 | 4.17 | 401 | 5948 |
| 50 | 75  | 0.020 | 3.75 | 0.0469 | 1.76 | 1.5 | 1.164 | 2.25 | 2.81 | 3.38 | 3.94 | 0.31 | 0.39 | 0.47 | 0.55 | 15  | 2109 |
| 50 | 100 | 0.027 | 5.00 | 0.0833 | 4.17 | 2.0 | 1.128 | 3.00 | 3.75 | 4.51 | 5.26 | 0.54 | 0.68 | 0.82 | 0.95 | 37  | 2812 |
| 50 | 125 | 0.033 | 6.25 | 0.130  | 8.14 | 2.5 | 1.101 | 3.75 | 4.69 | 5.63 | 6.57 | 0.83 | 1.04 | 1.25 | 1.46 | 73  | 3515 |
| 50 | 150 | 0.040 | 7.50 | 0.188  | 14.1 | 3.0 | 1.079 | 4.51 | 5.63 | 6.76 | 7.89 | 1.17 | 1.47 | 1.76 | 2.06 | 126 | 4218 |
| 50 | 175 | 0.047 | 8.75 | 0.255  | 22.3 | 3.5 | 1.061 | 5.26 | 6.57 | 7.89 | 9.20 | 1.57 | 1.97 | 2.36 | 2.76 | 200 | 4921 |
| 50 | 200 | 0.054 | 10.0 | 0.333  | 33.3 | 4.0 | 1.045 | 6.01 | 7.51 | 9.02 | 10.52 | 2.03 | 2.53 | 3.04 | 3.55 | 300 | 5625 |
| 50 | 225 | 0.060 | 11.3 | 0.422  | 47.5 | 4.5 | 1.032 | 6.76 | 8.45 | 10.14 | 11.83 | 2.53 | 3.17 | 3.80 | 4.44 | 427 | 6328 |
| 63 | 150 | 0.051 | 9.45 | 0.236  | 17.7 | 2.3 | 1.079 | 5.68 | 7.10 | 8.52 | 9.94 | 1.48 | 1.85 | 2.22 | 2.60 | 159 | 5315 |
| 63 | 175 | 0.059 | 11.0 | 0.322  | 28.1 | 2.7 | 1.061 | 6.62 | 8.28 | 9.94 | 11.60 | 1.98 | 2.48 | 2.98 | 3.48 | 253 | 6201 |
| 63 | 200 | 0.068 | 12.6 | 0.420  | 42.0 | 3.1 | 1.045 | 7.57 | 9.47 | 11.36 | 13.25 | 2.56 | 3.20 | 3.84 | 4.48 | 378 | 7087 |
| 63 | 225 | 0.076 | 14.2 | 0.532  | 59.8 | 3.5 | 1.032 | 8.52 | 10.65 | 12.78 | 14.91 | 3.19 | 3.99 | 4.79 | 5.59 | 538 | 7973 |
| 75 | 150 | 0.060 | 11.3 | 0.281  | 21.1 | 2.0 | 1.079 | 6.76 | 8.45 | 10.14 | 11.83 | 1.76 | 2.21 | 2.65 | 3.09 | 189 | 6328 |
| 75 | 175 | 0.070 | 13.1 | 0.383  | 33.5 | 2.3 | 1.061 | 7.89 | 9.86 | 11.83 | 13.81 | 2.36 | 2.96 | 3.55 | 4.14 | 301 | 7382 |
| 75 | 200 | 0.081 | 15.0 | 0.500  | 50.0 | 2.6 | 1.045 | 9.02 | 11.27 | 13.53 | 15.78 | 3.04 | 3.80 | 4.57 | 5.33 | 450 | 8437 |
| 75 | 225 | .091  | 16.9 | 0.633  | 71.2 | 3.0 | 1.032 | 10.14 | 12.68 | 15.22 | 17.75 | 3.80 | 4.75 | 5.71 | 6.66 | 640 | 9492 |

\* Self-weight based on 5.4 kN/m³.
Permissible long term shear stress = 0.82 N/mm² × $K_8$.
Permissible long term bending stress = 5.3 N/mm² × $K_7$ × $K_8$.

$E_{mean} = 9000$ N/mm².
$G = E_{mean}/16 = 562.5$ N/mm².

**Table 5.25   Load-sharing members: European Whitewood or Redwood. Regularized only on width. Grade GS or MGS**

| b (mm) | h (mm) | Self-wt* (kN/m) | Area (mm² × 10³) | Section modulus (mm³ × 10⁶) | Second moment of area (mm⁴ × 10⁶) | $\frac{h}{b}$ | $K_7$ | $\bar{F}_{vL}$ | $\bar{F}_{vM}$ | $\bar{F}_{vS}$ | $\bar{F}_{vVS}$ | $\bar{M}_L$ | $\bar{M}_M$ | $\bar{M}_S$ | $\bar{M}_{VS}$ | EI (kN.m²) | AG (kN) |
|---|---|---|---|---|---|---|---|---|---|---|---|---|---|---|---|---|---|
| | | | | | | | | Shear (kN) | | | | Bending (kN.m) | | | | Deflection | |
| 38 | 72 | 0.014 | 2.74 | 0.0328 | 1.18 | 1.8 | 1.169 | 1.64 | 2.05 | 2.46 | 2.87 | 0.22 | 0.27 | 0.33 | 0.39 | 10 | 1539 |
| 38 | 97 | 0.019 | 3.69 | 0.0596 | 2.89 | 2.5 | 1.132 | 2.21 | 2.77 | 3.32 | 3.87 | 0.39 | 0.49 | 0.59 | 0.68 | 26 | 2073 |
| 38 | 122 | 0.025 | 4.64 | 0.0943 | 5.75 | 3.2 | 1.104 | 2.78 | 3.48 | 4.18 | 4.87 | 0.60 | 0.75 | 0.91 | 1.06 | 51 | 2607 |
| 38 | 147 | 0.030 | 5.59 | 0.137 | 10.1 | 3.8 | 1.081 | 3.35 | 4.19 | 5.03 | 5.87 | 0.86 | 1.07 | 1.29 | 1.51 | 90 | 3142 |
| 38 | 170 | 0.034 | 6.46 | 0.183 | 15.6 | 4.4 | 1.064 | 3.88 | 4.85 | 5.82 | 6.79 | 1.13 | 1.41 | 1.70 | 1.98 | 140 | 3633 |
| 38 | 195 | 0.040 | 7.41 | 0.241 | 23.5 | 5.1 | 1.048 | 4.45 | 5.56 | 6.68 | 7.79 | 1.47 | 1.84 | 2.20 | 2.57 | 211 | 4168 |
| 38 | 220 | 0.045 | 8.36 | 0.307 | 33.7 | 5.7 | 1.034 | 5.02 | 6.28 | 7.54 | 8.79 | 1.84 | 2.31 | 2.77 | 3.23 | 303 | 4702 |
| 47 | 72 | 0.018 | 3.38 | 0.0406 | 1.46 | 1.5 | 1.169 | 2.03 | 2.54 | 3.05 | 3.56 | 0.27 | 0.34 | 0.41 | 0.48 | 13 | 1903 |
| 47 | 97 | 0.024 | 4.56 | 0.0737 | 3.57 | 2.0 | 1.132 | 2.74 | 3.42 | 4.11 | 4.79 | 0.48 | 0.60 | 0.72 | 0.85 | 32 | 2564 |
| 47 | 122 | 0.030 | 5.73 | 0.117 | 7.11 | 2.5 | 1.104 | 3.44 | 4.31 | 5.17 | 6.03 | 0.75 | 0.93 | 1.12 | 1.31 | 64 | 3225 |
| 47 | 147 | 0.037 | 6.91 | 0.169 | 12.4 | 3.1 | 1.081 | 4.15 | 5.19 | 6.23 | 7.27 | 1.06 | 1.33 | 1.60 | 1.86 | 111 | 3886 |
| 47 | 170 | 0.043 | 7.99 | 0.226 | 19.2 | 3.6 | 1.064 | 4.80 | 6.00 | 7.20 | 8.40 | 1.40 | 1.75 | 2.10 | 2.45 | 173 | 4494 |
| 47 | 195 | 0.049 | 9.17 | 0.298 | 29.0 | 4.1 | 1.048 | 5.51 | 6.88 | 8.26 | 9.64 | 1.82 | 2.27 | 2.73 | 3.18 | 261 | 5155 |
| 47 | 220 | 0.055 | 10.3 | 0.379 | 41.7 | 4.6 | 1.034 | 6.21 | 7.77 | 9.32 | 10.88 | 2.28 | 2.85 | 3.43 | 4.00 | 375 | 5816 |
| 50 | 72 | 0.019 | 3.60 | 0.0432 | 1.56 | 1.4 | 1.169 | 2.16 | 2.70 | 3.24 | 3.78 | 0.29 | 0.36 | 0.44 | 0.51 | 13 | 2025 |
| 50 | 97 | 0.026 | 4.85 | 0.0784 | 3.80 | 1.9 | 1.132 | 2.91 | 3.64 | 4.37 | 5.10 | 0.51 | 0.64 | 0.77 | 0.90 | 34 | 2728 |
| 50 | 122 | 0.032 | 6.10 | 0.124 | 7.57 | 2.4 | 1.104 | 3.66 | 4.58 | 5.50 | 6.41 | 0.79 | 0.99 | 1.19 | 1.39 | 68 | 3431 |
| 50 | 147 | 0.039 | 7.35 | 0.180 | 13.2 | 2.9 | 1.081 | 4.41 | 5.52 | 6.62 | 7.73 | 1.13 | 1.41 | 1.70 | 1.98 | 119 | 4134 |
| 50 | 170 | 0.045 | 8.50 | 0.241 | 20.5 | 3.4 | 1.064 | 5.11 | 6.38 | 7.66 | 8.94 | 1.49 | 1.86 | 2.24 | 2.61 | 184 | 4781 |
| 50 | 195 | 0.052 | 9.75 | 0.317 | 30.9 | 3.9 | 1.048 | 5.86 | 7.32 | 8.79 | 10.26 | 1.93 | 2.42 | 2.90 | 3.38 | 278 | 5484 |
| 50 | 220 | 0.059 | 11.0 | 0.403 | 44.4 | 4.4 | 1.034 | 6.61 | 8.26 | 9.92 | 11.57 | 2.43 | 3.04 | 3.64 | 4.25 | 399 | 6187 |
| 63 | 147 | 0.050 | 9.26 | 0.227 | 16.7 | 2.3 | 1.081 | 5.56 | 6.96 | 8.35 | 9.74 | 1.43 | 1.78 | 2.14 | 2.50 | 150 | 5209 |
| 63 | 170 | 0.057 | 10.7 | 0.303 | 25.8 | 2.6 | 1.064 | 6.44 | 8.05 | 9.66 | 11.27 | 1.88 | 2.35 | 2.82 | 3.29 | 232 | 6024 |
| 63 | 195 | 0.066 | 12.3 | 0.399 | 38.9 | 3.0 | 1.048 | 7.38 | 9.23 | 11.08 | 12.92 | 2.44 | 3.05 | 3.66 | 4.27 | 350 | 6910 |
| 63 | 220 | 0.074 | 13.9 | 0.508 | 55.9 | 3.4 | 1.034 | 8.33 | 10.41 | 12.50 | 14.58 | 3.06 | 3.83 | 4.59 | 5.36 | 503 | 7796 |
| 75 | 147 | 0.059 | 11.0 | 0.270 | 19.9 | 1.9 | 1.081 | 6.62 | 8.28 | 9.94 | 11.60 | 1.70 | 2.12 | 2.55 | 2.98 | 178 | 6201 |
| 75 | 170 | 0.068 | 12.8 | 0.361 | 30.7 | 2.2 | 1.064 | 7.66 | 9.58 | 11.50 | 13.41 | 2.24 | 2.80 | 3.36 | 3.92 | 276 | 7171 |
| 75 | 195 | 0.078 | 14.6 | 0.475 | 46.3 | 2.6 | 1.048 | 8.79 | 10.99 | 13.19 | 15.39 | 2.90 | 3.63 | 4.35 | 5.08 | 417 | 8226 |
| 75 | 220 | 0.089 | 16.5 | 0.605 | 66.6 | 2.9 | 1.034 | 9.92 | 12.40 | 14.88 | 17.36 | 3.64 | 4.56 | 5.47 | 6.38 | 598 | 9281 |

* Self-weight based on 5.4 kN/m³.

**Table 5.26**   Load-sharing members: Hem-Fir CLS. No. 1 and No. 2 Joist and Plank Grade

| $b$ (mm) | $h$ (mm) | Self-wt* (kN/m) | Area (mm² × 10³) | Section modulus (mm³ × 10⁶) | Second moment of area (mm⁴ × 10⁶) | $\frac{h}{b}$ | $K_7$ | Shear (kN) $\bar{F}_{vL}$ | $\bar{F}_{vM}$ | $\bar{F}_{vS}$ | $\bar{F}_{vVS}$ | Bending (kN.m) $\bar{M}_L$ | $\bar{M}_M$ | $\bar{M}_S$ | $\bar{M}_{VS}$ | Deflection $EI$ (kN.m²) | $AG$ (kN) |
|---|---|---|---|---|---|---|---|---|---|---|---|---|---|---|---|---|---|
| 38 | 140 | 0.028 | 5.32 | 0.124 | 8.68 | 3.6 | 1.087 | 2.76 | 3.46 | 4.15 | 4.84 | 0.83 | 1.03 | 1.24 | 1.45 | 91 | 3491 |
| 38 | 184 | 0.037 | 6.99 | 0.214 | 19.72 | 4.8 | 1.055 | 3.64 | 4.55 | 5.46 | 6.37 | 1.39 | 1.74 | 2.09 | 2.43 | 207 | 4588 |
| 38 | 235 | 0.048 | 8.93 | 0.350 | 41.09 | 6.1 | 1.027 | 4.64 | 5.81 | 6.97 | 8.13 | 2.21 | 2.76 | 3.31 | 3.87 | 431 | 5860 |

* Self-weight based on 5.4 kN/m³.
Permissible long term shear stress = 0.71 N/mm² × $K_8$.
Permissible long term bending stress = 5.6 N/mm² × $K_7$ × $K_8$.

$E_{mean} = 10\,500$ N/mm².
$G = E_{mean}/16 = 656.3$ N/mm².

**Table 5.27**   Load-sharing members: Spruce-Pine-Fir CLS. No. 1 and No. 2 Joist and Plank Grade

| $b$ (mm) | $h$ (mm) | Self-wt* (kN/m) | Area (mm² × 10³) | Section modulus (mm³ × 10⁶) | Second moment of area (mm⁴ × 10⁶) | $\frac{h}{b}$ | $K_7$ | Shear (kN) $\bar{F}_{vL}$ | $\bar{F}_{vM}$ | $\bar{F}_{vS}$ | $\bar{F}_{vVs}$ | Bending (kN.m) $\bar{M}_L$ | $\bar{M}_M$ | $\bar{M}_S$ | $\bar{M}_{VS}$ | Deflection $EI$ (kN.m²) | $AG$ (kN) |
|---|---|---|---|---|---|---|---|---|---|---|---|---|---|---|---|---|---|
| 38 | 140 | 0.028 | 5.32 | 0.124 | 8.68 | 3.6 | 1.087 | 2.65 | 3.31 | 3.97 | 4.64 | 0.83 | 1.03 | 1.24 | 1.45 | 78 | 2992 |
| 38 | 184 | 0.037 | 6.99 | 0.214 | 19.72 | 4.8 | 1.055 | 3.48 | 4.35 | 5.23 | 6.10 | 1.39 | 1.74 | 2.09 | 2.43 | 177 | 3933 |
| 38 | 235 | 0.048 | 8.93 | 0.350 | 41.09 | 6.1 | 1.027 | 4.45 | 5.56 | 6.67 | 7.79 | 2.21 | 2.76 | 3.31 | 3.87 | 369 | 5023 |

* Self-weight based on 5.4 kN/m³.
Permissible long term shear stress = 0.68 N/mm² × $K_8$.
Permissible long term bending stress = 5.6 N/mm² × $K_7$ × $K_8$.

$E_{mean} = 9000$ N/mm².
$G = E_{mean}/16 = 562.5$ N/mm².

## 5.10   GEOMETRICAL PROPERTIES OF PLYWOOD

BS 5268 bases the geometrical properties of plywood on the minimum thicknesses of veneers permitted by the relevant product standards. They are said to apply to both the dry and wet exposure conditions. Refer to §1.9 for information on qualities and layups.

Tables 5.28–5.36 give geometrical properties on the same basis as BS 5268: Part 2. The properties are on the 'full area' method.


## 5.11   STRENGTH CAPACITIES OF PLYWOOD

Tables 5.28–5.36 give the strength capacities of many of the available plywoods. These are on the 'full area' basis and for the dry exposure condition. For 'wet' conditions it is necessary to multiply by the $K_{36}$ factor from BS 5268: Part 2.

The grade stresses are generally at the lower 5 percentile level. The moduli are generally mathematical mean values. A certain amount of balancing between the strength properties of plywoods has been necessary in BS 5268 due to different test methods having been used in various countries to establish test results. Values are at the 15% moisture content level. $E$ values do not include an allowance for shear modulus.

The duration of load factor ($K_3$) has the same values as for solid timber.

## Table 5.28 Canadian Softwood plywood, unsanded, all grades

Properties and capacities per metre width based on minimum thickness. Bending as in Fig. 4.27.
Where two or more layups make up one thickness the values given in this table are for the weakest layup, and the number of plies is indicated (by 3, 4, 5 etc).
In BS 5268 separate stresses are given for each layup.

| Thickness nominal (mm) | minimum (mm) | No. of plies | Area (mm² × 10³) | Section modulus $Z_X$ (mm³ × 10³) | Second moment of area $I_X$ (mm⁴ × 10³) | Face grain parallel to span $\sigma_{m,adm}$ (N/mm²) | $\bar{M}_{XL}$ (kN.m) | $\bar{M}_{XM}$ (kN.m) | $E$ (N/mm²) | $EI_X$ (kN.m²) | Face grain perpendicular to span $\sigma_{m,adm}$ (N/mm²) | $\bar{M}_{XL}$ (kN.m) | $\bar{M}_{XM}$ (kN.m) | $E$ (N/mm²) | $EI_X$ (kN.m²) |
|---|---|---|---|---|---|---|---|---|---|---|---|---|---|---|---|
| 7.5 | 7.0 | 3 | 7.0 | 8.17 | 28.6 | 7.69 | 0.062 | 0.078 | 7600 | 0.217 | 2.28 | 0.018 | 0.023 | 650 | 0.018 |
| 9.5 | 9.0 | 3 | 9.0 | 13.5 | 60.7 | 7.16 | 0.096 | 0.120 | 7100 | 0.430 | 2.25 | 0.030 | 0.037 | 600 | 0.036 |
| 12.5 | 12.0 | 3, 4, 5 | 12.0 | 24.0 | 144 | 6.99 (4) | 0.167 | 0.209 | 6900 (4) | 0.993 | 2.26 (3) | 0.054 | 0.067 | 600 (3) | 0.086 |
| 15.5 | 15.0 | 5 | 15.0 | 37.5 | 281 | 6.49 | 0.243 | 0.304 | 6100 | 1.714 | 3.14 | 0.117 | 0.147 | 1750 | 0.491 |
| 18.5 | 18.0 | 5, 6, 7 | 18.0 | 54.0 | 486 | 6.37 (7) | 0.343 | 0.429 | 5400 (5) | 2.624 | 3.33 (6) | 0.179 | 0.224 | 2100 (6) | 1.020 |
| 20.5 | 20.0 | 5, 6, 7 | 20.0 | 66.7 | 667 | 5.50 (6) | 0.366 | 0.458 | 5150 (6) | 3.435 | 3.35 (6) | 0.223 | 0.279 | 2250 (6) | 1.500 |

## Table 5.29 Canadian Douglas Fir plywood, unsanded, all grades

Properties and capacities per metre width based on minimum thickness. Bending as in Fig. 4.27.
Where two or more layups make up one thickness the values given in this table are for the weakest layup, and the number of plies is indicated (by 3, 4, 5 etc).
In BS 5268 separate stresses are given for each layup.

| Thickness nominal (mm) | minimum (mm) | No. of plies | Area (mm² × 10³) | Section modulus $Z_X$ (mm³ × 10³) | Second moment of area $I_X$ (mm⁴ × 10³) | Face grain parallel to span $\sigma_{m,adm}$ (N/mm²) | $\bar{M}_{XL}$ (kN.m) | $\bar{M}_{XM}$ (kN.m) | $E$ (N/mm²) | $EI_X$ (kN.m²) | Face grain perpendicular to span $\sigma_{m,adm}$ (N/mm²) | $\bar{M}_{XL}$ (kN.m) | $\bar{M}_{XM}$ (kN.m) | $E$ (N/mm²) | $EI_X$ (kN.m²) |
|---|---|---|---|---|---|---|---|---|---|---|---|---|---|---|---|
| 7.5 | 7.0 | 3 | 7.0 | 8.17 | 28.6 | 11.60 | 0.094 | 0.118 | 12350 | 0.353 | 2.78 | 0.022 | 0.028 | 850 | 0.024 |
| 9.5 | 9.0 | 3 | 9.0 | 13.5 | 60.7 | 10.70 | 0.144 | 0.180 | 11450 | 0.695 | 2.76 | 0.037 | 0.046 | 800 | 0.048 |
| 12.5 | 12.0 | 3, 4, 5 | 12.0 | 24.0 | 144 | 9.39 (4) | 0.225 | 0.281 | 10050 (4) | 1.447 | 2.77 (3) | 0.066 | 0.083 | 850 (3) | 0.122 |
| 15.5 | 15.0 | 5 | 15.0 | 37.5 | 281 | 9.72 | 0.364 | 0.455 | 9800 | 2.753 | 3.41 | 0.127 | 0.159 | 2150 | 0.604 |
| 18.5 | 18.0 | 5, 6, 7 | 18.0 | 54.0 | 486 | 8.63 (5) | 0.466 | 0.582 | 8700 (5) | 4.228 | 3.58 (6) | 0.193 | 0.241 | 2500 (6) | 1.215 |
| 20.5 | 20.0 | 5, 6, 7 | 20.0 | 66.7 | 667 | 8.14 (6) | 0.542 | 0.678 | 8200 (6) | 5.469 | 3.58 (6) | 0.238 | 0.298 | 2650 (6) | 1.767 |

**Table 5.30   Canadian Douglas Fir plywood, sanded, all grades**

Properties and capacities per metre width based on minimum thickness. Bending as in Fig. 4.27.
Where two or more layups make up one thickness the values given in this table are for the weakest layup, and the number of plies is indicated (by 3, 4, 5 etc).
In BS 5268 separate stresses are given for each layup.

| Thickness | | No. of plies | Area | Section modulus $Z_X$ | Second moment of area $I_X$ | Face grain parallel to span | | | | | Face grain perpendicular to span | | | | |
|---|---|---|---|---|---|---|---|---|---|---|---|---|---|---|---|
| nominal (mm) | minimum (mm) | | $(\text{mm}^2 \times 10^3)$ | $(\text{mm}^3 \times 10^3)$ | $(\text{mm}^4 \times 10^3)$ | $\sigma_{m,adm}$ $(\text{N/mm}^2)$ | $\bar{M}_{XL}$ (kN.m) | $\bar{M}_{XM}$ (kN.m) | $E$ $(\text{N/mm}^2)$ | $EI_X$ $(\text{kN.m}^2)$ | $\sigma_{m,adm}$ $(\text{N/mm}^2)$ | $\bar{M}_{XL}$ (kN.m) | $\bar{M}_{XM}$ (kN.m) | $E$ $(\text{N/mm}^2)$ | $EI_X$ $(\text{kN.m}^2)$ |
| 6 | 5.5 | 3 | 5.5 | 5.04 | 13.9 | 11.20 | 0.056 | 0.070 | 11850 | 0.164 | 3.28 | 0.016 | 0.020 | 1250 | 0.017 |
| 8 | 7.5 | 3 | 7.5 | 9.38 | 35.2 | 9.85 | 0.092 | 0.115 | 10450 | 0.367 | 3.05 | 0.028 | 0.035 | 1100 | 0.038 |
| 11 | 10.5 | 3, 4, 5 | 10.5 | 18.4 | 96.5 | 8.11 (4) | 0.149 | 0.186 | 8650 (4) | 0.601 | 2.96 (3) | 0.054 | 0.068 | 1000 (3) | 0.069 |
| 14 | 13.5 | 5 | 13.5 | 30.4 | 205 | 8.06 | 0.245 | 0.306 | 8100 | 1.660 | 3.99 | 0.121 | 0.151 | 2750 | 0.563 |
| 17 | 16.5 | 5, 6, 7 | 16.5 | 45.4 | 374 | 7.11 (6) | 0.322 | 0.403 | 7050 (5) | 2.636 | 4.11 (6) | 0.186 | 0.233 | 3150 (6) | 1.178 |
| 19 | 18.5 | 5, 6, 7 | 18.5 | 57.0 | 528 | 6.57 (6) | 0.374 | 0.468 | 6600 (6) | 3.484 | 4.06 (6) | 0.231 | 0.289 | 3250 (6) | 1.716 |

**Table 5.31  Finnish Coniferous plywood, sanded, all grades**

Properties and capacities per metre width based on minimum thickness. Bending as in Fig. 4.27. Where two or more layups make up one thickness the values given in this table are for the weakest layup, and the number of plies is indicated (by 3, 4, 5 etc). In BS 5268 separate stresses are given for each layup.

| Thickness | | No. of plies | Area (mm² × 10³) | Section modulus $Z_X$ (mm³ × 10³) | Second moment of area $I_X$ (mm⁴ × 10³) | Face grain parallel to span | | | | | Face grain perpendicular to span | | | | |
|---|---|---|---|---|---|---|---|---|---|---|---|---|---|---|---|
| nominal (mm) | minimum (mm) | | | | | $\sigma_{m,adm}$ (N/mm²) | $\bar{M}_{XL}$ (kN.m) | $\bar{M}_{XM}$ (kN.m) | $E$ (N/mm²) | $EI_X$ (kN.m²) | $\sigma_{m,adm}$ (N/mm²) | $\bar{M}_{XL}$ (kN.m) | $\bar{M}_{XM}$ (kN.m) | $E$ (N/mm²) | $EI_X$ (kN.m²) |
| 6.5 | 6.1 | 5 | 6.1 | 6.20 | 18.9 | 8.28 | 0.051 | 0.064 | 7600 | 0.143 | 3.93 | 0.024 | 0.030 | 1800 | 0.034 |
| 9 | 8.6 | 5, 7 | 8.6 | 12.3 | 53.0 | 6.89(5) | 0.084 | 0.105 | 6300(5) | 0.333 | 4.60(5) | 0.056 | 0.070 | 2400(5) | 0.127 |
| 12 | 10.9 | 7, 9 | 10.9 | 19.8 | 108 | 6.13(7) | 0.121 | 0.151 | 5600(9) | 0.604 | 4.00(9) | 0.079 | 0.099 | 1900(9) | 0.205 |
| 13.5 | 12.4 | 5, 7 | 12.4 | 25.6 | 159 | 5.89(7) | 0.150 | 0.188 | 5550(5) | 0.882 | 4.20(5) | 0.107 | 0.134 | 1950(5) | 0.310 |
| 15 | 13.9 | 7, 11 | 13.9 | 32.2 | 224 | 6.77(11) | 0.217 | 0.272 | 5400(11) | 1.209 | 4.40(7) | 0.141 | 0.177 | 2000(7) | 0.448 |
| 16.5 | 15.5 | 7 | 15.5 | 40.0 | 310 | 5.56 | 0.222 | 0.278 | 5400 | 1.674 | 4.51 | 0.180 | 0.225 | 2250 | 0.697 |
| 18 | 16.6 | 9, 13 | 16.6 | 45.9 | 382 | 5.78(9) | 0.265 | 0.331 | 5350(9) | 2.043 | 4.51(9) | 0.207 | 0.258 | 2500(9) | 0.955 |
| 19.5 | 18.1 | 7, 9 | 18.1 | 54.6 | 494 | 5.07(9) | 0.276 | 0.346 | 5200(7) | 2.568 | 4.51(7) | 0.246 | 0.307 | 2550(7) | 1.259 |
| 21 | 19.8 | 15 | 19.8 | 65.3 | 647 | 6.16 | 0.402 | 0.502 | 5100 | 3.299 | 4.51 | 0.294 | 0.368 | 2850 | 1.843 |
| 22 | 21.0 | 9 | 21.0 | 73.5 | 772 | 5.35 | 0.393 | 0.491 | 5000 | 3.860 | 4.51 | 0.331 | 0.414 | 2850 | 2.200 |
| 23 | 22.3 | 11, 17 | 22.3 | 82.9 | 924 | 6.15(11) | 0.509 | 0.637 | 4950(11) | 4.573 | 4.51(11) | 0.373 | 0.467 | 3050(11) | 2.818 |
| 25 | 23.7 | 9, 11 | 23.7 | 93.6 | 1110 | 4.54(11) | 0.424 | 0.531 | 4900(9) | 5.439 | 4.51(9) | 0.422 | 0.527 | 3100(9) | 3.441 |
| 26 | 25.2 | 19 | 25.2 | 106 | 1330 | 5.87 | 0.622 | 0.777 | 4900 | 6.517 | 4.51 | 0.478 | 0.597 | 3150 | 4.189 |
| 27 | 26.4 | 11, 13 | 26.4 | 116 | 1530 | 5.21(11) | 0.604 | 0.755 | 4800(11) | 7.344 | 4.51(11) | 0.523 | 0.653 | 3200 | 4.896 |

**Table 5.32   Finnish  Birch-faced plywood, sanded, all grades**

Properties and capacities per metre width based on minimum thickness. Bending as in Fig. 4.27.
Where two or more layups make up one thickness the values given in this table are for the weakest layup, and the number of plies is indicated (by 3, 4, 5 etc).
In BS 5268 separate stresses are given for each layup.

| Thickness | | No. of plies | Area | Section modulus $Z_X$ | Second moment of area $I_X$ | Face grain parallel to span | | | | | Face grain perpendicular to span | | | | |
| --- | --- | --- | --- | --- | --- | --- | --- | --- | --- | --- | --- | --- | --- | --- | --- |
| nominal (mm) | minimum (mm) | | (mm²×10³) | (mm³×10³) | (mm⁴×10³) | $\sigma_{m,adm}$ (N/mm²) | $\overline{M}_{XL}$ (kN.m) | $\overline{M}_{XM}$ (kN.m) | $E$ (N/mm²) | $EI_X$ (kN.m²) | $\sigma_{m,adm}$ (N/mm²) | $\overline{M}_{XL}$ (kN.m) | $\overline{M}_{XM}$ (kN.m) | $E$ (N/mm²) | $EI_X$ (kN.m²) |
| 6.5 | 6.1 | 5 | 6.1 | 6.20 | 18.9 | 18.20 | 0.112 | 0.141 | 10300 | 0.194 | 6.42 | 0.039 | 0.049 | 2750 | 0.051 |
| 9 | 8.8 | 7 | 8.8 | 12.9 | 56.8 | 15.90 | 0.205 | 0.256 | 9500 | 0.539 | 9.33 | 0.120 | 0.150 | 3800 | 0.215 |
| 12 | 11.5 | 9 | 11.5 | 22.0 | 127 | 14.30 | 0.314 | 0.393 | 8900 | 1.130 | 9.83 | 0.216 | 0.270 | 4300 | 0.546 |
| 15 | 13.9 | 9, 11 | 13.9 | 32.2 | 224 | 14.20(9) | 0.457 | 0.571 | 8550(11) | 1.915 | 9.95(9) | 0.320 | 0.400 | 4300 | 0.963 |
| 18 | 17.1 | 11, 13 | 17.1 | 48.7 | 417 | 12.80(13) | 0.623 | 0.779 | 8050 | 3.356 | 9.75(13) | 0.474 | 0.593 | 4200 | 1.751 |
| 21 | 19.9 | 15 | 19.9 | 66.0 | 657 | 12.40 | 0.818 | 1.023 | 7550 | 4.960 | 9.39 | 0.619 | 0.774 | 4150 | 2.726 |
| 22 | 21.1 | 13 | 21.1 | 74.2 | 783 | 13.20 | 0.979 | 1.224 | 7400 | 5.794 | 9.35 | 0.693 | 0.867 | 4150 | 3.249 |
| 23 | 22.7 | 17 | 22.7 | 85.9 | 975 | 12.10 | 1.039 | 1.299 | 6300 | 6.142 | 9.30 | 0.798 | 0.998 | 4150 | 4.046 |
| 25 | 24.4 | 15 | 24.4 | 99.2 | 1210 | 12.90 | 1.279 | 1.599 | 7150 | 8.651 | 9.20 | 0.912 | 1.140 | 4150 | 5.021 |
| 26 | 25.2 | 19 | 25.2 | 106 | 1330 | 11.70 | 1.240 | 1.550 | 6950 | 9.243 | 9.10 | 0.964 | 1.205 | 4100 | 5.453 |
| 29 | 27.9 | 17 | 27.9 | 130 | 1810 | 12.60 | 1.638 | 2.047 | 6850 | 12.398 | 9.05 | 1.176 | 1.470 | 4050 | 7.330 |

**Table 5.33  Finnish Birch plywood, sanded, all grades**
Properties and capacities per metre width based on minimum thickness. Bending as in Fig. 4.27.

| Thickness | | No. of plies | Area | Section modulus $Z_X$ | Second moment of area $I_X$ | Face grain parallel to span | | | | | Face grain perpendicular to span | | | | |
|---|---|---|---|---|---|---|---|---|---|---|---|---|---|---|---|
| nominal (mm) | minimum (mm) | | (mm²× 10³) | (mm³× 10³) | (mm⁴× 10³) | $\sigma_{m,adm}$ (N/mm²) | $\bar{M}_{XL}$ (kN.m) | $\bar{M}_{XM}$ (kN.m) | $E$ (N/mm²) | $EI_X$ (kN.m²) | $\sigma_{m,adm}$ (N/mm²) | $\bar{M}_{XL}$ (kN.m) | $\bar{M}_{XM}$ (kN.m) | $E$ (N/mm²) | $EI_X$ (kN.m²) |
| 6.5 | 6.1 | 5 | 6.1 | 6.20 | 18.9 | 19.00 | 0.117 | 0.147 | 10750 | 0.203 | 9.79 | 0.060 | 0.075 | 4000 | 0.075 |
| 9 | 8.8 | 7 | 8.8 | 12.9 | 56.8 | 18.30 | 0.236 | 0.295 | 9900 | 0.562 | 11.50 | 0.148 | 0.185 | 5350 | 0.303 |
| 12 | 11.5 | 9 | 11.5 | 22.0 | 127 | 17.60 | 0.387 | 0.484 | 9250 | 1.174 | 12.40 | 0.272 | 0.341 | 5800 | 0.736 |
| 15 | 14.3 | 11 | 14.3 | 34.1 | 244 | 16.90 | 0.576 | 0.720 | 8800 | 2.147 | 12.20 | 0.416 | 0.520 | 5950 | 1.451 |
| 18 | 17.1 | 13 | 17.1 | 48.7 | 417 | 16.30 | 0.793 | 0.992 | 8600 | 3.586 | 12.10 | 0.589 | 0.736 | 5950 | 2.481 |
| 21 | 19.9 | 15 | 19.9 | 66.0 | 657 | 15.90 | 1.049 | 1.311 | 8550 | 5.617 | 12.00 | 0.792 | 0.990 | 6000 | 3.942 |
| 24 | 22.5 | 17 | 22.5 | 84.4 | 949 | 15.70 | 1.325 | 1.656 | 8500 | 8.066 | 11.90 | 1.004 | 1.255 | 6050 | 5.741 |
| 27 | 25.2 | 19 | 25.2 | 106 | 1330 | 15.60 | 1.653 | 2.067 | 8450 | 11.238 | 11.80 | 1.250 | 1.563 | 6050 | 8.046 |

**Table 5.34   American Constructional and Industrial plywood, unsanded C–D grade (as specified in BS 5268 : Part 2) and C–C grade**

Properties and capacities per metre width based on minimum thickness. Bending as in Fig. 4.27.
Where two or more layups make up one thickness the values given in this table are for the weakest layup, and the number of plies is indicated (by 3, 4, 5 etc).
In BS 5268 separate stresses are given for each layup.

| Thickness | | No. of plies | Area | Section modulus $Z_X$ | Second moment of area $I_X$ | Face grain parallel to span | | | | | Face grain perpendicular to span | | | | |
|---|---|---|---|---|---|---|---|---|---|---|---|---|---|---|---|
| nominal (mm) | minimum (mm) | | $(mm^2 \times 10^3)$ | $(mm^3 \times 10^3)$ | $(mm^4 \times 10^3)$ | $\sigma_{m,adm}$ $(N/mm^2)$ | $\overline{M}_{XL}$ $(kN.m)$ | $\overline{M}_{XM}$ $(kN.m)$ | $E$ $(N/mm^2)$ | $EI_X$ $(kN.m^2)$ | $\sigma_{m,adm}$ $(N/mm^2)$ | $\overline{M}_{XL}$ $(kN.m)$ | $\overline{M}_{XM}$ $(kN.m)$ | $E$ $(N/mm^2)$ | $EI_X$ $(kN.m^2)$ |
| **C–D grade** | | | | | | | | | | | | | | | |
| 8 | 7.1 | 3 | 7.1 | 8.40 | 29.8 | 7.07 | 0.059 | 0.074 | 9850 | 0.293 | 2.17 | 0.018 | 0.022 | 750 | 0.022 |
| 9.5 | 8.7 | 3 | 8.7 | 12.6 | 54.9 | 6.26 | 0.078 | 0.098 | 8800 | 0.483 | 2.02 | 0.025 | 0.031 | 600 | 0.032 |
| 12.5 | 11.9 | 3, 4, 5 | 11.9 | 23.6 | 140 | 6.39(3) | 0.150 | 0.188 | 8600(5) | 1.204 | 2.16(3) | 0.050 | 0.063 | 750(3) | 0.105 |
| 16 | 15.1 | 4, 5 | 15.1 | 38.0 | 287 | 4.66(4) | 0.177 | 0.221 | 6600(4) | 1.894 | 2.50(5) | 0.095 | 0.118 | 1650(5) | 0.473 |
| 19 | 18.3 | 5, 6, 7 | 18.3 | 55.8 | 511 | 5.42(5) | 0.302 | 0.378 | 7100(5) | 3.628 | 2.81(5) | 0.156 | 0.195 | 1950(5) | 0.996 |
| **C–C grade** | | | | | | | | | | | | | | | |
| 8 | 7.1 | 3 | | As above | | 11.40 | 0.095 | 0.119 | 10750 | 0.320 | 3.50 | 0.029 | 0.036 | 900 | 0.026 |
| 9.5 | 8.7 | 3 | | | | 10.20 | 0.128 | 0.160 | 9600 | 0.527 | 3.24 | 0.040 | 0.051 | 700 | 0.038 |
| 12.5 | 11.9 | 3, 4, 5 | | | | 10.40(3) | 0.245 | 0.306 | 9350(5) | 1.309 | 3.49(3) | 0.082 | 0.102 | 900(3) | 0.126 |
| 16 | 15.1 | 4, 5 | | | | 7.63(4) | 0.289 | 0.362 | 7200(4) | 2.066 | 4.13(5) | 0.156 | 0.196 | 1900(5) | 0.545 |
| 19 | 18.3 | 5, 6, 7 | | | | 8.69(5) | 0.484 | 0.606 | 7750(5) | 3.960 | 4.62(5) | 0.257 | 0.322 | 2300(5) | 1.175 |

BS 5268 recommends that C–D grade should not be used for prolonged use in damp or wet conditions.
For panels less than 600 mm wide bending, tension and compression stresses should be reduced in proportion to their width, taking no reduction at 600 mm, and 50% reduction at 200 mm and less.

**Table 5.35  American Constructional and Industrial plywood, sanded, A–C grade and B–C grade**

Properties and capacities per metre width based on minimum thickness. Bending as in Fig. 4.27.
Where two or more layups make up one thickness the values given in this table are for the weakest layup, and the number of plies is indicated (by 3, 4, 5 etc).
In BS 5268 separate stresses are given for each layup.

| Thickness | | No. of plies | Area (mm²×10³) | Section modulus $Z_X$ (mm³×10³) | Second moment of area $I_X$ (mm⁴×10³) | Face grain parallel to span | | | | | Face grain perpendicular to span | | | | |
| --- | --- | --- | --- | --- | --- | --- | --- | --- | --- | --- | --- | --- | --- | --- | --- |
| nominal (mm) | minimum (mm) | | | | | $\sigma_{m,adm}$ (N/mm²) | $\bar{M}_{XL}$ (kN.m) | $\bar{M}_{XM}$ (kN.m) | $E$ (N/mm²) | $EI_X$ (kN.m²) | $\sigma_{m,adm}$ (N/mm²) | $\bar{M}_{XL}$ (kN.m) | $\bar{M}_{XM}$ (kN.m) | $E$ (N/mm²) | $EI_X$ (kN.m²) |
| 9.5 | 9.1 | 3 | 9.1 | 13.8 | 62.8 | 9.21 | 0.127 | 0.158 | 8700 | 0.546 | 7.37 | 0.101 | 0.127 | 3400 | 0.213 |
| 12.5 | 12.3 | 4, 5 | 12.3 | 25.2 | 155 | 7.70(4) | 0.194 | 0.242 | 7250(4) | 1.123 | 6.96(5) | 0.175 | 0.219 | 4400(5) | 0.682 |
| 16 | 15.5 | 5 | 15.5 | 40.0 | 310 | 7.82 | 0.312 | 0.391 | 6950 | 2.154 | 7.90 | 0.316 | 0.395 | 5400 | 1.674 |
| 19 | 18.6 | 5, 6 | 18.6 | 57.7 | 536 | 6.95(5) | 0.401 | 0.501 | 6200(5) | 3.323 | 8.62(6) | 0.497 | 0.621 | 6200(6) | 3.323 |

For panels less than 600 mm wide bending, tension and compression stresses should be reduced in proportion to their width, taking no reduction at 600 mm, and 50% reduction at 200 mm and less.

**Table 5.36   Swedish P30 Softwood plywood, unsanded and sanded**

Properties and capacities per metre width based on minimum thickness. Bending as in Fig. 4.27.
Where two or more layups make up one thickness the values given in this table are for the weakest layup, and the number of plies is indicated (by 3, 4, 5 etc).
In BS 5268 separate stresses are given for each layup.

| Thickness | | No. of plies | Area (mm² × 10³) | Section modulus $Z_X$ (mm³ × 10³) | Second moment of area $I_X$ (mm⁴ × 10³) | Face grain parallel to span | | | | | Face grain perpendicular to span | | | | |
| nominal (mm) | minimum (mm) | | | | | $\sigma_{m,adm}$ (N/mm²) | $\bar{M}_{XL}$ (kN.m) | $\bar{M}_{XM}$ (kN.m) | $E$ (N/mm²) | $EI_X$ (kN.m²) | $\sigma_{m,adm}$ (N/mm²) | $\bar{M}_{XL}$ (kN.m) | $\bar{M}_{XM}$ (kN.m) | $E$ (N/mm²) | $EI_X$ (kN.m²) |
|---|---|---|---|---|---|---|---|---|---|---|---|---|---|---|---|
| **Unsanded** | | | | | | | | | | | | | | | |
| 7.5 | 6.8 | 3 | 6.8 | 7.71 | 26.2 | 9.35 | 0.072 | 0.090 | 10050 | 0.263 | 2.88 | 0.022 | 0.027 | 800 | 0.020 |
| 9.5 | 8.7 | 3 | 8.7 | 12.6 | 54.9 | 9.42 | 0.118 | 0.148 | 10050 | 0.551 | 2.62 | 0.033 | 0.041 | 600 | 0.032 |
| 12.5 | 11.9 | 5 | 11.9 | 23.6 | 140 | 9.12 | 0.215 | 0.269 | 9300 | 1.302 | 4.38 | 0.103 | 0.129 | 2700 | 0.378 |
| 16 | 15.1 | 5 | 15.1 | 38.0 | 287 | 10.10 | 0.383 | 0.479 | 10650 | 3.056 | 4.39 | 0.166 | 0.208 | 2650 | 0.760 |
| 19 | 18.2 | 7 | 18.2 | 55.2 | 502 | 7.78 | 0.429 | 0.536 | 8100 | 4.066 | 4.41 | 0.243 | 0.304 | 3000 | 1.506 |
| 22 | 21.4 | 7, 9 | 21.4 | 76.3 | 817 | 7.94(7) | 0.605 | 0.757 | 8150(9) | 6.658(9) | 4.95(7) | 0.377 | 0.472 | 3600(7) | 2.941 |
| 25 | 24.7 | 9 | 24.7 | 102 | 1260 | 6.80 | 0.693 | 0.867 | 6900 | 8.694 | 6.07 | 0.619 | 0.773 | 5000 | 6.300 |
| **Sanded** | | | | | | | | | | | | | | | |
| 12.5 | 12.0 | 5 | 12.0 | 24.0 | 144 | 7.49 | 0.179 | 0.224 | 7600 | 1.094 | 5.26 | 0.126 | 0.157 | 3600 | 0.518 |
| 16 | 15.7 | 7 | 15.7 | 41.1 | 322 | 7.61 | 0.312 | 0.390 | 7700 | 2.479 | 5.48 | 0.225 | 0.281 | 4200 | 1.352 |
| 19 | 18.2 | 7 | 18.2 | 55.2 | 502 | 7.09 | 0.391 | 0.489 | 7400 | 3.714 | 4.88 | 0.269 | 0.336 | 3600 | 1.807 |
| 22 | 22.2 | 9 | 22.2 | 82.1 | 912 | 7.08 | 0.581 | 0.726 | 7150 | 6.520 | 5.53 | 0.454 | 0.567 | 4750 | 4.332 |
| 25 | 25.2 | 11 | 25.2 | 106 | 1330 | 6.64 | 0.703 | 0.879 | 6750 | 8.977 | 5.91 | 0.626 | 0.783 | 5200 | 6.916 |

**Chapter Six**

# Beams of Simple Composites

## 6.1 INTRODUCTION

As in Chapter 5 the method chosen to explain the design of simple composites is to give examples. Simple composites are defined as components of more than one member which do not qualify as glulam. The examples in this chapter are limited to components of two or three members.

Some simple composites such as Tee sections are not usually encountered as beams, although they can be used to advantage as compression members, particularly in wind girders, and are discussed here for completeness.

## 6.2 COMPOSITE ACTION

When the joint between two members in a simple composite beam is taking horizontal shear, as in the cases sketched in Figs 6.1–6.3, it is better structurally to surface the sections and make the joint a glue joint. In this way full

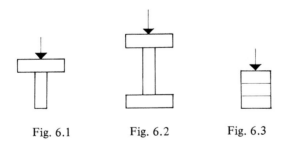

Fig. 6.1          Fig. 6.2          Fig. 6.3

composite action can be obtained with no slip between the members. If a nailed joint is substituted for a glue joint, there is no need to surface the sections therefore each section is slightly larger. However, full composite action will not be achieved even if the nails are spaced close together. This type of construction may suit a particular manufacturer, and in these cases a test is recommended to determine the strength and the stiffness (see Chapter 26 on prototype testing).

When a joint between two members in a simple composite is not taking horizontal shear, as in the cases sketched in Figs 6.4 and 6.5, a glue joint is not necessary and the sections need not be surfaced unless for appearance (although, from § 4.4.1 it can be seen that slightly higher stresses can be used with a glued composite because $K_{27}$ values are larger than $K_8$ values). A nailed or bolted joint is usually adequate to link the members. The joint is only being stressed if the load is not applied absolutely equally to each member or if the difference in stiffness between members tends to cause

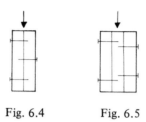

Fig. 6.4          Fig. 6.5

each to deflect to a different extent. With this type of member it is desirable to make individual pieces flush top so that the loading is applied equally. Regularizing of the depth is therefore advantageous.

## 6.3   TWO-MEMBER COMPOSITE BEAM, ON STRONG AXIS, NAILED TOGETHER

Check the design of the beam sketched in Fig. 6.6 supporting 4 kN long term UDL on a design span of 4.0 m. Assume lateral restraint adequate to allow full bending stress to develop. The beam is not part of a load-sharing system.

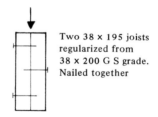

Two 38 x 195 joists
regularized from
38 x 200 G S grade.
Nailed together

Fig. 6.6

$$\text{Total load} = 4\,\text{kN}$$

$$M = \frac{4 \times 4}{8} = 2\,\text{kN.m}$$

From Table 5.14   $Z_X = 2 \times 0.241 \times 10^6 = 0.482 \times 10^6\,\text{mm}^3$

$$\sigma_{m,a,par} = \frac{2.0 \times 10^6}{0.482 \times 10^6} = 4.15\,\text{N/mm}^2$$

$\sigma_{m,adm,par}$ = grade bending stress (from Table 2.2) = 5.30
$\times K_8$ load-sharing factor (from Fig. 2.9) = 1.10
$\times K_7$ depth factor (from Table 4.4) = 1.049
= 6.10 N/mm²

The beam is not in a load-sharing system of four or more members therefore the value of $E$ to be used is the statistical minimum for two members acting together. See Fig. 2.3 for value of $K_9$ for two members.

$$E_2 = E_{min} \times K_9 = 6000 \times 1.14 = 6840 \, \text{N/mm}^2$$

$$I_X = 2 \, (23.5 \times 10^6) = 47.0 \times 10^6 \, \text{mm}^4$$

$$\therefore \, EI_X = 321 \, \text{kN.m}^2$$

Bending deflection $= \dfrac{5FL^3}{384EI} = \dfrac{5 \times 4 \times 4^3}{384 \times 321} = 0.0104 \, \text{m} = 10.4 \, \text{mm}$

With $L/h = 4000/195 = 20.5$, from Fig. 4.26 $\delta_t/\delta_m = 1.037$.

$\therefore$ Total deflection $= 10.4 \times 1.037 = 10.8 \, \text{mm}$
(including shear)

Permissible deflection $= 0.003 \times 4 \times 10^3 = 12 \, \text{mm}$

$$\tau_a = \frac{3F_v}{2 \times A} = \frac{3 \times 2 \times 10^3}{2 \times 2 \times 38 \times 195} = 0.202 \, \text{N/mm}^2$$

$$\tau_{adm} = \tau_g \times K_8 = 0.82 \times 1.10 = 0.902 \, \text{N/mm}^2$$

The section is adequate but bearing would have to be checked.


## 6.4  TWO-MEMBER COMPOSITE BEAM, ON STRONG AXIS, GLUED TOGETHER

If the two pieces of the beam shown in Fig. 6.6 are surfaced to 35 × 194 mm (Table 1.9) and glued together, one effect is that the actual stresses and deflection will be increased by approximately 9%. The only justification for this extra work coupled with loss of performance could be an improvement in appearance.


## 6.5  THREE-MEMBER COMPOSITE BEAM, HORIZONTAL MEMBERS, GLUED TOGETHER

Calculate the stresses and deflection of the beam sketched in Fig. 6.7 supporting a 2 kN long term UDL on a design span of 4.0 m. Assume lateral restraint adequate to allow full bending stress to develop. The beam is not part of a load-sharing system. As this beam is made from three members

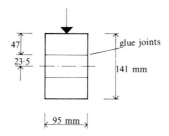

Fig. 6.7

rather than four it cannot be classified as glulam, and therefore has to be calculated by modifying the stresses for solid timber. Assume SS grade Whitewood members, and bonding pressure during manufacture to be by clamps rather than nailing. Assume that the individual members are planed to 47 × 97 from 50 × 100 mm then glued together and the width surfaced to 95 mm. If any member is jointed in length this should be with a glued joint (see Chapter 19) of at least the required strength.

$$M = \frac{2 \times 4}{8} = 1.0\,\text{kN.m}$$

$$Z_X = \frac{95 \times 141^2}{6} = 315 \times 10^3\,\text{mm}^3$$

$$I_X = 22.2 \times 10^6\,\text{mm}^4$$

$$\sigma_{m,a,par} = \frac{1.0 \times 10^6}{315 \times 10^3} = 3.17\,\text{N/mm}^2$$

The permissible bending stress is the product of the long term grade stress for SS Whitewood and the designer's interpretation of the extent to which the load-sharing factor $K_8$ (or even $K_{27}$–see Fig. 2.9) and the depth factor $K_7$ (see Fig. 4.4) are applicable to this case. Because only one member is stressed at the extreme fibres (in bending tension and bending compression) the authors are inclined to disregard $K_8$, but regard use of $K_7$ (of 1.086) to be justified.

$$\therefore\ \sigma_{m,adm,par} = 7.50 \times 1.086 = 8.15\,\text{N/mm}^2$$

When considering bearing stresses, because bearing is on one member, the authors are inclined to disregard $K_8$. This agrees with the philosophy of how $K_{18}$ and $K_{24}$ are used with horizontally glued laminated members.

The beam is not part of a load-sharing system of four or more members therefore the value of $E$ to be used is that appropriate for three members acting together (see Fig. 2.3). The value of $K_9$ to take for three members acting together is 1.21 therefore the $E$ value for these three SS Whitewood members is:

$$E_{min} \times K_9 = 7000 \times 1.21 = 8470\,\text{N/mm}^2$$

Certain designers might argue that, because the centre laminate contributes so little to the stiffness of the beam, the $E$ value appropriate to two members instead of three should be taken. However, to take three, fits in with the general philosophy in BS 5268 for glulam beams with unequal numbers of laminates.

$$EI_X = 8.47 \times 22.2 = 188 \, \text{kN.m}^2$$

$$\text{Bending deflection} = \frac{5 \times 2 \times 4^3}{384 \times 188} = 0.0089 \, \text{m} = 8.9 \, \text{mm}$$

$L/h = 4000/141 = 28.4$ therefore, from §4.15.8, $\delta_t/\delta_m = 1.019$.

$$\therefore \text{ Total deflection} = 8.9 \times 1.019 = 9.1 \, \text{mm}$$

$$\text{Permissible deflection} = 0.003 \times 4000 = 12 \, \text{mm}$$

Maximum shear stress parallel to grain at neutral axis is:

$$\tau_a = \frac{3F_v}{2A} = \frac{3 \times 1 \times 10^3}{2 \times 95 \times 141} = 0.112 \, \text{N/mm}^2$$

The permissible shear stress is the product of the long term grade stress for SS Whitewood and the designer's interpretation of the extent to which the load-sharing factor $K_8$ is applicable to this case. Because only one member is stressed at the level of the neutral axis the authors are inclined to disregard $K_8$ in this case.

$$\tau_{adm} = 0.82 \, \text{N/mm}^2$$

The stress along the glue lines is away from the neutral axis and is therefore less than $0.82 \, \text{N/mm}^2$. From §4.11.1 it can be seen that the stress on the glue lines is:

$$= \frac{1 \times 10^3 \times 47 \times 95 \times 47}{22.2 \times 10^6 \times 95} = 0.10 \, \text{N/mm}^2$$

The permissible stress at the glue line is the shear parallel to grain stress in the timber without any reduction for stress concentrations.

## 6.6   TEE-SHAPED COMPOSITE BEAM, GLUED

Tee sections from two solid sections (usually equal) are frequently used as compression members, particularly in wind girder systems in industrial applications. Their value lies not only in the improved axial compression capacity compared to a rectangular section of a similar area, but also in simplifying the end jointing to trusses etc. Although not often used purely as a beam, the design method is given here for completeness.

Fig. 6.8

It is usual to use two pieces of the same cross-section to form the composite Tee profile, so that a section has the general dimensions shown in Figs 6.8 and 6.9. Frequently, the vertical solid element is curtailed at a point of support to simplify fixings through the wide face of the horizontal element (see Fig. 6.8). If this detail is adopted, the shear capacity should be limited to that of a solid of dimensions $t \times h$ and the vertical component should be mitred to reduce stress concentrations.

Although the proportions of stiffness are not contributed equally by each element of the Tee section (i.e. with thinner sections the vertical element contributes about 55% to the stiffness of the section while for thicker sections the vertical element contributes about 65%) it is customary to modify the minimum value of $E$ to that appropriate to two members acting together. Composite action is achieved by a glue line between the elements which is usually assembled by glue/nailing therefore permissible stresses for the glue line must be reduced by $K_{70} = 0.9$ (§4.14.1).

Figure 6.9 gives properties for a selected range of Tee sections, using 38 mm and 50 mm basic thicknesses of timber surfaced on both faces prior to assembly into the composite unit.

### 6.6.1   Example of design of a Tee-shaped beam

Calculate the stresses and deflection of a $_{47}T_{97}$ section (Fig. 6.9) carrying 2.4 kN total UDL on a span of 3 m about its $XX$ axis. The shape is such as to ensure that the full compressive bending stress is realised before any lateral instability takes place if the table of the Tee is on the compression side (as in this case). For cases where the stalk of the Tee is on the compression side, a conservative assessment for lateral stability is to ensure that the proportion of the stalk member as a separate entity complies with the requirements of §4.6.1. If there is a joint in the length of either piece it should be a glued joint of at least the required strength.

The beam is not part of a load-sharing system.

$$\text{Maximum } M = \frac{2.4 \times 3}{8} = 0.90 \, \text{kN.m}^2$$

$$\sigma_{m,a,par} = \frac{0.90 \times 10^6}{0.192 \times 10^6} = 4.69 \, \text{N/mm}^2$$

$$\text{Maximum shear} = F_v = 1.2 \, \text{kN}$$

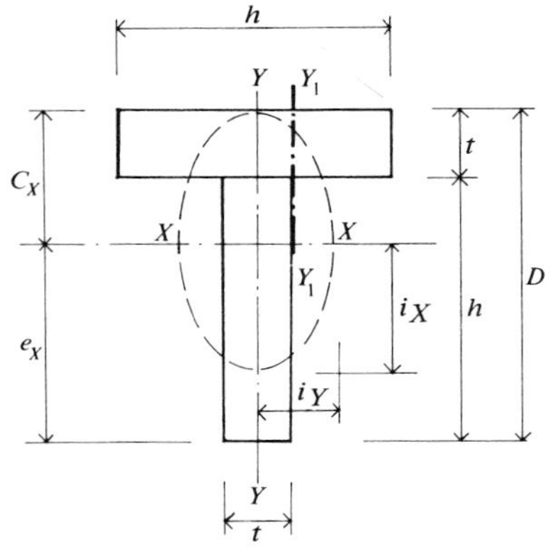

$$A = 2ht$$
$$C_X = 0.75t + 0.25h$$
$$e_X = 0.25t + 0.75h$$
$$I_Y = \frac{ht^3}{12} + \frac{th^3}{12}$$
$$I_X = \frac{ht(t+h)^2}{8} + I_Y$$
$$i_X = \sqrt{(I_X/A)}$$
$$i_Y = \sqrt{(I_Y/A)}$$
$$S_X = 0.5e_X^2\, t$$
$$S_{Y1} = \frac{t(h-t)^2}{8} \text{ at axis } Y_1Y_1$$

| Section | D (mm) | h (mm) | t (mm) | Centre of gravity $C_X$ (mm) | $e_X$ (mm) | Area $A$ (mm²×10³) | First moment of area $S_X$ (mm³×10⁶) | $S_{Y1}$ (mm²×10⁶) | Second moment of area $I_X$ (mm⁴×10⁶) | $I_Y$ (mm⁴×10⁶) | Section modulus $Z_X$ max (mm³×10⁶) | min (mm³×10⁶) | $Z_Y$ (mm³×10⁶) | Radius of gyration $i_X$ (mm) | $i_Y$ (mm) |
|---|---|---|---|---|---|---|---|---|---|---|---|---|---|---|---|
| $_{35}T_{97}$ | 132 | 97 | 35 | 50.50 | 81.50 | 6.79 | 0.116 | 0.0168 | 10.4 | 3.00 | 0.206 | 0.128 | 0.062 | 39.1 | 21.0 |
| $_{35}T_{122}$ | 157 | 122 | 35 | 56.75 | 100.25 | 8.54 | 0.176 | 0.0331 | 18.9 | 5.73 | 0.333 | 0.188 | 0.094 | 47.0 | 25.9 |
| $_{35}T_{147}$ | 182 | 147 | 35 | 63.00 | 119.00 | 10.3 | 0.248 | 0.0549 | 31.1 | 9.79 | 0.494 | 0.261 | 0.133 | 55.0 | 30.8 |
| $_{35}T_{170}$ | 205 | 170 | 35 | 68.75 | 136.00 | 11.9 | 0.325 | 0.0797 | 46.2 | 14.9 | 0.672 | 0.339 | 0.176 | 62.3 | 35.4 |
| $_{35}T_{195}$ | 230 | 195 | 35 | 75.00 | 155.00 | 13.7 | 0.420 | 0.112 | 67.5 | 22.3 | 0.899 | 0.435 | 0.229 | 70.3 | 40.4 |
| $_{35}T_{220}$ | 255 | 220 | 35 | 81.30 | 174.00 | 15.4 | 0.528 | 0.150 | 94.4 | 31.8 | 1.16 | 0.543 | 0.289 | 78.3 | 45.5 |
| $_{47}T_{97}$ | 144 | 97 | 47 | 59.50 | 84.50 | 9.12 | 0.168 | 0.0147 | 16.2 | 4.41 | 0.273 | 0.192 | 0.091 | 42.2 | 22.0 |
| $_{47}T_{122}$ | 169 | 122 | 47 | 65.75 | 103.25 | 11.5 | 0.250 | 0.0330 | 28.6 | 8.17 | 0.436 | 0.277 | 0.134 | 50.0 | 26.7 |
| $_{47}T_{147}$ | 194 | 147 | 47 | 72.00 | 122.00 | 13.8 | 0.350 | 0.0587 | 46.6 | 14.1 | 0.648 | 0.382 | 0.192 | 58.1 | 32.0 |
| $_{47}T_{170}$ | 217 | 170 | 47 | 77.80 | 139.00 | 16.0 | 0.456 | 0.0888 | 67.7 | 20.7 | 0.871 | 0.486 | 0.244 | 65.1 | 36.0 |
| $_{47}T_{195}$ | 242 | 195 | 47 | 84.00 | 158.00 | 18.3 | 0.587 | 0.129 | 97.8 | 30.7 | 1.16 | 0.619 | 0.315 | 73.1 | 40.9 |
| $_{47}T_{220}$ | 267 | 220 | 47 | 90.30 | 177.00 | 20.7 | 0.734 | 0.176 | 136.0 | 43.6 | 1.50 | 0.768 | 0.396 | 81.0 | 45.9 |

Fig. 6.9: *Geometrical properties of Tee sections.*

Maximum shear stress occurs on axis $XX$.

From § 4.11.1:

$$\tau_a = \frac{F_v A_u \bar{y}}{I_X t} = \frac{F_v S_X}{I_X t}$$

$$= \frac{1200 \times (0.168 \times 10^6)}{(16.2 \times 10^6) \times 47} = 0.265 \text{ N/mm}^2$$

Because $C_X$ is fairly close to the value of $t$, the value of $\tau_a$ above may be regarded as a conservative estimate of the stress at the glue line. The bending and shear stresses can be satisfied using GS grade Whitewood. In the opinion of the authors a value of 1.10 for $K_8$ should not be used in deriving the permissible stresses, because at the levels for which stress is computed there is no load sharing. However, in calculating deflection, because the members have similar stiffness it seems reasonable to use $E_2$ (see § 6.6) using a $K_9$ factor of 1.14 (see Fig. 2.3). It seems reasonable to take a conservative value for $K_7$ based on the depth of the stalk (97 mm). Hence $K_7$ is 1.132 (Table 4.4).

$$E_2 I = 6000 \times 1.14 \times 16.2 \times 10^6$$

$$= 111 \times 10^9 \text{ N.mm}^2 = 111 \text{ kN.m}^2$$

$$\text{Bending deflection} = \frac{5 \times 2.4 \times 3^3}{384 \times 111} = 0.0076 \text{ m} = 7.6 \text{ mm}$$

From § 4.15.7, for a simply supported span:

$$\text{Shear deflection} = \frac{K_{\text{form}} M_0}{AG}$$

For a Tee section the form factor $K_{\text{form}}$ may be taken as:

$$K_{\text{form}} = \frac{\text{total cross-sectional area}}{\text{cross-sectional area of web}}$$

For simplicity, assume that the area of the stalk of the Tee is the web area, therefore $K_{\text{form}} = 2.0$ for all the Tee sections shown in Fig. 6.9. $G = E/16 = 375 \text{ N/mm}^2$. Therefore:

$$\text{Shear deflection} = \frac{2 \times 0.9 \times 10^6}{(9.12 \times 10^3) \times 375} = 0.53 \text{ mm}$$

Therefore total deflection = 8.13 mm, compared to a permissible deflection of 0.003 × 3 m = 9 mm.

## 6.7 I-SHAPED COMPOSITE BEAM, GLUED

Beams of I shapes of three pieces of solid timber are not common but can be useful in certain cases particularly if plywood is not available. If there is a joint in the length in any piece it should be a glued joint of at least the required strength.

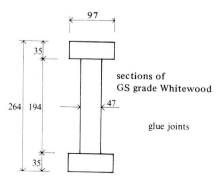

sections of
GS grade Whitewood

glue joints

Fig. 6.10

Calculate the stresses and deflection of the I beam sketched in Fig. 6.10 supporting 8 kN long term loading on a design span of 4.0 m. The beam is not part of a load-sharing system but is restrained laterally.

$$I_X = \frac{97 \times 264^3}{12} - \frac{50 \times 194^3}{12} = 118.3 \times 10^6 \, \text{mm}^4$$

$$Z_X = \frac{118.3 \times 10^6}{132} = 0.895 \times 10^6 \, \text{mm}^3$$

$$M = \frac{8 \times 4}{8} = 4 \, \text{kN.m}$$

$$\sigma_{m,a,par} = \frac{4 \times 10^6}{0.895 \times 10^6} = 4.47 \, \text{N/mm}^2$$

The long term permissible bending stress is the product of the grade stress and the designer's interpretation of the extent to which the load-sharing factor $K_8$ and the depth factor $K_7$ are applicable to this section. Because only one member is stressed at the levels for which bending (and shear) stresses are calculated the authors are inclined to disregard $K_8$ in this case. Regarding depth factor $K_7$, on balance, they are inclined to adopt it for the full depth of 264 mm for which depth it has the value 1.014 (see Table 4.4). (Particularly for deeper beams a case could be made for using a $K_7$ value related to the thickness of the 'flange'.)

$$\sigma_{m,adm,par} = 5.30 \times 1.014 = 5.37 \, \text{N/mm}^2$$

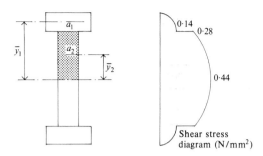

Fig. 6.11

The distribution of shear stress (see Fig. 6.11) is as follows:

Shear stress at neutral axis:

$$= \frac{F_v(a_1\bar{y}_1 + a_2\bar{y}_2)}{Ib}$$

$$= \frac{4.0 \times 10^3 [(35 \times 97 \times 114.5) + (47 \times 97 \times 48.5)]}{118.3 \times 10^6 \times 47}$$

$$= 0.44 \text{ N/mm}^2$$

Shear stress at glue line $= \dfrac{4.0 \times 10^3 (35 \times 97 \times 114.5)}{118.3 \times 10^6 \times 97} = 0.14 \text{ N/mm}^2$
($b = 97$ mm)

Shear stress at glue line $= 0.14 \times 97/47 = 0.28 \text{ N/mm}^2$
($b = 47$ mm)

The beam is not part of a load-sharing system therefore the value of $E$ (and $K_9$) to be used is the one appropriate for three members acting together (see Fig. 2.3). Although the web member is on the strong axis and the top and bottom members are on their weak axis, all three members contribute approximately the same amount of stiffness and justify the use of $E_3$ rather than $E_2$ in the deflection calculations.

For GS grade $E_3$ (from Table 2.2) is $6000 \times 1.21 = 7260 \text{ N/mm}^2$.

$$EI_X = 859 \text{ kN.m}$$

$$\text{Bending deflection} = \frac{5 \times 8 \times 4^3}{384 \times 859} = 0.0078 \text{ m} = 7.8 \text{ mm}$$

From §4.15.7, for a simply supported span:

$$\text{Shear deflection} = \frac{K_{\text{form}} M_0}{AG}$$

For a solid I section the form factor $K_{\text{form}}$ may be taken as:

$$K_{\text{form}} = \frac{\text{total cross-sectional area}}{\text{cross-sectional area of web}}$$

In this case:

$$K_{\text{form}} = \frac{15\,908}{9118} = 1.74$$

$$G = \frac{E}{16} = 375\,\text{N/mm}^2$$

$$\text{Shear deflection} = \frac{1.74 \times 4 \times 10^6}{15\,908 \times 375} = 1.17\,\text{mm}$$

Therefore the total deflection is 8.97 mm compared to a permissible deflection of 0.003 × 4 m = 12 mm.

For a solid I section, shear deflection may be simplified to:

$$\text{Shear deflection} = \frac{M_0}{AG}$$

where $A$ = the area of the web.

Hence:

$$\text{Shear deflection} = \frac{4 \times 10^6}{47 \times 194 \times 375} = 1.17\,\text{mm as above}$$

# Chapter Seven
# Glulam Beams

## 7.1 INTRODUCTION

A horizontally laminated glulam section is one manufactured by gluing together at least four laminations with their grain essentially parallel. The laminations are usually machined from timber of 38–50 mm basic thicknesses although thinner laminations are used to suit certain curved details. Clause 47.1 of BS 5268:Part 2 limits the thickness of laminates to 50 mm, although BS 4169 gives a limit of 47 mm. Manufacture must be in accordance with BS 4169 which requires laminations to be dried to within 3% of the average moisture content in service or between 12–15% moisture content before being machined and assembled. When radio-frequency curing of the adhesive is used the lower moisture content applies. The reduction in thickness when machining laminations is usually 3–5 mm. Assembly is normally carried out by tightening clamps at right angles to the glue lines, and holding the pressure until curing of the adhesive is complete.

The section is machined after assembly to remove any deviations caused by individual laminations slipping sideways during the clamping (Fig. 7.1) and to obtain a machined surface.

It is easy to build in a camber at the time of assembly, although it should be realised that it is normal for the standard glulam sections which one can obtain from stockists to be manufactured without a camber.

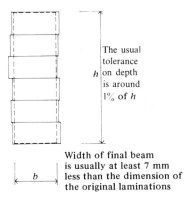

The usual tolerance on depth is around 1% of $h$

Width of final beam is usually at least 7 mm less than the dimension of the original laminations

Fig. 7.1

Glulam sections are built-up from thin members, and it is therefore possible to manufacture complicated shapes, curving laminates within limits. The designs in this chapter (indeed in this manual) are limited to beams with parallel, mono-pitch or duo-pitch profiles. The method of building up laminations for beams with sloping tops is indicated in Fig. 7.2. Although the grain of the laminations which occur near the sloping surface is not parallel to this surface, it is permissible to consider these laminations as full strength in the design.

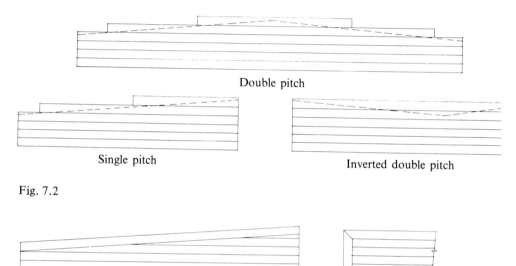

Double pitch

Single pitch

Inverted double pitch

Fig. 7.2

Fig. 7.3

When it is necessary, for appearance, to fit a lamination parallel to the sloping surface or the end of the beam (Fig. 7.3), provided it is correctly glued into place, it may be considered to add to the strength of the beam.

Under no circumstances should two prefabricated part-sections which have been manufactured separately be glued together as sketched in Fig. 7.4 and treated as a fully composite beam. Each part is almost certain to have too much inertia to be held by one glue line once the clamps are removed. It has been proved by experience, however, that a section which, for example, is too deep for the sanding machine, may be manufactured initially with a 'dry joint', the two pieces being separated after the glue has been

Fig. 7.4

cured; then each piece is sanded (not on the dry joint), glued, re-assembled and clamped. When feasible, the method of adding to a glulam section should, however, be by one laminate at a time, because great care and manufacturing expertise are required to obtain a successful member by the method explained above.

## 7.2    TIMBER STRESS GRADES FOR GLULAM

BS 4978 describes three stress grades for softwood for particular use with horizontally laminated glulam members. They are denoted LA, LB and LC. For vertically laminated glued members the stress grades for solid timber (e.g. GS, SS etc. to BS 4978) are those to be used in design and manufacture. For horizontally and vertically laminated members of hardwood, the HS grade from BS 5756 is to be used in design and manufacture. Only softwood is covered in this manual.

Grade LA is the strongest of the grades, but LC is not a particularly weak grade. It is possible to manufacture a glulam section using the same grade throughout or to mix the grades LA with LB, or LB with LC, as indicated in Fig. 7.5. No allowance is made in BS 5268 for a mixture of LA and LC.

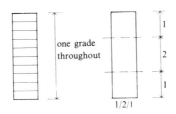

The higher grade must occupy at least the outer zones
(i.e. 25% each of depth)

Fig. 7.5

LB and LC grades are the more common in Redwood or Whitewood, because the slope of grain restriction of LA is rather exacting and can lead to a high reject rate. Also it is practically impossible to provide end joints in laminates of sufficient strength to develop the full strength of LA grade (see §7.5). Therefore strength properties are not given in this chapter for beams using LA grade.

At present LA, LB and LC grades are only visually graded and can not be graded by machine even though, in the Nordic area, machine stress grading of laminations is common.

## 7.3 PERMISSIBLE STRESSES FOR HORIZONTALLY OR VERTICALLY LAMINATED BEAMS

### 7.3.1 Horizontally laminated beams

The grades LA, LB and LC are intended to be used when designing members with horizontal laminations as shown in Fig. 7.6. With horizontal laminations the method of arriving at the permissible stresses is to multiply the relevant SS grade stress of the timber species being used by modifications factors $K_{15}$–$K_{19}$ or $K_{21}$–$K_{25}$ according to the nature of the stress, the number of laminations, and the grade or grade mix (see Tables 7.1 and 7.2).

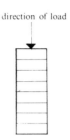

direction of load

Fig. 7.6

For beams the grade bending stress must be multiplied by depth factor $K_7$ as for solid timber (see Fig. 4.4 and Table 4.4).

Load-duration factor $K_3$ applies.

Wane is not permitted with glulam therefore the permissible stress for compression perpendicular to the grain is the SS grade stress for the species multiplied by 1.33 ($K_{18}$ or $K_{24}$) and the load-duration factor $K_3$.

The $E$ value to be taken with beams of four or more laminations is the $E_{\text{mean}}$ value for SS grade of the species multiplied either by $K_{20}$ or $K_{26}$ (see Tables 7.1 and 7.2). It seems strange, particularly now that BS 5268: Part 2 includes values of $E_2$, $E_3$ etc. for solid timber (i.e. $K_9$ and $K_{27}$), that BS 5268 gives values of $K_{20}$ or $K_{26}$ which do not vary with the number of laminations.

Glulam beams are rarely part of a load-sharing system but, even when they are, the modified value of $E_{\text{mean}}$ should be used rather than $E_{\text{mean}}$ on its own.

The values of modification factors $K_{15}$–$K_{20}$ as presented in BS 5268 for glulam of one grade are given in Table 7.1, and the corresponding factors $K_{21}$–$K_{26}$ for glulam of mixed grades in Table 7.2.

Care must be taken to provide end joints of adequate strength along each laminate (see § 7.5).

The additional modification factors listed on page 173 apply to glulam beams of horizontal laminations:

**Table 7.1   Modification factors $K_{15}$–$K_{20}$ for single-grade glued laminated members and horizontally glued laminated beams**

| | | Values of the modification factors | | | | | |
|---|---|---|---|---|---|---|---|
| Grade | No. of laminations | $K_{15}*$ (bending parallel to grain) | $K_{16}*$ (tension parallel to grain) | $K_{17}*$ (compression parallel to grain) | $K_{18}*†$ (compression perpendicular to grain) | $K_{19}*$ (shear parallel to grain) | $K_{20}‡$ (modulus of elasticity) |
| LA | 4 or more | 1.85 | 1.85 | 1.15 | 1.33 | 2.00 | 1.00 |
| LB | 4 | 1.26 | 1.26 | 1.04 | 1.33 | 2.00 | 0.90 |
| | 5 | 1.34 | 1.34 | | | | |
| | 7 | 1.39 | 1.39 | | | | |
| | 10 | 1.43 | 1.43 | | | | |
| | 15 | 1.48 | 1.48 | | | | |
| | 20 or more | 1.52 | 1.52 | | | | |
| LC | 4 | 0.74 | 0.74 | 0.92 | 1.33 | 2.00 | 0.80 |
| | 5 | 0.82 | 0.82 | | | | |
| | 7 | 0.91 | 0.91 | | | | |
| | 10 | 0.98 | 0.98 | | | | |
| | 15 | 1.05 | 1.05 | | | | |
| | 20 or more | 1.11 | 1.11 | | | | |

Interpolation for intermediate number of laminations is permitted.
\* Applied to the corresponding grade stress for SS grade of the species.
† Assumes no wane.
‡ Applied to the SS grade mean modulus of elasticity (and the shear modulus) of the species.

**Table 7.2   Modification factors $K_{21}$–$K_{26}$ for combined-grade glued laminated members and horizontally glued laminated beams**

| | | Values of the modification factors | | | | | |
|---|---|---|---|---|---|---|---|
| Combination** of laminating grades | No. of laminations | $K_{21}*$ (bending parallel to grain) | $K_{22}*$ (tension parallel to grain) | $K_{23}*$ (compression parallel to grain) | $K_{24}*†$ (compression perpendicular to grain) | $K_{25}*$ (shear parallel to grain) | $K_{26}‡$ (modulus of elasticity) |
| LA/LB/LA | 4 or more | 1.76 | 1.76 | 1.09 | 1.33 | 2.00 | 0.93 |
| LB/LC/LB | 4 | 1.26 | 1.26 | 0.97 | 1.33 | 2.00 | 0.85 |
| | 8 | 1.32 | 1.32 | | | | |
| | 12 | 1.37 | 1.37 | | | | |
| | 16 | 1.41 | 1.41 | | | | |
| | 20 or more | 1.45 | 1.45 | | | | |

Interpolation for intermediate number of laminations is permitted.
 \* Applied to the corresponding grade stress for SS grade of the species.
 † Assumes no wane.
 ‡ Applied to the SS grade mean modulus of elasticity (and the shear modulus) of the species.
 ** Not less than the outer quarter of depth, top and bottom, must consist of the superior grade (see Fig. 7.5).

$K_4$    factor for bearing stresses
$K_5$    factor for beams notched at the ends (see §4.12)
$K_{33}, K_{34}$ and $K_{35}$    factors for curved laminated members.

The lateral support and the maximum depth-to-breadth ratio must either comply with the figures tabulated in §4.6.1, or one of the special design methods given in Chapter 9 must be used to check the beam.

### 7.3.2    Vertically laminated beams

Even with glued beams formed from four or more vertical laminations as sketched in Fig. 7.7 the special glulam grades LA, LB and LC do *not* apply.

direction of load

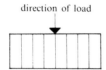

Fig. 7.7

The grades which apply are the ordinary stress grades for solid timber from BS 4978. When a designer is considering a beam with four or more vertical laminations, it is often because a horizontally laminated member with bending about the $XX$ axis is also being called upon to take bending on its $YY$ axis. If the beam is made from horizontal laminations of LB grade it is normally safe to assume that these can be taken as SS grade, as vertical laminates. Likewise LC grade (or a mixture of LB or LC) may be considered to be GS grade, as vertical laminates. Alternatively of course, the grading can be checked against the GS, SS etc. rules.

The permissible stresses for bending and shear parallel to the grain are calculated by multiplying the *grade* stress for the assumed stress grade (GS, SS etc.) by the relevant factor ($K_{27}$) from BS 5268 (see Table 7.3) for the number of laminations. Note that the number $N$ starts at two (as for $K_8$), and not at four as for horizontally laminated members.

The $E$ value is calculated by multiplying the minimum $E$ value for the stress grade by factor $K_{28}$ (see Table 7.3). Note that the number starts at two (as for $K_9$).

Load sharing applies to bearing stresses therefore $K_{29}$ (see Table 7.3) is 1.10 (therefore $K_8$ should not be applied as well as $K_{29}$).

The additional factors listed below apply with beams of vertical laminations:

$K_3$    load-duration factor
$K_4$    factor for bearing stresses
$K_5$    factor for beams notched at ends (also see §4.12)
$K_7$    depth factor.

**Table 7.3** Modification factors $K_{27}$–$K_{29}$ for vertically glued laminated members

| Number of laminations | Values of modification factors | | |
|---|---|---|---|
| | $K_{27}$* (bending, tension and shear parallel to grain) | $K_{28}$ (modulus of elasticity†) (and compression parallel to grain*) | $K_{29}$*‡ (compression perpendicular to grain) |
| 2 | 1.11 | 1.14 | |
| 3 | 1.16 | 1.21 | |
| 4 | 1.19 | 1.24 | |
| 5 | 1.21 | 1.27 | |
| 6 | 1.23 | 1.29 | 1.10 |
| 7 | 1.24 | 1.30 | |
| 8 or more | 1.25 | 1.32 | |

* Applied to the corresponding grade stress of the stress grade of the species.
† For $E$ value is applied to the minimum modulus of elasticity of the stress grade of the species.
‡ If no wane is present, $K_{29}$ is shown in BS 5268: Part 2: 1984 as being 1.33, however it should probably be 1.33 × 1.10 for load sharing. Regardless of the stress grade used, it is applied to the SS stress grade for the species.

Lateral restraint is unlikely to be a problem but should be checked, remembering that any beam tends to buckle only at right angles to the direction of bending.

### 7.3.3 Glulam beams with bending about both axes

With bending about both axes the distribution of bending stresses is as shown in Fig. 7.8.

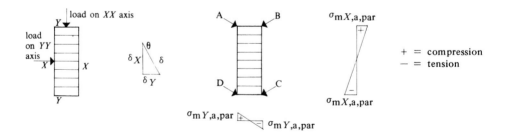

Fig. 7.8

At point A the two compression bending stress ratios are additive.
At point C the two tensile bending stress ratios are additive.
At points B and D the two stress ratios are combined algebraically.

$$\text{The permissible bending stress about } XX = \sigma_{mX,adm,par}$$
$$\text{The permissible bending stress about } YY = \sigma_{mY,adm,par}$$

At point A the bending stresses are combined thus:

$$\frac{\sigma_{mX,a,par}}{\sigma_{mX,adm,par}} + \frac{\sigma_{mY,a,par}}{\sigma_{mY,adm,par}} \leqslant 1.0$$

The deflection of the beam $\delta = \sqrt{(\delta_X^2 + \delta_Y^2)}$ and will take place at an angle $\theta$ to the direction of loading on the $XX$ axis where $\tan \theta = \delta_Y/\delta_X$.

## 7.4   APPEARANCE GRADES FOR GLULAM MEMBERS

AMD 3453 to BS 4169 replaced the three previous appearance classifications, Economy, Industrial, Architectural, by:

Regularized
Planed
Sanded.

The term 'Regularized' has a different definition, although a similar intention, to the same word used in BS 4471 in relation to solid timber. The definition of the classifications is:

**7.2.1** *Regularized.* Not less than 50% of the surface sawn or planed to remove the protruding laminations. Surface defects not made good or filled.
   This classification is suitable for use in industrial buildings or in similar utilitarian situations, where appearance is not of prime importance. Finishing treatment, when specified, would usually be of the opaque or pigmented type.

**7.2.2** *Planed.* Fully planed surface free from glue stains. Significant knot holes, fissures, skips in planing, voids and similar defects on exposed surfaces shall be filled or made good with glued inserts.
   This classification is suitable for most applications other than where varnish or similar non-reflective finish is specified.

**7.2.3** *Sanded.* Exposed surfaces fully planed, with knot holes, fissures, voids and similar defects filled or made good with glued inserts, and sanded. Normal secondary sanding marks are acceptable. Outside laminations shall be selected with reasonable care to match grain and colour at end joints where practicable, and shall be free from loose knots and open knot holes. Reasonable care shall be exercised in matching the direction of grain and colour of glued inserts.
   This classification is recommended for use where appearance is a prime consideration and where it is desired to apply a varnish or similar finish.

Although these classifications make reference to finishing, the manufacturer will not varnish etc. or provide protective covering unless such requirements are added to the specification.

Laminations will contain the natural characteristics of the species within the grading limits for the individual laminations, including fissures on the edges and faces of laminations.

Fig. 7.9

The corners of members may be 'pencil rounded' (Fig. 7.9) to avoid chipped edges.

The limiting sizes of permitted characteristics and defects for individual laminations are given in BS 4978. However, no limits are stated for fissures which may appear on the sides of completed glulam members. From an appearance point of view this can be a matter for individual specification or discussion. From a structural aspect, such fissures are unlikely to be a worry and the effect on the shear parallel to grain stress can be calculated by normal formulae, substituting $b_{net}$ for $b$ in the formula $\tau = F_v S_X / I_X b$ (see Fig. 7.10).

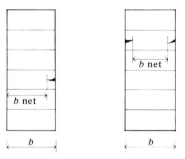

Fig. 7.10

It can be seen that the $K_{19}$ and $K_{25}$ factors of 2.00, applied to SS grade shear stress (with which 50% fissures are allowed) imply no fissures in the glulam members.

## 7.5  JOINTS IN LAMINATIONS

### 7.5.1  Joints in horizontal laminations

With structural glulam the end jointing of individual laminations is carried out almost certainly by finger jointing or scarf joints. If a butt joint is used, then the laminate in which it occurs has to be disregarded in the stress calculations, and this usually makes the member uneconomical. Finger joints and scarf joints may be considered not to reduce the $EI$ value of the member.

Finger joints can be used on either axis of laminates and have equal strength whichever way they are cut.

There are two ways of establishing if the strength of an end joint in a lamination is adequate for its function. The simple way is to provide a joint with an efficiency in bending which compares with the efficiency in bending of the laminate. The second way is to calculate the strength required of the joint in bending, tension or compression (or combinations of bending and tension, or bending and compression), and to provide a joint of adequate strength.

When a horizontally laminated member is made from LB or LC laminates, providing the end joint has an efficiency in bending of at least:

70% for LB laminates, or
55% for LC laminatates,

BS 5268:Part 2 (Clause 20) permits the joint to be used without a further design check.

The concept of a joint efficiency dates from the idea of stress grades 40, 50, 65 and 75 as described initially in CP 112:1967, and to the idea of a 'basic' stress. Now that 'basic' stress is omitted from BS 5268:Part 2 there is less relevance in referring to the efficiency of a joint. When a finger joint is said to have an efficiency in bending of 70% this refers to the efficiency compared to defect-free timber (i.e. the 'old' basic stress) but, because no value is given in BS 5268:Part 2 for this, if the designer has to check the strength of a finger joint, another method must be used. The method given in BS 5268 is to relate to the SS grade of the same species as laminates LB or LC. This is discussed below.

Guidance on the efficiency in bending, tension and compression of various finger joints is given in BS 5291 and in BS 5268:Part 2 (Appendix F). Table 7.4 of this manual gives an extract for those finger joints most likely

**Table 7.4**

| Finger profile (mm) | | | %<br>Efficiency rating<br>in bending<br>and tension | %<br>Efficiency rating<br>in compression |
|---|---|---|---|---|
| length $l$ | pitch $p$ | tip width $t$ | | |
| 50 | 12.0 | 2.0 | 75 | 83 |
| 32 | 6.2 | 0.5 | 75 | 92 |
| 20 | 6.2 | 1.0 | 65 | 84 |
| 15 | 3.8 | 0.5 | 75 | 87 |
| 12.5 | 4.0 | 0.7 | 65 | 82 |
| 12.5 | 3.0 | 0.5 | 65 | 83 |
| 10 | 3.7 | 0.6 | 65 | 84 |
| 10 | 3.8 | 0.6 | 65 | 84 |
| 7.5 | 2.5 | 0.2 | 65 | 92 |

to be used. See Fig. 19.3 for a diagram of the two types of finger joints which are used structurally.

Note that, although the efficiency in tension is the same as in bending, the efficiency in compression is higher. The efficiencies in bending and tension have been established by test. Tests in compression tend to indicate an efficiency of 100% but, rather than quote 100%, the following formula is used:

$$\frac{p - t}{p} \times 100\%$$

See Fig. 19.3 for an explanation of $p$ and $t$.

Guidance on the efficiency of a scarf joint used in individual laminations of a glued laminated softwood member is given in BS 5268:Part 2 (Appendix F). These values are given in Fig. 7.11 of this manual with an explanation of slope. BS 5268:Part 2 gives efficiencies in compression which are usually 100% but this seems somewhat doubtful. A compression force in a scarf jointed member will have a similar shearing effect on the glue line as

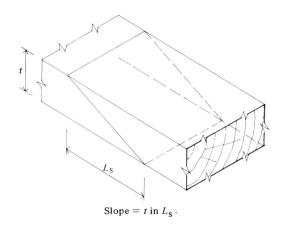

Slope $= t$ in $L_s$ .

|  | % Efficiency rating for a scarf slope of: | | | |
|---|---|---|---|---|
| Grade | 1 in 6 | 1 in 8 | 1 in 10 | 1 in 12 |
| | in bending and tension | | | |
| LA | 69 | 77 | 84 | 88 |
| LB | 67 | 75 | 81 | 85 |
| LC | 50 | 65 | 68 | 72 |
| | in compression (but see text) | | | |
| LA | 100 | 100 | 100 | 100 |
| LB | 100 | 100 | 100 | 100 |
| LC | 80 | 95 | 100 | 100 |

Fig. 7.11: *Plain scarf joints in laminations.*

a tension force although, in a glulam section, the joint is contained by at least one other laminate (see Fig. 7.12). Certainly if a scarf joint is called upon to take a compression force the authors feel that the efficiency in compression should be considered as not more than that in tension or bending. BS 5268 intends scarf joints to be used only in glued laminated members.

From the efficiency ratings shown in Table 7.4 and Fig. 7.11 it can be calculated that it is usually not possible to use LA laminates if they have to be jointed, unless they are purely in compression, or unless a design penalty is accepted due to the level of efficiency of the joint.

When a designer is establishing a glulam section before he knows who is to be the manufacturer, the effect of joint efficiency on his choice of grade and working stress is one which he cannot ignore and one on which he may have to be conservative. Most manufacturers of structural glulam can justify end joint efficiencies of at least 70% in bending, but this must be checked. Preferably the designer should check with the chosen manufacturer to ensure that tests have been carried out to establish the strength of the joint to be used.

End joints should be staggered in adjacent laminations, and this will ensure that the actual joint efficiency is larger than the quoted value because, even in the outer lamination, as well as the strength of the joint in isolation, the adjacent lamination will act as a splice plate which will increase the strength of the joint (Fig. 7.12). BS 4169 calls for 'excessive grouping of finger joints in adjacent laminations' to be avoided.

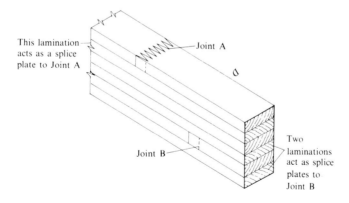

Fig. 7.12

When designing with LA laminates, or when designing with LB or LC laminates and unable to provide an end joint of 70% or 55% efficiency respectively, the designer is required to check the strength of the joint which is to be used against the actual strength required. To assist with this, BS 5268 : Part 2 gives coefficients which, for softwood, are:

$K_{30} = 1.85$ for bending parallel to grain

$K_{31} = 1.85$ for tension parallel to grain

$K_{32} = 1.15$ for compression parallel to grain.

The best way to explain how these factors apply is to give examples, as below. The factors are probably on the conservative side and may be reviewed in future editions of BS 5268.

Take the case of a horizontally laminated beam 540 mm deep of twelve laminates and calculate the permissible extreme fibre stress in bending under medium term loading. The beam is LB European Whitewood throughout. The only finger joint available has an efficiency in bending of 65% (i.e. less than the 70% required for 'blanket approval' with LB grade).

The permissible bending stress in the member as limited by the joint being used is calculated as:

> The SS grade stress for the species $(7.50 \text{ N/mm}^2)$
> X the relevant load-duration factor $K_3$ (1.25)
> X the relevant moisture-content factor $K_2$ (1.00 for dry exposure)
> X the modification factor for depth of *member* $K_7$ (0.893 for 540 mm)
> X the ratio for efficiency in bending of the joint (0.65 in this case)
> X factor $K_{30}$ (1.85)
> $= 7.50 \times 1.25 \times 1.00 \times 0.893 \times 0.65 \times 1.85 = 10.07 \text{ N/mm}^2$

Note that if the laminates had not been jointed the permissible bending stress for this member would be:

> The SS grade stress for the species
> X the relevant load-duration factor
> X the relevant moisture-content factor
> X the modification factor for depth of member
> X the relevant value of $K_{15}$ (1.45 in this case)
> $= 12.10 \text{ N/mm}^2$

Even if a 70% efficient joint were to be used, the permissible stress based on $K_{30}$ is still only $10.81 \text{ N/mm}^2$ which indicates that the value of 1.85 for $K_{30}$ is at least 10% on the conservative side.

The designer may be surprised to see that the depth factor $K_7$ is to be applied in calculating the permissible bending stress in a joint in the member, but the authors have checked that this is the intention of Clause 20 of BS 5268:Part 2.

When using an LB/LC/LB beam it is almost certain that a joint of 55% efficiency would be available to use in the LC laminates, and hence no design check would be necessary. However, from inspection of Fig. 7.5, it is obvious from proportions of the beam, that the joint in the LB laminates would be more critical than the joint in the centre LC laminates.

The case of finger joints in a member taking compression and bending is discussed in § 17.7 but it is convenient to cover here the case of a glulam beam taking axial tension as well as bending.

Repeat the example of the 540 mm deep, twelve laminates of LB White-wood section carrying medium term loading. Assume that the only finger joint available is one with an efficiency in bending and tension of 65% (i.e. less than the 70% required for 'blanket approval' with LB grade).

As before, the permissible bending stress is 10.07 N/mm².
The permissible tension stress in the member as limited by the joint being used is calculated as:

The SS grade stress for the species (4.50 N/mm²)
X the relevant load-duration factor $K_3$ (1.25)
X the relevant moisture-content factor $K_2$ (1.00 for dry exposure)
X the modification factor for width of tension member $K_{14}$
$\qquad$ (0.937 for 540 mm – see Fig. 4.4 and Table 4.4)
X the ratio for efficiency of the joint (0.65 in this case)
X factor $K_{31}$ (1.85)
= 4.50 X 1.25 X 1.00 X 0.937 X 0.65 X 1.85 = 6.34 N/mm²

For the purpose of this example assume actual (applied) bending and tension stresses of 5.00 and 3.00 N/mm² respectively.

$$\frac{\sigma_{m,a,par}}{\sigma_{m,adm,par}} + \frac{\sigma_{t,a,par}}{\sigma_{t,adm,par}} \leqslant 1.00$$

$$\frac{5.00}{10.07} + \frac{3.00}{6.34} = 0.50 + 0.47 = 0.97 < 1.0$$

∴ Section is adequate despite the 65% efficient joint.

### 7.5.2 Joints in vertical laminations

With a member made up of vertical laminates the stress grades which apply are those for solid timber (e.g. GS, SS etc.). As with joints in horizontal laminates the designer has the option of providing a joint of a certain efficiency for a particular stress grade or of comparing the strength of a joint of a certain efficiency with the strength required for the actual stresses in the member.

If a joint of at least the efficiency in bending given in Table 7.5 is provided in the relevant stress grade there is no requirement in BS 5268 : Part 2 for a further design check.

If it is not convenient or possible to provide a joint of the efficiency given in Table 7.5 (or Table 19.3), the joint provided may be adequate if the timber is not fully stressed or if part or all of the stress is due to compression. In § 19.7.1 and Table 19.4 the authors suggest a method of comparing the

**Table 7.5**   **Finger joint efficiency in bending for various stress grades**

| Joint efficiency in bending (%) | Stress grades | Grading rules |
|:---:|:---|:---|
| 50 | M50 and GS/MGS | BS 4978 |
|  | No. 1, No. 2 and No. 3 Joist and Plank | NLGA/NGRDL |
|  | No. 1, No. 2 and No. 3 Structural Light Framing | NLGA/NGRDL |
|  | Construction and Standard Light Framing | NLGA/NGRDL |
|  | Utility, Stud | NLGA/NGRDL |
| 60 | SS/MSS | BS 4978 |
| 65 | Select Structural Joist and Plank | NLGA/NGRDL |
|  | Select Structural Light Framing | NLGA/NGRDL |
| 75 | M75 | BS 4978 |

actual stress at a joint with the required stress. This is equally applicable to joints in vertical laminations and therefore is not repeated here.

## 7.6   CHOICE OF GLUE FOR GLULAM

The tendency is to use a WBP adhesive (see Chapter 19) in the manufacture of glulam both for gluing the laminations and for the finger joints or scarf joints, but the designer should not insist on a WBP adhesive being used when conditions do not require it. Much glulam is used in conditions under which the moisture content of the timber will not rise above 18% and the tempera-ture will not exceed 50°C. In low hazard situations BS 5268:Part 2 (Table 85) permits use of a BR or even an MR adhesive to BS 1204:Part 1. See Chapter 19 for further information.

## 7.7   PRESERVATIVE TREATMENT

BS 4169 should be consulted on preservation of glulam but it should be realised that currently (1984) it is being amended to take account of the latest experience.

There are two stages at which preservation can be carried out. Either individual laminations can be preserved before assembly or the member can be treated after assembly and after all notching etc. has been carried out. Usually a manufacturer prefers to adopt the latter method and to treat with an organic-solvent rather than use a water-borne treatment. The specifier is encouraged to discuss each case with the manufacturer because space and size of preservation plant available, weight of components etc. all have a bearing on the ability of a manufacturer to preserve members.

If individual laminations are preserved by a water-borne method (either before or after end jointing) the laminates will have to be kiln dried before being assembled into a member. Even if machined before being treated they will probably have to be re-machined to obtain a surface suitable for gluing and this will remove some of the preserved timber. The adhesive and the preservative must be compatible. The manufacturer of each must be consulted.

If individual laminations are preserved by an organic-solvent process after being machined this will have no effect on the moisture content, nor will it lead to deterioration of the surface. There is some doubt as to whether or not a water repellent or other additive can be included in the preservative. Certainly the preservative and the adhesive must be compatible.

Currently there is disagreement on whether or not timber pre-treated at the sawmill by the boron diffusion process can be glued. The treatment process is not practised to any extent. Few mills offer it but, if the designer is called upon to use timber treated in this way, it would be prudent to obtain the latest information on whether or not it can be glued.

If completed members are to be preserved, the specifier must ensure that equipment is available for the size and weight of members. If a water-borne process is used it is likely that a certain amount of surface deterioration will occur, and either time must be allowed for air drying or cost must be included for kiln drying (which could also lead to further surface deterioration (fissures). An organic-solvent process is usually preferable.

If there is a requirement to use certain fire-retardant treatments it may not be possible to glue at all after treatment. If treatment is to be carried out on completed members it will probably be necessary to delay treatment until the glue is fully cured (which may be seven days or more) and compatibility must be checked.

Whitewood is 'resistant' to water-borne preservatives whereas Redwood is easy to preserve. Both can be treated by organic-solvent processes.

## 7.8 STANDARD SIZES

Several attempts have been made at both national and international level to agree a range of standard sizes. Some manufacturers prefer to use 45 mm finished laminations machined from 50 mm. Some prefer 33.3 mm from 38 mm. The latter allows 100 mm increments of depth, but requires more glue lines and machining. The tendency is for European (including Nordic) manufacturers to standardise on 45 mm increments of depth when manufacturing straight members.

The tables of suggested sizes and properties in this chapter are based on 45 mm finished laminations.

Width of 65 mm in depths of 180 to 495 mm in 45 increments.
90 mm in depths of 180 to 720 mm in 45 increments.
115 mm in depths of 180 to 900 mm in 45 increments.
135 mm in depths of 180 to 900 mm in 45 increments.
160 mm in depths of 180 to 900 mm in 45 increments.
185 mm in depths of 180 to 900 mm in 45 increments.

## 7.9 TABLES OF PROPERTIES AND CAPACITIES OF STANDARD SIZES IN EUROPEAN WHITEWOOD AND REDWOOD OF LB AND LC GRADES

### 7.9.1 Introduction

The most popular grade used by fabricators is LB grade, as this grade is compatible with the available economic commercial grades, and can be end jointed without loss of strength efficiency.

LC grade is less popular, perhaps due to its less attractive appearance in a component which is often selected for its appearance value, or perhaps because it requires the same expenditure of labour and an equal degree of quality control in manufacture as LB grade, but has less strength capacity.

It is unusual to find a manufacturer offering LA grade or adopting combined grade laminations unless specially requested to do so. Consequently the remainder of this discussion relates only to LB and LC single-grade sections.

A limited range of the smaller sections with no camber is available from stock. These are usually of LB grade.

The net laminate thickness for other than curved work such as portal frames is usually 45 mm for maximum economy in manufacture. This thickness would normally be used for straight members and those provided with a nominal camber to off-set all or part of the anticipated dead load deflection.

As the depth $h$ of the section increases, $K_7$ the depth factor reduces, whereas $K_{15}$ the modification factor for number of laminations increases. These two factors tend to counteract each other giving a reasonably consistent value for the permissible bending stress at all beam depths. The effect of change in $K_7$ and $K_{15}$ can be seen in Table 7.6.

Capacities of sections are given in Table 7.7 for LB grade and in Table 7.8 for LC grade determined from the following stress considerations.

*Shear capacity*

$$\text{Shear capacity} = \overline{F_v} = \frac{2bh\tau_{adm}}{3}$$

**Table 7.6** Permissible long term bending stresses for horizontally laminated beams of LB and LC grades

| No. of laminations | $h$ (mm) | Coefficients from and interpolated from Table 22 of BS 5268 : Part 2 | | | | | | Long term $\sigma_{m, adm}$ | |
|---|---|---|---|---|---|---|---|---|---|
| | | $K_{15}$ | | $K_7$ | $K_{15} \times K_7$ | | | | |
| | | LB | LC | | | LB | LC | LB | LC |
| 4 | 180 | 1.260 | 0.740 | 1.058 | 1.332 | 0.782 | | 9.99 | 5.87 |
| 5 | 225 | 1.340 | 0.820 | 1.032 | 1.383 | 0.846 | | 10.37 | 6.34 |
| 6 | 270 | 1.365 | 0.865 | 1.012 | 1.380 | 0.875 | | 10.35 | 6.56 |
| 7 | 315 | 1.390 | 0.910 | 0.994 | 1.382 | 0.904 | | 10.36 | 6.78 |
| 8 | 360 | 1.403 | 0.933 | 0.964 | 1.352 | 0.899 | | 10.14 | 6.74 |
| 9 | 405 | 1.417 | 0.956 | 0.940 | 1.332 | 0.898 | | 9.99 | 6.74 |
| 10 | 450 | 1.430 | 0.980 | 0.921 | 1.316 | 0.902 | | 9.87 | 6.76 |
| 11 | 495 | 1.440 | 0.994 | 0.905 | 1.303 | 0.899 | | 9.77 | 6.74 |
| 12 | 540 | 1.450 | 1.008 | 0.893 | 1.294 | 0.899 | | 9.70 | 6.74 |
| 13 | 585 | 1.460 | 1.022 | 0.882 | 1.287 | 0.901 | | 9.65 | 6.76 |
| 14 | 630 | 1.470 | 1.036 | 0.873 | 1.283 | 0.904 | | 9.62 | 6.78 |
| 15 | 675 | 1.480 | 1.050 | 0.866 | 1.281 | 0.909 | | 9.61 | 6.82 |
| 16 | 720 | 1.488 | 1.062 | 0.860 | 1.279 | 0.913 | | 9.59 | 6.84 |
| 17 | 765 | 1.496 | 1.074 | 0.855 | 1.278 | 0.918 | | 9.59 | 6.88 |
| 18 | 810 | 1.504 | 1.086 | 0.850 | 1.278 | 0.923 | | 9.59 | 6.92 |
| 19 | 855 | 1.512 | 1.098 | 0.846 | 1.279 | 0.929 | | 9.59 | 6.97 |
| 20 | 900 | 1.520 | 1.110 | 0.843 | 1.281 | 0.935 | | 9.61 | 7.01 |

where $b$ = breadth (thickness) of section

$\quad h$ = depth of section

$\tau_{adm}$ = permissible shear stress = $\tau_g$ for SS grade $\times K_{19}$ (long term)

$\qquad\qquad = 0.82 \times 2.00$

$\qquad\qquad = 1.64 \text{ N/mm}^2$ for long term stresses

and applies to both LB and LC members.

*Moment capacity*

$\qquad$ Moment capacity about $XX$ axis = $\bar{M}$ = $\sigma_{m, adm} \times Z$

where $\sigma_{m, adm}$ = permissible long term bending stress

$\qquad\qquad = \sigma_{m, g}$ for SS grade $\times K_{15} \times K_7$

$\quad Z$ = section modulus

$\qquad = bh^2/6.$

*Deflection properties*
About $XX$ axis:

$$\text{Bending rigidity} = E_N I_X$$

where $E_N = E_{mean}$ for SS grade $\times K_{20}$
   $I_X$ = second moment of area
   $= bh^3/12$.

$K_{20}$ is constant at 0.9 and 0.8 for LB and LC grades respectively.
   Therefore

$$E_N = 10\,500 \times 0.9$$
$$= 9450 \text{ N/mm}^2 \text{ for LB grade}$$

and

$$E_N = 10\,500 \times 0.8$$
$$= 8400 \text{ N/mm}^2 \text{ for LC grade}$$

About $YY$ axis:

$$\text{Bending rigidity} = E_N I_Y$$

where $E_N = E_{min}$ for SS grade $\times K_{28}$ for LB grade $\Big\}$ see § 7.3.2
   $= E_{min}$ for GS grade $\times K_{28}$ for LC grade
   $I_Y = hb^3/12$.

*Shear rigidity*
From § 4.15.8 it can be seen that for shear deflection about $XX$ axis:

$$AG_X = \frac{3EI_X}{4h^2}$$

Similarly, about the $YY$ axis:

$$AG_Y = \frac{3EI_Y}{4b^2}$$

### 7.9.2   LB grade beams

The properties and sectional capacities for horizontally laminated LB grade beams are given in Table 7.7.

### 7.9.3   LC grade beams

The properties and sectional capacities for horizontally laminated LC grade beams are given in Table 7.8.

**Table 7.7  Properties and sectional capacities for horizontally lamined LB grade beams**

| b (mm) | h (mm) | No. of lam. | Self-wt (kN/m) | $\frac{h}{b}$ | Area $A$ (mm²×10³) | $Z_X$ (mm³×10⁶) | $Z_Y$ (mm³×10⁶) | $I_X$ (mm⁴×10⁶) | $I_Y$ (mm⁴×10⁶) | $\bar{F}_{vL}$ (kN) | $\bar{F}_{vM}$ (kN) | $\bar{F}_{vS}$ (kN) | $\bar{F}_{vVS}$ (kN) | $\bar{M}_L$ (kN.m) | $\bar{M}_M$ (kN.m) | $\bar{M}_S$ (kN.m) | $\bar{M}_{VS}$ (kN.m) | $EI_X$ (kN.m²) | $EI_Y$ (kN.m²) | $AG_X$ (kN) | $AG_Y$ (kN) |
|---|---|---|---|---|---|---|---|---|---|---|---|---|---|---|---|---|---|---|---|---|---|
| 65 | 180 | 4 | 0.06 | 2.76 | 11.7 | 0.351 | 0.126 | 31 | 4.1 | 12.7 | 15.9 | 19.1 | 22.3 | 3.5 | 4.3 | 5.2 | 6.1 | 298 | 35 | 6910 | 6347 |
| 65 | 225 | 5 | 0.07 | 3.46 | 14.6 | 0.548 | 0.158 | 61 | 5.1 | 15.9 | 19.9 | 23.9 | 27.9 | 5.6 | 7.1 | 8.5 | 9.9 | 583 | 45 | 8637 | 8126 |
| 65 | 270 | 6 | 0.09 | 4.15 | 17.5 | 0.789 | 0.190 | 106 | 6.1 | 19.1 | 23.9 | 28.7 | 33.5 | 8.1 | 10.2 | 12.2 | 14.3 | 1007 | 55 | 10365 | 9904 |
| 65 | 315 | 7 | 0.11 | 4.84 | 20.4 | 1.074 | 0.221 | 169 | 7.2 | 22.3 | 27.9 | 33.5 | 39.1 | 11.1 | 13.9 | 16.7 | 19.4 | 1599 | 65 | 12093 | 11645 |
| 65 | 360 | 8 | 0.12 | 5.53 | 23.4 | 1.404 | 0.253 | 252 | 8.2 | 25.5 | 31.9 | 38.3 | 44.7 | 14.2 | 17.8 | 21.4 | 24.9 | 2388 | 76 | 13820 | 13513 |
| 65 | 405 | 9 | 0.14 | 6.23 | 26.3 | 1.776 | 0.285 | 359 | 9.2 | 28.7 | 35.9 | 43.1 | 50.3 | 17.7 | 22.2 | 26.6 | 31.0 | 3400 | 85 | 15548 | 15202 |
| 65 | 450 | 10 | 0.15 | 6.92 | 29.2 | 2.193 | 0.316 | 493 | 10.2 | 31.9 | 39.9 | 47.9 | 55.9 | 21.6 | 27.0 | 32.5 | 37.9 | 4664 | 95 | 17275 | 16891 |
| 65 | 495 | 11 | 0.17 | 7.61 | 32.1 | 2.654 | 0.348 | 656 | 11.3 | 35.1 | 43.9 | 52.7 | 61.5 | 25.9 | 32.4 | 38.9 | 45.4 | 6208 | 104 | 19003 | 18581 |
| 90 | 180 | 4 | 0.08 | 2.00 | 16.2 | 0.486 | 0.243 | 43 | 10.9 | 17.7 | 22.1 | 26.5 | 30.9 | 4.9 | 6.1 | 7.3 | 8.5 | 413 | 94 | 9568 | 8788 |
| 90 | 225 | 5 | 0.10 | 2.50 | 20.2 | 0.759 | 0.303 | 85 | 13.6 | 22.1 | 27.6 | 33.2 | 38.7 | 7.8 | 9.8 | 11.8 | 13.7 | 807 | 121 | 11960 | 11251 |
| 90 | 270 | 6 | 0.13 | 3.00 | 24.3 | 1.093 | 0.364 | 147 | 16.4 | 26.5 | 33.2 | 39.8 | 46.4 | 11.3 | 14.1 | 16.9 | 19.8 | 1395 | 148 | 14352 | 13714 |
| 90 | 315 | 7 | 0.15 | 3.50 | 28.3 | 1.488 | 0.425 | 234 | 19.1 | 30.9 | 38.7 | 46.4 | 54.2 | 15.4 | 19.2 | 23.1 | 26.9 | 2215 | 174 | 16744 | 16124 |
| 90 | 360 | 8 | 0.17 | 4.00 | 32.4 | 1.944 | 0.486 | 349 | 21.8 | 35.4 | 44.2 | 53.1 | 61.9 | 19.7 | 24.6 | 29.6 | 34.5 | 3306 | 202 | 19136 | 18711 |
| 90 | 405 | 9 | 0.19 | 4.50 | 36.4 | 2.460 | 0.546 | 498 | 24.6 | 39.8 | 49.8 | 59.7 | 69.7 | 24.6 | 30.7 | 36.9 | 43.0 | 4708 | 227 | 21528 | 21049 |
| 90 | 450 | 10 | 0.21 | 5.00 | 40.5 | 3.037 | 0.607 | 683 | 27.3 | 44.2 | 55.3 | 66.4 | 77.4 | 30.0 | 37.5 | 45.0 | 52.5 | 6458 | 252 | 23920 | 23388 |
| 90 | 495 | 11 | 0.24 | 5.50 | 44.5 | 3.675 | 0.668 | 909 | 30.0 | 48.7 | 60.8 | 73.0 | 85.2 | 35.9 | 44.9 | 53.9 | 62.8 | 8596 | 277 | 26312 | 25727 |
| 90 | 540 | 12 | 0.26 | 6.00 | 48.6 | 4.374 | 0.729 | 1180 | 32.8 | 53.1 | 66.4 | 79.7 | 92.9 | 42.4 | 53.0 | 63.6 | 74.2 | 11160 | 303 | 28704 | 28066 |
| 90 | 585 | 13 | 0.28 | 6.50 | 52.6 | 5.133 | 0.789 | 1501 | 35.5 | 57.5 | 71.9 | 86.3 | 100.7 | 49.5 | 61.9 | 74.3 | 86.7 | 14189 | 328 | 31096 | 30405 |
| 90 | 630 | 14 | 0.30 | 7.00 | 56.7 | 5.953 | 0.850 | 1875 | 38.2 | 61.9 | 77.4 | 92.9 | 108.4 | 57.3 | 71.6 | 85.9 | 100.3 | 17722 | 353 | 33488 | 32744 |
| 90 | 675 | 15 | 0.32 | 7.50 | 60.7 | 6.834 | 0.911 | 2306 | 41.0 | 66.4 | 83.0 | 99.6 | 116.2 | 65.7 | 82.1 | 98.5 | 114.9 | 21797 | 378 | 35880 | 35083 |
| 90 | 720 | 16 | 0.34 | 8.00 | 64.8 | 7.776 | 0.972 | 2799 | 43.7 | 70.8 | 88.5 | 106.2 | 123.9 | 74.6 | 93.2 | 111.9 | 130.6 | 26453 | 404 | 38272 | 37422 |
| 115 | 180 | 4 | 0.11 | 1.56 | 20.7 | 0.621 | 0.396 | 55 | 22.8 | 22.6 | 28.2 | 33.9 | 39.6 | 6.2 | 7.8 | 9.3 | 10.9 | 528 | 198 | 12225 | 11229 |
| 115 | 225 | 5 | 0.13 | 1.95 | 25.8 | 0.970 | 0.495 | 109 | 28.5 | 28.2 | 35.3 | 42.4 | 49.5 | 10.0 | 12.5 | 15.0 | 17.6 | 1031 | 253 | 15282 | 14376 |
| 115 | 270 | 6 | 0.16 | 2.34 | 31.0 | 1.397 | 0.595 | 188 | 34.2 | 33.9 | 42.4 | 50.9 | 59.4 | 14.4 | 18.0 | 21.7 | 25.3 | 1782 | 309 | 18338 | 17523 |
| 115 | 315 | 7 | 0.19 | 2.73 | 36.2 | 1.901 | 0.694 | 299 | 39.9 | 39.6 | 49.5 | 59.4 | 69.3 | 19.7 | 24.6 | 29.5 | 34.4 | 2830 | 363 | 21395 | 20602 |
| 115 | 360 | 8 | 0.22 | 3.13 | 41.4 | 2.484 | 0.793 | 447 | 45.6 | 45.2 | 56.5 | 67.8 | 79.2 | 25.2 | 31.5 | 37.8 | 44.1 | 4225 | 421 | 24451 | 23908 |
| 115 | 405 | 9 | 0.25 | 3.52 | 46.5 | 3.143 | 0.892 | 636 | 51.3 | 50.9 | 63.6 | 76.3 | 89.1 | 31.4 | 39.2 | 47.1 | 54.9 | 6016 | 474 | 27508 | 26897 |
| 115 | 450 | 10 | 0.27 | 3.91 | 51.7 | 3.881 | 0.991 | 873 | 57.0 | 56.5 | 70.7 | 84.8 | 99.0 | 38.3 | 47.9 | 57.5 | 67.0 | 8252 | 526 | 30564 | 29885 |
| 115 | 495 | 11 | 0.30 | 4.30 | 56.9 | 4.696 | 1.091 | 1162 | 62.7 | 62.2 | 77.7 | 93.3 | 108.9 | 45.9 | 57.3 | 68.8 | 80.3 | 10984 | 579 | 33621 | 32874 |

**Table 7.7** (*contd*)

| b (mm) | h (mm) | No. of lam. | Self-wt (kN/m) | $\frac{h}{b}$ | Area $A$ (mm² × 10³) | $Z_X$ (mm³ × 10⁶) | $Z_Y$ (mm³ × 10⁶) | $I_X$ (mm⁴ × 10⁶) | $I_Y$ (mm⁵ × 10⁶) | $\bar{F}_{vL}$ (kN) | $\bar{F}_{vM}$ (kN) | $\bar{F}_{vS}$ (kN) | $\bar{F}_{vVS}$ (kN) | $\bar{M}_L$ (kN.m) | $\bar{M}_M$ (kN.m) | $\bar{M}_S$ (kN.m) | $\bar{M}_{VS}$ (kN.m) | $EI_X$ (kN.m²) | $EI_Y$ (kN.m²) | $AG_X$ (kN) | $AG_Y$ (kN) |
|---|---|---|---|---|---|---|---|---|---|---|---|---|---|---|---|---|---|---|---|---|---|
| 115 | 540 | 12 | 0.33 | 4.69 | 62.1 | 5.589 | 1.190 | 1509 | 68.4 | 67.8 | 84.8 | 101.8 | 118.8 | 54.2 | 67.8 | 81.3 | 94.9 | 14260 | 632 | 36677 | 35862 |
| 115 | 585 | 13 | 0.36 | 5.08 | 67.2 | 6.559 | 1.289 | 1918 | 74.1 | 73.5 | 91.9 | 110.3 | 128.7 | 63.3 | 79.1 | 95.0 | 110.8 | 18130 | 685 | 39734 | 38851 |
| 115 | 630 | 14 | 0.39 | 5.47 | 72.4 | 7.607 | 1.388 | 2396 | 79.8 | 79.2 | 99.0 | 118.8 | 138.6 | 73.2 | 91.5 | 109.8 | 128.1 | 22644 | 737 | 42790 | 41839 |
| 115 | 675 | 15 | 0.41 | 5.86 | 77.6 | 8.732 | 1.487 | 2947 | 85.5 | 84.8 | 106.0 | 127.3 | 148.5 | 83.9 | 104.9 | 125.9 | 146.9 | 27852 | 790 | 45847 | 44828 |
| 115 | 720 | 16 | 0.44 | 6.26 | 82.8 | 9.936 | 1.587 | 3576 | 91.2 | 90.5 | 113.1 | 135.7 | 158.4 | 95.3 | 119.2 | 143.0 | 166.8 | 33802 | 843 | 48903 | 47817 |
| 115 | 765 | 17 | 0.47 | 6.65 | 87.9 | 11.216 | 1.686 | 4290 | 96.9 | 96.1 | 120.2 | 144.2 | 168.3 | 107.5 | 134.4 | 161.3 | 188.2 | 40544 | 895 | 51960 | 50805 |
| 115 | 810 | 18 | 0.50 | 7.04 | 93.1 | 12.575 | 1.785 | 5092 | 102.6 | 101.8 | 127.3 | 152.7 | 178.2 | 120.6 | 150.7 | 180.9 | 211.0 | 48128 | 948 | 55016 | 53794 |
| 115 | 855 | 19 | 0.53 | 7.43 | 98.3 | 14.011 | 1.884 | 5989 | 108.3 | 107.5 | 134.3 | 161.2 | 188.1 | 134.4 | 168.1 | 201.7 | 235.3 | 56603 | 1001 | 58073 | 56782 |
| 115 | 900 | 20 | 0.55 | 7.82 | 103.5 | 15.525 | 1.983 | 6986 | 114.0 | 113.1 | 141.4 | 169.7 | 198.0 | 149.2 | 186.5 | 223.8 | 261.1 | 66020 | 1053 | 61129 | 59771 |
| 135 | 180 | 4 | 0.13 | 1.33 | 24.3 | 0.729 | 0.546 | 65 | 36.9 | 26.5 | 33.2 | 39.8 | 46.4 | 7.3 | 9.1 | 10.9 | 12.7 | 620 | 320 | 14352 | 13182 |
| 135 | 225 | 5 | 0.16 | 1.66 | 30.3 | 1.139 | 0.683 | 128 | 46.1 | 33.2 | 41.5 | 49.8 | 58.1 | 11.8 | 14.7 | 17.7 | 20.6 | 1210 | 410 | 17940 | 16877 |
| 135 | 270 | 6 | 0.19 | 2.00 | 36.4 | 1.640 | 0.820 | 221 | 55.3 | 39.8 | 49.8 | 59.7 | 69.7 | 16.9 | 21.2 | 25.4 | 29.7 | 2092 | 499 | 21528 | 20571 |
| 135 | 315 | 7 | 0.22 | 2.33 | 42.5 | 2.232 | 0.956 | 351 | 64.5 | 46.4 | 58.1 | 69.7 | 81.3 | 23.1 | 28.9 | 34.7 | 40.4 | 3322 | 587 | 25116 | 24186 |
| 135 | 360 | 8 | 0.26 | 2.66 | 48.6 | 2.916 | 1.093 | 524 | 73.8 | 53.1 | 66.4 | 79.7 | 92.9 | 29.6 | 37.0 | 44.3 | 51.7 | 4960 | 682 | 28704 | 28066 |
| 135 | 405 | 9 | 0.29 | 3.00 | 54.6 | 3.690 | 1.230 | 747 | 83.0 | 59.7 | 74.7 | 89.6 | 104.6 | 36.9 | 46.1 | 55.3 | 64.5 | 7062 | 767 | 32292 | 31574 |
| 135 | 450 | 10 | 0.32 | 3.33 | 60.7 | 4.556 | 1.366 | 1025 | 92.2 | 66.4 | 83.0 | 99.6 | 116.2 | 45.0 | 56.2 | 67.5 | 78.7 | 9687 | 852 | 35880 | 35083 |
| 135 | 495 | 11 | 0.36 | 3.66 | 66.8 | 5.513 | 1.503 | 1364 | 101.4 | 73.0 | 91.3 | 109.5 | 127.8 | 53.9 | 67.3 | 80.8 | 94.3 | 12894 | 937 | 39468 | 38591 |
| 135 | 540 | 12 | 0.39 | 4.00 | 72.9 | 6.561 | 1.640 | 1771 | 110.7 | 79.7 | 99.6 | 119.5 | 139.4 | 63.6 | 79.6 | 95.5 | 111.4 | 16740 | 1023 | 43056 | 42099 |
| 135 | 585 | 13 | 0.42 | 4.33 | 78.9 | 7.700 | 1.776 | 2252 | 119.9 | 86.3 | 107.9 | 129.5 | 151.1 | 74.3 | 92.9 | 111.5 | 130.1 | 21283 | 1108 | 46644 | 45608 |
| 135 | 630 | 14 | 0.45 | 4.66 | 85.0 | 8.930 | 1.913 | 2813 | 129.1 | 92.9 | 116.2 | 139.4 | 162.7 | 85.9 | 107.4 | 128.9 | 150.4 | 26583 | 1193 | 50232 | 49116 |
| 135 | 675 | 15 | 0.49 | 5.00 | 91.1 | 10.251 | 2.050 | 3459 | 138.3 | 99.6 | 124.5 | 149.4 | 174.3 | 98.5 | 123.1 | 147.8 | 172.4 | 32696 | 1278 | 53820 | 52624 |
| 135 | 720 | 16 | 0.52 | 5.33 | 97.2 | 11.664 | 2.187 | 4199 | 147.6 | 106.2 | 132.8 | 159.4 | 185.9 | 111.9 | 139.9 | 167.9 | 195.9 | 39680 | 1364 | 57408 | 56133 |
| 135 | 765 | 17 | 0.55 | 5.66 | 103.2 | 13.167 | 2.323 | 5036 | 156.8 | 112.9 | 141.1 | 169.3 | 197.5 | 126.2 | 157.8 | 189.4 | 221.0 | 47595 | 1449 | 60996 | 59641 |
| 135 | 810 | 18 | 0.59 | 6.00 | 109.3 | 14.762 | 2.460 | 5978 | 166.0 | 119.5 | 149.4 | 179.3 | 209.2 | 141.5 | 176.9 | 212.3 | 247.7 | 56498 | 1534 | 64584 | 63149 |
| 135 | 855 | 19 | 0.62 | 6.33 | 115.4 | 16.448 | 2.597 | 7031 | 175.3 | 126.1 | 157.7 | 189.2 | 220.8 | 157.8 | 197.3 | 236.8 | 276.3 | 66448 | 1619 | 68172 | 66657 |
| 135 | 900 | 20 | 0.65 | 6.66 | 121.5 | 18.225 | 2.733 | 8201 | 184.5 | 132.8 | 166.0 | 199.2 | 232.4 | 175.1 | 218.9 | 262.7 | 306.5 | 77501 | 1705 | 71760 | 70166 |

| | | | | | | | | | | | | | | | | | | | | | |
|---|---|---|---|---|---|---|---|---|---|---|---|---|---|---|---|---|---|---|---|---|---|
| 160 | 180 | 4 | 0.15 | 1.12 | 28.8 | 0.768 | 0.864 | 77 | 61.4 | 31.4 | 39.3 | 47.2 | 55.1 | 8.6 | 10.7 | 12.9 | 15.1 | 734 | 533 | 17010 | 15624 |
| 160 | 225 | 5 | 0.19 | 1.40 | 36.0 | 0.960 | 1.350 | 151 | 76.8 | 39.3 | 49.2 | 59.0 | 68.8 | 14.0 | 17.5 | 21.0 | 24.5 | 1435 | 682 | 21262 | 20002 |
| 160 | 270 | 6 | 0.23 | 1.68 | 43.2 | 1.152 | 1.944 | 262 | 92.1 | 47.2 | 59.0 | 70.8 | 82.6 | 20.1 | 25.1 | 30.2 | 35.2 | 2480 | 832 | 25515 | 24381 |
| 160 | 315 | 7 | 0.27 | 1.96 | 50.4 | 1.344 | 2.646 | 416 | 107.5 | 55.1 | 68.8 | 82.6 | 96.4 | 27.4 | 34.2 | 41.1 | 47.9 | 3938 | 978 | 29767 | 28665 |
| 160 | 360 | 8 | 0.31 | 2.25 | 57.6 | 1.536 | 3.456 | 622 | 122.8 | 62.9 | 78.7 | 94.4 | 110.2 | 35.0 | 43.8 | 52.6 | 61.3 | 5878 | 1135 | 34020 | 33264 |
| 160 | 405 | 9 | 0.34 | 2.53 | 64.8 | 1.728 | 4.374 | 885 | 138.2 | 70.8 | 88.5 | 106.2 | 123.9 | 43.7 | 54.6 | 65.5 | 76.5 | 8370 | 1277 | 38272 | 37422 |
| 160 | 450 | 10 | 0.38 | 2.81 | 72.0 | 1.920 | 5.400 | 1215 | 153.6 | 78.7 | 98.4 | 118.0 | 137.7 | 53.3 | 66.6 | 80.0 | 93.3 | 11481 | 1419 | 42525 | 41580 |
| 160 | 495 | 11 | 0.42 | 3.09 | 79.2 | 2.112 | 6.534 | 1617 | 168.9 | 86.5 | 108.2 | 129.8 | 151.5 | 63.8 | 79.8 | 95.8 | 111.7 | 15282 | 1561 | 46777 | 45738 |
| 160 | 540 | 12 | 0.46 | 3.37 | 86.4 | 2.304 | 7.776 | 2099 | 184.3 | 94.4 | 118.0 | 141.6 | 165.3 | 75.4 | 94.3 | 113.2 | 132.0 | 19840 | 1703 | 51030 | 49896 |
| 160 | 585 | 13 | 0.50 | 3.65 | 93.6 | 2.496 | 9.126 | 2669 | 199.6 | 102.3 | 127.9 | 153.5 | 179.0 | 88.1 | 110.1 | 132.2 | 154.2 | 25225 | 1845 | 55282 | 54054 |
| 160 | 630 | 14 | 0.54 | 3.93 | 100.8 | 2.688 | 10.584 | 3333 | 215.0 | 110.2 | 137.7 | 165.3 | 192.8 | 101.9 | 127.3 | 152.8 | 178.3 | 31505 | 1986 | 59535 | 58212 |
| 160 | 675 | 15 | 0.58 | 4.21 | 108.0 | 2.880 | 12.150 | 4100 | 230.4 | 118.0 | 147.6 | 177.1 | 206.6 | 116.8 | 146.0 | 175.2 | 204.4 | 38750 | 2128 | 63787 | 62370 |
| 160 | 720 | 16 | 0.62 | 4.50 | 115.2 | 3.072 | 13.824 | 4976 | 245.7 | 125.9 | 157.4 | 188.9 | 220.4 | 132.6 | 165.8 | 199.0 | 232.1 | 47029 | 2270 | 68040 | 66528 |
| 160 | 765 | 17 | 0.66 | 4.78 | 122.4 | 3.264 | 15.606 | 5969 | 261.1 | 133.8 | 167.2 | 200.7 | 234.1 | 149.6 | 187.0 | 224.5 | 261.9 | 56409 | 2412 | 72292 | 70686 |
| 160 | 810 | 18 | 0.69 | 5.06 | 129.6 | 3.456 | 17.496 | 7085 | 276.4 | 141.6 | 177.1 | 212.5 | 247.9 | 167.8 | 209.7 | 251.7 | 293.6 | 66961 | 2554 | 76545 | 74844 |
| 160 | 855 | 19 | 0.73 | 5.34 | 136.8 | 3.648 | 19.494 | 8333 | 291.8 | 149.5 | 186.9 | 224.3 | 261.7 | 187.1 | 233.9 | 280.6 | 327.4 | 78753 | 2696 | 80797 | 79002 |
| 160 | 900 | 20 | 0.77 | 5.62 | 144.0 | 3.840 | 21.600 | 9720 | 307.2 | 157.4 | 196.8 | 236.1 | 275.5 | 207.6 | 259.5 | 311.4 | 363.3 | 91854 | 2838 | 85050 | 83160 |
| 185 | 180 | 4 | 0.17 | 0.97 | 33.3 | 1.026 | 0.999 | 89 | 94.9 | 36.4 | 45.5 | 54.6 | 63.7 | 10.0 | 12.5 | 15.0 | 17.5 | 849 | 824 | 19667 | 18065 |
| 185 | 225 | 5 | 0.22 | 1.21 | 41.6 | 1.283 | 1.560 | 175 | 118.7 | 45.5 | 56.8 | 68.2 | 79.6 | 16.1 | 20.2 | 24.2 | 28.3 | 1659 | 1055 | 24584 | 23127 |
| 185 | 270 | 6 | 0.26 | 1.45 | 49.9 | 1.540 | 2.247 | 303 | 142.4 | 54.6 | 68.2 | 81.9 | 95.5 | 23.2 | 29.0 | 34.9 | 40.7 | 2867 | 1286 | 29501 | 28190 |
| 185 | 315 | 7 | 0.31 | 1.70 | 58.2 | 1.796 | 3.059 | 481 | 166.2 | 63.7 | 79.6 | 95.5 | 111.4 | 31.7 | 39.6 | 47.5 | 55.4 | 4553 | 1512 | 34418 | 33143 |
| 185 | 360 | 8 | 0.35 | 1.94 | 66.6 | 2.053 | 3.996 | 719 | 189.9 | 72.8 | 91.0 | 109.2 | 127.4 | 40.5 | 50.6 | 60.8 | 70.9 | 6797 | 1755 | 39335 | 38461 |
| 185 | 405 | 9 | 0.40 | 2.18 | 74.9 | 2.310 | 5.057 | 1024 | 213.6 | 81.9 | 102.3 | 122.8 | 143.3 | 50.5 | 63.1 | 75.8 | 88.4 | 9678 | 1974 | 44252 | 43269 |
| 185 | 450 | 10 | 0.44 | 2.43 | 83.2 | 2.566 | 6.243 | 1404 | 237.4 | 91.0 | 113.7 | 136.5 | 159.2 | 61.6 | 77.0 | 92.5 | 107.9 | 13275 | 2193 | 49169 | 48076 |
| 185 | 495 | 11 | 0.49 | 2.67 | 91.5 | 2.823 | 7.554 | 1869 | 261.1 | 100.1 | 125.1 | 150.1 | 175.2 | 73.8 | 92.3 | 110.7 | 129.2 | 17670 | 2413 | 54086 | 52884 |
| 185 | 540 | 12 | 0.53 | 2.91 | 99.9 | 3.080 | 8.991 | 2427 | 284.9 | 109.2 | 136.5 | 163.8 | 191.1 | 87.2 | 109.0 | 130.9 | 152.7 | 22940 | 2632 | 59003 | 57692 |
| 185 | 585 | 13 | 0.58 | 3.16 | 108.2 | 3.336 | 10.551 | 3086 | 308.6 | 118.3 | 147.9 | 177.4 | 207.0 | 101.9 | 127.3 | 152.8 | 178.3 | 29166 | 2852 | 63920 | 62499 |
| 185 | 630 | 14 | 0.62 | 3.40 | 116.5 | 3.593 | 12.237 | 3854 | 332.4 | 127.4 | 159.2 | 191.1 | 222.9 | 117.8 | 147.2 | 176.7 | 206.2 | 36428 | 3071 | 68837 | 67307 |
| 185 | 675 | 15 | 0.67 | 3.64 | 124.8 | 3.850 | 14.048 | 4741 | 356.1 | 136.5 | 170.6 | 204.7 | 238.9 | 135.0 | 168.8 | 202.5 | 236.3 | 44805 | 3290 | 73754 | 72115 |
| 185 | 720 | 16 | 0.71 | 3.89 | 133.2 | 4.107 | 15.984 | 5754 | 379.8 | 145.6 | 182.0 | 218.4 | 254.8 | 153.4 | 191.7 | 230.1 | 268.4 | 54377 | 3510 | 78671 | 76923 |
| 185 | 765 | 17 | 0.76 | 4.13 | 141.5 | 4.363 | 18.044 | 6901 | 403.6 | 154.7 | 193.4 | 232.1 | 270.7 | 173.0 | 216.3 | 259.5 | 302.8 | 65223 | 3729 | 83588 | 81730 |
| 185 | 810 | 18 | 0.80 | 4.37 | 149.8 | 4.620 | 20.229 | 8193 | 427.3 | 163.8 | 204.7 | 245.7 | 286.7 | 194.0 | 242.5 | 291.0 | 339.5 | 77424 | 3949 | 88505 | 86538 |
| 185 | 855 | 19 | 0.85 | 4.62 | 158.1 | 4.877 | 22.539 | 9635 | 451.1 | 172.9 | 216.1 | 259.4 | 302.6 | 216.3 | 270.4 | 324.5 | 378.6 | 91058 | 4168 | 93422 | 91346 |
| 185 | 900 | 20 | 0.89 | 4.86 | 166.5 | 5.133 | 24.975 | 11238 | 474.8 | 182.0 | 227.5 | 273.0 | 318.5 | 240.0 | 300.0 | 360.0 | 420.1 | 106206 | 4387 | 98339 | 96153 |

**Table 7.8**   Properties and sectional capacities for horizontally laminated LC grade beams. European Whitewood and Redwood

| $b$ (mm) | $h$ (mm) | No. of lam. | Self-wt (kN/m) | $\frac{h}{b}$ | Area $A$ (mm² × 10³) | $Z_X$ (mm³ × 10⁶) | $Z_Y$ (mm³ × 10⁶) | $I_X$ (mm⁴ × 10⁶) | $I_Y$ (mm⁴ × 10⁶) | $\bar{F}_{vL}$ (kN) | $\bar{F}_{vM}$ (kN) | $\bar{F}_{vS}$ (kN) | $\bar{F}_{vVS}$ (kN) | $\bar{M}_L$ (kN.m) | $\bar{M}_M$ (kN.m) | $\bar{M}_S$ (kN.m) | $\bar{M}_{VS}$ (kN.m) | $EI_X$ (kN.m²) | $EI_Y$ (kN.m²) | $AG_X$ (kN) | $AG_Y$ (kN) |
|---|---|---|---|---|---|---|---|---|---|---|---|---|---|---|---|---|---|---|---|---|---|
| | | | | | | Section modulus | | Second moment of area | | Shear capacity XX axis | | | | Moment capacity XX axis | | | | Deflection bending | | Deflection shear | |
| 65 | 180 | 4 | 0.06 | 2.76 | 11.7 | 0.351 | 0.126 | 31 | 4.1 | 12.7 | 15.9 | 19.1 | 22.3 | 2.0 | 2.5 | 3.0 | 3.6 | 265 | 30 | 6142 | 5440 |
| 65 | 225 | 5 | 0.07 | 3.46 | 14.6 | 0.548 | 0.158 | 61 | 5.1 | 15.9 | 19.9 | 23.9 | 27.9 | 3.4 | 4.3 | 5.2 | 6.0 | 518 | 39 | 7678 | 6965 |
| 65 | 270 | 6 | 0.09 | 4.15 | 17.5 | 0.789 | 0.190 | 106 | 6.1 | 19.1 | 23.9 | 28.7 | 33.5 | 5.1 | 6.4 | 7.7 | 9.0 | 895 | 47 | 9213 | 8489 |
| 65 | 315 | 7 | 0.11 | 4.84 | 20.4 | 1.074 | 0.221 | 169 | 7.2 | 22.3 | 27.9 | 33.5 | 39.1 | 7.2 | 9.1 | 10.9 | 12.7 | 1422 | 56 | 10749 | 9981 |
| 65 | 360 | 8 | 0.12 | 5.53 | 23.4 | 1.404 | 0.253 | 252 | 8.2 | 25.5 | 31.9 | 38.3 | 44.7 | 9.4 | 11.8 | 14.2 | 16.5 | 2122 | 65 | 12285 | 11583 |
| 65 | 405 | 9 | 0.14 | 6.23 | 26.3 | 1.776 | 0.285 | 359 | 9.2 | 28.7 | 35.9 | 43.1 | 50.3 | 11.9 | 14.9 | 17.9 | 20.9 | 3022 | 73 | 13820 | 13030 |
| 65 | 450 | 10 | 0.15 | 6.92 | 29.2 | 2.193 | 0.316 | 493 | 10.2 | 31.9 | 39.9 | 47.9 | 55.9 | 14.8 | 18.5 | 22.2 | 25.9 | 4146 | 81 | 15356 | 14478 |
| 65 | 495 | 11 | 0.17 | 7.61 | 32.1 | 2.654 | 0.348 | 656 | 11.3 | 35.1 | 43.9 | 52.7 | 61.5 | 17.9 | 22.3 | 26.8 | 31.3 | 5518 | 89 | 16891 | 15926 |
| 90 | 180 | 4 | 0.08 | 2.00 | 16.2 | 0.486 | 0.243 | 43 | 10.9 | 17.7 | 22.1 | 26.5 | 30.9 | 2.8 | 3.5 | 4.2 | 4.9 | 367 | 81 | 8505 | 7533 |
| 90 | 225 | 5 | 0.10 | 2.50 | 20.2 | 0.759 | 0.303 | 85 | 13.6 | 22.1 | 27.6 | 33.2 | 38.7 | 4.8 | 6.0 | 7.2 | 8.4 | 717 | 104 | 10631 | 9644 |
| 90 | 270 | 6 | 0.13 | 3.00 | 24.3 | 1.093 | 0.364 | 147 | 16.4 | 26.5 | 33.2 | 39.8 | 46.4 | 7.1 | 8.9 | 10.7 | 12.5 | 1240 | 126 | 12757 | 11755 |
| 90 | 315 | 7 | 0.15 | 3.50 | 28.3 | 1.488 | 0.425 | 234 | 19.1 | 30.9 | 38.7 | 46.4 | 54.2 | 10.1 | 12.6 | 15.1 | 17.6 | 1969 | 149 | 14883 | 13820 |
| 90 | 360 | 8 | 0.17 | 4.00 | 32.4 | 1.944 | 0.486 | 349 | 21.8 | 35.4 | 44.2 | 53.1 | 61.9 | 13.1 | 16.3 | 19.6 | 22.9 | 2939 | 173 | 17010 | 16038 |
| 90 | 405 | 9 | 0.19 | 4.50 | 36.4 | 2.460 | 0.546 | 498 | 24.6 | 39.8 | 49.8 | 59.7 | 69.7 | 16.5 | 20.7 | 24.8 | 29.0 | 4185 | 194 | 19136 | 18042 |
| 90 | 450 | 10 | 0.21 | 5.00 | 40.5 | 3.037 | 0.607 | 683 | 27.3 | 44.2 | 55.3 | 66.4 | 77.4 | 20.5 | 25.6 | 30.8 | 35.9 | 5740 | 216 | 21262 | 20047 |
| 90 | 495 | 11 | 0.24 | 5.50 | 44.5 | 3.675 | 0.668 | 909 | 30.0 | 48.7 | 60.8 | 73.0 | 85.2 | 24.8 | 31.0 | 37.2 | 43.4 | 7641 | 238 | 23388 | 22052 |
| 90 | 540 | 12 | 0.26 | 6.00 | 48.6 | 4.374 | 0.729 | 1180 | 32.8 | 53.1 | 66.4 | 79.7 | 92.9 | 29.5 | 36.8 | 44.2 | 51.6 | 9920 | 259 | 25515 | 24057 |
| 90 | 585 | 13 | 0.28 | 6.50 | 52.6 | 5.133 | 0.789 | 1501 | 35.5 | 57.5 | 71.9 | 86.3 | 100.7 | 34.7 | 43.3 | 52.0 | 60.7 | 12612 | 281 | 27641 | 26061 |
| 90 | 630 | 14 | 0.30 | 7.00 | 56.7 | 5.953 | 0.850 | 1875 | 38.2 | 61.9 | 77.4 | 92.9 | 108.4 | 40.4 | 50.5 | 60.6 | 70.7 | 15752 | 303 | 29767 | 28066 |
| 90 | 675 | 15 | 0.32 | 7.50 | 60.7 | 6.834 | 0.911 | 2306 | 41.0 | 66.4 | 83.0 | 99.6 | 116.2 | 46.6 | 58.2 | 69.9 | 81.5 | 19375 | 324 | 31893 | 30071 |
| 90 | 720 | 16 | 0.34 | 8.00 | 64.8 | 7.776 | 0.972 | 2799 | 43.7 | 70.8 | 88.5 | 106.2 | 123.9 | 53.2 | 66.5 | 79.8 | 93.2 | 23514 | 346 | 34020 | 32076 |
| 115 | 180 | 4 | 0.11 | 1.56 | 20.7 | 0.621 | 0.396 | 55 | 22.8 | 22.6 | 28.2 | 33.9 | 39.6 | 3.6 | 4.5 | 5.4 | 6.3 | 469 | 169 | 10867 | 9625 |
| 115 | 225 | 5 | 0.13 | 1.95 | 25.8 | 0.970 | 0.495 | 109 | 28.5 | 28.2 | 35.3 | 42.4 | 49.5 | 6.1 | 7.6 | 9.2 | 10.7 | 916 | 217 | 13584 | 12322 |
| 115 | 270 | 6 | 0.16 | 2.34 | 31.0 | 1.397 | 0.595 | 188 | 34.2 | 33.9 | 42.4 | 50.9 | 59.4 | 9.1 | 11.4 | 13.7 | 16.0 | 1584 | 264 | 16301 | 15020 |
| 115 | 315 | 7 | 0.19 | 2.73 | 36.2 | 1.901 | 0.694 | 299 | 39.9 | 39.6 | 49.5 | 59.4 | 69.3 | 12.9 | 16.1 | 19.3 | 22.5 | 2516 | 311 | 19018 | 17659 |
| 115 | 360 | 8 | 0.22 | 3.13 | 41.4 | 2.484 | 0.793 | 447 | 45.6 | 45.2 | 56.5 | 67.8 | 79.2 | 16.7 | 20.9 | 25.1 | 29.3 | 3755 | 361 | 21735 | 20493 |
| 115 | 405 | 9 | 0.25 | 3.52 | 46.5 | 3.143 | 0.892 | 636 | 51.3 | 50.9 | 63.6 | 76.3 | 89.1 | 21.1 | 26.4 | 31.7 | 37.0 | 5347 | 406 | 24451 | 23054 |
| 115 | 450 | 10 | 0.27 | 3.91 | 51.7 | 3.881 | 0.991 | 873 | 57.0 | 56.5 | 70.7 | 84.8 | 99.0 | 26.2 | 32.8 | 39.4 | 45.9 | 7335 | 451 | 27168 | 25616 |

| | | | | | | | | | | | | | | | | | | | | | |
|---|---|---|---|---|---|---|---|---|---|---|---|---|---|---|---|---|---|---|---|---|---|
| 115 | 495 | 11 | 0.30 | 4.30 | 56.9 | 1.091 | 4.696 | 1162 | 62.7 | 62.2 | 77.7 | 93.3 | 108.9 | 31.6 | 39.6 | 47.5 | 55.4 | 9763 | 496 | 29885 | 28177 |
| 115 | 540 | 12 | 0.33 | 4.69 | 62.1 | 1.190 | 5.589 | 1509 | 68.4 | 67.8 | 84.8 | 101.8 | 118.8 | 37.7 | 47.1 | 56.5 | 65.9 | 12675 | 542 | 32602 | 30739 |
| 115 | 585 | 13 | 0.36 | 5.08 | 67.2 | 1.289 | 6.559 | 1918 | 74.1 | 73.5 | 91.9 | 110.3 | 128.7 | 44.3 | 55.4 | 66.5 | 77.6 | 16116 | 587 | 35319 | 33301 |
| 115 | 630 | 14 | 0.39 | 5.47 | 72.4 | 1.388 | 7.607 | 2396 | 79.8 | 79.2 | 99.0 | 118.8 | 138.6 | 51.6 | 64.5 | 77.4 | 90.3 | 20128 | 632 | 38036 | 35862 |
| 115 | 675 | 15 | 0.41 | 5.86 | 77.6 | 1.487 | 8.732 | 2947 | 85.5 | 84.8 | 106.0 | 127.3 | 148.5 | 59.5 | 74.4 | 89.3 | 104.2 | 24757 | 677 | 40753 | 38424 |
| 115 | 720 | 16 | 0.44 | 6.26 | 82.8 | 1.587 | 9.936 | 3576 | 91.2 | 90.5 | 113.1 | 135.7 | 158.4 | 68.0 | 85.0 | 102.0 | 119.1 | 30046 | 722 | 43470 | 40986 |
| 115 | 765 | 17 | 0.47 | 6.65 | 87.9 | 1.686 | 11.216 | 4290 | 96.9 | 96.1 | 120.2 | 144.2 | 168.3 | 77.2 | 96.5 | 115.8 | 135.1 | 36039 | 767 | 46186 | 43547 |
| 115 | 810 | 18 | 0.50 | 7.04 | 93.1 | 1.785 | 12.575 | 5092 | 102.6 | 101.8 | 127.3 | 152.7 | 178.2 | 87.0 | 108.8 | 130.6 | 152.4 | 42781 | 813 | 48903 | 46109 |
| 115 | 855 | 19 | 0.53 | 7.43 | 98.3 | 1.884 | 14.011 | 5989 | 108.3 | 107.5 | 134.3 | 161.2 | 188.1 | 97.6 | 122.0 | 146.5 | 170.9 | 50314 | 858 | 51620 | 48670 |
| 115 | 900 | 20 | 0.55 | 7.82 | 103.5 | 1.983 | 15.525 | 6986 | 114.0 | 113.1 | 141.4 | 169.7 | 198.0 | 108.9 | 136.2 | 163.4 | 190.7 | 58684 | 903 | 54337 | 51232 |
| 135 | 180 | 4 | 0.13 | 1.33 | 24.3 | 0.546 | 0.729 | 65 | 36.9 | 26.5 | 33.2 | 39.8 | 46.4 | 4.2 | 5.3 | 6.4 | 7.4 | 551 | 274 | 12757 | 11299 |
| 135 | 225 | 5 | 0.16 | 1.66 | 30.3 | 0.683 | 1.139 | 128 | 46.1 | 33.2 | 41.5 | 49.8 | 58.1 | 7.2 | 9.0 | 10.8 | 12.6 | 1076 | 351 | 15946 | 14466 |
| 135 | 270 | 6 | 0.19 | 2.00 | 36.4 | 0.820 | 1.640 | 221 | 55.3 | 39.8 | 49.8 | 59.7 | 69.7 | 10.7 | 13.4 | 16.1 | 18.8 | 1860 | 428 | 19136 | 17632 |
| 135 | 315 | 7 | 0.22 | 2.33 | 42.5 | 0.956 | 2.232 | 351 | 64.5 | 46.4 | 58.1 | 69.7 | 81.3 | 15.1 | 18.9 | 22.7 | 26.5 | 2953 | 503 | 22325 | 20730 |
| 135 | 360 | 8 | 0.26 | 2.66 | 48.6 | 1.093 | 2.916 | 524 | 73.8 | 53.1 | 66.4 | 79.7 | 92.9 | 19.6 | 24.5 | 29.5 | 34.4 | 4408 | 584 | 25515 | 24057 |
| 135 | 405 | 9 | 0.29 | 3.00 | 54.6 | 1.230 | 3.690 | 747 | 83.0 | 59.7 | 74.7 | 89.6 | 104.6 | 24.8 | 31.0 | 37.3 | 43.5 | 6277 | 657 | 28704 | 27064 |
| 135 | 450 | 10 | 0.32 | 3.33 | 60.7 | 1.366 | 4.556 | 1025 | 92.2 | 66.4 | 83.0 | 99.6 | 116.2 | 30.8 | 38.5 | 46.2 | 53.9 | 8611 | 730 | 31893 | 30071 |
| 135 | 495 | 11 | 0.36 | 3.66 | 66.8 | 1.503 | 5.513 | 1364 | 101.4 | 73.0 | 91.3 | 109.5 | 127.8 | 37.2 | 46.5 | 55.8 | 65.1 | 11461 | 803 | 35083 | 33078 |
| 135 | 540 | 12 | 0.39 | 4.00 | 72.9 | 1.640 | 6.561 | 1771 | 110.7 | 79.7 | 99.6 | 119.5 | 139.4 | 44.2 | 55.3 | 66.4 | 77.4 | 14880 | 876 | 38272 | 36085 |
| 135 | 585 | 13 | 0.42 | 4.33 | 78.9 | 1.776 | 7.700 | 2252 | 119.9 | 86.3 | 107.9 | 129.5 | 151.1 | 52.0 | 65.0 | 78.0 | 91.1 | 18919 | 949 | 41461 | 39092 |
| 135 | 630 | 14 | 0.45 | 4.66 | 85.0 | 1.913 | 8.930 | 2813 | 129.1 | 92.9 | 116.2 | 139.4 | 162.7 | 60.6 | 75.7 | 90.9 | 106.0 | 23629 | 1023 | 44651 | 42099 |
| 135 | 675 | 15 | 0.49 | 5.00 | 91.1 | 2.050 | 10.251 | 3459 | 138.3 | 99.6 | 124.5 | 149.4 | 174.3 | 69.9 | 87.4 | 104.8 | 122.3 | 29063 | 1096 | 47840 | 45106 |
| 135 | 720 | 16 | 0.52 | 5.33 | 97.2 | 2.187 | 11.664 | 4199 | 147.6 | 106.2 | 132.8 | 159.4 | 185.9 | 79.8 | 99.8 | 119.8 | 139.8 | 35271 | 1169 | 51030 | 48114 |
| 135 | 765 | 17 | 0.55 | 5.66 | 103.2 | 2.323 | 13.167 | 5036 | 156.8 | 112.9 | 141.1 | 169.3 | 197.5 | 90.2 | 113.3 | 135.9 | 158.6 | 42307 | 1242 | 54219 | 51121 |
| 135 | 810 | 18 | 0.59 | 6.00 | 109.3 | 2.460 | 14.762 | 5978 | 166.0 | 119.5 | 149.4 | 179.3 | 209.2 | 102.2 | 127.8 | 153.3 | 178.9 | 50221 | 1315 | 57408 | 54128 |
| 135 | 855 | 19 | 0.62 | 6.33 | 115.4 | 2.597 | 16.448 | 7031 | 175.3 | 126.1 | 157.7 | 189.2 | 220.8 | 114.6 | 143.3 | 171.9 | 200.6 | 59064 | 1388 | 60598 | 57135 |
| 135 | 900 | 20 | 0.65 | 6.66 | 121.5 | 2.733 | 18.225 | 8201 | 184.5 | 132.8 | 166.0 | 199.2 | 232.4 | 127.9 | 159.9 | 191.8 | 223.8 | 68890 | 1461 | 63787 | 60142 |
| 160 | 180 | 4 | 0.15 | 1.12 | 28.8 | 0.768 | 0.864 | 77 | 61.4 | 31.4 | 39.3 | 47.2 | 55.1 | 5.0 | 6.3 | 7.6 | 8.8 | 653 | 457 | 15120 | 13392 |
| 160 | 225 | 5 | 0.19 | 1.40 | 36.0 | 0.960 | 1.350 | 151 | 76.8 | 39.3 | 49.2 | 59.0 | 68.8 | 8.5 | 10.7 | 12.8 | 14.9 | 1275 | 585 | 18900 | 17145 |
| 160 | 270 | 6 | 0.23 | 1.68 | 43.2 | 1.152 | 1.944 | 262 | 92.1 | 47.2 | 59.0 | 70.8 | 82.6 | 12.7 | 15.9 | 19.1 | 22.3 | 2204 | 713 | 22680 | 20898 |
| 160 | 315 | 7 | 0.27 | 1.96 | 50.4 | 1.344 | 2.646 | 416 | 107.5 | 55.1 | 68.8 | 82.6 | 96.4 | 17.9 | 22.4 | 26.9 | 31.4 | 3500 | 838 | 26460 | 24570 |
| 160 | 360 | 8 | 0.31 | 2.25 | 57.6 | 1.536 | 3.456 | 622 | 122.8 | 62.9 | 78.7 | 94.4 | 110.2 | 23.3 | 29.1 | 34.9 | 40.8 | 5225 | 973 | 30240 | 28512 |
| 160 | 405 | 9 | 0.34 | 2.53 | 64.8 | 1.728 | 4.374 | 885 | 137.7 | 70.8 | 88.5 | 106.2 | 123.9 | 29.4 | 36.8 | 44.2 | 51.6 | 7440 | 1094 | 34020 | 32076 |
| 160 | 450 | 10 | 0.38 | 2.81 | 72.0 | 1.920 | 5.400 | 1215 | 153.6 | 78.7 | 98.4 | 118.0 | 137.7 | 36.5 | 45.6 | 54.8 | 63.9 | 10206 | 1216 | 37800 | 35640 |
| 160 | 495 | 11 | 0.42 | 3.09 | 79.2 | 2.112 | 6.534 | 1617 | 168.9 | 86.5 | 108.2 | 129.8 | 151.5 | 44.0 | 55.1 | 66.1 | 77.1 | 13584 | 1338 | 41580 | 39204 |
| 160 | 540 | 12 | 0.46 | 3.37 | 86.4 | 2.304 | 7.776 | 2099 | 184.3 | 94.4 | 118.0 | 141.6 | 165.3 | 52.4 | 65.5 | 78.7 | 91.8 | 17635 | 1459 | 45360 | 42768 |
| 160 | 585 | 13 | 0.50 | 3.65 | 93.6 | 2.496 | 9.126 | 2669 | 199.6 | 102.3 | 127.7 | 153.5 | 179.0 | 61.7 | 77.1 | 92.5 | 107.9 | 22422 | 1581 | 49140 | 46332 |
| 160 | 630 | 14 | 0.54 | 3.93 | 100.8 | 2.688 | 10.584 | 3333 | 215.0 | 110.2 | 137.7 | 165.3 | 192.8 | 71.8 | 89.7 | 107.7 | 125.6 | 28005 | 1703 | 52920 | 49896 |
| 160 | 675 | 15 | 0.58 | 4.21 | 108.0 | 2.880 | 12.150 | 4100 | 230.4 | 118.0 | 147.6 | 177.1 | 206.6 | 82.8 | 103.5 | 124.3 | 145.0 | 34445 | 1824 | 56700 | 53460 |

**Table 7.8** (contd)

| b (mm) | h (mm) | No. of lam. | Self-wt (kN/m) | $\frac{h}{b}$ | Area $A$ (mm² × 10³) | $Z_X$ (mm³ × 10⁶) | $Z_Y$ (mm³ × 10⁶) | $I_X$ (mm⁴ × 10⁶) | $I_Y$ (mm⁴ × 10⁶) | $\overline{F}_{vL}$ (kN) | $\overline{F}_{vM}$ (kN) | $\overline{F}_{vS}$ (kN) | $\overline{F}_{vVS}$ (kN) | $\overline{M}_L$ (kN.m) | $\overline{M}_M$ (kN.m) | $\overline{M}_S$ (kN.m) | $\overline{M}_{VS}$ (kN.m) | $EI_X$ (kN.m²) | $EI_Y$ (kN.m²) | $AG_X$ (kN) | $AG_Y$ (kN) |
|---|---|---|---|---|---|---|---|---|---|---|---|---|---|---|---|---|---|---|---|---|---|
| 160 | 720 | 16 | 0.62 | 4.50 | 115.2 | 13.824 | 3.072 | 4976 | 245.7 | 125.9 | 157.4 | 188.9 | 220.4 | 94.6 | 118.3 | 142.0 | 165.7 | 41803 | 1946 | 60480 | 57024 |
| 160 | 765 | 17 | 0.66 | 4.78 | 122.4 | 15.606 | 3.264 | 5969 | 261.1 | 133.8 | 167.2 | 200.7 | 234.1 | 107.4 | 134.3 | 161.1 | 188.0 | 50142 | 2068 | 64260 | 60588 |
| 160 | 810 | 18 | 0.69 | 5.06 | 129.6 | 17.496 | 3.456 | 7085 | 276.4 | 141.6 | 177.1 | 212.5 | 247.9 | 121.1 | 151.4 | 181.7 | 212.0 | 59521 | 2189 | 68040 | 64152 |
| 160 | 855 | 19 | 0.73 | 5.34 | 136.8 | 19.494 | 3.648 | 8333 | 291.8 | 149.5 | 186.9 | 224.3 | 261.7 | 135.8 | 169.8 | 203.8 | 237.8 | 70002 | 2311 | 71820 | 67716 |
| 160 | 900 | 20 | 0.77 | 5.62 | 144.0 | 21.600 | 3.840 | 9720 | 307.2 | 157.4 | 196.8 | 236.1 | 275.5 | 151.6 | 189.5 | 227.4 | 265.3 | 81648 | 2433 | 75600 | 71280 |
| 185 | 180 | 4 | 0.17 | 0.97 | 33.3 | 0.999 | 1.026 | 89 | 94.9 | 36.4 | 45.5 | 54.6 | 63.7 | 5.8 | 7.3 | 8.7 | 10.2 | 755 | 706 | 17482 | 15484 |
| 185 | 225 | 5 | 0.22 | 1.21 | 41.6 | 1.560 | 1.283 | 175 | 118.7 | 45.5 | 56.8 | 68.2 | 79.6 | 9.9 | 12.3 | 14.8 | 17.3 | 1475 | 904 | 21853 | 19823 |
| 185 | 270 | 6 | 0.26 | 1.45 | 49.9 | 2.247 | 1.540 | 303 | 142.4 | 54.6 | 68.2 | 81.9 | 95.5 | 14.7 | 18.4 | 22.1 | 25.8 | 2548 | 1102 | 26223 | 24163 |
| 185 | 315 | 7 | 0.31 | 1.70 | 58.2 | 3.059 | 1.796 | 481 | 166.2 | 63.7 | 79.6 | 95.5 | 111.4 | 20.7 | 25.9 | 31.1 | 36.3 | 4047 | 1296 | 30594 | 28409 |
| 185 | 360 | 8 | 0.35 | 1.94 | 66.6 | 3.996 | 2.053 | 719 | 189.9 | 72.8 | 91.0 | 109.2 | 127.4 | 26.9 | 33.7 | 40.4 | 47.1 | 6041 | 1504 | 34965 | 32967 |
| 185 | 405 | 9 | 0.40 | 2.18 | 74.9 | 5.057 | 2.310 | 1024 | 213.6 | 81.9 | 102.3 | 122.8 | 143.3 | 34.0 | 42.6 | 51.1 | 59.6 | 8602 | 1692 | 39335 | 37087 |
| 185 | 450 | 10 | 0.44 | 2.43 | 83.2 | 6.243 | 2.566 | 1404 | 237.4 | 91.0 | 113.7 | 136.5 | 159.2 | 42.2 | 52.8 | 63.3 | 73.9 | 11800 | 1880 | 43706 | 41208 |
| 185 | 495 | 11 | 0.49 | 2.67 | 91.5 | 7.554 | 2.823 | 1869 | 261.1 | 100.1 | 125.1 | 150.1 | 175.2 | 50.9 | 63.7 | 76.4 | 89.2 | 15706 | 2068 | 48076 | 45329 |
| 185 | 540 | 12 | 0.53 | 2.91 | 99.9 | 8.991 | 3.080 | 2427 | 284.9 | 109.2 | 136.5 | 163.8 | 191.1 | 60.6 | 75.8 | 91.0 | 106.1 | 20391 | 2256 | 52447 | 49450 |
| 185 | 585 | 13 | 0.58 | 3.16 | 108.2 | 10.551 | 3.336 | 3086 | 308.6 | 118.3 | 147.9 | 177.4 | 207.0 | 71.3 | 89.1 | 107.0 | 124.8 | 25926 | 2444 | 56818 | 53571 |
| 185 | 630 | 14 | 0.62 | 3.40 | 116.5 | 12.237 | 3.593 | 3854 | 332.4 | 127.4 | 159.2 | 191.1 | 222.9 | 83.0 | 103.8 | 124.5 | 145.3 | 32381 | 2632 | 61188 | 57692 |
| 185 | 675 | 15 | 0.67 | 3.64 | 124.8 | 14.048 | 3.850 | 4741 | 356.1 | 136.5 | 170.6 | 204.7 | 238.9 | 95.8 | 119.7 | 143.7 | 167.6 | 39827 | 2820 | 65559 | 61813 |
| 185 | 720 | 16 | 0.71 | 3.89 | 133.2 | 15.984 | 4.107 | 5754 | 379.8 | 145.6 | 182.0 | 218.4 | 254.8 | 109.4 | 136.8 | 164.2 | 191.6 | 48335 | 3008 | 69930 | 65934 |
| 185 | 765 | 17 | 0.76 | 4.13 | 141.5 | 18.044 | 4.363 | 6901 | 403.6 | 154.7 | 193.4 | 232.1 | 270.7 | 124.2 | 155.3 | 186.3 | 217.4 | 57976 | 3196 | 74300 | 70054 |
| 185 | 810 | 18 | 0.80 | 4.37 | 149.8 | 20.229 | 4.620 | 8193 | 427.3 | 163.8 | 204.7 | 245.7 | 286.7 | 140.1 | 175.1 | 210.1 | 245.1 | 68821 | 3384 | 78671 | 74175 |
| 185 | 855 | 19 | 0.85 | 4.62 | 158.1 | 22.539 | 4.877 | 9635 | 451.1 | 172.9 | 216.1 | 259.4 | 302.6 | 157.1 | 196.4 | 235.6 | 274.9 | 80940 | 3572 | 83041 | 78296 |
| 185 | 900 | 20 | 0.89 | 4.86 | 166.5 | 24.975 | 5.133 | 11238 | 474.8 | 182.0 | 227.5 | 273.0 | 318.5 | 175.3 | 219.1 | 262.9 | 306.7 | 94405 | 3760 | 87412 | 82417 |

## 7.10 TYPICAL DESIGNS

### 7.10.1 Typical design of an uncambered laminated beam

Determine a suitable glulam section in LB grade to support a uniformly distributed load of 4 kN/m dead load (including self-weight) and 2.4 kN/m medium term imposed over a design span of 6.0 m. Assume full lateral restraint of the section.

$$\text{Long term load} = F_L = 24 \text{ kN}$$
$$\text{Medium term load} = F_M = 38.4 \text{ kN}$$

$$\text{Long term shear} = F_{vL} = 12 \text{ kN}$$
$$\text{Medium term shear} = F_{vM} = 19.2 \text{ kN}$$

$$\text{Long term moment} = M_L = 18 \text{ kN.m}$$
$$\text{Medium term moment} = M_M = 28.8 \text{ kN.m}$$

For laminated beams assume depth of section will be $L/16$. From Fig. 4.26, $\delta_t/\delta_m = 1.060$ and a first approximation to $EI_X$ may be taken as:

$$EI_X = 4.6 F_M L^2 \qquad \text{(see § 5.3.2)}$$

where $F_M$ = total load (kN).

$$\text{Required } EI_X = 4.6 \times 38.4 \times 6^2 = 6359 \text{ kN.m}^2$$

From Table 7.7, try 90 × 450 mm section LB grade for which:

$$\overline{F}_{vL} = 44.2 \text{ kN} \quad > 12 \text{ kN}$$
$$\overline{F}_{vM} = 55.3 \text{ kN} \quad > 19.2 \text{ kN}$$
$$\overline{M}_L = 30.0 \text{ kN.m} > 18 \text{ kN.m}$$
$$\overline{M}_M = 37.5 \text{ kN.m} > 28.8 \text{ kN.m}$$
$$EI_X = 6458 \text{ kN.m}^2$$
$$AG_X = 23\,920 \text{ kN}$$

$$\text{Total deflection} = \frac{5 \times 38.4 \times 6^3}{384 \times 6458} + \frac{28.8}{23\,920}$$

$$= 0.0167 + 0.0012 = 0.0179 \text{ m}$$

Allowable deflection = 0.003 × 6 = 0.018 m which is satisfactory.

### 7.10.2 Typical design of a cambered laminated beam

Determine a suitable glulam section and camber in LB grade glulam to support a uniformly distributed load of 2.5 kN/m dead load (including self-

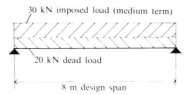

Fig. 7.13

weight) and 3.75 kN/m medium term imposed over a design span of 8.0 m as shown in Fig. 7.13 assuming full lateral restraint to the section.

Also determine a suitable size in LC grade and compare the material content.

| | | |
|---|---|---|
| Long term load | $= F_L$ | $= 20\,kN$ |
| Medium term load | $= F_M$ | $= 50\,kN$ |

| | | |
|---|---|---|
| Long term shear | $= F_{vL}$ | $= 10\,kN$ |
| Medium term shear | $= F_{vM}$ | $= 25\,kN$ |

| | | |
|---|---|---|
| Long term moment | $= M_L$ | $= 20\,kN.m$ |
| Medium term moment | $= M_M$ | $= 50\,kN.m$ |

As a first approximation take $EI_X$ as:

$$EI_X = 4.6 F_i\, L^2 \qquad \text{(see § 5.3.2)}$$

where $F_i$ = imposed load (kN).

$$\text{Required } EI_X = 4.6 \times 30 \times 8^2 = 8832\,kN.m^2$$

From Table 7.7, try 90 × 540 mm section LB grade for which:

$$\bar{F}_{vL} = 53.1 \quad kN \quad > 10\,kN$$
$$\bar{F}_{vM} = 66.4 \quad kN \quad > 25\,kN$$
$$\bar{M}_L = 42.4 \quad kN.m > 20\,kN.m$$
$$\bar{M}_M = 53.0 \quad kN.m > 50\,kN.m$$
$$EI_X = 11\,160\,kN.m^2$$
$$AG_X = 28\,704\,kN$$

$$\text{Total deflection} = \frac{5 \times 50 \times 8^3}{384 \times 11\,160} + \frac{50}{28\,704}$$

$$= 0.0299 + 0.0017 = 0.0316\,m$$

By proportion:

Imposed load deflection $= 0.0316 \times 30/50 = 0.0190\,m$
Dead load deflection $\quad= 0.0316 \times 20/50 = 0.0126\,m.$

Therefore provide say 15 mm camber.

Allowable imposed load deflection $= 0.003 \times 8 = 0.024$ m which is satisfactory.

From Table 7.8 a 90 $\times$ 630 mm LC grade section would be required as determined by the medium term moment capacity. The LC grade choice would require 17% more timber than the LB grade section and would be 90 mm deeper. The design favours the LB grade beam unless there is a unit cost economy sufficient to off-set the extra material. If the beam has to be sanded, varnished or stained, there will also be higher finishing cost for the LC grade which has 14% more surface area.

## 7.11   THE CALCULATION OF DEFLECTION AND BENDING STRESS OF GLULAM BEAMS WITH TAPERED PROFILES

### 7.11.1   Introduction

The calculation of bending deflection and shear deflection for glulam beams with tapering profiles by the normal stain energy method is extremely time-consuming, as is the calculation of the position and value of the maximum bending stress. In §§ 7.11.2, 7.11.4 and 7.11.6, formulae and coefficients are given for simply supported beams of symmetrical duo-pitch, mono-pitch and inverted duo-pitch to cover most cases of normal loading. These coefficients facilitate the calculation of bending deflection and shear deflection at mid span, and maximum bending stress.

The formulae for bending deflection at mid span are established from the general formula:

$$\delta_m = \int_0^L \frac{M_X m_X \, dx}{EI_X}$$

where $M_X =$ the moment at $X$ due to the applied loading

$m_X =$ the moment at $X$ due to unit load applied at mid span.

In addition, the varying nature of the second moment of area is taken into account by introducing a ratio

$$\frac{H}{h} = n$$

where $H =$ the mid span depth

$h =$ the minimum end depth.

The formulae for shear deflection at mid span are established from the general formula:

$$\delta_v = K_{form} \int_0^L \frac{F_{vX} F_{vu} \, dx}{(AG)_X}$$

where $K_{form}$ = a form factor equal to 1.2 for a solid rectangle

$F_{vX}$ = the vertical shear at $X$ due to the applied loading

$F_{vu}$ = the vertical shear at $X$ due to unit load applied at mid span.

In addition, the varying nature of the area is taken into account by introducing a ratio $H/h = n$, where $H$ = the mid span depth and $h$ = the minimum end depth.

## 7.11.2   Formulae for calculating bending deflection of glulam beams with tapered profiles

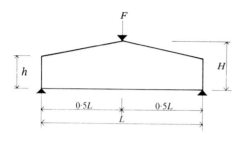

Central bending deflection is:

$$\delta_{m1} = \frac{5FL^3 \Delta_{m1}}{384EI_h}$$

where $\Delta_{m1} = 4.8 \left[ \dfrac{1}{n-1} \right]^3 \left[ \log_e n - \dfrac{3n^2 - 4n + 1}{2n^2} \right]$ as tabulated

$I_h$ = the second moment of area at depth $h$

$n = H/h$.

Fig. 7.14

| Values of $\Delta_{m1}$ for values of $n$ from 1.0 to 3.9 | | | |
|---|---|---|---|
| $n$ | 1.0 | 2.0 | 3.0 |
| 0 | 1.6 | 0.327 | 0.126 |
| 0.05 | 1.433 | | |
| 0.10 | 1.291 | 0.292 | 0.116 |
| 0.15 | 1.167 | | |
| 0.20 | 1.060 | 0.262 | 0.108 |
| 0.25 | 0.966 | | |
| 0.30 | 0.883 | 0.236 | 0.100 |
| 0.35 | 0.810 | | |
| 0.40 | 0.746 | 0.213 | 0.0933 |
| 0.45 | 0.688 | | |
| 0.50 | 0.637 | 0.194 | 0.0870 |
| 0.55 | 0.590 | | |
| 0.60 | 0.549 | 0.177 | 0.0813 |
| 0.65 | 0.511 | | |
| 0.70 | 0.477 | 0.162 | 0.0762 |
| 0.75 | 0.446 | | |
| 0.80 | 0.418 | 0.149 | 0.0714 |
| 0.85 | 0.392 | | |
| 0.90 | 0.369 | 0.137 | 0.0670 |
| 0.95 | 0.347 | | |

Example: for $n = 1.90$,
$\Delta_{m1} = 0.369$

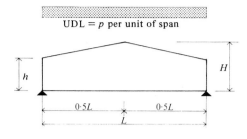

Values of $\Delta_{m2}$ for values of $n$ from 1.0 to 3.9

| $n$ | 1.0 | 2.0 | 3.0 |
|---|---|---|---|
| 0 | 1.000 | 0.218 | 0.0869 |
| 0.05 | 0.900 | | |
| 0.10 | 0.814 | 0.195 | 0.0806 |
| 0.15 | 0.739 | | |
| 0.20 | 0.673 | 0.176 | 0.0749 |
| 0.25 | 0.616 | | |
| 0.30 | 0.565 | 0.159 | 0.0698 |
| 0.35 | 0.520 | | |
| 0.40 | 0.481 | 0.144 | 0.0651 |
| 0.45 | 0.445 | | |
| 0.50 | 0.413 | 0.132 | 0.0609 |
| 0.55 | 0.384 | | |
| 0.60 | 0.358 | 0.120 | 0.0571 |
| 0.65 | 0.334 | | |
| 0.70 | 0.313 | 0.111 | 0.0535 |
| 0.75 | 0.293 | | |
| 0.80 | 0.276 | 0.102 | 0.0503 |
| 0.85 | 0.259 | | |
| 0.90 | 0.244 | 0.0939 | 0.0474 |
| 0.95 | 0.231 | | |

Central bending deflection is:

$$\delta_{m2} = \frac{5pL^4 \Delta_{m2}}{384EI_h}$$

where $\Delta_{m2} = 2.4\left[\frac{1}{n-1}\right]^3 \left[\frac{2n+1}{n-1}\log_e n\right.$

$$\left. - \frac{8n^2 - 3n + 1}{2n^2}\right] \text{ as tabulated}$$

$I_h$ = the second moment of area at depth $h$

$n$ = $H/h$.

Fig. 7.15

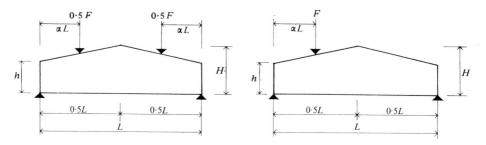

Central bending deflection is:

$$\delta_{m3} = \frac{5FL^3 \Delta_{m3}}{384EI_h}$$

where    $\Delta_{m3} = 4.8 \left[\dfrac{1}{n-1}\right] \left[\dfrac{\log_e N}{n-1} - \dfrac{\alpha}{N} - \dfrac{\alpha(2n-1)}{n^2}\right]$ as tabulated

$I_h$ = the second moment of area at depth $h$

$n = H/h$

$N = 2\alpha(n-1) + 1.$

| | | | | Values of $\Delta_{m3}$ for values of $n$ from 1.0 to 3.9 | | | | | | | | |
|---|---|---|---|---|---|---|---|---|---|---|---|---|
| | $\alpha = 0.1$ | | | $\alpha = 0.2$ | | | $\alpha = 0.3$ | | | $\alpha = 0.4$ | | |
| $n$ | 1.0 | 2.0 | 3.0 | 1.0 | 2.0 | 3.0 | 1.0 | 2.0 | 3.0 | 1.0 | 2.0 | 3.0 |
| 0 | 0.474 | 0.115 | 0.0495 | 0.909 | 0.209 | 0.0860 | 1.27 | 0.276 | 0.109 | 1.51 | 0.315 | 0.122 |
| 0.10 | 0.390 | 0.104 | 0.0462 | 0.745 | 0.188 | 0.0799 | 1.03 | 0.247 | 0.101 | 1.22 | 0.281 | 0.113 |
| 0.20 | 0.327 | 0.0946 | 0.0432 | 0.621 | 0.170 | 0.0745 | 0.854 | 0.223 | 0.0942 | 1.01 | 0.252 | 0.105 |
| 0.30 | 0.278 | 0.0862 | 0.0405 | 0.525 | 0.154 | 0.0695 | 0.717 | 0.201 | 0.0877 | 0.840 | 0.228 | 0.0973 |
| 0.40 | 0.239 | 0.0789 | 0.0380 | 0.449 | 0.141 | 0.0650 | 0.610 | 0.183 | 0.0818 | 0.711 | 0.206 | 0.0907 |
| 0.50 | 0.208 | 0.0725 | 0.0357 | 0.388 | 0.129 | 0.0609 | 0.524 | 0.166 | 0.0765 | 0.608 | 0.187 | 0.0846 |
| 0.60 | 0.182 | 0.0668 | 0.0337 | 0.338 | 0.118 | 0.0571 | 0.454 | 0.152 | 0.0716 | 0.525 | 0.171 | 0.0791 |
| 0.70 | 0.161 | 0.0617 | 0.0318 | 0.297 | 0.109 | 0.0537 | 0.397 | 0.139 | 0.0671 | 0.457 | 0.156 | 0.0741 |
| 0.80 | 0.143 | 0.0570 | 0.0300 | 0.263 | 0.100 | 0.0506 | 0.349 | 0.128 | 0.0631 | 0.401 | 0.144 | 0.0695 |
| 0.90 | 0.128 | 0.0531 | 0.0284 | 0.234 | 0.0927 | 0.0477 | 0.310 | 0.118 | 0.0593 | 0.354 | 0.132 | 0.0653 |
| $N$ | $0.2n + 0.8$ | | | $0.4n + 0.6$ | | | $0.6n + 0.4$ | | | $0.8n + 0.2$ | | |

Fig. 7.16

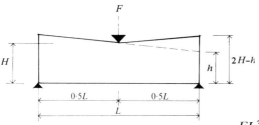

As this case is not particularly common values of $n$ for values of $\Delta_{m4}$ are not tabulated. The same applies to the next case.

Central bending deflection is:  $\delta_{m4} = \dfrac{FL^3}{16EI_h} \left[\dfrac{1}{n-1}\right]^3 \left[\log_e \dfrac{2n-1}{n} - \dfrac{n^2-1}{2n^2}\right]$

Fig. 7.17

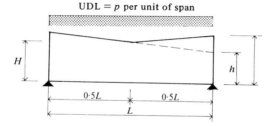

Central bending deflection is:

$$\delta_{m5} = \frac{pL^4}{32EI_h}\left[\frac{1}{n-1}\right]^3\left[\frac{6n^2-n+1}{2n^2}-\frac{4n-1}{n-1}\times\log_e\frac{2n-1}{n}\right]$$

Fig. 7.18

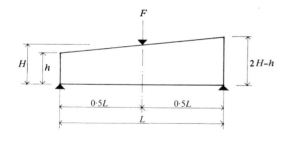

Central bending deflection is:

$$\delta_{m6} = \frac{5FL^3\Delta_{m6}}{384EI_h}$$

where $\Delta_{m6} = 2.4\left[\frac{1}{n-1}\right]^3\left[\log_e(2n-1)\right.$

$$\left.-\frac{2(n-1)}{n}\right]$$ as tabulated

$I_h$ = the second moment of area at depth $h$

$n$ = $H/h$.

| | Values of $\Delta_{m6}$ for values of $n$ from 1.0 to 3.9 | | |
|---|---|---|---|
| $n$ | 1.0 | 2.0 | 3.0 |
| 0 | 1.60 | 0.237 | 0.0828 |
| 0.05 | 1.38 | | |
| 0.10 | 1.21 | 0.208 | 0.0761 |
| 0.15 | 1.06 | | |
| 0.20 | 0.942 | 0.184 | 0.0702 |
| 0.25 | 0.839 | | |
| 0.30 | 0.752 | 0.164 | 0.0649 |
| 0.35 | 0.678 | | |
| 0.40 | 0.613 | 0.147 | 0.0601 |
| 0.45 | 0.557 | | |
| 0.50 | 0.508 | 0.132 | 0.0558 |
| 0.55 | 0.465 | | |
| 0.60 | 0.427 | 0.120 | 0.0519 |
| 0.65 | 0.393 | | |
| 0.70 | 0.363 | 0.109 | 0.0484 |
| 0.75 | 0.336 | | |
| 0.80 | 0.312 | 0.0989 | 0.0452 |
| 0.85 | 0.290 | | |
| 0.90 | 0.271 | 0.0904 | 0.0423 |
| 0.95 | 0.253 | | |

Fig. 7.19

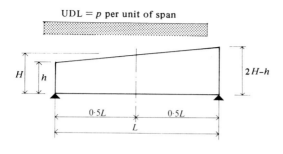

UDL = $p$ per unit of span

$2H-h$

$H$   $h$

$0.5L$    $0.5L$

$L$

Central bending deflection is:

$$\delta_{m7} = \frac{5pL^4 \Delta_{m7}}{384EI_h}$$

where    $\Delta_{m7} = 1.2 \left[\frac{1}{n-1}\right]^4 \left[6n \log_e n \right.$

$$\left. - (4n-1)\log_e(2n-1) - \frac{(n-1)^2}{n}\right]$$

$I_h$ = the second moment of area at depth $h$

$n = H/h$.

| $n$ | 1.0 | 2.0 | 3.0 |
|-----|-----|-----|-----|
| 0 | 1.000 | 0.153 | 0.0553 |
| 0.05 | 0.865 | | |
| 0.10 | 0.756 | 0.135 | 0.0510 |
| 0.15 | 0.666 | | |
| 0.20 | 0.590 | 0.120 | 0.0472 |
| 0.25 | 0.527 | | |
| 0.30 | 0.473 | 0.107 | 0.0437 |
| 0.35 | 0.427 | | |
| 0.40 | 0.387 | 0.0965 | 0.0406 |
| 0.45 | 0.353 | | |
| 0.50 | 0.322 | 0.0872 | 0.0378 |
| 0.55 | 0.296 | | |
| 0.60 | 0.272 | 0.0790 | 0.0352 |
| 0.65 | 0.251 | | |
| 0.70 | 0.232 | 0.0719 | 0.0329 |
| 0.75 | 0.215 | | |
| 0.80 | 0.200 | 0.0657 | 0.0308 |
| 0.85 | 0.187 | | |
| 0.90 | 0.174 | 0.0602 | 0.0289 |
| 0.95 | 0.163 | | |

Values of $\Delta_{m7}$ for values of $n$ from 1.0 to 3.9

Fig. 7.20

### 7.11.3    Example of calculating bending deflection of a tapered glulam beam

Consider a 135 mm wide LC grade Whitewood laminated beam tapered and loaded as in Fig. 7.21. Calculate the mid span bending deflection (a) by considering elemental strips and using the strain energy method, and (b) by using the coefficients from Fig. 7.14.

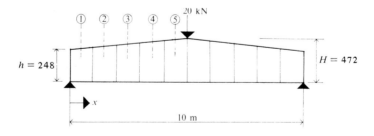

Fig. 7.21

*Method (a)*

Divide the beam into a convenient number of pieces, in this case ten equal elements of 1 m width. Then:

$$\text{Deflection} = 2 \times \Sigma_1^5 \frac{M_X m_X \, \delta s}{E I_X}$$

where $M$ = the moment at mid point of elements 1 to 5 due to the 20 kN
point load

$m$ = the unit moment at mid points of elements 1 to 5 due to unit
load at mid span

$EI = EI$ at mid points of elements 1 to 5

$\delta s$ = increment of span (equal to 1 m in this particular example).

At a distance $x$ from the left hand support up to mid span:

$$M_X = 10x \qquad m_X = 0.5x \qquad \delta s = 1.0 \, \text{m}$$

By tabulation, consider one half of the beam, taking the $EI$ value of each element from Fig. 7.21 appropriate to the mid position depth of each element.

From Table 7.9:

$$\text{Deflection} = 2 \times 0.03309 \times 1.0 = 0.0662 \, \text{m (i.e. 66.2 mm)}$$

**Table 7.9**

| Element | $x$ (m) | $M_X$ (kN.m) | $m_X$ (m) | $h_X$ (mm) | $E I_X$ (kN.m²) | $\frac{M_X m_X}{E I_X}$ |
|---------|---------|--------------|-----------|------------|-----------------|-------------------------|
| 1 | 0.5 | 5 | 0.25 | 270 | 1860 | 0.00067 |
| 2 | 1.5 | 15 | 0.75 | 315 | 2953 | 0.00381 |
| 3 | 2.5 | 25 | 1.25 | 360 | 4408 | 0.00709 |
| 4 | 3.5 | 35 | 1.75 | 405 | 6277 | 0.00976 |
| 5 | 4.5 | 45 | 2.25 | 450 | 8611 | 0.01176 |
| | | | | | | $\Sigma = 0.03309$ |

*Method (b)*

Use coefficients from Fig. 7.14 and the $E$ value appropriate to the minimum height:

$$n = \frac{H}{h} = \frac{472}{248} = 1.90 \qquad \therefore \Delta_{m1} = 0.369$$

$$I_h = \frac{135 \times 248^3}{12} = 171.6 \times 10^6 \, \text{mm}^4$$

From Tables 7.1 and 2.2, $E = 0.80 \times 10\,500 = 8400 \, \text{N/mm}^2$.

$$\therefore EI_h = \frac{8400 \times 171.6 \times 10^6}{10^9} = 1441 \text{ kN.m}^2$$

$$\text{Deflection (see Fig. 7.14)} = \frac{5 \times 20 \times 10^3 \times 0.369}{384 \times 1441}$$

$$= 0.0667 \text{ m (i.e. 66.7 mm)}$$

which is within 1% of the value calculated by the much lengthier method (a).

In this example, the end and mid span depths were chosen to give $h_X$ (see Table 7.9) equal to standard depths, and thereby simplify the tabular work in method (a). In practice, method (b) is further simplified by choosing a standard depth at $h$.

### 7.11.4   Formulae for calculating shear deflection of glulam beams with tapered profiles

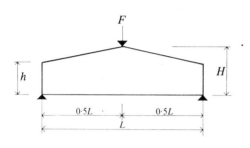

Central shear deflection is:

$$\delta_{v1} = \frac{K_{form} M_0 \Delta_{v1}}{GA_h}$$

where $M_0 = FL/4$

$\Delta_{v1} = \dfrac{\log_e n}{n-1}$ as tabulated

$A_h$ = the area at depth $h$
$K_{form}$ = the form factor (1.2 for a rectangular section)
$n = H/h$.

Fig. 7.22

Values of $\Delta_{v1}$ for values of $n$ from 1.0 to 3.9

| $n$ | 1.0 | 2.0 | 3.0 |
|---|---|---|---|
| 0 | 1.000 | 0.693 | 0.549 |
| 0.05 | 0.976 | | |
| 0.10 | 0.953 | 0.674 | 0.539 |
| 0.15 | 0.932 | | |
| 0.20 | 0.912 | 0.657 | 0.529 |
| 0.25 | 0.893 | | |
| 0.30 | 0.874 | 0.641 | 0.519 |
| 0.35 | 0.857 | | |
| 0.40 | 0.841 | 0.625 | 0.510 |
| 0.45 | 0.826 | | |
| 0.50 | 0.811 | 0.611 | 0.501 |
| 0.55 | 0.797 | | |
| 0.60 | 0.783 | 0.597 | 0.493 |
| 0.65 | 0.770 | | |
| 0.70 | 0.758 | 0.584 | 0.485 |
| 0.75 | 0.746 | | |
| 0.80 | 0.735 | 0.572 | 0.477 |
| 0.85 | 0.724 | | |
| 0.90 | 0.713 | 0.560 | 0.469 |
| 0.95 | 0.703 | | |

Example: for $n = 2.50$,
$\Delta_{v1} = 0.611$.

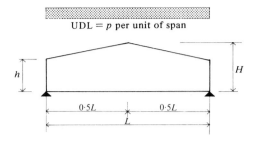

UDL = $p$ per unit of span

| | Values of $\Delta_{v2}$ for values of $n$ from 1.0 to 3.9 | | |
|---|---|---|---|
| $n$ | 1.0 | 2.0 | 3.0 |
| 0 | 1.000 | 0.773 | 0.648 |
| 0.05 | 0.984 | | |
| 0.10 | 0.968 | 0.757 | 0.638 |
| 0.15 | 0.953 | | |
| 0.20 | 0.939 | 0.742 | 0.629 |
| 0.25 | 0.926 | | |
| 0.30 | 0.913 | 0.729 | 0.620 |
| 0.35 | 0.900 | | |
| 0.40 | 0.888 | 0.715 | 0.611 |
| 0.45 | 0.877 | | |
| 0.50 | 0.866 | 0.703 | 0.603 |
| 0.55 | 0.855 | | |
| 0.60 | 0.844 | 0.691 | 0.595 |
| 0.65 | 0.834 | | |
| 0.70 | 0.825 | 0.679 | 0.587 |
| 0.75 | 0.815 | | |
| 0.80 | 0.806 | 0.668 | 0.580 |
| 0.85 | 0.797 | | |
| 0.90 | 0.789 | 0.658 | 0.573 |
| 0.95 | 0.781 | | |

Central shear deflection is:

$$\delta_{v2} = \frac{K_{\text{form}} M_0 \Delta_{v2}}{GA_h}$$

where $M_0 = pL^2/8$

$$\Delta_{v2} = \left[\frac{2}{n-1}\right]\left[\left(\frac{n}{n-1}\right)\log_e n - 1\right]$$

$A_h$ = the area at depth $h$
$K_{\text{form}}$ = the form factor (1.2 for a rectangular section)
$n = H/h$.

Example: for $n = 2.60$,
$\Delta_{v2} = 0.691$.

Fig. 7.23

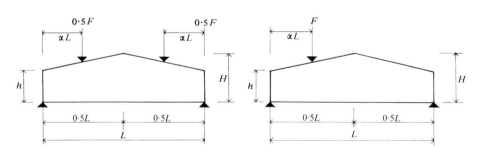

0·5F    0·5F    F
$\alpha L$

Central bending deflection is:

$$\delta_{v3} = \frac{K_{\text{form}} FL}{GA_h}\left[\frac{\log_e N}{n-1}\right]$$

where $N = 2\alpha(n-1) + 1$
$n = H/h$
$A_h$ = the area at depth $h$
$K_{\text{form}}$ = the form factor (1.2 for a rectangular section).

Fig. 7.24

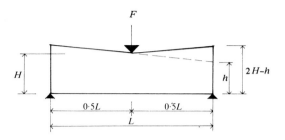

Central bending deflection is:

$$\delta_{v4} = \frac{K_{form}\,FL}{4GA_h\,(n-1)}\left[\log_e\left(\frac{2n-1}{n}\right)\right]$$

where $n = H/h$
$\quad A_h$ = the area at depth $h$
$\quad K_{form}$ = the form factor (1.2 for a rectangular section).

Fig. 7.25

---

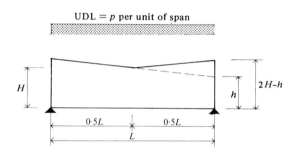

Central bending deflection is:

$$\delta_{v5} = \frac{K_{form}\,pL^2}{4GA_h\,(n-1)}\left[1 - \frac{n}{n-1}\log_e\frac{2n-1}{n}\right]$$

where $n = H/h$
$\quad A_h$ = the area at depth $h$
$\quad K_{form}$ = the form factor (1.2 for a rectangular section).

Fig. 7.26

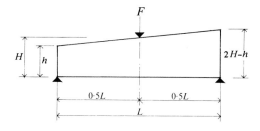

Fig. 7.27

Central shear deflection is:

$$\delta_{v6} = \frac{K_{form}\, M_0\, \Delta_{v6}}{GA_h}$$

where  $\Delta_{v6} = \dfrac{1}{2(n-1)} \log_e (2n-1)$

as tabulated

$M_0$ = the moment at mid span
$A_h$ = the area at depth $h$
$K_{form}$ = the form factor (1.2 for a rectangular section)
$n = H/h$.

| | Values of $\Delta_{v6}$ for values of $n$ from 1.0 to 3.9 | | |
|---|---|---|---|
| $n$ | 1.0 | 2.0 | 3.0 |
| 0 | 1.000 | 0.549 | 0.402 |
| 0.05 | 0.953 | | |
| 0.10 | 0.912 | 0.529 | 0.392 |
| 0.15 | 0.874 | | |
| 0.20 | 0.841 | 0.510 | 0.383 |
| 0.25 | 0.811 | | |
| 0.30 | 0.783 | 0.493 | 0.374 |
| 0.35 | 0.758 | | |
| 0.40 | 0.735 | 0.477 | 0.366 |
| 0.45 | 0.713 | | |
| 0.50 | 0.693 | 0.462 | 0.358 |
| 0.55 | 0.674 | | |
| 0.60 | 0.657 | 0.448 | 0.351 |
| 0.65 | 0.641 | | |
| 0.70 | 0.625 | 0.436 | 0.344 |
| 0.75 | 0.611 | | |
| 0.80 | 0.597 | 0.424 | 0.337 |
| 0.85 | 0.584 | | |
| 0.90 | 0.572 | 0.413 | 0.330 |
| 0.95 | 0.560 | | |

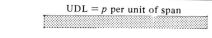

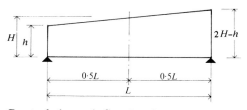

Central shear deflection is:

$$\delta_{v7} = \frac{K_{form}\, M_0\, \Delta_{v7}}{GA_h}$$

where  $\Delta_{v7} = \dfrac{n}{(n-1)^2} \log_e \dfrac{n^2}{2n-1}$

as tabulated

$M_0$ = the moment at mid span
$A_h$ = the area at depth $h$
$K_{form}$ = the form factor (1.2 for a rectangular section)
$n = H/h$.

| | Values of $\Delta_{v7}$ for values of $n$ from 1.0 to 3.9 | | |
|---|---|---|---|
| $n$ | 1.0 | 2.0 | 3.0 |
| 0 | 1.000 | 0.575 | 0.441 |
| 0.05 | 0.953 | | |
| 0.10 | 0.913 | 0.557 | 0.432 |
| 0.15 | 0.877 | | |
| 0.20 | 0.845 | 0.539 | 0.423 |
| 0.25 | 0.816 | | |
| 0.30 | 0.790 | 0.524 | 0.415 |
| 0.35 | 0.767 | | |
| 0.40 | 0.745 | 0.509 | 0.407 |
| 0.45 | 0.725 | | |
| 0.50 | 0.707 | 0.496 | 0.400 |
| 0.55 | 0.689 | | |
| 0.60 | 0.673 | 0.483 | 0.393 |
| 0.65 | 0.658 | | |
| 0.70 | 0.645 | 0.472 | 0.386 |
| 0.75 | 0.631 | | |
| 0.80 | 0.619 | 0.461 | 0.379 |
| 0.85 | 0.607 | | |
| 0.90 | 0.596 | 0.450 | 0.373 |
| 0.95 | 0.585 | | |

Fig. 7.28

### 7.11.5   Example of calculating shear deflection of a tapered glulam beam

Calculate the shear deflection at mid span for the beam profile and loading condition illustrated in Fig. 7.21, using the $\Delta_v$ coefficients as in Fig. 7.22.

From Fig. 7.22, with $n = 1.90$, $\Delta_{v1} = 0.713$.

The modulus of rigidity $G$ is taken as $E/16$.

From Tables 7.1 and 2.2, $E = 0.80 \times 10\,500 = 8400\,\text{N/mm}^2$ therefore $G = 525\,\text{N/mm}^2$.

The shear ridigity $GA_h$ at depth $h$ of 248 mm is:

$$= \frac{525 \times 135 \times 248}{10^3} = 17\,580\,\text{kN}$$

The mid span moment $M_0 = 50\,\text{kN.m}$

The shear deflection is:

$$\delta_{v1} = \frac{K_{\text{form}}\,M_0\,\Delta_{v1}}{GA_h} = \frac{1.2 \times 50 \times 0.713}{17\,580} = 0.0024\,\text{m} \quad (2.4\,\text{mm})$$

This is 3.6% of the bending deflection.

### 7.11.6   Formulae for calculating position and value of maximum bending stress of glulam beams with tapered profiles

For a simply supported beam of the tapering profiles as shown in Fig. 7.29 with uniform loading, it can be shown that the position of the maximum bending stress occurs at:

$$x_m = \frac{Lh}{2H}$$

where $x_m$, $L$, $h$ and $H$ are as shown in Fig. 7.29.

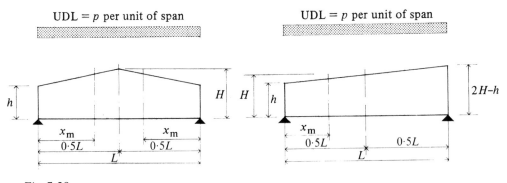

Fig. 7.29

It can also be shown that:

$$\text{Maximum bending stress} = \frac{M_0}{Z_h\,(2n - 1)}$$

where $M_0$ = the moment at mid span
$\quad Z_h$ = the section modulus at depth $h$
$\quad n = H/h$.

By transposing, one arrives at an equation which will permit the designer to avoid time-consuming trial and error in the design of taper beams:

$$n = \tfrac{1}{2}\left[\frac{M_0}{\overline{m}_h} + 1\right]$$

where $\overline{m}_h$ is the moment of resistance at depth $h$.

For a simply supported beam of the tapering profile shown in Fig. 7.30 with a central point load, it can be shown that the position of the maximum bending stress occurs at:

$$x_m = \frac{L}{2(n-1)} \qquad (n = H/h)$$

where $x_m$, $L$, $h$ and $H$ are as shown in Fig. 7.30.

The value of $x_m$ is for values of $n$ greater than 2. For cases with $n$ less than 2 the maximum bending stress occurs at mid span. It can be shown that:

$$\text{Maximum bending stress} = \frac{M_0}{4Z_h\,(n-1)}$$

where $M_0$ = the moment at mid span
$\quad Z_h$ = the section modulus at depth $h$
$\quad n = H/h$.

By transposing one arrives at an equation which will permit the designer to avoid time-consuming trial and error in the design of taper beams.

$$n = \frac{M_0}{4\overline{m}_h} + 1$$

where $\overline{m}_h$ is the moment of resistance at depth $h$.

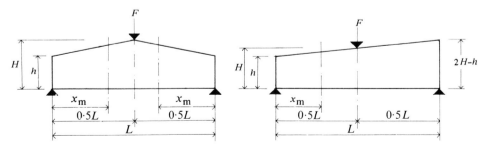

Fig. 7.30

**Chapter Eight**
# Ply Web Beams

## 8.1  INTRODUCTION

Ply web beams are usually of I or Box profile utilising plywood in the web construction and softwood in the flanges. Each flange may consist of one or several members glued together, placed with either their $XX$ or $YY$ axes horizontal. Some of the possible variations are sketched in Fig. 8.1. The profiles marked with a P are patented.

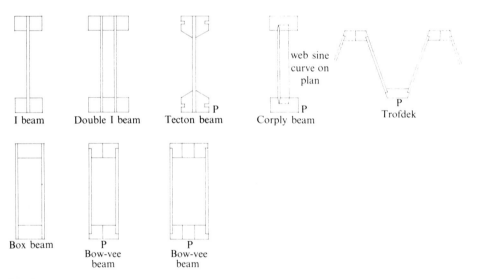

Fig. 8.1

Ply web beams are particularly suitable and economical for roof spans of 6–20 m when spaced at about 1.2 m centres. The closeness of the spacing permits simple, cheap secondary systems of cladding and ceilings to be used.

Whereas it is very usual to expose a glulam beam, the main use for ply web beams is with a ceiling fixed to the lower flange, so that the appearance of the beams is not important. With Box beams, it may be possible for sanding at an extra cost to lead to an acceptable appearance if the beams are to be left exposed, but I sections are not usually suitable for being left exposed unless finished, for example, with an aggregate-loaded paint which can cover glue marks, excess beads of glue, and nail heads and holes.

It is usual to have glue joints between webs and flanges. Assembly of beams is usually by glue/nailing although some beams are assembled purely by clamping, with no nails being used in the flange-to-web connection. The I beam with the curved web (Corply) sketched in Fig. 8.1 has no nails in the flange-to-web connections. The flanges of Corply can use sawn timber sizes. Flanges of the other sections would be processed.

Although only plywood webs are discussed in this chapter, BS 5268: Part 2 does give strength values for tempered hardboard and this is used in the construction of web beams. Note has to be taken of the different coefficients and particularly to the service environment.

Ply web beams can be manufactured with a camber.

See §§7.6, 7.7 and Chapter 19 for guidance on gluing and preservation.

## 8.2 DISCUSSION ON METHODS OF ANALYSIS

### 8.2.1 Bending stress and deflection

Several methods of analysis have been proved by test and years of usage to give suitable designs. Five methods are discussed in §8.4 in relation to bending stress. The resistance to deflection is very often the main design criterion. Shear deflection must be taken into account (§8.6).

### 8.2.2 Ply web

With regard to the contribution to strength and stiffness made by the ply web, there are two distinct areas of opinion regarding the design methods. The traditional North American method is the 'parallel ply method' in which only those plies which are parallel to the direction of stress are considered to resist stress, whereas BS 5268:Part 2 gives permissible stresses related to the full cross-sectional area of the plywood. As one would expect, the stresses permitted on the full cross-section are less than those for parallel plies only. Either method should normally lead to the same section being used, which is hardly surprising when one considers that both sets of stresses are derived from the same type of tests on actual plywood.

The parallel ply method is probably favoured by those engineers who have a pure approach to design, as it does relate more correctly to the actual stresses in individual plies. The permissible stresses are related to a veneer strength (applicable to the strong direction), and the geometrical properties of the plywood overall are computed from the thickness and disposition of parallel plies only, disregarding those veneers which are perpendicular to the direction of stress.

With the full area method the calculation of geometrical properties is less complex, but permissible stresses differ for each thickness of plywood and different layup of veneers. This is the method of presentation selected by BS 5268:Part 2, consequently the full area method is used in this chapter.

It must be appreciated that a ply web beam is a composite beam with flanges and webs of different materials, and consequently, differing $E$ values. On a few occasions, the $E$ values of the flange material and the plywood may be close enough to justify the designer taking the section as monolithic and therefore calculating geometrical properties without any modification for differing $E$ values. However, the more normal situation involves differing $E$ values, and therefore the use of straightforward geometrical properties may not lead to a correct or sufficiently accurate design. The method of dealing with differing materials in design is explained in this chapter.

### 8.2.3 Degree of load sharing

Clause 14.10 of BS 5268:Part 2:1984 clarifies the doubts which existed in CP 112:Part 2 on the degree of load sharing which could be assumed to occur within one built-up beam such as a ply web beam, and in a system of beams which are usually spaced at centres around 1.2 m. As far as deflection is concerned the recommendations agree with the principles outlined in the First Edition of this manual. However, as far as bending stress is concerned, it has now been agreed that the number of pieces to consider as acting together is the number per flange rather than in cross-section. For example, referring to Fig. 8.1, the number of pieces $N$ acting against deflection in the I beam as sketched is four, whereas the number of pieces considered to act together against stress in each flange is two. For the Corply beam and the Box beam, the number $N$ resisting deflection is two and, resisting stress in each flange is one.

Of course, if built-up beams are spaced at centres not exceeding 610 mm it can be argued that $E_{mean}$ can be used. Normally, however, the factors to use are $K_{27}$, $K_{28}$ and $K_{29}$ (numerically the same as for vertically glued laminated beams).

For example, once a system of ply web beams spaced at 1.2 m centres is complete, including continuous noggings fixed at the level of the bottom flanges, it is pretty certain that this is a load-sharing system. Even if this is not taken into account in choosing the $E$ value and value of $\sigma_{m,adm}$ to assume in calculations, it can be an important consideration with regard to the use of finger joints in flanges in view of the wording of Clause 47.4.1 of BS 5268:Part 2 (see §19.7.3).

## 8.3  CONSTRUCTION OF PLY WEB BEAMS

### 8.3.1  Timber and plywood

Although the designer appears to have the choice of a number of species of timber and plywood from which to construct a ply web beam, in practice this choice is limited.

The timber used in the flanges is usually European Whitewood with Redwood as an alternative, and Douglas Fir-Larch occasionally used. However, sawn sizes of Douglas Fir-Larch are imported green and therefore the cost and time for kiln drying in the UK has to be allowed for in designs if Canadian or USA timber is specified. In choosing the stress grade for flange material it can be advantageous to choose a grade which can be selected both visually or by machine. If that point is agreed to be relevant, then the choice becomes either General Structural or Special Structural of BS 4978 (plus the added proviso that there must be no wane in the completed beam).

If the sections will usually be machine stress graded with visual grading used only as a standby, it is probably preferable to base designs (particularly if standard in-company designs) on Special Structural. The yield from normal productions of machine stress-graded European Whitewood and Redwood should be high. If, however, most of the sections will be visually stress graded, consideration should be given to using General Structural, particularly if each section is quite small.

When stress grading a small section to the SS rules, it is quite normal for a 'margin condition' to occur (§2.4.2), therefore the total knot area ratio must be limited to $\frac{1}{5}$ which, in real terms, means that only very small knots are permitted say in a quarter flange of 35 × 60 mm finished size.

When designing using rules from BS 4978 it is necessary to remember that currently (1984) the smallest size permitted is 35 × 60 and the minimum depth is 60 mm. (Thus 44 × 44 mm is not permitted, although this situation is under review.)

In the further discussion and examples in this chapter, SS grade European Whitewood is assumed for flanges, although GS is usually adequate for stiffeners, even load-bearing stiffeners.

The plywood used should be bonded with an exterior quality resin adhesive.

When the webs are highly stressed the designer will probably wish to use Finnish all-Birch plywood, but continuous supplies can be difficult to obtain when required, and it would probably be wise to design on the basis of Finnish Birch-faced plywood. This has alternate plies of birch and softwood (see §1.9.3) but, except for 6.5 mm thicknesses, the two outer veneers are of birch. It is more readily available than the all-birch plywood. The IV/IV face quality is usually adequate if the beams are hidden, but a superior face quality may be desirable if the beams are exposed. (The same stresses apply to all face qualities.)

The main alternatives are Douglas Fir-faced plywoods or one of the other softwood plywoods referred to in §1.9 (but take note of quality). Although Finnish Birch-faced plywood has a panel shear stress about 15% lower than all-Birch plywood it is still 50% higher than the value allocated to any softwood plywood quoted in BS 5268:Part 2. For those layups where two outer veneers are of birth a higher rolling shear stress is permissible compared to the case where only one outer veneer on each side is birch (i.e. 6.5 mm thick plywood).

The lower grades of plywood may be acceptable if beams are not to be exposed, but are unlikely to be acceptable for exposed beams unless an aggregate loaded finish is to be applied. If in doubt about appearance, the specifier should examine a specimen parcel of plywood sheets. The designer should also consider whether or not the face is suitable for gluing, and consider the integrity of sheets for use in an engineered component.

The plywood is normally orientated so that web joints occur only at 2440 (or 2400) mm centres rather than 1220 mm, with a make-up piece at the centre of the beam. Therefore, with Finnish plywood the face grain is perpendicular to span and with most others it is parallel to span. The permissible stresses vary with direction of face grain.

### 8.3.2   Joints

The structural interaction between flanges and web or webs is achieved by a continuous glue joint. Although the glue faces can be held together during curing by pressure clamping, the more common method is to obtain glue-line pressure by either nails or staples. Staples would be used only in fixing plywood to softwood, as with a Box beam, and not for fixing softwood to softwood.

Clause 47.3 of BS 5268:Part 2 gives requirements where the bonding pressure is to be obtained by nailing, and a new British Standard (BS 6446) on the manufacture of glued structural components of timber and wood based panels should be printed in 1984. In general if plywood (or tempered hardboard) is not more than 12.5 mm thick, nails along the grain (of the softwood flanges) should occur at centres not in excess of 100 mm and, for thicker plywood, not in excess of 150 mm. Obviously, however, there is a difference between plywood sandwiched between two quarter flanges in the case of an I beam, and the case of a Box beam.

Glue squeeze-out must be apparent. Usually a WBP glue is used although, if the service conditions permit, a BR or MR glue may be acceptable (but take account of delivery and erection conditions). See Chapter 19.

Stiffeners and web joints are normally glued to the webs in a similar way to the flange-to-web connection.

### 8.3.3 Camber

The majority of ply web beams are built with a camber to off-set dead load deflection. This usually requires a camber in the order of 3–4 mm per 2.4 m chord of camber curve. The top and bottom lines of the plywood will have been cut straight but, in a beam which is not exposed, the manufacturer may not find it necessary to machine off any part of the plywood which falls outside the beam profile. With exposed beams, particularly Box beams, it is usually necessary to fix the plywood to encompass the final curve of the beam completely and then trim to the final manufactured shape.

## 8.4 CALCULATION OF BENDING STRESSES

### 8.4.1 Discussion of various methods

Most designers agree that, at the design level, a straight-line distribution can be assumed for the calculation of the actual (applied) stresses in the extreme fibres of the flanges of a ply web beam due to bending. However, there can be disagreement as to whether the stress in the extreme fibre should be taken as bending tension and bending compression (as one would assume in the elastic design of a steel welded plate girder) or as pure (or approaching pure) tension or compression. In designs to CP 112 : Part 2 there was no great need to resolve this point because the same values were given for permissible bending and tension stresses and, even though compression stresses were lower, most discussion centred on the tension side. In BS 5268 : Part 2 permissible tension stresses are considerably lower than those given for bending and therefore there is more need to resolve the point.

Because there is a solid web (rather than a lattice) there seems no reason, other than perhaps for a quick estimate design, to assume pure tension in one flange and pure compression in the other. The authors have considerable sympathy with methods which assume the extreme fibre stress to be a bending stress. Certainly experience in practice, and with controlled tests to destruction in laboratories, shows designs based on this principle to have adequate factors of safety.

However, in view of the doubts which have been expressed, the authors have evolved a design method (Method 5 below) in which combined bending and tension is considered to occur in the 'tension' flange and combined bending and compression is considered to occur in the 'compression' flange. This seems to offer a sensible compromise in that, for a shallow beam where the amount of flange material in cross-section approximates to that for a solid beam, the bending effect is significant whereas, for a deep beam, the tension or compression in the flanges is more significant.

However, particularly in view of the lower stresses being quoted for permissible tension stresses, the authors leave the choice of which permissible stress to take to individual designers and the approving authority.

Five basic methods of calculating the stress in the flanges are discussed below. The titles given to them have no universal significance.

*Method 1. The homogenous beam method*
This method is used by some engineers and is analogous to the design of rolled-steel beams. The method is the simplest for those engineers familiar with steel design, because the ply web beam is treated as a homogeneous section for the purpose of calculating bending stresses. The modulus $Z$ of the section is calculated by normal geometrical means, and the maximum moment capacity $\bar{M}$ of the section is given by the formula:

$$\bar{M} = \sigma_{\mathrm{adm}} Z$$

and the actual extreme fibre stress $= M/Z$.

The choice of whether or not to assume bending stress (or tension and compression) is still open. This method over-estimates the strength of the beam in bending (if permissible bending stress is taken) (§8.4.2) and hence should be used only with caution.

*Method 2. The lever-arm method*
This is an approximate method, often used to give an estimate design or a first check in anticipation of a full design check, and neglects the contribution made by the web in resisting bending. The section is considered as a compression flange with area $A_c$ and a tension flange with area $A_t$ having their centres of area distance $h_a$ apart (Fig. 8.2).

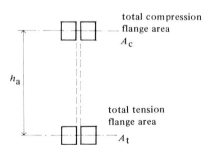

Fig. 8.2

Once again the designer is faced with the choice of which permissible stress to assume. (The 'compression' flange must be adequately restrained laterally.)

$$\text{Moment capacity } \bar{M} = \sigma_{\mathrm{adm}} \times h_a \times A_t \text{ (or } A_c)$$

Even though this method disregards the contribution made by the web in resisting bending, it can lead to a slight over-estimation of the bending strength of the beam (if permissible bending stress is taken).

*Method 3. The flange modulus method*

This method has much to commend it, although it usually under-estimates the bending strength of the beam. The method is similar to Method 2 in that the contribution of the web to the bending strength is disregarded, but the extreme fibre bending stress is more accurately assessed.

The modulus $Z_f$ of the beam flanges only is taken as $I_f/y_h$

where $I_f$ = the second moment of area of the flanges only, about the neutral
axis

    $y_h$ = the distance from the neutral axis to the extreme fibre being
considered ($h/2$ for an extreme fibre of a symmetrical beam).

$$\text{Moment capacity } \bar{M} = \sigma_{adm} Z_f = \sigma_{adm} \frac{I_f}{y_h}$$

As with the previous methods the designer is faced with the choice of which permissible stress to assume.

Apart from the simplicity of this method, disregarding the contribution of the web to the bending strength reduces the design of the web splices (as discussed in § 8.9.2) to a consideration of shear forces only.

*Method 4. The composite material method*

With this method the beam is considered as a composite of two materials and the web is included in the calculation of moment resistance. In the case of bending about the major $XX$ axis, the total moment is taken by each element of the beam in proportion to its $EI_X$ value, in accordance with the theory of composite elements. The designer is still faced with the choice of which permissible stress to assume for the flange material.

$$M = M_f + M_w$$

where $M_f$ = the moment taken by the flanges
    $M_w$ = the moment taken by the webs.

$$M_f = M\left[\frac{E_{fN} \, I_{Xf}}{EI_X}\right] \quad \text{and} \quad M_w = M\left[\frac{E_w I_{Xw}}{EI_X}\right] \tag{8.1}$$

where $E_{fN}$ = the $E$ value of the flanges relevant to the number of flange
pieces in cross-section

    $I_{Xf}$ = the second moment of area of the flanges about $XX$

    $I_{Xw}$ = the second moment of area of the plywood web/s about $XX$

    $E_w$ = the $E$ value of the plywood web/s

    $EI_X$ = the summation of $E_{fN} \, I_{Xf}$ and $E_w \, I_{Xw}$.

The moment capacity of a section symmetrical about the $XX$ axis is then the lesser of the values of:

$$\overline{M} = \frac{2EI_X \, \sigma_{f,adm}}{E_{fN} \, h} \quad \text{or} \quad \overline{M} = \frac{2EI_X \, \sigma_{w,adm}}{E_w \, h}$$

where $\sigma_{f,adm}$ = the permissible stress in bending of the flanges
$\sigma_{w,adm}$ = the permissible stress in bending of the web
$h$ = the depth of the beam.

In most cases the moment capacity is limited by the flange stress. Values of permissible stress for Finnish Birch-faced plywood placed vertically are tabulated in Table 8.1. Because values of bending on edge are not available for plywoods the lesser of the permissible stresses for compression or tension may be taken. As the compression values are usually evolved from tests with the compression edges not fully restrained laterally, the permissible stresses are likely to be conservative when used for ply web beam designs (and hence compression values are smaller than tension).

**Table 8.1**

|  | $\sigma_{adm}$ (N/mm²) for Finnish Birch-faced ply – face grain vertical–perp. to span. Nominal thicknesses (minimum thickness in brackets) in millimetres. | | | | | | | | | | |
|---|---|---|---|---|---|---|---|---|---|---|---|
| Duration of loading | 6.5 (6.1) | 9 (8.8) | 12 (11.5) | 15 (13.9) | 18 (17.1) | 21 (19.9) | 22 (21.1) | 23 (22.7) | 25 (24.4) | 26 (25.2) | 29 (27.9) |
| Long | 3.51 | 5.33 | 5.40 | 5.43 | 5.30 | 5.20 | 5.19 | 5.18 | 5.17 | 5.00 | 4.98 |
| Medium | 4.39 | 6.66 | 6.75 | 6.79 | 6.63 | 6.50 | 6.49 | 6.48 | 6.46 | 6.25 | 6.23 |
| Short | 5.27 | 8.00 | 8.10 | 8.15 | 7.95 | 7.80 | 7.79 | 7.77 | 7.76 | 7.50 | 7.47 |
| Very short | 6.14 | 9.33 | 9.45 | 9.50 | 9.28 | 9.10 | 9.08 | 9.07 | 9.05 | 8.75 | 8.72 |

*Method 5. The combined stress method*
With this method the beam is considered as a composite of two materials and the web is included in the calculation of moment resistance. The applied stress in one flange is a combination of bending and pure tension, and in the other is a combination of bending and pure compression.

As with Method 4 the total moment is shared between flanges and web/s in proportion to their respective *EI* values. A check is required both on the extreme fibre stress in the flanges and the extreme fibre stress in the web. Web stresses will be the same as those established with Method 4.

With a symmetrical section, the 'tension' flange is usually the critical flange for the timber, the 'compression' edge for the plywood.

Consider the tension flange in Fig. 8.3:

The stress at the extreme fibre due to the applied moment $M_f$ taken by the flanges is:

$$\sigma_{m,a} + \sigma_{t,a} = \frac{M_f \, h}{2I_f}$$

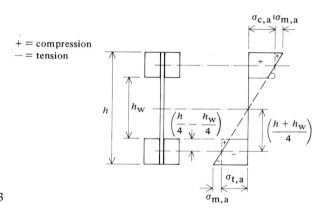

Fig. 8.3

$\therefore$ By proportion, from Fig. 8.3:

$$\frac{M_f h}{2I_f (h/2)} = \frac{\sigma_{t,a}}{(h + h_w)/4}$$

$$\therefore \quad \sigma_{t,a} = \frac{M_f (h + h_w)}{4I_f}$$

Also by proportion, from Fig. 8.3:

$$\frac{M_f h}{2I_f (h/2)} = \frac{\sigma_{m,a}}{(h - h_w)/4}$$

$$\therefore \quad \sigma_{m,a} = \frac{M_f (h - h_w)}{4I_f}$$

The flange section must be proportioned so that:

$$\frac{\sigma_{m,a}}{\sigma_{m,adm}} + \frac{\sigma_{t,a}}{\sigma_{t,adm}} \leqslant 1.0$$

Therefore, substituting the formulae above:

$$\frac{M_f}{4I_f} \left[ \frac{(h - h_w)}{\sigma_{m,adm}} + \frac{(h + h_w)}{\sigma_{t,adm}} \right] \leqslant 1.0$$

In the case of all the softwoods listed in Table 9 of BS 5268 : Part 2 stress graded to BS 4978, the grade tension stress is set at 60% of the grade bending stress, and, providing that, in the case of I beams, $b/2$ does not exceed $h_f$ and, in the case of Box beams, $b$ does not exceed $h_f$, $K_{14}$ is numerically equal to $K_7$ in widths/depths up to 300 mm, and $K_{27}$ is the same for bending as tension. Therefore, for these cases:

$$\frac{\sigma_{t,adm}}{\sigma_{m,adm}} = 0.6$$

Therefore:

$$\frac{M_f}{4I_f}\left[\frac{(h-h_w)}{\sigma_{m,adm}} + \frac{(h+h_w)}{0.6\sigma_{m,adm}}\right] \leqslant 1.0$$

$$\therefore \qquad \frac{M_f(4h+h_w)}{6\,I_f\,\sigma_{m,adm}} \leqslant 1.0$$

The moment capacity of the section, as set by the tension flange is:

$$\bar{M} = \frac{6I_f\,\sigma_{m,adm}\,(EI_X)}{(4h+h_w)\,(E_{fN}\,I_f)} = \frac{6\sigma_{m,adm}\,(EI_X)}{(4h+h_w)\,E_{fN}} \tag{8.2}$$

and the moment capacity, as set by the web, is the same as for Method 4:

$$\bar{M} = \frac{2EI_X\,\sigma_{w,adm}}{E_w\,h}$$

It is worth noting that the base design code produced by working group W18 within CIB (Conseil International du Bâtiment) shows the extreme fibre stress as being limited to the permissible bending stress, but calls for the stress parallel to grain at the centre of area of each flange to be limited, as appropriate, to the permissible tension or compression stress. It hardly seems likely that a stress could change from pure bending to pure tension within half the depth of a flange but, in practice, the method may give similar results to Method 5 as explained above.

### 8.4.2   Example of comparing the five methods

Compare the long term bending moment capacity for the beam sketched in Fig. 8.4 by each of the five methods discussed in § 8.4.1. No allowance is made for load sharing other than $K_{27}$, $E_N$ ($K_{28}$) and $K_{29}$.

$E_{fN}$ for flanges is $E_{min} \times K_{28}$ (for $N = 4$) = 7000 $\times$ 1.24 = 8680 N/mm$^2$ and $E_w$ = 5750 N/mm$^2$ (Table 5.32).

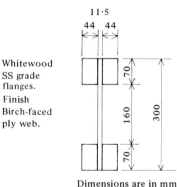

Whitewood
SS grade
flanges.

Finish
Birch-faced
ply web.

Dimensions are in mm

$$I_{Xf} = \frac{88}{12}(300^3 - 160^3) = 168 \times 10^6 \, mm^4$$

$$I_{Xw} = \frac{11.5 \times 300^3}{12} = 25.9 \times 10^6 \, mm^4$$

$$I_X = 193.9 \times 10^6 \, mm^4$$

$$\begin{aligned} EI_X &= E_{fN} I_{Xf} + E_w I_{Xw} \\ &= 8680(168 \times 10^6) + 5750(25.9 \times 10^6) = (1458 + 149)\,10^9 \, N.mm^2 \\ &= 1607 \, kN.m^2 \end{aligned}$$

Depth factor $K_7$ for 70 mm = 1.17. Width factor $K_{14}$ for 70 mm = 1.17 (Fig. 4.4).

*Method 1*

$$\bar{M} = \sigma_{adm} Z \text{ and } Z = \frac{193.9 \times 10^6}{150} = 1.29 \times 10^6 \, mm^3.$$

If the designer decides that the relevant permissible stress is bending then:

$$\sigma_{adm} = \sigma_{g,adm} \times K_7 \times K_{27} \,(\text{for } N = 2) = 7.50 \times 1.17 \times 1.11 = 9.74 \, N/mm^2$$

and

$$\bar{M} = 9.74 \times 1.29 \times 10^6 = 12.56 \times 10^6 \, N.mm$$

If, however, the designer decides that the relevant permissible stress is pure tension or compression whichever is the lesser:

$$\sigma_{t,adm} = \sigma_{t,g,adm} \times K_{14} \times K_{27} = 4.50 \times 1.17 \times 1.11 = 5.844 \, N/mm^2$$

and

$$\sigma_{c,adm} = \sigma_{c,g,adm} \times K_{28} = 7.90 \times 1.14 = 9.006 \, N/mm^2$$
$$\text{(if laterally restrained)}$$

Then:

$$\bar{M} = 5.844 \times 1.29 \times 10^6 = 7.54 \times 10^6 \, N.mm$$

(i.e. only 60% of the value if bending stress is taken).

*Method 2*

$$A_c = A_t = 2 \times 44 \times 70 = 6160 \, mm^2$$
$$h_a = 300 - 35 - 35 = 230 \, mm$$
$$\begin{aligned} \bar{M} &= \sigma_{adm} \times A_c \times h_a \\ &= 9.74 \times 6160 \times 230 = 13.80 \times 10^6 \, N.mm \text{ if bending stress is taken} \end{aligned}$$

or

$$= 5.844 \times 6160 \times 230 = 8.28 \times 10^6 \, N.mm \text{ if tension stress is taken.}$$

*Method 3*

$$I_f = 168 \times 10^6 \, \text{mm}^4$$

$$Z_f = \frac{168 \times 10^6}{150} = 1.12 \times 10^6 \, \text{mm}^4$$

$$\overline{M} = \sigma_{adm} \times Z_f$$
$$= 9.74 \times 1.12 \times 10^6 = 10.91 \times 10^6 \, \text{N.mm if bending stress is taken}$$

or

$$= 5.844 \times 1.12 \times 10^6 = 6.55 \times 10^6 \, \text{N.mm if tension stress is taken.}$$

*Method 4*
$\overline{M}$ is the lesser of:

$$\frac{2EI_X \, \sigma_{f,adm}}{E_{fN} h} = \frac{2 \times 1607 \times 10^9 \times \sigma_{f,adm}}{8680 \times 300} = 1.234 \times 10^6 \times \sigma_{f,adm} \, \text{N.mm}$$

and

$$\frac{2EI_X \, \sigma_{w,adm}}{E_w h} = \frac{2 \times 1607 \times 10^9 \times \sigma_{w,adm}}{5750 \times 300} = 1.863 \times 10^6 \times \sigma_{w,adm} \, \text{N.mm}$$

If the bending stress is taken as the permissible stress in the flanges, the moment capacity based on the tension flange becomes:

$$1.234 \times 9.74 \times 10^6 = 12.02 \times 10^6 \, \text{N.mm (for flanges)}$$

In calculating the moment capacity of the beam based on the extreme fibre stress in the web, it can be seen from Table 45 of BS 5268: Part 2 that, for 12 mm Finnish Birch-faced plywood with face grain perpendicular to line of action of the load, the compression stress (5.40 N/mm²) is less than the tension stress (8.06 N/mm²). (Also see Table 8.1.) Therefore, in the case of the web, the extreme fibre stress on the compression side is the critical edge. Therefore the moment capacity of the beam based on the web is:

$$1.863 \times 5.40 \times 10^6 = 10.06 \times 10^6 \, \text{N.mm (for web)}$$

which shows that the web would be the limiting part of the moment capacity of this beam with this method.

If the tension stress is taken as the permissible stress in the flanges, the calculations become:

$$1.234 \times 5.844 \times 10^6 = 7.21 \times 10^6 \, \text{N.mm (for flanges)}$$

and

$$1.863 \times 5.40 \times 10^6 = 10.06 \times 10^6 \, \text{N.mm (for web)}$$

which shows that the tension flange would be the limiting part of this beam, with this method.

*Method 5*
$\overline{M}$ is the lesser of

$$\frac{2EI_X\,\sigma_{f,adm}}{E_{fN}\,h} \quad \text{and} \quad \frac{2EI_X\,\sigma_{w,adm}}{E_w\,h}$$

In the special case (see §8.4.1) of the timbers stress graded to BS 4978 with $\sigma_{t,adm} = 0.6\sigma_{m,adm}$, and with $K_{14} = K_7$, the first formula becomes:

$$\overline{M} = \frac{6\sigma_{m,adm}\,(EI_X)}{(4h + h_w)\,E_{fN}}$$

With a 70 mm deep flange of SS Whitewood, $\sigma_{t,adm} = 0.6\sigma_{m,adm}$, $K_7 = K_{14}$, and $K_{27}$ applies equally to bending and tension, therefore, for the tension flange:

$$\overline{M} = \frac{6 \times (7.50 \times 1.17 \times 1.11) \times 1607 \times 10^9}{(1200 + 160) \times 8680} = 7.95 \times 10^6\,\text{N.mm}$$
$$\text{(based on 'tension' flange)}$$

$\overline{M}$ based on web is (as Method 4) $= 10.06 \times 10^6\,\text{N.mm}$.

These figures show that the 'tension' flange, taking combined tension and bending, would be the limiting part of this beam. Note, however that the capacity of $7.95 \times 10^6$ is 10% larger than the value of $7.21 \times 10^6\,\text{N.mm}$ derived from Method 4 if the permissible stress in the lower flange is based on pure tension.

The authors are aware that a designer who has not previously designed or checked a ply web beam could be concerned by the number of alternative design methods discussed above. However, particuarly if involved in checking calculations produced by another designer, the engineer may encounter any of the methods. Particularly with deep beams, purists are of the opinion that the stress in the lower flange approximates closer to pure tension than bending, therefore would probably prefer use of Method 5 (or even Method 4 taking the permissible stress as $\sigma_{t,adm}$). However, based on experience and a few tests to destruction of actual beams, the authors are still of the opinion that Method 4 taking the permissible stress as $\sigma_{m,adm}$ is a reasonable assumption. Nevertheless, in view of the comment made in the third sentence of this paragraph, the capacity values given in Tables 8.5 and 8.6 are based on Method 5 assuming combined bending and tension in the lower flange.

If Methods 4 or 5 are used the web splices must be adequate to transmit both shear and bending moment (§ 8.9.2).

If the designer wishes to carry out a check on the design of Method 5, one way is to consider the unity factor at the extreme fibres in the lower flange. When the actual moment is numerically equal to the moment capacity, the actual extreme fibre stress is the same as the permissible stress, therefore:

$$M = \frac{2EI_X\,\sigma_{f,a}}{E_{fN}\,h} = 7.95 \times 10^6 = \frac{2 \times 1607 \times 10^9 \times \sigma_{f,a}}{8680 \times 300}$$

from which:

$$\sigma_{f,a} = 6.441 \text{ N/mm}^2$$

By proportion, from Fig. 8.4:

$$\frac{\sigma_{m,a}}{35} = \frac{\sigma_{t,a}}{115} \quad \text{and} \quad \sigma_{m,a} + \sigma_{t,a} = 6.441$$

from which:

$$\sigma_{m,a} = 1.503 \text{ N/mm}^2 \quad \text{and} \quad \sigma_{t,a} = 4.938 \text{ N/mm}^2$$

$$\sigma_{t,adm} = 4.50 \times 1.17 \times 1.11 = 5.844 \text{ N/mm}^2$$
$$\sigma_{m,adm} = 7.50 \times 1.17 \times 1.11 = 9.74 \text{ N/mm}^2$$

$$\frac{1.503}{9.74} + \frac{4.938}{5.844} = 0.155 + 0.845 = 1.00 \qquad \text{Check.}$$

If the designer wishes to check on the extreme fibre stress in the web this can be carried out by modifying the formula above changing $\bar{M}$ and $\sigma_{w,adm}$ to $M$ and $\sigma_{w,a}$.

$$M = \frac{2EI_X\,\sigma_{w,a}}{E_w\,h} = 7.95 \times 10^6 = \frac{2 \times 1607 \times 10^9 \times \sigma_{w,a}}{5750 \times 300}$$

from which

$$\sigma_{w,a} = 4.27 \text{ N/mm}^2$$

$$\sigma_{w,adm} = 5.40 \text{ N/mm}^2 \text{ (see Table 8.1).}$$

## 8.5   CALCULATION OF BENDING DEFLECTION

### 8.5.1   Introduction

Bending deflection is calculated in a similar way to solid beams, except that the sectional rigidity $EI$ must take account of the differing $E$ values of the flange and web. About the major $XX$ axis, the sectional rigidity is:

$$EI_X = E_{fN}\,I_{Xf} + E_w\,I_{Xw}$$

where $E_{fN}$ = the $E$ value of the flanges relevant to the number of pieces in the flanges ($E_{min} \times K_{28}$)

$I_{Xf}$ = the second moment of area of the flanges about the $XX$ axis

$E_w$ = the $E$ value of the plywood webs

$I_{Xw}$ = the second moment of area of the webs about the $XX$ axis.

Similarly the sectional rigidity about the minor $YY$ axis is:

$$EI_Y = E_{fN}I_{Yf} + E_w I_{Yw}$$

### 8.5.2   Example of calculating bending deflection

Calculate the sectional rigidity about the $XX$ axis for the beam shown in Fig. 8.4 (SS grade Whitewood and Finnish Birch-faced plywood web, face grain perpendicular to span) and then determine the bending deflection caused by a UDL of 1.6 kN/m on a simply supported span of 6.7 m.

$$
\begin{aligned}
E_{fN} &= 7000 \times 1.24 = 8680 \text{ N/mm}^2 \\
E_w &= 5750 \text{ N/mm}^2 \text{ (see Table 8.7)} \\
I_{Xf} &= 168 \times 10^6 \text{ mm}^4 \\
I_{Xw} &= 25.9 \times 10^6 \text{ mm}^4 \\
EI_X &= 1607 \text{ kN.m}^2
\end{aligned}
\quad \left. \right\} \text{ from § 8.4.2}
$$

The bending deflection $\delta_m = \dfrac{5 \times 1.6 \times 6.7^4}{384 \times 1607} = 0.0261$ m, say 26 mm.

## 8.6   CALCULATION OF SHEAR DEFLECTION

### 8.6.1   Introduction

The shear deflection of a ply web beam about the $YY$ axis is small as a percentage of the total, whereas the shear deflection about the $XX$ axis is likely to be a significant proportion of the total deflection (frequently 15% or more), and must be taken into account in determining deflection and camber of ply web beams.

For simple spans the shear deflection at mid span for a section of constant $EI_X$ is given in § 4.15.7 as:

$$\delta_v = \frac{K_{form} M_0}{GA}$$

For an I or Box section having flanges and webs of uniform thickness throughout the span, Roark (1956) gives the formula for the section constant $K_{form}$ (the form factor) as:

$$K_{form} = \left[ 1 + \frac{3(D_2^2 - D_1^2)D_1}{2D_2^3} \left( \frac{t_2}{t_1} - 1 \right) \right] \frac{4D_2^2}{10r^2}$$

where $D_1$ = distance from neutral axis to the nearest surface of the flange
$D_2$ = distance from neutral axis to extreme fibre
$t_1$ = thickness of web (or webs, in Box beams)
$t_2$ = width of flange (including web thickness)
$r$ = radius of gyration of section with respect to the neutral axis
$= \sqrt{(I_X/A)}$.

When transposed into the terms of Fig. 8.5 this becomes (omitting intermediate steps):

$$K_{form} = \frac{Ah^2}{10I_X}\left[1 + \frac{3(h^2 - h_w^2)h_w}{2h^3}\left(\frac{b}{t}\right)\right]$$

where $A$ = the area of the full section
$\quad\ b$ = the total flange width (excluding thickness of web).

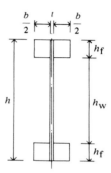

Fig. 8.5

If $\alpha$ is made equal to $h_f/h$ so that $h_w = h(1 - 2\alpha)$ then:

$$K_{form} = \frac{Ah^2}{10I_X}\left[1 + 6\alpha(1 - \alpha)(1 - 2\alpha)\left(\frac{b}{t}\right)\right]$$

or

$$K_{form} = \frac{Ah^2 K_v}{10I_X}$$

where $K_v = 1 + 6\alpha(1 - \alpha)(1 - 2\alpha)\left(\frac{b}{t}\right)$

Hence shear deflection is:

$$\delta_v = \frac{K_{form}M_0}{GA} = \frac{Ah^2 K_v M_0}{10I_X GA} = \frac{h^2 K_v M_0}{10GI_X} \tag{8.3}$$

Note that for a solid section $\alpha = 0.5$, hence $K_v = 1.0$
and:

$$\delta_v = \frac{h^2 M_0}{10G(bh^3/12)} = \frac{1.2M_0}{GA}$$

which agrees with the form factor $K_{form} = 1.2$ given in §4.15.7 for solid sections.

There is little if any inaccuracy in adopting Roark's recommended approximation that $K_{form}$ may be taken as unity if $A$ is taken as the area of the web or webs only. The shear deflection may then be simplified to:

$$\delta_v = \frac{M_0}{G_w A_w} \tag{8.4}$$

where $M_0$ = the bending moment at mid span
$G_w$ = the modulus of rigidity of the webs
$A_w$ = the area of the webs.

$G_w A_w$ is the 'sectional rigidity in shear'.

### 8.6.2 Example of calculating shear deflection

Calculate the shear deflection of the beam shown in Fig. 8.4 (SS grade Whitewood and Finnish Birch-faced plywood web) caused by a UDL of 1.6 kN/m on a simply supported span of 6.7 m (as §8.5.2).

From equation (8.3) and §8.6.1,

$$\delta_v = \frac{h^2 K_v M_0}{10 G_w I_X}$$

$h = 300\,\text{mm} \qquad \alpha = 70/300 \qquad \therefore K_v = 5.38\ (\text{see }\S\,8.6.1)$

$$M_0 = \frac{1.6 \times 6.7^2}{8} = 8.98\,\text{kN.m} = 8.98 \times 10^6\,\text{N.mm}$$

$G_w = 650\,\text{N/mm}^2\ (\text{Table 8.6})$
$I_X = 193.9 \times 10^6\,\text{mm}^4\ (\S\,8.4.2)$

$$\delta_v = \frac{300^2 \times 5.38 \times 8.98 \times 10^6}{10 \times 650 \times 193.9 \times 10^6} = 3.45\,\text{mm}$$

From equation (8.4):

$$\delta_v = \frac{M_0}{G_w A_w}$$

with $M_0 = 8.98\,\text{kN.m}$
$G_w = 650\,\text{N/mm}^2$
$A_w = 11.5 \times 300 = 3450\,\text{mm}^2$
$G_w A_w = 2243\,\text{kN}$

$$\delta_v = \frac{8.98}{2243} = 0.004\,\text{m} = 4\,\text{mm}$$

The answer produced by equation (8.4) is similar to that from the more complex equation (8.3) and the authors feel that this simplification is justified. Note that the shear deflection of 4 mm is 13% of the total deflection of $26 + 4 = 30\,\text{mm}$ in this particular case and $\delta_v$ is 15% of $\delta_m$.

## 8.7 CALCULATION OF PANEL SHEAR STRESS

### 8.7.1 Introduction

Panel shear stress is the traditional term used for horizontal shear stress in a plywood web, not to be confused with 'rolling shear stress' as discussed in §8.8.2. It is appropriate to stresses about the $XX$ axis only of beams as sketched in Fig. 8.6. For a beam symmetrical about both axes, the maximum panel shear stress occurs at the $XX$ axis. The symbol used in this manual for panel shear stress is $\tau_p$.

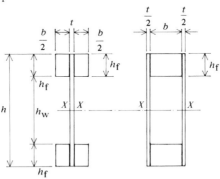

Fig. 8.6

The panel shear stress at the $XX$ axis for I and Box beams as shown in Fig. 8.6 is:

$$\tau_p = \frac{F_v S_X}{I_X t} \tag{8.5}$$

where $F_v$ = the external shear at the section

$S_X$ = the first moment of area of the flange and web or webs above the $XX$ axis

$I_X$ = the second moment of area of the section about the $XX$ axis

$t$ = the total web thickness at the $XX$ axis.

Referring to Fig. 8.6:

$$S_X = S_{Xf} \text{ of flange} + S_{Xw} \text{ of web}$$

$$= b\left(\frac{h - h_w}{2}\right)\left(\frac{h + h_w}{4}\right) + \frac{th^2}{8}$$

$$= \frac{b}{8}(h^2 - h_w^2) + \frac{th^2}{8}$$

$$\therefore \tau_p = \frac{F_v}{8I_X}\left[\frac{b}{t}(h^2 - h_w^2) + h^2\right] \tag{8.6}$$

### 8.7.2 Example of calculating panel shear stress

Calculate the panel shear stress for the I section shown in Fig. 8.4 for a shear of 5360 N. $I_X = 193.9 \times 10^6 \text{ mm}^4$.

$$\tau_p = \frac{5360}{8 \times 193.9 \times 10^6} \left[ \frac{88}{11.5} (300^2 - 160^2) + 300^2 \right]$$

$$= 2.01 \text{ N/mm}^2$$

The permissible panel shear stress (long term) for 12 mm nominal thickness of Finnish Birch-faced plywood is 3.00 N/mm$^2$ (Table 8.7) whether face grain is parallel or perpendicular to the direction of the load. The load-duration factor $K_3$ applies, but load-sharing factors $K_{27}$, $K_{28}$ and $K_{29}$ are not generally applied even if two or more sheets of plywood act together.

## 8.8 STRESS AT FLANGE-TO-WEB JOINT

### 8.8.1 Flange-to-web shear stress

The shear stress between flange and web for I and Box beams as shown in Fig. 8.6 is:

$$\tau_r = \frac{F_v S_{Xf}}{I_X (nh_f)} \tag{8.7}$$

where $F_v$ = the external shear at the section

$S_{Xf}$ = the first moment of area of the flanges above the XX axis

$I_X$ = the second moment of area of the section about the XX axis

$n$ = the number of glue-line contact faces for the (total) flange being considered = 2 for the section shown in Fig. 8.6.

Referring to Fig. 8.6:

$$S_{Xf} = b \left( \frac{h - h_w}{2} \right) \left( \frac{h + h_w}{4} \right)$$

$$nh_f = 2h_f = h - h_w$$

$$\therefore \tau_r = \frac{F_v b (h + h_w)}{8I_X} \tag{8.8}$$

Note that equation (8.7) is obtained by modifying the formulae in §4.11.1. The load per unit length in the flange at the position of maximum shear is derived and divided by the contact area of the adhesive to give an average stress in the glue line.

### 8.8.2    Rolling shear stress

The flange-to-web shear in a glued ply web beam is frequently referred to as 'rolling shear'. Figures 8.7 and 8.8 represent a plan on the top flange of an I and Box beam, and Fig. 8.9 an idealised magnification of the junction between web and flange.

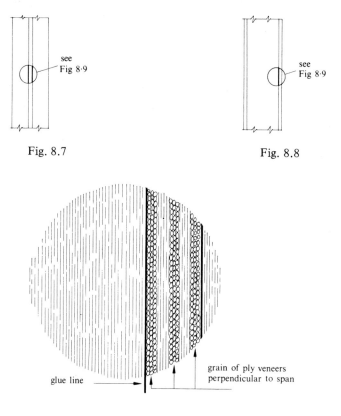

Fig. 8.7                                                              Fig. 8.8

Fig. 8.9

The term 'rolling shear' is frequently used because it best describes the appearance of the failure which can result if the ultimate stress at the junction between web and flange is exceeded. In transferring horizontal shear forces from web to flange, a 'rolling action' takes place. If the face grain of the plywood runs perpendicular to the general grain direction of the timber in the flange, the rolling takes place at this interface. If the face grain of the plywood runs parallel to the general grain direction of the timber in the flange, the rolling will take place between the face veneer of the plywood and the next veneer into the plywood. (This is why Birch-faced plywood has the two outer veneers of birch, except for the thinnest layup). To avoid this rolling action, the rolling shear must be limited by providing sufficient glued contact depth between flange and web.

With plywoods with a stronger face veneer two values are given for rolling shear. If the line of action of load is parallel to the face grain, 'rolling' will take place at the first glue line in the plywood and the lesser value for rolling shear is relevant. If the line of action of load is perpendicular to face grain, the greater value is relevant. However, some plywoods have the stronger veneer on only one face (not on the 'back') therefore this point must be considered before using the greater value.

Rolling shear in ply web beams applies only to the plywood, not to the solid timber. Any glue lines to the solid timber flanges in ply web beams are stressed parallel to grain.

The permissible rolling shear stress for the plywood may be increased by modifying the tabulated values for the plywood by the load-sharing factor $K_3$. However, it is unlikely that any load-sharing factors such as $K_{27}$ can be applied. BS 5268 : Part 2 (Clause 30) requires certain permissible rolling shear stresses, such as those at the detail being discussed (see Fig. 4.24) to be multiplied by modification factor $K_{37} = 0.5$ to take account of possible stress concentrations. Also, if the bonding pressure is obtained by nails or staples, a further reduction factor of 0.9 ($K_{70}$) is to be taken (Clause 47.4.2 of BS 5268 : Part 2).

## 8.8.3 Relation between panel and rolling shear

For the usual proportions of ply web sections and·glue-line depth, rolling shear is not critical, but it is worth considering the relation which exists between panel shear and rolling shear for a given shear force. See equations (8.5) and (8.7). By transposing:

$$F_v = \frac{\tau_p I_X t}{S} = \frac{\tau_r I_X n h_f}{S_f}$$

For a given external shear $F_v$, rolling shear and panel shear will be equally critical if:

$$\frac{h_f}{t} = \frac{\tau_p S_f}{n \tau_r S}$$

However, $S_f$ is less than $S$ and therefore one can state that, providing the panel shear stress is not greater than permitted, then rolling shear will never be critical if:

$$\frac{h_f}{t} \geqslant \frac{\tau_p}{n \tau_r}$$

Values of $\tau_p/n\tau_r$ are given in Table 8.2 for $n = 2$, from which it can be seen that in general the total flange/web contact depth should be about three to seven times the total web thickness. If a beam with $n = 2$ is at maximum panel shear, then providing the value of $\tau_p/n\tau_r$ is not less than the value in Table 8.2, rolling shear will not be critical.

**Table 8.2**   Values of $\tau_p/n\tau_r = \tau_p/2\tau_r$ where $n = 2$.

| Plywood | Grade | Nominal thickness (mm) | $\dfrac{\tau_p}{2\tau_r}$ * |
|---|---|---|---|
| Canadian Softwood plywood | Select tight face<br>Select sheathing<br>Sheathing grade } | All | 3.03–4.92 |
| Finnish Birch-faced plywood | All | 6.5 | 7.4 |
| | All | 9–29 incl | 5.55 |

\* 0.5 factor taken into account; 0.9 factor taken into account.

### 8.8.4   Calculation of rolling shear stress

Calculate the rolling shear stress for the beam shown in Fig. 8.4 for a vertical external shear of 5360 N.

From equation (8.8):

$$\tau_r = \frac{5360 \times 88\,(300 + 160)}{8 \times 193.9 \times 10^6} = 0.14\,\text{N/mm}^2$$

The permissible rolling shear stress

$$= 0.60\,(\text{from Table 8.7}) \times 0.5 \times 0.9\,(\S\,8.8.2)$$
$$= 0.27\,\text{N/mm}^2$$

### 8.9   WEB SPLICES

### 8.9.1   Introduction

Plywood sheets usually have a longer side of 2.4 m (or 2.44 m: see § 1.9) and consequently web splices are normally required at 2.4 m intervals along the beam. The web splices described in this chapter are of plywood of the same species and thickness as the web. With an I section the splice plate is placed on both sides of the web. With a Box beam it may be possible to have the splice on only the inside of each web.

By special arrangement it is occasionally possible to purchase long lengths of plywood scarf jointed, or for the manufacturer of a standard range of beams to install a scarf jointing machine. In these cases the strength of the joints should be proved for the particular application.

Certain manufacturers use softwood instead of plywood for splices, particularly if the splice is only being called upon to take vertical shear and not a bending moment. They are not described in this chapter, but if they are used they should be carefully dried before installation. If the softwood splices are called upon to take a bending moment, the designer should realise that this entails rolling shear at the interface, and that the permissible glue-line stresses have to be reduced accordingly. A softwood splice can also be designed to act as a stiffener or as part of a stiffener.

### 8.9.2 Stresses in ply web splices

Once the designer has decided which method to use in calculating bending stresses on the complete cross-section (§ 8.4.1) this will determine if the web splices have to be designed for vertical shear only, or vertical shear plus a bending moment.

*Vertical shear only on web splice*
When a web splice is required only to transmit a pure shear $F_v$, on the assumption that the web is not transmitting a proportion of the bending moment (see § 8.4.1), the splice plate should be designed to transmit a force $F_v$ applied at the centroid of the area of the splice plate at each side of the joint. It is usual to resist the resulting stresses with a glued connection as shown in Fig. 8.10.

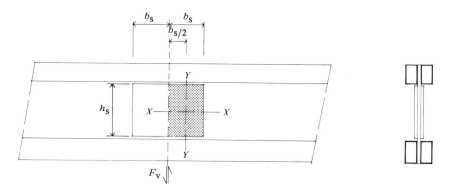

Fig. 8.10

Shear $F_v$ must be resisted at an eccentricity $b_s/2$ resulting in a combination of shear and bending on the glue area.

The shear on the glue area due to vertical shear can be calculated on the assumption of an average distribution without concentration of stress.

$$\therefore \; \tau_v = \frac{F_v}{nb_s h_s} \tag{8.10}$$

where $n$ is the number of glue lines.

The permissible stress is the rolling shear stress of the plywood modified by the load-duration factor $K_3$. If the bonding pressure is achieved by nails or staples the permissible stress should be reduced by 0.9 ($K_{70}$). The moment due to the eccentricity of shear force $F_v$ is $0.5 F_v b_s$. The method of resisting moment on the splice plate is described below (but for the case where the moment is $M_w + 0.5 F_v b_s$).

*Vertical shear and moment on web splice*

If the method of calculating bending stresses on the complete cross-section (§ 8.4.1) requires the web to transmit a proportion of the bending, the total moment on the glue area (Fig. 8.10) is $M_w + 0.5 F_v b_s$ and the maximum stress due to this moment occurs at the extreme fibre of the splice plate and is:

$$\sigma_m (\text{max}) = \frac{(M_w + 0.5 F_v b_s) a_{\text{max}}}{n I_p}$$

where $n$ = number of glue lines

$a_{\text{max}}$ = distance from the centre of area of the glue to the extreme fibre
$= 0.5 \sqrt{(b_s^2 + h_s^2)}$ (see Fig. 8.11)
$I_p$ = the polar second moment of area of the glue
$= I_X + I_Y$.

$$\therefore \ \sigma_m (\text{max}) = \frac{6 (M_w + 0.5 F_v b_s)}{n b_s h_s \sqrt{(b_s^2 + h_s^2)}} \qquad (8.11)$$

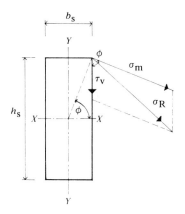

Fig. 8.11

The maximum resulting stress $\sigma_R$ acting on the splice plate is the vector summation of $\tau_v$ and $\sigma_m$. This can be arrived at by plotting $\tau_v$ and $\sigma_m$ to scale or from:

$$\sigma_R = \sqrt{(\sigma_m^2 + \tau_v^2 + 2\sigma_m \tau_v \cos \phi)} \qquad (8.12)$$

where $\phi$ is the angle between $\tau_v$ and $\sigma_m$ and also the angle between a line drawn from the centre of area of the glue to an extreme corner (Fig. 8.11).

$$\cos \phi = \frac{0.5 b_s}{a_{\text{max}}} = \frac{b_s}{\sqrt{(b_s^2 + h_s^2)}} = \frac{1}{\sqrt{[1 + (h_s/b_s)^2]}} \qquad (8.13)$$

The permissible stress is the rolling shear stress modified for load duration. If the bonding pressure is achieved by nails or staples the permissible stress should be multiplied by 0.9 ($K_{70}$).

An estimate design of web splices taking moment and shear is to assume that the upper and lower 40% of plate depth acts to resist the moment while the remaining 20% resists shear only. The force in the upper portion of the plate (shown hatched in Fig. 8.12) can be taken as the moment in the web

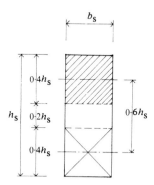

Fig. 8.12

(neglecting moment due to eccentricity from shear) divided by the lever arm $0.6h_s$. Width $b_s$ can be chosen to limit rolling shear to either:

$$b_s = \frac{M_w}{0.24h_s^2 \tau_r} \tag{8.14}$$

or

$$b_s = \frac{F_v}{h_s \tau_r}$$

In addition to the check on the glue area it is necessary to check the strength in bending of the plywood splices on the line of the gap in the web by dividing the moment by the $Z$ of the plywood.

### 8.9.3 Example of checking splice plates

Provide a web splice for the beam shown in Fig. 8.4 to resist an external shear at the splice position of 2680 N and a bending moment on the total section of 6.73 kN.m, each of medium term duration.

The permissible medium term rolling shear (no load sharing) is:

$$0.60 \times 0.9 \times 1.25 = 0.675 \, \text{N/mm}^2 \, \text{(Table 8.7)}$$

from § 8.4.2:

$$E_w I_w = 5750 \times 25.9 \times 10^6 = 149 \times 10^9 \, \text{N.mm}^2$$
$$EI_X = 1607 \times 10^9 \, \text{N.mm}^2$$

from equation (8.1):

$$M_w = 6.73 \times 10^6 \left[ \frac{149 \times 10^9}{1607 \times 10^9} \right] = 0.624 \times 10^6 \, \text{N.mm}$$

An approximate value for $b_s$ can be obtained from equation (8.14):

$$b_s = \frac{0.624 \times 10^6}{0.24 \times 160^2 \times 0.624} = 163 \, \text{mm} \qquad \text{say } 170 \, \text{mm}$$

This approximation is now checked in detail.

From equation (8.10):

$$\tau_v = \frac{2680}{2 \times 170 \times 160} = 0.0493 \, \text{N/mm}^2$$

from equation (8.11):

$$\sigma_m = \frac{6\,(0.624 \times 10^6 + 0.5 \times 2680 \times 170)}{2 \times 170 \times 160 \sqrt{(170^2 + 160^2)}} = 0.402 \, \text{N/mm}^2$$

from equation (8.13):

$$\cos \phi = \frac{1}{\sqrt{[1 + (160/170)^2]}} = 0.728$$

from equation (8.12):

$$\sigma_R = \sqrt{[0.402^2 + 0.0493^2 + (2 \times 0.402 \times 0.0493 \times 0.728)]}$$
$$= 0.44 \, \text{N/mm}^2 < 0.675 \, \text{N/mm}^2 \quad \text{(acceptable)}$$

A total width of splice plate of $2b_s = 340 \, \text{mm}$ is adequate.

To complete the design of a splice plate, the bending stress in the plywood plates on the line of the joint should be checked for the bending moment in the web only. Assume same plywood as web.

$$\text{The modulus of the plates only} = \frac{th_s^2}{6} = \frac{2 \times 11.5 \times 160^2}{6}$$
$$= 0.098 \times 10^6 \, \text{mm}^3$$

$$\therefore \text{Applied bending stress} = \frac{0.624 \times 10^6}{0.098 \times 10^6} = 6.37 \, \text{N/mm}^2$$

The permissible medium term bending stress (from Table 8.1) $=$ 6.75 N/mm$^2$.

## 8.10 WEB STIFFENERS

### 8.10.1 Introduction

BS 5268:Part 2:1984 gives no guidance on the stabilising of plywood webs of Box or I section beams, and manufacturers of ply web beams often have to revert to tests to prove the chosen centres, or to place stiffeners much closer than probably required. Tests on certain proprietary beam systems have given adequate factors of safety even with the thinnest structural plywood available (beams of 1.2 m overall depth with stiffeners placed at 2.4 m centres), but obviously a satisfactory design method is desirable because, for any given thickness and species of plywood, there is a critical shear stress at which buckling of the web will occur. The calculation of this critical stress and its confirmation by laboratory testing is an aspect of timber design on which little information is available.

The CIB Structural Timber Design Code produced by working group W18 does state that a buckling investigation is not necessary if the webs are made from structural plywood and the clear depth of the web ($h_w$) is less than $2h_{max}$ (where $h_{max}$ is as given below) and the shear force ($F_v$) is not greater than the values given below. It is assumed that load-bearing stiffeners are provided at support positions and under any concentrated load.

$$\text{For } h_w \leqslant h_{max}: \quad F_v \leqslant \tau_{v,adm} \, t \, (h_w + h_f)$$

$$\text{For } h_{max} \leqslant h_w \leqslant 2h_{max}: F_v \leqslant \tau_{v,adm} \, t \, h_{max} \left(1 + \frac{h_f}{h_w}\right)$$

where $t$ = the thickness of one web.

$$\text{For plywood with } \psi < 0.5: \quad h_{max} = (20 + 50\psi)\,t$$
$$\text{For plywood with } \psi \geqslant 0.5: \quad h_{max} = 45t$$

where $\psi$ = the ratio between the bending stiffness of a strip of unit width cut perpendicularly to the beam axis, and the bending stiffness of a corresponding strip cut parallel to the longitudinal direction of the beam.

Where a buckling investigation is necessary the CIB document gives one based on 'linear elastic theory for perfect plates simply supported along the flanges and web stiffeners'. It is not detailed here.

### 8.10.2 Canadian design method for Douglas Fir ply webs

The spacing and design for Douglas Fir plywood is treated empirically in the COFI publication *Fir Plywood Web Beam Design*. The following notes are a summary of the method, which gives a conservative spacing for stiffeners suitable for plywood with the face grain running parallel to the span. Although produced for Douglas Fir plywood it may be used for Douglas Fir-faced plywood.

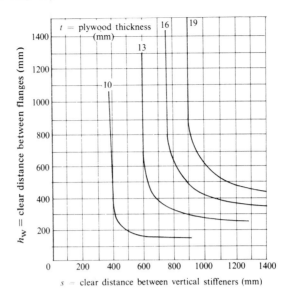

Fig. 8.13

The recommended basic clear distance between vertical stiffeners $s$ for combinations of web thickness $t$ and the clear distance between flanges $h_w$ can be taken from Fig. 8.13. The basic spacing may be increased if the actual panel shear stress is less than the permissible according to equation (8.15) except that the value of $p$ must never be taken as less than 50% and the actual allowable improved clear spacing between stiffeners must not exceed $3s$.

$$s' = s \left[ 1 + \frac{100 - p}{25} \right]$$   (8.15)

where $s'$ = the allowable improved clear stiffener spacing

$p$ = percentage of allowable panel shear stress occurring in the panel.

$p$ must not be taken as less than 50%. $s' \not> 3s$ ($s$ from Fig. 8.13).

### 8.10.3   A method for Finnish Birch-faced plywood

At present there are no recommendations for the spacing of stiffeners in Finnish Birch-faced (or all-Birch plywood) comparable to the method given in § 8.10.2 for Douglas Fir, and the authors have developed a method based upon the more general recommendations outlined by Hanson (1948).

The procedure assumes that the plywood has the face grain perpendicular to span, under which circumstances experience shows that horizontal stiffeners are more effective in limiting web buckling than are the more conventional vertical stiffeners. This aspect is reflected in the method.

Hanson states that the critical panel shear stress for the commencement of web buckling is given by the equation:

$$\text{Critical shear stress} = \frac{32t^2 \sqrt[4]{(E_{tra}E_{par}^3)}}{12h_w^2}$$

where $t$ = web thickness (mm)

$\quad h_w$ = depth between flanges or horizontal stiffeners (mm)

$\quad E_{tra}$ = $E$ (bending) perpendicular to grain

$\quad E_{par}$ = $E$ (bending) parallel to grain.

If a factor of safety of 2.25 is applied to the critical shear stress and a material stress coefficient $K_m$ is introduced, then the permissible panel shear stress is given as:

$$\tau_{p,adm} = 1.18K_m \left(\frac{t}{h_w}\right)^2 \tag{8.16}$$

where $K_m = \sqrt[4]{(E_{tra}E_{par}^3)}$ N/mm$^2$.

Transposing equation (8.16) provides an immediate guide to the maximum slenderness ratio $h_w/t$ acceptable to the full development of panel shear.

$$\frac{h_w}{t} = \sqrt{\left(\frac{1.18K_m}{\tau_{p,adm}}\right)}$$

For various thicknesses of plywood, $K_m$ and $h_w/t$ are determined in Table 8.3 for medium term duration of load. (Medium term leads to lower values than long term.)

To summarise the values in Table 8.3, note that webs of Finnish Birch-faced plywood (face grain perpendicular to span) develop the full shear capacity of the plywood without buckling, provided that the web slenderness ratio $h_w/t$ does not exceed 42–48, this requirement being independent of any vertical stiffeners. However, it has to be said that certain manufacturers of proprietary ply web beam systems have proved (by test) $h_w/t$ ratios as high as 150 to work in certain cases.

Compare the slenderness figures in Table 8.3 with the values obtained from the method given in the CIB document (§ 8.10.1).

Consider the 12 mm nominal thickness with face grain vertical (i.e. perpendicular to the beam axis).

$$\psi = \frac{8900}{4300} = 2.07 \qquad \therefore h_{max} = 45t = 518 \text{ mm}$$

The ratio for $h_{max}/t = 45$ which is similar to the slenderness values given in Table 8.3, however a clear depth of 2 × $h_{max}$ is permitted providing that the shear stress is not exceeded.

Referring back to the example shown in Fig. 8.4 and the calculation of panel shear stress in § 8.7.2 for a shear of 5360 N (assume long term):

**Table 8.3   Determination of maximum web slenderness, Finnish Birch-faced plywood**

| Thickness $t$ | | | | | medium term | |
| nom. (mm) | min (mm) | $E_{par}$ (N/mm$^2$) | $E_{tra}$ (N/mm$^2$) | $K_m$ (N/mm$^2$) | $\tau_{p,adm}$ (N/mm$^2$) | $\dfrac{h_w}{t}$ |
|---|---|---|---|---|---|---|
| 6.5 | 6.1 | 10 300 | 2750 | 7404 | 3.75 | 48.3 |
| 9 | 8.8 | 9 500 | 3800 | 7555 | 3.75 | 48.8 |
| 12 | 11.5 | 8 900 | 4300 | 7420 | 3.75 | 48.3 |
| 15 | 13.9 | 8 550 | 4300 | 7200 | 3.75 | 47.6 |
| 18 | 17.1 | 8 050 | 4200 | 6842 | 3.75 | 46.4 |
| 21 | 19.9 | 7 550 | 4150 | 6501 | 3.75 | 45.2 |
| 22 | 21.1 | 7 400 | 4150 | 6404 | 3.75 | 44.9 |
| 23 | 22.7 | 6 300 | 4150 | 5676 | 3.75 | 42.3 |
| 25 | 24.4 | 7 150 | 4150 | 6241 | 3.75 | 44.3 |
| 26 | 25.2 | 6 950 | 4100 | 6091 | 3.75 | 43.8 |
| 29 | 27.9 | 6 850 | 4050 | 6007 | 3.75 | 43.5 |

$$h_w < h_{max}$$
$$\therefore \text{ shear capacity } = \tau_{v,adm} t (h_w + h_f) = 3.00 \times 11.5 (160 + 70)$$
$$= 7935 \, \text{N} > 5360 \, \text{N which is acceptable.}$$

The conclusion from a check by the CIB method is that in this case the web does not require intermediate stiffeners for the applied shear.

### 8.10.4   Design of non-load-bearing stiffeners

Very little work has been carried out on the design of non-load-bearing stiffeners, but beams which have been tested show that very small stiffeners are adequate. For practical reasons of straightness of timber, fixing etc. it can be said that any reasonable size stiffener will usually be adequate. Stiffeners would normally be at least 50 mm basic along length of beam, extend close to the outside line (in cross-section) of the flanges, be glued to the web, and fit between the flanges (although a tight fit to the flanges is not essential because no concentrated load is being transferred to non-load-bearing stiffeners).

### 8.10.5   Design of load-bearing stiffeners

Load-bearing stiffeners will occur at points of support and under concentrated loads. In effect they are short columns. A typical end stiffener is sketched in Fig. 8.14.

As well as the cross-section of the timber in the stiffener, it is likely that part of the length of web will contribute to strength. For simplicity it is suggested that only the area of web enclosed by the stiffener is included, and that it be treated as timber; therefore one has a stiffener with cross-section of $B \times S_w$.

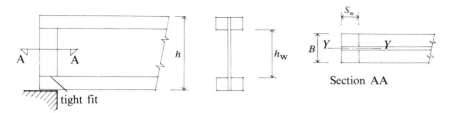

Fig. 8.14

It is essential that the detailing of the stiffener is arranged so that the concentrated load for which it is provided is transferred (usually by bearing) into the stiffener, and is then transferred, through the glue line/s between stiffener and web, into the web. The end stiffener shown in Fig. 8.14 is resisting the concentrated load of the reaction at the support and is transferring the reaction into the web. Therefore it is necessary for there to be tight fit bearing between the stiffener and the top of the lower flange. If there had been a concentrated load applied to the top of the beam and it required a load-bearing stiffener, this stiffener would have to have been fitted tightly to the underside of the top flange.

Sufficient bearing area must be provided. This is usually dictated by bearing across the grain (§ 4.10). Usually wane is excluded.

The radius of gyration out of the plane $YY$ of the beam $= \sqrt{(I_{Ys}/BS_w)}$, where $I_{Ys}$ is the inertia of the stiffener about $YY$. As the stiffener is restrained along the beam by the web, the slenderness ratio in that direction does not have to be considered. The effective length of the stiffener can usually be taken as $h_w$ providing the top and bottom of the ends of the beam are laterally restrained. In certain cases the designer may decide that the effective length can be $0.85h_w$ or $0.7h_w$. The stiffener is then designed as a simple column and permissible and actual stresses compared. General Structural grade will usually be adequate.

## 8.11   HOLES OR SLOTS IN PLY WEB BEAMS

The passage of pipes, ducting or other services through the webs of beams requires a consideration of the resulting stresses in the beam at the position of the hole.

Consider the rectangle slot in the web of a beam as shown in Fig. 8.15. The bending stresses at the extreme fibres a and b at any section BB are given as:

$$\sigma_{ma} = \frac{M_A}{Z_A} + \frac{F_{vA} X (EI)_u}{Z_u [(EI)_u + (EI)_L]} \quad \text{and} \quad \sigma_{mb} = \frac{M_A}{Z_A} + \frac{F_{vA} X (EI)_L}{Z_u [(EI)_u + (EI)_L]}$$

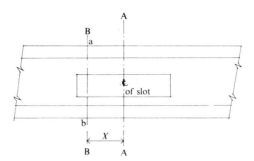

Fig. 8.15

where $M_A$ = the moment at AA (mid length of slot)
  $F_{vA}$ = the shear at AA
   $X$ = the distance from section AA to BB
   $Z_A$ = the section modulus of full section at AA
   $Z_u$ = the section modulus of section at BB above slot
   $Z_L$ = the section modulus of section at BB below slot
  $(EI)_u$ = the sectional rigidity of section at BB above slot
  $(EI)_L$ = the sectional rigidity of section at BB below slot

The formulae above are based on the assumption that the moment $M_A$ produces a bending stress given by the first part of each formula, while the shear is divided between the upper and the lower portion of the beam in proportion to the sectional rigidity of the upper and lower portion of the beam.

For ply web beams it is usually convenient to form a slot symmetrical about the mid depth of section, and then to reinforce the perimeter of the slot. With symmetry the above equations simplify to:

$$\sigma_{ma} = \sigma_{mb} = \frac{M_A}{Z_A} + \frac{F_{vA} X}{2Z_u}$$

In the case of simple span beams, a slot is frequently provided at mid span, where there is no shear under a symmetrical loading condition, and the slot may be of the clear depth $h_w$ between flanges providing that the remaining flange section is adequate to resist the moment $M_A$. It is necessary, however, to check the slot for excessive stress due to unbalanced loadings.

Figure 8.16 shows a typical unbalanced loading condition for a simple span beam with a mid span slot of length $X$ on each side of the centre line of the beam. Because the dead load is placed with progressively balanced layers of elements such as claddings, felt, chippings etc., it can be assumed that the maximum shear at beam mid span occurs only if the imposed loading is unbalanced.

Take the case of a beam stressed to its full permissible bending stress under the action of a UDL over the full span from dead and imposed loading.

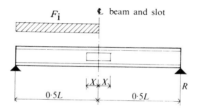

Fig. 8.16

To consider the worst case of shear at the centre line, the imposed loading should be curtailed as shown in Fig. 8.16.

The shear at $R$ from $F_i$ is $F_i/4$, and the moment at mid span is $F_i L/8$. The removal of $F_i$ from the right hand side gives a reduction in the bending stress of:

$$\frac{M_A}{Z_A} = \frac{F_i L}{8 Z_A}$$

whereas the eccentricity of shear on the slotted section increases the extreme fibre stress of the remaining section above (or below) the slot by:

$$\frac{F_{vA} X}{2 Z_u} = \frac{F_i X}{8 Z_u}$$

Thus the section will be restored to its original permissible bending stress when:

$$\frac{M_A}{Z_A} = \frac{F_{vA} X}{2 Z_u}$$

i.e. when:

$$\frac{F_i L}{8 Z_A} = \frac{F_i X}{8 Z_u}$$

or:

$$\frac{L}{Z_A} = \frac{X}{Z_u}$$

Thus the maximum value of $X$ is $Z_u L/Z_A$. Naturally, if the section is not fully stressed under maximum possible load, the distance $X$ can be increased, and therefore $2X$ is a conservative first estimate of a potential slot size.

## 8.12  LATERAL STABILITY

The discussion on ply web beams so far in this chapter has been limited to cases with the compression flange assumed to be fully restrained. The designer must however check on lateral restraint.

**Table 8.4**

| $I_X/I_Y$ | Degree of restraint* |
|---|---|
| Up to 5 | No lateral restraint is required |
| 5–10 | The ends of beams should be held in position at the bottom flange at the supports |
| 10–20 | Beams should be held in line at the ends |
| 20–30 | One edge should be held in line |
| 30–40 | The beam should be restrained by bridging or other bracing at intervals of not more than 2.4 m |
| More than 40 | The compression flange should be fully restrained |

\* Also see note in second paragraph of §4.6.2.

BS 5268:Part 2:1984 permits a ply web beam to be considered as being fully restrained providing the ratio $I_X/I_Y$ is in accordance with certain values as given in Table 8.4. These ratios are based on practical experience and cannot be justified theoretically. They are conservative. When, for architectural or other reasons, the designer is faced with a condition in which these ratios are not satisfied he should refer to Chapter 9 of this manual, where other methods of justifying full restraint, or of calculating for partial restraint are discussed in detail. (Also see §4.6.2.)

## 8.13   PLY WEB BEAMS BENDING ABOUT THE MINOR AXIS

### 8.13.1   Introduction

The designer may have to check the suitability of a ply web beam either for bending about the minor axis only or for combined $XX$ and $YY$ axis bending. This item deals only with bending about the minor axis but the combining of stress ratios is a simple matter providing the designer takes care to combine stresses or stress ratios algebraically at a point (as discussed in §§7.3.3 and 14.7.1 on column designs).

Bending deflection may be calculated using $EI_Y$ similarly to $EI_X$ in §8.5.1. For most designs of ply web beams it is satisfactory to calculate shear deflection for I beams about the $YY$ axis by assuming that only the flanges resist deflection and have a thickness of $h_f$ and a depth of $B$, and $K_{form} = 1.2$. For a Box beam (as shown in Fig. 8.18) it is on the safe side to calculate shear deflection about the $YY$ axis by assuming that only the flanges resist deflection but have a depth of $b + t$ (= 100.2 mm in Fig. 8.18) and a thickness of $h_f = 70$.

In considering bending stresses it is assumed that the flanges contain no joints weaker than dictated by the grade stress. There is no need to discuss various bending analysis methods as in §8.4.1 for beams on the strong axis. For an I section beam the contribution of the plywood is insignificant as therefore is the type of web splice used when bending on the minor axis is considered. However, for a Box beam the bending contribution from the plywood can be quite considerable and requires correctly designed splice plates at joints (if the plywood is considered to add to the moment capacity).

A beam requires lateral stability only at right angles to the direction of bending. Due to the proportions of a ply web beam it is safe to assume full lateral stability when it is being bent on its minor axis.

### 8.13.2 Example of an I section bending about the minor axis

Calculate the deflection, bending stress and shear stress of the beam shown in Fig. 8.17 (same as Fig. 8.4) loaded on the $YY$ axis by its self-weight of 0.06 kN/m on a design span of 6.7 m. Flanges are SS Whitewood and plywood is Finnish Birch-faced.

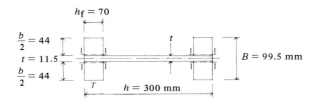

Fig. 8.17

$$I_{Yf} = \frac{2h_f}{12}[B^3 - t^3] = \frac{2 \times 70}{12}[99.5^3 - 11.5^3]$$

$$= 11.47 \times 10^6 \, mm^4$$

$$I_w = \frac{ht^3}{12} = \frac{300 \times 11.5^3}{12} = 0.038 \times 10^6 \, mm^4$$

$$EI_Y = E_{fN}I_{Yf} + E_w I_w$$
$$= [(8680 \times 11.47) + (4300 \times 0.038)] \, 10^6$$
$$= [99.6 + 0.2] \, 10^9 = 99.8 \times 10^9 \, N.mm^2$$
$$= 99.8 \, kN.m^2$$

$$\text{Bending deflection} = \frac{5 \times 0.06 \times 6.7^4}{384 \times 99.8} = 0.016 \, m = 16 \, mm$$

(Note that the plywood is bending flat rather than in tension or compression.)

Note that the effect of the web could be disregarded and the designer could simplify the design by assuming two softwood sections $h_f$ mm wide and $b + t$ deep.

(Obviously in the case of a Box beam the effect of the plywood should not be similarly disregarded, and the plywood will be more in compression and tension than bending.)

$$\text{Shear deflection (approx.)} = \frac{K_{\text{form}} \times M_0}{AG} = \frac{1.2 \times 0.06 \times 6.7^2 \times 10^6 \times 20}{99.5 \times 70 \times 8680 \times 8}$$

$$= 0.13 \text{ mm}$$

$$\text{Bending moment due to self-weight} = \frac{0.06 \times 6.7^2}{8} = 0.337 \text{ kN.m}$$

$$Z_Y = \frac{2I_Y}{B} = \frac{2 \times 11.47 \times 10^6}{99.5}$$

$$= 0.231 \times 10^6 \text{ mm}^3$$

$$\text{Bending stress} = \frac{0.337 \times 10^6}{0.231 \times 10^6} = 1.46 \text{ N/mm}^2$$

In deciding on the permissible long term bending stress there is probably some doubt as to whether the appropriate $K_7$ value should be taken for a depth of 44 mm or 99.5 mm. On balance the value for 99.5 mm is taken in this case. Depending on the method of applying load, $K_{27}$ may be relevant. It is taken here for $N = 2$ therefore $K_{27} = 1.11$.

$$\sigma_{m,\text{adm}} = 7.50 \times 1.129 \times 1.11 = 9.41 \text{ N/mm}^2 \qquad \text{(Table 4.4)}$$

The shear stress at the $YY$ axis (for I beams) is:

$$\tau_v = \frac{F_v S}{I_Y h}$$

However, for I beams, the greater shear stress occurs at plane 1–1 shown in Fig. 8.17:

$$\tau_{v1} = \frac{F_v b \ (b/2 + t)}{4 I_Y}$$

This stress $\tau_{v1}$ should not exceed the permissible horizontal shear stress in the softwood or the permissible rolling shear stress in the plywood 'bending flat'.

*Example*
Calculate $\tau_{v1}$ at plane 1–1 for a shear of 201 N resulting from self-weight.

$$I_Y = (11.47 + 0.038) \times 10^6 \text{ mm}^4$$
$$= 11.5 \times 10^6 \text{ mm}^4$$

$$\tau_{v1} = \frac{201 \times 88 \times (44 + 11.5)}{4 \times 11.5 \times 10^6} = 0.021 \text{ N/mm}^2$$

The permissible long term horizontal shear stress in the softwood at plane 1–1 (shear parallel to grain) is $0.82 \text{ N/mm}^2$ (Table 2.2) multiplied by $K_{27}$ depending on the way the load is applied.

The permissible long term rolling shear 'bending flat' in the plywood at plane 1–1 is $0.60\ \text{N/mm}^2$ (see Table 8.7) $\times 0.5$ ($K_{37}$ for stress concentration) $\times 0.9$ ($K_{70}$ assuming bonding pressure is by nails) $= 0.27\ \text{N/mm}^2$.

Although the beam is not over-stressed due to the static action of self-weight, the magnitude of the deflection and the stresses indicates the care required during storage, handling and erection. If ply web beams are stored flat (in the factory or on site) they should be supported at fairly close centres (say 2 to 3 m) partly for reasons of stress, partly to prevent damage or distortion due to the self-weight of any beams stored above them.

### 8.13.3 Example of a Box beam bending about the minor axis

Calculate the maximum bending stresses in the plywood ($E_w = 4650\ \text{N/mm}^2$ from Table 8.7) and the timber ($E_{fN} = 8680\ \text{N/mm}^2$ for $N = 4$) for the beam shown in Fig. 8.18, taking a bending moment on the $YY$ axis of $1.0\ \text{kN.m}$ medium term (see standard tables, Table 8.5). (Total UDL $= 1.19\ \text{kN}$ on design span of 6.72 m.)

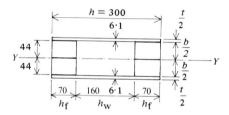

Fig. 8.18

$$I_{Xf} = \frac{140 \times 88^3}{12} = 7.95 \times 10^6\ \text{mm}^4$$

$$I_{Yw} = \frac{300}{12}(100.1^3 - 88^3) = 8.04 \times 10^6\ \text{mm}^4$$

$$E_{fN}\,I_{Yf} = 8680 \times 7.95 \times 10^6 = 69.0 \times 10^9\ \text{N.mm}^2$$
$$E_w\,I_{Yw} = 4650 \times 8.04 \times 10^6 = 37.4 \times 10^9\ \text{N.mm}^2$$
$$(EI_Y) = 106.4\ \text{kN.m}^2$$

The bending stress at a distance $y$ from the $YY$ axis is given by the formula:

$$\sigma_{m,a} = \frac{MEy}{(EI_Y)}$$

where $E$ is the value for the element being considered.

For the plywood at $y = 50.1$ mm:

$$\sigma_{m,a} = \frac{(1.0 \times 10^6) \times 4650 \times 50.1}{106.4 \times 10^9} = 2.19 \text{ N/mm}^2$$

The permissible stress (medium term) $= 3.51 \times 1.25 = 4.39$ N/mm$^2$ (the plywood is mainly in tension or compression).

For the timber at $y = 44$ mm:

$$\sigma_{m,a} = \frac{(1.0 \times 10^6) \times 8680 \times 44}{106.4 \times 10^9} = 3.59 \text{ N/mm}^2$$

The permissible stress (medium term)

$$= 7.50 \times 1.25 \times K_7 \text{ (for 88 mm depth)} \times K_{27} \text{ (for two pieces)}$$
$$= 7.50 \times 1.25 \times 1.144 \times 1.11 = 11.9 \text{ N/mm}^2$$

$$\text{Bending deflection} = \frac{5 \times 1.19 \times 6.72^3}{384 \times 106.9} = 0.044 \text{ m} = 44 \text{ mm}$$

$$\text{Shear deflection} = \frac{K_{form} \times M_0}{AG}$$

$$= \frac{1.2 \times 1.0 \times 10^6 \times 16}{(2 \times 70 \times 100.6) \times 8680} = 0.16 \text{ mm}$$

As can be seen, the shear deflection is very small and could have been disregarded.

Shear stress is maximum at the $YY$ axis, at which there is a glue line (but no 'rolling' shear because the stress is parallel to grain). $F_v = 595$ N.

$$\tau_{v,a} = \frac{F_v S}{2 h_f I_Y}$$

where $S =$ the first moment of area of the section above the $YY$ axis
$$= (140 \times 44 \times 22) + (6.1 \times 300 \times 47.2)$$
$$= 0.222 \times 10^6 \text{ mm}^3$$
$I_Y = I_{Yf} + I_{Yw}$
$$= (7.95 + 8.11) \times 10^6 \text{ mm}^4$$
$$= 16.06 \times 10^6 \text{ mm}^4$$

$$\therefore \tau_{v,a} = \frac{595 \times 0.222 \times 10^6}{140 \times 16.06 \times 10^6} = 0.059 \text{ N/mm}^2$$

The permissible shear stress parallel to grain, medium term

$$= 0.82 \times 1.25 \times 0.9 \text{ (if bonding pressure is by nails)}$$
$$= 0.92 \text{ N/mm}^2 \text{ (Note that the 0.5 reduction factor } K_{37} \text{ does not apply.)}$$

The value 0.92 N/mm$^2$ may be increased by $K_{27}$ (for two pieces acting together $K_{27} = 1.11$) depending on the way the load is applied.

## 8.14 STANDARD SECTIONS

### 8.14.1 Tables of strength capacities of Box and I sections

To aid the designer in the choice of a suitable or trial beam section, geometrical properties and section capacities are given in Tables 8.5 and 8.6 for a range of Box and I sections with SS grade flanges of European Whitewood (and Redwood) and Finnish Birch-faced plywood webs (face grain perpendicular to span).

The designer must note that the moment capacities are calculated in accordance with Method 5.

$E$ values are calculated taking account of $K_{28}$.

Permissible bending stresses are calculated taking account of $K_7$ and $K_{14}$, and $K_{27}$.

The assumption in preparing the tables is that the compression flange is fully restrained.

Capacities are given for long term and medium term loading.

### 8.14.2 Example of using the tables of standard sections

Using the standard tabulated sections (Table 8.5) choose a Box section suitable to carry a uniform load of 10 kN (dead) and 12 kN (imposed) over a simple span of 9 m, assuming full restraint to the compression flange. The beam is to be cambered and the deflection due to imposed loading is not to exceed 0.003 × span.

Assume that the beam is not in a load-sharing system. Note that the standard section tables (Tables 8.5 and 8.6) are based on Method 5, the 'Combined Stress' method, and glued-nailed assembly.

$$
\begin{aligned}
\text{Long term load} &= 10\,\text{kN} \\
\text{Long term shear} &= 5\,\text{kN} \leqslant \bar{F}_{vXL} \\
\text{Long term moment} &= 11.2\,\text{kN.m} \leqslant \bar{M}_{XL} \\
\text{Medium term load} &= 22\,\text{kN} \\
\text{Medium term shear} &= 11\,\text{kN} \leqslant \bar{F}_{vXM} \\
\text{Medium term moment} &= 24.7\,\text{kN.m} \leqslant \bar{M}_{XM}
\end{aligned}
$$

A good first approximation for $EI_X$ to include for shear deflection in cambered ply web beams is:

$$EI_X \simeq 5.5\,F_i L^2\,\text{kN.m}^2$$

where $F_i$ = total imposed load (kN)
  $L$ = span (m).

In this case:

$$\text{Required } EI_X \simeq 5.5 \times 12 \times 9^2 = 5350\,\text{kN.m}^2$$

**Table 8.5   Geometrical properties and sectional capacities for a selected range of Box beams–glued-nailed assembly**

Flanges of SS grade Whitewood. Webs of 6.5 mm Finnish Birch-faced plywood (face grain perpendicular to span). $E_w = 4650$ N/mm², $G_w = 650$ N/mm².

**Geometrical Properties**

flanges of 2, 3 or 4 pieces of 44 × 70

| Section size (mm) $B \times h$ | Weight (kN/m) | Internal dimensions (mm) $b \times h_w$ | Total area (mm² × 10³) $A_f$ | $A_w$ | First moment of area (mm³ × 10⁶) XX $S_{Xf}$ | $S_{Xw}$ | YY $S_{Yf}$ | $S_{Yw}$ | Second moment of area (mm⁴ × 10⁶) XX $I_{Xf}$ | $I_{Xw}$ | YY $I_{Yf}$ | $I_{Yw}$ | Rolling shear ratio $\dfrac{h_f}{t}$ | Stability ratio $\dfrac{I_X}{I_Y}$ |
|---|---|---|---|---|---|---|---|---|---|---|---|---|---|---|
| 100.2 × 300 | 0.09 | 88 × 160 | 12.3 | 3.66 | 0.70 | 0.137 | 0.135 | 0.086 | 167 | 27 | 7.9 | 8.1 | 5.7 | 12.1 |
| 100.2 × 400 | 0.10 | 88 × 260 | 12.3 | 4.88 | 1.01 | 0.244 | 0.135 | 0.114 | 340 | 65 | 7.9 | 10.8 | 5.7 | 21.6 |
| 144.2 × 300 | 0.13 | 132 × 160 | 18.4 | 3.66 | 1.06 | 0.137 | 0.304 | 0.126 | 251 | 27 | 26.8 | 17.4 | 5.7 | 6.3 |
| 144.2 × 400 | 0.13 | 132 × 260 | 18.4 | 4.88 | 1.52 | 0.244 | 0.304 | 0.168 | 510 | 65 | 26.8 | 23.2 | 5.7 | 11.4 |
| 144.2 × 500 | 0.14 | 132 × 360 | 18.4 | 6.10 | 1.98 | 0.381 | 0.304 | 0.210 | 861 | 127 | 26.8 | 29.1 | 5.7 | 17.6 |
| 144.2 × 600 | 0.15 | 132 × 460 | 18.4 | 7.32 | 2.44 | 0.549 | 0.304 | 0.252 | 1305 | 219 | 26.8 | 34.9 | 5.7 | 24.6 |
| 188.2 × 600 | 0.18 | 176 × 460 | 24.6 | 7.32 | 3.26 | 0.549 | 0.542 | 0.333 | 1740 | 219 | 63.6 | 60.7 | 5.7 | 15.7 |
| 188.2 × 700 | 0.19 | 176 × 560 | 24.6 | 8.54 | 3.88 | 0.747 | 0.542 | 0.388 | 2454 | 348 | 63.6 | 70.8 | 5.7 | 20.8 |
| 188.2 × 800 | 0.20 | 176 × 660 | 24.6 | 9.76 | 4.49 | 0.976 | 0.542 | 0.444 | 3292 | 520 | 63.6 | 80.9 | 5.7 | 26.3 |
| 188.2 × 900 | 0.21 | 176 × 760 | 24.6 | 10.98 | 5.11 | 1.235 | 0.542 | 0.499 | 4253 | 741 | 63.6 | 91.0 | 5.7 | 32.2 |

**(Table 8.5)** *contd*

**Section Capacities: no load sharing**

| Section size (mm) $B \times h$ | No. of flange elements $N$ | $E_{min} \times K_{28} = E$ for flanges (N/mm²) $E_{fN}$ | Sectional rigidity | | | | | Shear capacity (kN) | | | | Moment capacity (kN.m)* | | | | On $XX$ axis $K_7 = 1.17$ |
|---|---|---|---|---|---|---|---|---|---|---|---|---|---|---|---|---|
| | | | Bending (kN.m²) | | | | Shear (kN) | $XX$ | | $YY$ | | $XX$† | | $YY$ | | |
| | | | $XX$ | | | $YY$ | $XX$ | long | medium | long | medium | long | medium | long | medium | |
| | | | $E_{fN}I_{xf}$ | $E_w I_{xw}$ | $EI_X$ | $EI_Y$ | $A_w G_w$ | $\overline{F}_{vXL}$ | $\overline{F}_{vXM}$ | $\overline{F}_{vYL}$ | $\overline{F}_{vYM}$ | $\overline{M}_{XL}$ | $\overline{M}_{XM}$ | $\overline{M}_{YL}$ | $\overline{M}_{YM}$ | $K_{27}$ |
| 100.2 × 300 | 4 | 8680 | 1457 | 127 | 1585 | 106 | 2379 | 7.8 | 9.7 | 7.4 | 9.3 | 7.7 | 9.6 | 1.6 | 2.0 | 1.11 |
| 100.2 × 400 | 4 | 8680 | 2955 | 302 | 3257 | 119 | 3172 | 11.3 | 14.1 | 7.7 | 9.6 | 11.6 | 14.5 | 1.7 | 2.2 | 1.11 |
| 144.2 × 300 | 6 | 9030 | 2275 | 127 | 2402 | 323 | 2379 | 7.4 | 9.3 | 10.6 | 13.2 | 11.3 | 14.1 | 3.3 | 4.2 | 1.16 |
| 144.2 × 400 | 6 | 9030 | 4611 | 302 | 4913 | 350 | 3172 | 10.7 | 13.3 | 10.9 | 13.6 | 16.8 | 21.1 | 3.6 | 4.5 | 1.16 |
| 144.2 × 500 | 6 | 9030 | 7781 | 590 | 8372 | 377 | 3965 | 14.1 | 17.6 | 11.2 | 14.0 | 22.6 | 28.2 | 3.9 | 4.9 | 1.16 |
| 144.2 × 600 | 6 | 9030 | 11786 | 1021 | 12808 | 404 | 4758 | 17.6 | 22.0 | 11.4 | 14.3 | 28.5 | 35.6 | 4.2 | 5.2 | 1.16 |
| 188.2 × 600 | 8 | 9240 | 16081 | 1021 | 17102 | 869 | 4758 | 17.0 | 21.2 | 14.6 | 18.3 | 37.0 | 46.2 | 6.9 | 8.7 | 1.19 |
| 188.2 × 700 | 8 | 9240 | 22683 | 1621 | 24305 | 917 | 5551 | 20.4 | 25.6 | 14.9 | 18.6 | 44.7 | 55.9 | 7.3 | 9.1 | 1.19 |
| 188.2 × 800 | 8 | 9240 | 30424 | 2420 | 32845 | 964 | 6344 | 24.0 | 30.0 | 15.1 | 18.9 | 52.5 | 65.6 | 7.7 | 9.6 | 1.19 |
| 188.2 × 900 | 8 | 9240 | 39304 | 3446 | 42750 | 1011 | 7137 | 27.6 | 34.6 | 15.3 | 19.1 | 60.5 | 75.6 | 8.1 | 10.1 | 1.19 |

\* Design Method 5 used to determine moment capacity.

† For moment capacity on $XX$ axis, note that $K_{14}$ will vary and, because $b$ is greater that $h_f$, $K_{14}$ and $K_7$ are not the same (see Method 5 in §8.4.1):

with $N = 4$, $b = 88$, $K_{14} = 1.144$

with $N = 6$, $b = 132$, $K_{14} = 1.095$

with $N = 8$, $b = 176$, $K_{14} = 1.060$.

## Table 8.6   Geometrical properties and sectional capacities for a selected range of I beams–glued-nailed assembly

Flanges of SS grade Whitewood. Webs of 12 mm Finnish Birch-faced plywood (face grain perpendicular to span). $E_W = 5750$ N/mm², $G_W = 650$ N/mm².

**Geometrical Properties**

| Section size (mm) B × h | Weight (kN/m) | Internal dimensions (mm) b × h_w | Total area (mm² × 10³) | | First moment of area (mm³ × 10⁶) | | | | Second moment of area (mm⁴ × 10⁶) | | | | Rolling shear ratio | Stability ratio |
|---|---|---|---|---|---|---|---|---|---|---|---|---|---|---|
| | | | | | XX | | YY | | XX | | YY | | | |
| | | | $A_f$ | $A_w$ | $S_{Xf}$ | $S_{Xw}$ | $S_{Yf}$ | $S_{Yw}$ | $I_{Xf}$ | $I_{Xw}$ | $I_{Yf}$ | $I_{Yw}$ | $\dfrac{h_f}{t}$ | $\dfrac{I_X}{I_Y}$ |
| 99.5 × 300 | 0.09 | 88 × 160 | 12.3 | 3.45 | 0.70 | 0.129 | 0.170 | 0.004 | 167 | 25 | 11.4 | 0.038 | 6.0 | 16.8 |
| 99.5 × 400 | 0.10 | 88 × 260 | 12.3 | 4.60 | 1.01 | 0.230 | 0.170 | 0.006 | 340 | 61 | 11.4 | 0.050 | 6.0 | 34.8 |
| 99.5 × 500 | 0.10 | 88 × 360 | 12.3 | 5.75 | 1.32 | 0.359 | 0.170 | 0.008 | 574 | 119 | 11.4 | 0.063 | 6.0 | 60.1 |
| 99.5 × 600 | 0.11 | 88 × 460 | 12.3 | 6.90 | 1.63 | 0.517 | 0.170 | 0.009 | 870 | 207 | 11.4 | 0.073 | 6.0 | 93.2 |
| 151.5 × 500 | 0.15 | 140 × 360 | 19.6 | 5.75 | 2.10 | 0.359 | 0.399 | 0.008 | 914 | 119 | 40.5 | 0.063 | 6.0 | 25.4 |
| 151.5 × 600 | 0.15 | 140 × 460 | 19.6 | 6.90 | 2.59 | 0.517 | 0.399 | 0.009 | 1384 | 207 | 40.5 | 0.076 | 6.0 | 39.1 |
| 151.5 × 700 | 0.16 | 140 × 560 | 19.6 | 8.05 | 3.08 | 0.704 | 0.399 | 0.011 | 1952 | 328 | 40.5 | 0.088 | 6.0 | 56.1 |
| 151.5 × 800 | 0.17 | 140 × 660 | 19.6 | 9.20 | 3.57 | 0.920 | 0.399 | 0.013 | 2619 | 490 | 40.5 | 0.101 | 6.0 | 76.5 |

**(Table 8.6)** *contd*

**Section Capacities: no load sharing**

| Section size (mm) $B \times h$ | No. of flange elements $N$ | $E_{min} \times K_{28} = E$ for flanges (N/mm²) $E_{fN}$ | Sectional rigidity Bending (kN.m²) XX $E_{fN}I_{Xf}$ | $E_w I_{Xw}$ | $EI_X$ | YY $EI_y$ | Shear (kN) XX $A_w G_w$ | Shear capacity (kN) XX long $\overline{F}_{vXL}$ | medium $\overline{F}_{vXM}$ | YY long $\overline{F}_{vYL}$ | medium $\overline{F}_{vYM}$ | Moment capacity (kN.m)* On XX axis XX† long $\overline{M}_{XL}$ | medium $\overline{M}_{XM}$ | YY long $\overline{M}_{YL}$ | medium $\overline{M}_{YM}$ | $K_7 = K_{14} = 1.17$ $K_{27}$ |
|---|---|---|---|---|---|---|---|---|---|---|---|---|---|---|---|---|
| 99.5 × 300 | 4 | 8680 | 1457 | 148 | 1606 | 99 | 2242 | 7.7 | 9.6 | 2.53 | 3.16 | 7.9 | 9.9 | 2.1 | 2.7 | 1.11 |
| 99.5 × 400 | 4 | 8680 | 2955 | 352 | 3307 | 99 | 2990 | 11.1 | 13.9 | 2.53 | 3.16 | 11.9 | 14.9 | 2.1 | 2.7 | 1.11 |
| 99.5 × 500 | 4 | 8680 | 4986 | 688 | 5675 | 99 | 3737 | 14.2 | 17.7 | 2.53 | 3.16 | 16.1 | 20.2 | 2.2 | 2.7 | 1.11 |
| 99.5 × 600 | 4 | 8680 | 7553 | 1190 | 8743 | 100 | 4485 | 17.2 | 21.6 | 2.53 | 3.16 | 20.5 | 25.7 | 2.2 | 2.7 | 1.11 |
| 151.5 × 500 | 8 | 9240 | 8445 | 688 | 9134 | 375 | 3737 | 13.9 | 17.3 | 3.84 | 4.80 | 25.6 | 32.0 | 5.0 | 6.3 | 1.16 |
| 151.5 × 600 | 8 | 9240 | 12791 | 1190 | 13982 | 375 | 4485 | 17.3 | 21.7 | 3.84 | 4.80 | 32.3 | 40.4 | 5.0 | 6.3 | 1.16 |
| 151.5 × 700 | 8 | 9240 | 18043 | 1890 | 19934 | 375 | 5232 | 20.7 | 25.9 | 3.84 | 4.80 | 39.2 | 49.0 | 5.0 | 6.3 | 1.16 |
| 151.5 × 800 | 8 | 9240 | 24201 | 2821 | 27022 | 375 | 5980 | 23.8 | 29.8 | 3.84 | 4.80 | 46.3 | 57.8 | 5.0 | 6.3 | 1.16 |

* Design Method 5 used to determine moment capacity.
† For moment capacity on $XX$ axis $b/2$ does not exceed $h_f$ therefore $K_{14} = K_7$ (see Method 5 in §8.4.1).

From Table 8.5 it can be seen that a 144.2 × 500 mm Box beam would satisfy the requirements above. Usually the lightest weight beam which satisfies the requirements is the most economical.

Because the $EI_X$ value calculated above was only an approximation, the deflection must be checked. From Table 8.5, $EI_X = 8372\,\text{kN.m}^2$ and $A_w\,G_w = 3965\,\text{kN}$.

$$\text{Total deflection} = \delta_m + \delta_v$$

$$= \frac{5 \times 22 \times 9^3}{384 \times 8372} + \frac{24.7}{3965} = 0.0249 + 0.0062 = 0.0311\,\text{m}$$

(Note that shear deflection is 20% of the total deflection.)

$$\text{Deflection under imposed loading (by proportion)} = 0.0170\,\text{m}$$
$$(= 0.0019\,\text{span})$$
$$\text{Camber (by proportion) against dead loading} = 0.014\,\text{m}$$

Manufacturers would probably increase this by a few millimetres and use, say 18 mm camber.

## 8.15   STRENGTH PROPERTIES OF FINNISH BIRCH-FACED PLYWOOD

The strength properties of Finnish Birch-faced plywood are given in Table 8.7.

## 8.16   STRENGTH PROPERTIES OF CANADIAN DOUGLAS FIR-FACED PLYWOOD

The strength properties of Canadian Douglas Fir-faced plywood are given in Table 8.8.

## 8.17   BEAMS OF VARIABLE SECTION

### 8.17.1   Bending deflection

The design principles for bending deflection outlined in the earlier parts of this chapter also apply to I beams and Box beams having a linear variation in cross-section.

If one considers the tapered beam illustrated in Fig. 8.19, Tuma and Munshi (1971) show that the variation in $EI$ can be represented with sufficient accuracy by the equation:

$$(EI)_x = (EI)_{h_c}\left[1 + \left(\frac{H_c}{h_c} - 1\right)\frac{x}{s}\right]^a$$

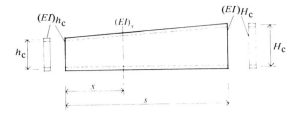

Fig. 8.19

where $(EI)_x$ = the sectional rigidity at a section, distance $x$ from the end of least depth

$(EI)_{h_c}$ = the sectional rigidity at end of least depth

$(EI)_{H_c}$ = the section rigidity at end with greater depth

$H_c$ = the depth between centre lines of flanges at end with greater depth

$h_c$ = the depth between centre lines of flanges at end with least depth

$x$ = the distance from end of least depth

$s$ = the distance along span between shallow and deep ends of beam

$$a = \frac{[\log_e (EI)_{H_c} - \log_e (EI)_{h_c}]}{[\log_e H_c - \log_e h_c]}$$

To simplify, let $n_1 = (EI)_{H_c}/(EI)_{h_c}$ and $n_2 = H_c/h_c$ then:

$$a = \frac{\log_e n_1}{\log_e n_2}$$

and:

$$(EI)_x = (EI)_{h_c} \left[1 + (n_2 - 1)\frac{x}{s}\right]^a$$

Although Tuma and Munshi present this formula with particular reference to steel sections where the depth of flange is small compared to the depth of section, the formula still has sufficient accuracy for timber sections, where the corresponding ratio of flange depth to section depth is usually much greater. With $h_f$ and $h$ as defined in Fig. 8.20 and assessed at the end with least depth, the above formula has the accuracy tabulated in Fig. 8.20.

Using the basic formula for bending deflection of a tapered section given in §7.11.1 and introducing the new definition of $(EI)_x$ leads to formulae for the bending deflection of ply web I and Box beams. Formulae for four simple cases are given in Fig. 8.21.

**Table 8.7**  Strength properties of Finnish Birch-faced plywood parallel and perpendicular to face grain. Sanded grades only. Long term duration stresses based on 'full area' method.

| | | Permissible stress (N/mm²) | | | | | | | | | | |
|---|---|---|---|---|---|---|---|---|---|---|---|---|
| Nominal thickness (mm) | | 6.5 | 9 | 12 | 15 | 18 | 21 | 22 | 23 | 25 | 26 | 29 |
| Minimum thickness (mm) | | 6.1 | 8.8 | 11.5 | 13.9 | 17.1 | 19.9 | 21.1 | 22.7 | 24.4 | 25.2 | 27.9 |
| No. of plies | | 5 | 7 | 9 | 9 or 11 | 11 or 13 | 15 | 13 | 17 | 15 | 19 | 17 |
| Bending stress (bending as Figs 4.27 and 4.28) $\sigma_m$ | par | 18.2 | 15.9 | 14.3 | 14.2 | 12.8 (13) | 12.4 | 13.2 | 12.1 | 12.9 | 11.7 | 12.6 |
| | perp | 6.42 | 9.33 | 9.83 | 9.95 (9) | 9.75 | 9.39 | 9.35 | 9.30 | 9.20 | 9.10 | 9.05 |
| Tension stress $\sigma_t$ | par | 9.67 | 7.76 | 6.76 | 6.40 (11) | 5.86 (13) | 5.62 | 6.15 | 5.42 | 5.94 | 5.25 | 5.77 |
| | perp | 3.51 | 7.07 | 8.06 | 7.50 (11) | 7.23 | 7.00 | 6.98 | 6.95 | 6.85 | 6.76 | 6.70 |
| Compression stress $\sigma_c$ | par | 7.53 | 6.18 | 5.83 | 5.82 (11) | 5.57 (13) | 5.46 | 6.18 | 5.38 | 6.23 | 5.35 | 6.10 |
| | perp | 4.20 | 5.33 | 5.40 | 5.43 (9) | 5.30 | 5.20 | 5.19 | 5.18 | 5.17 | 5.00 | 4.98 |
| Panel shear stress (par and perp) $\sigma_p$ | | 3.00 | 3.00 | 3.00 | 3.00 | 3.00 | 3.00 | 3.00 | 3.00 | 3.00 | 3.00 | 3.00 |
| Rolling shear stress‡ All thicknesses except 6.5 mm have two outer plies of birch each face $\tau_r$ | | 0.60 0.45* | 0.60 | 0.60 | 0.60 | 0.60 | 0.60 | 0.60 | 0.60 | 0.60 | 0.60 | 0.60 |
| Transverse shear stress (i.e. shear stress with the plywood bent like a plate – Figs 4.27 and 4.28) | par | 0.57 | 0.70 | 0.70 | 0.70 | 0.70 | 0.70 | 0.70 | 0.70 | 0.70 | 0.70 | 0.70 |
| | perp | 0.70 | 0.43 | 0.40 | 0.43 (9) | 0.42 (13) | 0.44 | 0.44 | 0.43 | 0.45 | 0.44 | 0.45 |
| Bearing on face | | 3.00 | 3.00 | 3.00 | 3.00 | 3.00 | 3.00 | 3.00 | 3.00 | 3.00 | 3.00 | 3.00 |
| Modulus of elasticity – bending $E$ | par | 10 300 | 9500 | 8900 | 8550 (11) | 8050 | 7500 | 7400 | 6300 | 7150 | 6950 | 6850 |
| | perp | 2 750 | 3800 | 4300 | 4300 | 4200 | 4150 | 4150 | 4150 | 4150 | 4100 | 4050 |

| Modulus of elasticity – tension and compression | $E$ | par | 7550 | 6950 | 6450 | 6150 (11) | 5950 | 5900 | 5850 | 5800 | 5800 | 5750 | 5700 |
|---|---|---|---|---|---|---|---|---|---|---|---|---|---|
| | | perp | 4650 | 5300 | 5750 | 5000 (11) | 5100 | 5100 | 5050 | 5000 | 5000 | 4900 | 4850 |
| Modulus of rigidity (for panel shear) | $G$ | | 650 | 650 | 650 | 650 | 650 | 650 | 650 | 650 | 650 | 650 | 650 |

Note: Where there are two layups for one thickness the lesser value of stress is given above (where the values are not the same). The relevant layup is indicated in brackets.

* Permissible stress at first glue line due to there being only one outer ply of birch.

† Not reduced by either 0.5 ($K_{37}$) or 0.9 ($K_{70}$ for bonding pressure by nails or staples).

'par' and 'perp' indicate direction of face grain.

**Table 8.8** Strength properties of Canadian Douglas Fir-faced plywood parallel and perpendicular to face grain. Unsanded grades: Select Tight Face, Sheathing and Select Sheathing. Long term duration stresses based on 'full area' method.

| | | Permissible stress (N/mm²) | | | | | |
|---|---|---|---|---|---|---|---|
| Nominal thickness (mm) | | 7.5 | 9.5 | 12.5 | 15.5 | 18.5 | 20.5 |
| Minimum thickness (mm) | | 7.0 | 9.0 | 12.0 | 15.0 | 18.0 | 20.0 |
| No. of plies | | 3 | 3 | 3, 4 or 5 | 5 | 5, 6 or 7 | 5, 6 or 7 |
| Bending stress $\sigma_m$ (bending as Figs 4.27 and 4.28) | par | 11.6 | 10.7 | 9.39 (4) | 9.72 | 8.63 (5) | 8.14 (6) |
| | perp | 2.78 | 2.76 | 2.77 (3) | 3.41 | 3.58 (6) | 3.58 (6) |
| Tension stress $\sigma_t$ | par | 5.13 | 4.05 | 3.09 (4) | 3.89 | 3.60 (5) | 3.60 (7) |
| | perp | 2.14 | 2.11 | 2.12 (3) | 2.01 | 1.70 (6) | 1.55 (6) |
| Compression stress $\sigma_c$ | par | 7.99 | 6.31 | 4.82 (4) | 5.32 | 5.00 (5) | 5.04 (7) |
| | perp | 3.22 | 3.15 | 3.17 (3) | 3.03 | 2.56 (6) | 2.33 (6) |
| Panel shear stress (par or perp) $\sigma_p$ | | 1.82 | 1.45 | 1.19 (3) | 1.43 | 1.18 (5) | 1.07 (5) |
| Rolling shear stress $\tau_r$ * | | 0.45 / 0.33 | 0.45 / 0.33 | 0.45 / 0.33 | 0.45 / 0.33 | 0.45 / 0.33 | 0.45 / 0.33 |
| Transverse shear stress (i.e. shear stress with the plywood bent like a plate – Figs 4.27 and 4.28) | par | 0.36 | 0.38 | 0.36 (3) | 0.42 | 0.38 (7) | 0.38 (7) |
| | perp | 0.68 | 0.68 | 0.23 (5) | 0.24 | 0.25 (5) | 0.26 (5) |
| Bearing on face | | 2.16 | 2.16 | 2.16 | 2.16 | 2.16 | 2.16 |
| Modulus of elasticity – bending $E$ | par | 12 350 | 11 450 | 10 050 (4) | 9800 | 8700 (5) | 8200 (6) |
| | perp | 850 | 800 | 850 (3) | 2150 | 2500 (6) | 2650 (6)(7) |

| | | | | | | | |
|---|---|---|---|---|---|---|---|
| Modulus of elasticity –tension and compression | $E$ | par | 10 550 | 8 350 | 6 350 (4) | 6900 | 6450 (5) | 6500 (7) |
| | | perp | 4 000 | 3 950 | 3 950 (3) | 3750 | 3200 (6) | 2900 (6) |
| Modulus of rigidity (for panel shear) | $G$ | | 660 | 630 | 600 (4)(5) | 590 | 570 | 570 (6)(7) |

Note: Where there are two or three layups for one thickness the lesser value of stress is given above (where the values are not the same). The relevant layup is indicated in brackets.

* Two values are given for rolling shear. The lower is for the first glue line (see §8.8.2).

† Not reduced by either 0.5 ($K_{37}$) or 0.9 ($K_{70}$ for bonding pressure by nails or staples).

'par' and 'perp' indicate direction of face grain.

| $\dfrac{h_f}{h}$ | Approximate accuracy (%) |
|---|---|
| 0.30 | 93 |
| 0.25 | 96 |
| 0.20 | 97.7 |

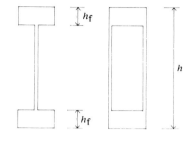

Fig. 8.20

| Loading arrangement | Formula for bending deflection |
|---|---|
| Case 1 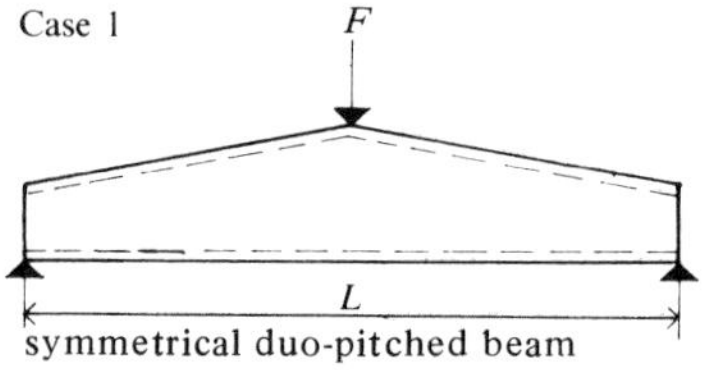      F <br><br> *symmetrical duo-pitched beam*   L | $\delta_{m1} = \dfrac{FL^3}{96\,(EI)_h}\left[\left(\dfrac{2}{n_2+1}\right)^a + \left(\dfrac{1}{n_2}\right)^a\right]$ |
| Case 2    P = load per unit length 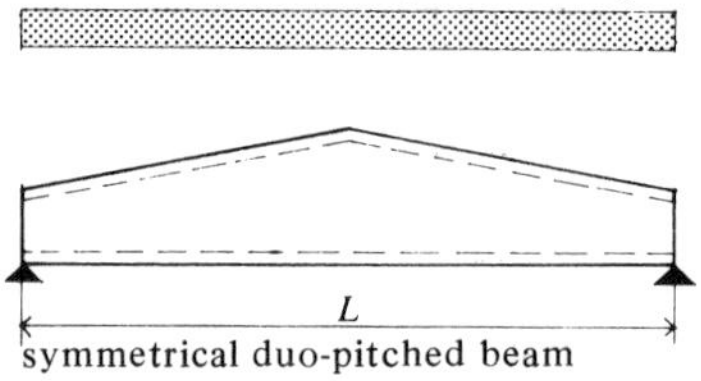 <br><br> *symmetrical duo-pitched beam*   L | $\delta_{m2} = \dfrac{pL^4}{192\,(EI)_h} \times$ <br><br> $\left[1{\cdot}5\left(\dfrac{2}{n_2+1}\right)^a + \left(\dfrac{1}{n_2}\right)^a\right]$ |
| Case 3 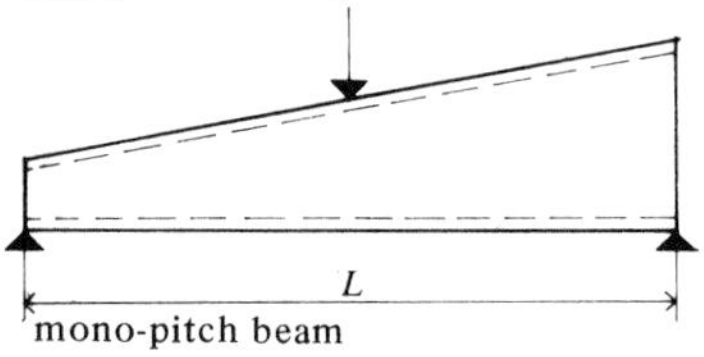     F <br><br> *mono-pitch beam*   L | $\delta_{m3} = \dfrac{FL^3}{192\,(EI)_h}\left[\left(\dfrac{4}{3+n_2}\right)^a + \right.$ <br><br> $\left. 2\left(\dfrac{2}{1+n_2}\right)^a + \left(\dfrac{4}{1+3n_2}\right)^a\right]$ |
| Case 4    p = load per unit length 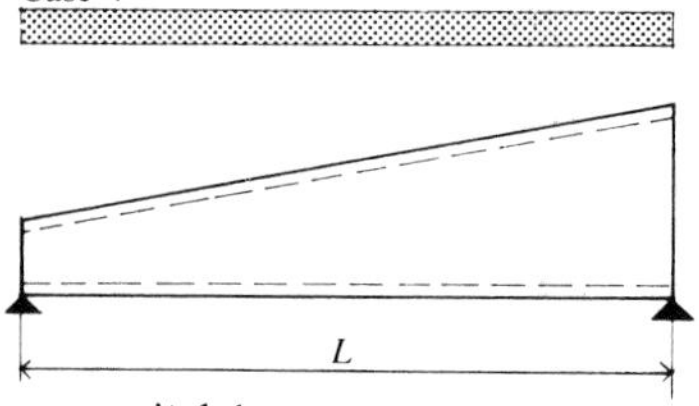 <br><br> *mono-pitch beam*   L | $\delta_{m4} = \dfrac{pL^4}{256\,(EI)_h}\left[\left(\dfrac{4}{3+n_2}\right)^a + \right.$ <br><br> $\left. \dfrac{4}{3}\left(\dfrac{2}{1+n_2}\right)^a + \left(\dfrac{4}{1+3n_2}\right)^a\right]$ |

Fig. 8.21

### 8.17.2 Shear deflection

The design principles for shear deflection of tapered solid sections introduced in §7.11.1 and formulated in §7.11.4 apply to ply web I and Box beams. The value of $K_{\text{form}}$ (the form factor) becomes unity, and $GA_h$ relates to the ply web only and is evaluated for some typical sections in Tables 8.5 and 8.6.

### 8.17.3 The location and magnitude of the maximum bending stress

Using the 'lever-arm' method of calculating bending stresses gives a linear variation of bending moment capacity with linear variation in beam depth, therefore, in the case of a central point load on a simple span, the maximum bending stress always occurs at mid span (Fig. 8.22).

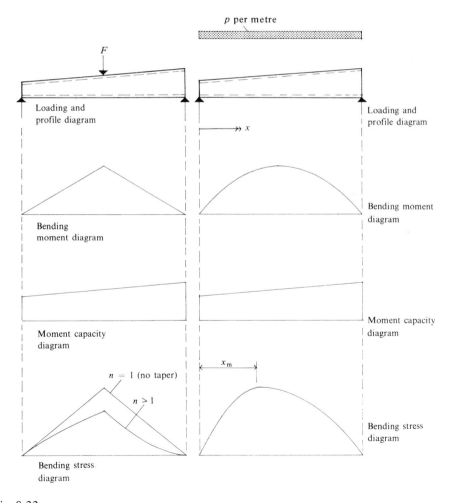

Fig. 8.22

In the case of a UDL, the maximum bending stress does not occur at mid span. By inspection of Fig. 8.22 it can be seen that the maximum bending stress occurs between $x = 0$ and $x = 0.5L$, in which range $M_x = 0.5p\,(Lx - x^2)$ at a distance $x_m$ from the end of least depth.

The corresponding moment capacity at any point $x$ is:

$$M_x = 0.5A_f h_x\, \sigma_{adm} \qquad \text{(Fig. 8.23)}$$

where $A_f$ = the total area of both flanges (assuming equal flange areas in top and bottom chord)

$\sigma_{adm}$ = the permissible stress

$$h_x = h_c\left[1 + \frac{x}{s}\,(n_2 - 1)\right]. \text{ (See Fig. 8.19 for position of } h_c.)$$

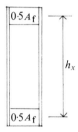

Fig. 8.23

Thus at any section a distance $x$ from the end of least depth, the actual stress in the flange is given by the formula:

$$\sigma_a = \frac{M_x}{0.5A_f h_x} = \frac{p\,(Lx - x^2)}{A_f h_c\,[1 + (x/s)(n_2 - 1)]}$$

which has a maximum value when $(Lx - x^2)/[1 + (x/s)(n_2 - 1)]$ is at a maximum (see Fig. 8.19) and $x_m$ is determined by differentiating this radical with respect to $x$.

For a mono-pitch beam with $s = L$, it can be shown that:

$$x_m = \left[\frac{\sqrt{n_2} - 1}{n_2 - 1}\right]L$$

For a duo-pitch beam with $s = 0.5L$ it can be shown that:

$$x_m = 0.5L\left[\frac{\sqrt{(2n_2 - 1)} - 1}{n_2 - 1}\right]$$

By substituting $x_m$ into the formula $\sigma_a = M_x/0.5A_f h_x$, one can obtain a direct formula for the evaluation of the maximum bending stress.

For a mono-pitch beam:

$$\sigma_{max} = \frac{pL^2}{A_f h_c}\left[\frac{n_2 + 1 - 2\sqrt{n_2}}{(n_2 - 1)^2}\right]$$

For a duo-pitch beam:

$$\sigma_{max} = \frac{pL^2}{2A_f h_c}\left[\frac{n_2 - \sqrt{(2n_2 - 1)}}{(n_2 - 1)^2}\right]$$

### 8.17.4   Design example. Mono-pitch beam

Calculate the maximum bending stress and mid span deflection of the beam in Fig. 8.24 when it is subjected to a total UDL of 4.5 kN/m.

$$E_f = 9240 \text{ N/mm}^2$$
$$E_w = 4650 \text{ N/mm}^2$$

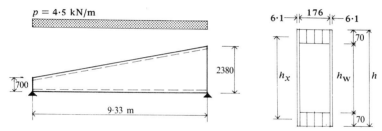

Fig. 8.24

Calculate the sectional rigidity at the 700 mm (least) depth:

$$EI_{700} = (EI)_{web} + (EI)_{flange}$$

$$= 4650\left[\frac{12.2}{12} \times 700^3\right] + 9240\left[\frac{176}{12}(700^3 - 560^3)\right]$$

$$= 24\,250 \times 10^9 \text{ N.mm}^2$$

First calculate bending stresses and bending deflection in a conventional tabular form, taking eight equal elements of span (Table 8.9). The *EI* values are calculated at the centre of each element in a similar manner to $EI_{700}$ above.

The mid span bending moment $M_0 = 48.96$ kN.m. $M_x$ is a parabolic proportion of the mid span bending moment.

i.e.
$$M_x = \frac{x(L - x)M_0}{(0.5L)^2}$$

From Table 8.9 maximum bending stress = 2.94 N/mm² and,

$$\delta_m = 1.1662 \times \Sigma \frac{M_x m}{EI} = 1.1662 \times 0.003\,094 = 0.003\,608 \text{ m}$$

**Table 8.9**

| Element | $x$ (m) | $Lx$ (m) | $M_x$ (kN.m) | $h$ (m) | $h_x$ (m) | $\sigma$ (N/mm²) | $EI$ (kN.m²) | $m$ | $\dfrac{M_x m}{EI}$ (m) |
|---|---|---|---|---|---|---|---|---|---|
| End 700 | 0 | 9.33 | 0 | 0.7 | 0.63 | – | 24 250 | – | – |
| 1 | 0.58 | 8.74 | 11.35 | 0.804 | 0.734 | 1.26 | 33 210 | 0.29 | 0.000099 |
| 2 | 1.74 | 7.58 | 29.54 | 1.013 | 0.942 | 2.55 | 55 620 | 0.87 | 0.000462 |
| 3 | 2.91 | 6.41 | 41.78 | 1.223 | 1.153 | 2.94 | 84 410 | 1.455 | 0.000720 |
| 4 | 4.08 | 5.25 | 47.98 | 1.434 | 1.364 | 2.86 | 119 930 | 2.04 | 0.000816 |
| 5 | 5.25 | 4.08 | 47.98 | 1.645 | 1.575 | 2.47 | 162 330 | 2.04 | 0.000603 |
| 6 | 6.41 | 2.91 | 41.78 | 1.853 | 1.783 | 1.90 | 211 120 | 1.455 | 0.000288 |
| 7 | 7.58 | 1.74 | 29.54 | 2.064 | 1.994 | 1.20 | 267 970 | 0.87 | 0.000096 |
| 8 | 8.74 | 0.58 | 11.35 | 2.273 | 2.203 | 0.42 | 331 850 | 0.29 | 0.000010 |
| End 2380 | 9.33 | 0 | 0 | 2.38 | 2.31 | – | 367 560 | – | – |

$$\Sigma \frac{M_x m}{EI} \qquad 0.003\,094$$

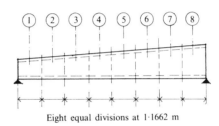

Eight equal divisions at 1·1662 m

The values of maximum bending stress and deflection have been established from a conventional method in order that the method below using coefficients can be compared.

*Alternative solution using coefficients*

$$\left.\begin{array}{l} (EI)_{h_c} = EI_{700} = 24\,250\,\text{kN.m}^2 \\ (EI)_{H_c} = EI_{2380} = 367\,560\,\text{kN.m}^2 \end{array}\right\} \quad \therefore\ n_1 = 15.16$$

$$\left.\begin{array}{l} h_c = 700 - 70 = 630\,\text{mm} \\ H_c = 2380 - 70 = 2310\,\text{mm} \end{array}\right\} \quad \therefore\ n_2 = 3.667$$

$$a = \frac{\log_e n_1}{\log_e n_2} = 2.09$$

and

$$\delta_m = \frac{pL^4}{256(EI)_{h_c}} \left[ \left(\frac{4}{3+n_2}\right)^a + \frac{4}{3}\left(\frac{2}{1+n_2}\right)^a + \left(\frac{4}{1+3n_2}\right)^a \right]$$

$$= \frac{4.5 \times 9.33^4}{256 \times 24\,250} \left[ \left(\frac{4}{6.667}\right)^{2.09} + \frac{4}{3}\left(\frac{2}{4.667}\right)^{2.09} + \left(\frac{4}{12}\right)^{2.09} \right]$$

$$= 0.003\,38\,\text{m}$$

$$= 3.38\,\text{mm}$$

The maximum bending stress for a single tapered (mono-pitch) section occurs at $x_m$, where:

$$x_m = \left[\frac{\sqrt{n_2}-1}{n_2-1}\right]L = \left[\frac{\sqrt{3.667}-1}{2.667}\right]9.33 = 3.2\,\text{m}$$

at $x_m = 3.2$ m:

$$M_x = \frac{3.2(9.33-3.2)48.96}{4.665^2} = 44.13\,\text{kN.m}$$

$$h_x = 630\left[1 + \frac{3.2}{9.33}(2.667)\right] = 1206\,\text{mm}$$

$$\therefore\ \sigma_a = \frac{M_x}{A_f h_x} = \frac{44.13 \times 10^6}{(70 \times 176)\,1206} = 2.97\,\text{N/mm}^2$$

which can be seen to lie between the tabulated values of 2.94 N/mm² and 2.86 N/mm² for elements 3 and 4 respectively.

This method is obviously faster and of similar accuracy. This example is principally presented as a comparison of two methods of solution, and in practice shear deflection must also be added to the bending deflection calculated above.

*Calculation of shear deflection using coefficients*
From Fig. 7.28 with $n = 2380/700 = 3.4$, $\Delta_{v7} = 0.407$, and $K_{form} = 1.0$ (§ 8.17.2).

$GA_h$ relates to the ply webs only, and at depth $h = 700$ mm, with 6.1 mm ply webs $GA_h = 650 \times 700 \times 12.2 \div 10^3 = 5551$ kN.
$M_0 = 48.96$ kN.m.

$$\delta_{v7} = \frac{K_{form}\,M_0\,\Delta_{v7}}{GA_h} = \frac{1.0 \times 48.96 \times 0.407}{5551} = 0.003\,59\,\text{m} = 3.59\,\text{mm}$$

Although this is a small deflection compared to the span (0.000 39 × span) it is more than the bending deflection.

## 8.17.5 Design example. Duo-pitch beam

Compare the bending and shear deflection for the beam in Fig. 8.25 as a principal member.

Section 700 mm tapering to 500 mm (see standard tables, Table 8.6).

$$\left.\begin{array}{l}\overline{EI}_{500} = \phantom{0}9\,130\,\text{kN.m}^2 \\ \overline{EI}_{700} = 19\,940\,\text{kN.m}^2\end{array}\right\}\ n_1 = 2.184 \qquad \therefore\ \log_e n_1 = 0.7812$$

$$\left.\begin{array}{l}h_c = 430\,\text{mm} \\ H_c = 630\,\text{mm}\end{array}\right\}\ n_2 = 1.465 \qquad \therefore\ \log_e n_2 = 0.3819$$

$$a = \frac{\log_e n_1}{\log_e n_2} = 2.05$$

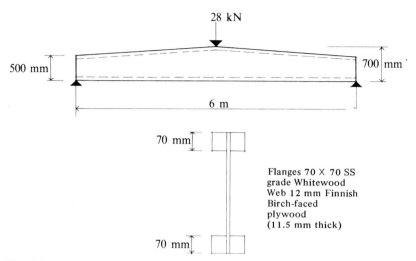

Fig. 8.25

Bending deflection is calculated from:

$$\delta_{m1} = \frac{FL^3}{96(EI)_h} \left[ \left(\frac{2}{n_2+1}\right)^a + \left(\frac{1}{n_2}\right)^a \right]$$

$$= \frac{28 \times 6^3}{96 \times 9130} \left[ \left(\frac{2}{2.465}\right)^{2.05} + \left(\frac{1}{1.465}\right)^{2.05} \right] = 0.007\,65 \text{ m}$$

Shear deflection is calculated from:

$$\delta_{v1} = \frac{K_{form} M \Delta_{v1}}{GA_h}$$

where $K_{form} = 1.0$ for ply web beam

$$n = \frac{H}{h} = \frac{700}{500} = 1.4$$

$$\therefore \Delta_{v1} = 0.841 \text{ (Fig. 7.22)}$$
$$GA_h = 3740 \text{ kN (Table 8.6)}$$

$$\delta_{v1} = \frac{42 \times 0.841}{3740} = 0.009\,44$$

Total deflection $= \delta_{m1} + \delta_{v1} = 0.0171$ m.

In this instance the shear deflection exceeds the bending deflection and is 55% of the total deflection.

## 8.18 SUMMARY

The authors are fully aware that any designer or engineer who has not previously designed or checked a ply web beam could be concerned by the length of this chapter and by the several methods put forward for calculating bending stresses. However, an engineer given the task of checking someone else's calculations may encounter any one of these methods and should be aware of the implications.

An engineer preparing ply web designs for the first time should decide which method to adopt in calculating bending stresses, which will also decide what method must be used to check web splices. Once this decision is taken the design becomes quite straightforward, and speed will come with practice, particularly if the coefficients for calculating deflection are used.

The authors emphasise that shear deflection must be taken into account as must the design of web splices and any load-bearing stiffeners. The spacing of non-load-bearing stiffeners must be considered, although their actual size is unlikely to be important.

It is good practice to build in a camber to off-set dead load deflection (except for cases similar to those discussed in Chapter 12).

# Chapter Nine
# Lateral Stability of Beams

## 9.1 LATERAL STABILITY TO PERMIT USE OF FULL GRADE BENDING STRESS

If the full grade bending stress (increased by the load-duration or load-sharing factors if applicable) is to be used in calculations, BS 5268:Part 2 gives limiting values of depth-to-breadth ratios for solid or glulam beams which must not be exceeded with various conditions of lateral restraint. These ratios are given in Table 17 of BS 5268:Part 2:1984 and are presented in Table 9.1 of this manual.

**Table 9.1  Solid and glulam beams**

| Degree of lateral support | Maximum depth-to-breadth ratio |
|---|---|
| No lateral support | 2 |
| Ends held in position | 3 |
| Ends held in position and members held in line, as by purlins or tie rods | 4 |
| Ends held in position and compression edge held in line, as by direct connection of sheathing, deck or joists | 5 |
| Ends held in position and compression edge held in line, as by direct connection of sheathing, deck or joists, together with adequate bridging or blocking spaced at intervals not exceeding six times the depth | 6 |
| Ends held in position and both edges firmly held in line | 7 |

Similar ratios are given for ply web Box and I sections, and they are given here in tabular form as Table 9.2. The ratios for ply web beams are quoted as $I_X/I_Y$. It will be noted that there is a similarity between the two tables, there being similar degrees of restraint required for $I_X/I_Y = (h/B)^2$.

The ratios for solid, glulam and ply web beams are empirical. They cannot be substantiated by theory, but are conservative and cover most of the cases encountered in practice. In addition to these empirical methods, BS 5268 permits checking by other methods to ensure that there is no risk of buckling under design load.

Table 9.2  Ply web Box and I beams

| $\dfrac{I_X}{I_Y}$ | Degree of lateral support |
|---|---|
| Up to 5 | No lateral support is required |
| 5–10 | The ends of beams should be held in position at the bottom flange at supports |
| 10–20 | Beams should be held in line at the ends |
| 20–30 | One edge should be held in line |
| 30–40 | The beam should be restrained by bridging or other bracing at intervals of not more than 2.4 m |
| More than 40 | The compression flange should be fully restrained |

## 9.2  PARTIAL LATERAL STABILITY. REDUCTION IN BENDING MOMENT CAPACITY

The simple, conservative limits detailed in § 9.1 cover the majority of cases a designer encounters, but there are cases where adequate restraint cannot be provided or where the designer wishes to use a more slender section for economical or architectural reasons. In any of these special cases it is necessary for the designer to see whether the reduced lateral stability leads to a reduction in bending strength capacity.

The lateral buckling of a beam depends not only on the depth-to-breadth ratio (or $I_X$ to $I_Y$ ratio) but also on:

(a) the geometrical and physical properties of the beam section,
(b) the nature of the applied loading,
(c) the position of the applied loading with respect to the neutral axis of the section, and
(d) the degree of restraint provided at the vertical supports and at points along the span.

§§ 9.3–9.6 give a design method which is applicable to solid, glulam and ply web beams and relate reductions in the bending moment capacity to partial lateral stability. § 9.7 gives a design method which is applicable to ply web beams (or lattice beams) only and treats the top flange as a horizontal column for the purpose of relating lateral restraint to permissible stress. § 9.8 gives a design method to establish the effective length of an unrestrained compression flange (as a horizontal column) due to restraint to the tension flange (the 'U frame' method).

## 9.3   RELATING PARTIAL LATERAL STABILITY TO BENDING MOMENT

The anticipated shape of the lateral buckling of the compression flange of a beam which is not fully restrained takes one of the forms sketched in plan in Fig. 9.1 or some intermediate position.

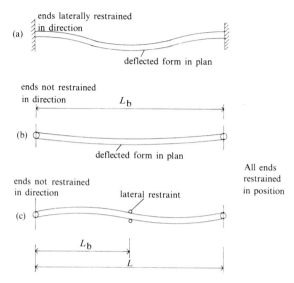

Fig. 9.1: *Forms of lateral buckling shown in plan.*

It is unusual and difficult to achieve full restraint in direction at the ends of beams as sketched at (a) in Fig. 9.1 and therefore, in the following design method, the assumption is made that all end joints have no directional restraint laterally, as sketched at (b) and (c) in Fig. 9.1. The ends of beams and any point of lateral restraint along the beam are assumed to be held in position laterally but not restrained in direction (i.e. no fixity).

In the following design method $L_b$ is the lateral unrestrained length of beam which is not necessarily the full span $L$. For solid, glulam, ply web Box beams and I beams which are symmetrical about the $XX$ and $YY$ axes, buckling occurs at a critical moment expressed as:

$$M_{crit} = \frac{C_1}{L_b} \left[ \frac{EI_Y\, GJ}{\alpha} \right]^{\frac{1}{2}} \left[ 1 - \frac{C_2 h_L}{L_b} \sqrt{\left( \frac{EI_Y}{GJ} \right)} \right] \tag{9.1}$$

where $EI_Y$ = effective bending stiffness about the $YY$ axis.

$\quad\quad GJ$ = effective torsional resistance.

$\quad\quad h_L$ = height from the $XX$ axis to the point of application of load, being positive above the neutral axis and negative below the neutral axis when the loading is in the direction shown in Fig. 9.2. If a load is placed on the top flange of a beam, there is a tipping action which increases the instability of the section (i.e. reduces the critical moment), whereas if a load is applied to the bottom flange there is a stabilising effect.

$\alpha = 1 - (EI_Y/EI_X)$, a factor taking account of the additional stability which occurs when a beam sags below the horizontal. It should not be applied to beams which are cambered. The designer may occasionally include $\alpha$ to obtain a slightly higher critical moment, but as most designs include camber, no allowance for $\alpha$ is taken in the design method given in §§ 9.3–9.6, where $\alpha$ is taken as 1.0.

$L_b$ = the laterally unrestrained length of beam which is not necessarily the full span between vertical supports.

$C_1$ and $C_2$ = constants determined by the nature of the applied loading and the conditions of effective restraint. There are a number of books and publications which derive the values of constants for many typical loading conditions. The cases most likely to be encountered are summarised in Table 9.3. When no external loading is carried between two points of lateral restraint (e.g. cases 6–9), then:

$$C_2 = 0 \quad \text{and} \quad C_1 = 5.5 - 3.3\beta + 0.94\beta^2$$

except that $C_1$ is not to be taken as more than 7.22 (which occurs at $\beta = -0.46$), where:

$$\beta = \frac{\text{smaller 'end' moment}}{\text{larger 'end' moment}}$$

The sign for $\beta$ is positive if one moment is anti-clockwise and the other clockwise; also if one moment is zero. If both moments are clockwise or both anti-clockwise, the sign for $\beta$ is negative.

The function:

$$\left[ 1 - \frac{C_2 h_L}{L_b} \sqrt{\left( \frac{EI_Y}{GJ} \right)} \right]$$

takes account of the point of application of the load above or below the $XX$ axis (Fig. 9.2). If the load is applied at the $XX$ axis, the whole function becomes unity.

The critical moment $M_{crit}$ must be reduced by a suitable factor of safety. A factor of 2.25 is assumed here, therefore the safe buckling moment $M_{buc}$ is found by dividing the critical moment by 2.25. Hence:

$$M_{buc} = M_{crit}/2.25$$

For a beam not to buckle under the design loading, the buckling moment must be equal to or greater than the maximum bending moment produced by the design loading on the $XX$ axis.

**Table 9.3   Values of $C_1$ and $C_2$ for selected common loading conditions**

| Case | External (or internal) loading arrangement between points of lateral restraint | $C_1$ | $C_2$ |
|------|------|------|------|
| 1 | | +3.55 | +1.41 |
| 2 | | +4.08 | +4.87 |
| 3 | | +4.24 | +1.73 |
| 4 | | +5.34 | +4.46 |
| 5 | | +3.27 | +2.64 |
| 6 | $M$ ⟵⟶ $M$ | +3.14 | +0.0 |
| 7 | $M$ ⟵⟶ $0.5M$ | +4.08 | +0.0 |
| 8 | $M$ ⟵⟶ | +5.5 | +0.0 |
| 9 | $M$ ⟵⟶ $0.46\,M$ ($\beta = -0.46$) | +7.22 | +0.0 |
| 10 | | +4.1 | +1.0 |
| 11 | | +6.42 | +1.8 |

To introduce some simplification into actual designs, it is convenient to introduce coefficients $N_1$ and $N_2$.

$$N_1 = \sqrt{\left(\frac{EI_Y\,GJ}{\alpha}\right)}\tag{9.2}$$

(simplifies with $\alpha =$ unity).

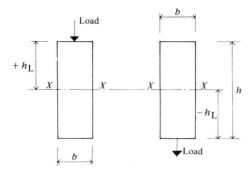

Fig. 9.2: *Positive and negative location of the applied load.*

$$N_2 = \sqrt{\left(\frac{EI_Y}{GJ}\right)} \tag{9.3}$$

$$\therefore M_{\text{buc}} = \frac{C_1 N_1}{2.25 L_b} \left(1 - \frac{C_2 h_L N_2}{L_b}\right) \tag{9.4}$$

The buckling moment $M_{\text{buc}}$ should not be increased to take account of any load-duration factor. If the beam is liable to buckle, it will be an instantaneous action dependent on the magnitude and not the duration of the loading.

It is unlikely that this design method will be used in a load-sharing system but, if it is, no increase in $M_{\text{buc}}$ should be made other than that which occurs due to the $E$ and $G$ values used in the calculation of $N_1$ and $N_2$, being those applicable to the number of pieces acting together (i.e. $E_{\text{min}}$ modified by $K_9$ or $K_{27}$, or $E_{\text{mean}}$ modified by $K_{20}$ or $K_{26}$).

## 9.4   PARTIALLY RESTRAINED SOLID AND GLULAM BEAMS

### 9.4.1   Design method

The design method is a continuation of that outlined in § 9.3. For a solid section the torsional constant $J$ may be taken as $\frac{1}{3}hb^3(1 - 0.63b/h)$. If $K_\lambda = \sqrt{(1 - 0.63b/h)}$ and the modulus of rigidity $G = E/16$ then:

$$N_1 = \frac{EI_Y K_\lambda}{2} \tag{9.5}$$

and

$$N_2 = \frac{2}{K_\lambda} \tag{9.6}$$

$$\therefore M_{\text{buc}} = \frac{C_1 EI_Y K_\lambda}{4.5 L_b} \left[1 - \frac{2h_L C_2}{K_\lambda L_b}\right] \tag{9.7}$$

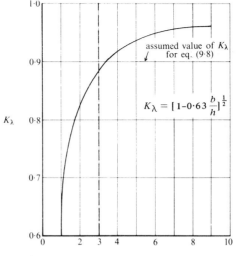

Fig. 9.3: *The relationship of $K_\lambda$ to $h/b$ for solid and glulam beams.*

The value of $K_\lambda$ approaches unity as the slenderness increases, particularly where $h/b$ is more than four. The relationship between $K_\lambda$ and $h/b$ for solid and glulam beams is given in Fig. 9.3.

In practice the load is most frequently applied at the top flange, and it is unlikely that a stability check will be necessary for a section having a depth-to-breadth ratio of less than 3 (where, from Table 9.1, it can be seen that only the ends would have to be held in position for the full grade bending stress to be permitted). With these assumptions $K_\lambda$ may be approximated to 0.9 and $h_L$ may be taken as $+h/2$.

A conservative simplification of equation (9.7) for solid and glulam beams is:

$$M_{\text{buc}} = \frac{C_1 E I_Y}{5L_b}\left[1 - \frac{C_2 h}{0.9L_b}\right] \tag{9.8}$$

### 9.4.2   Design of a glulam beam with no lateral restraint

A glulam beam spanning 9 m supports a medium term central point load of 10 kN (6 kN dead, 4 kN imposed) applied by a pin-ended post which gives no lateral restraint to the beam. Design a suitable glulam beam of LB grade Whitewood, checking lateral stability.

$$\text{Maximum shear} = \text{reaction} = 5\,\text{kN}$$

$$M_{MX} = \frac{10 \times 9}{4} = 22.5\,\text{kN.m}$$

Assuming that the beam is cambered to off-set dead load deflection, required $EI_X$ to limit bending deflection from imposed loading to $0.003L$

$$= \frac{FL^2}{48 \times 0.003} = \frac{4 \times 9^2}{48 \times 0.003} = 2250\,\text{kN.m}^2$$

Assume a 360 mm deep beam, therefore $L/h = 25$. From Fig. 4.26 bending deflection or required $EI_X$ value is increased by 1.025 to take account of shear deflection.

$$\therefore \ \text{Required } EI_X = 2250 \times 1.025 = 2306\,\text{kN.m}^2$$

From Table 7.7 it can be seen that a $90 \times 360$ mm beam of LB grade satisfies the requirements above.

*Stability check*
Lateral buckling will take the form of plan (b) in Fig. 9.1; therefore

$$L_b = L = 9\,\text{m}$$

The loading condition related to Table 9.3 is case 3; therefore:

$$C_1 = 4.24 \qquad \text{and} \qquad C_2 = 1.73$$

From Table 7.7 $E_N I_Y = 202\,\text{kN.m}^2$.

If one uses the more accurate equation (9.7), the $K_\lambda$ value (for $h/b = 4$) $= 0.92$ from Fig. 9.3, and with $h_L = +h/2 = +0.18$ m, the buckling moment is calculated as:

$$M_{\text{buc}} = \frac{4.24 \times 202 \times 0.92}{4.5 \times 9.0}\left[1 - \frac{2 \times 0.18 \times 1.73}{0.92 \times 9.0}\right] = 18.0\,\text{kN.m}$$

Or, from the simplified equation, (9.8):

$$M_{\text{buc}} = \frac{4.24 \times 202}{5 \times 9}\left[1 - \frac{1.73 \times 0.36}{0.9 \times 9}\right] = 17.6\,\text{kN.m}$$

(i.e. $2\tfrac{1}{2}\%$ less than 18.0 kN.m).

$M_{\text{buc}}$ is less than $M_{MX}$ therefore this section is unstable and the process has to be repeated using a less slender section. Try a $115 \times 315$ mm beam having $E_N I_Y = 363\,\text{kN.m}^2$. From equation (9.8):

$$M_{\text{buc}} = \frac{4.24 \times 363}{5 \times 9}\left[1 - \frac{1.73 \times 0.315}{0.9 \times 9}\right] = 31.9\,\text{kN.m} > 22.5\,\text{kN.m}$$

which is acceptable.

### 9.4.3   Design of a glulam beam with central lateral restraint

A glulam beam spanning 9 m carries a medium term central point load of 10 kN (6 kN dead, 4 kN imposed) applied by a secondary beam which also gives restraint to the beam at mid span, the loading being applied at mid height of the main beam (i.e. $h_L = 0$ in equation (9.7)). Design a suitable section of LB grade Whitewood checking lateral stability.

$$\text{Maximum shear} = 5\,\text{kN}$$

$$M_{MX} = \frac{10 \times 9}{4} = 22.5\,\text{kN.m}$$

Assume that the beam is cambered to off-set dead load deflection.

Required $EI_X$ to limit bending and shear deflection from imposed loading $= 2306\,\text{kN.m}^2$ as the example in §9.4.2.

Try a 90 × 360 mm LB Whitewood beam. Carry out a stability check using equation (9.7). Lateral buckling will take the form of plan (c) in Fig. 9.1 and $L_b = L/2 = 4.5\,\text{m}$. The loading condition related to Table 9.3 is case 8, therefore $C_1 = 5.5$ and $C_2 = 0.0$.

Because $h_L = 0$, the part in brackets in equation (9.7) becomes unity, $K_\lambda = 0.92$ (from Fig. 9.3 for $h/b = 4$), therefore:

$$M_{buc} = \frac{5.5 \times 202 \times 0.92}{4.5 \times 4.5} = 50.5\,\text{kN.m} > 22.5\,\text{kN.m}$$

which is acceptable.

### 9.4.4  Comparison design of a glulam beam with restrained/unrestrained conditions

A glulam beam of 12 m span is loaded on the top flange by point loads of 6 kN (2 kN dead, 4 kN medium term imposed) as shown in Fig. 9.4. Design a suitable glulam beam of LC grade if the method of applying the point loads restrains the beam, and compare the case if the beam is not restrained.

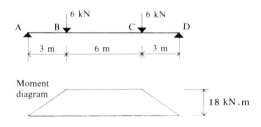

Fig. 9.4

The first part of the design is common to both the restrained and unrestrained conditions: to determine a beam adequate to resist shear and the bending moment about the $XX$ axis and deflection under imposed loading only (beam cambered) assuming full restraint.

$$\text{Maximum shear} = 6\,\text{kN}$$
$$M_{MX} = 18\,\text{kN.m}$$

From Table 4.15 required $EI_X$ to limit bending deflection under imposed loading to $0.003L$

$$= 4.34 \times (2 \times 1.10 \times 4)\, 12^2 = 5500 \text{ kN.m}^2 \text{ (assuming camber)}$$

Assume a 450 mm deep beam, therefore $L/h = 26.7$.
    From Fig. 4.26, bending deflection or required $EI_X$ value is increased by 1.022 to take account of shear deflection.

$$\therefore \text{ Required } EI_X = 1.022 \times 5500 = 5620 \text{ kN.m}^2$$

From Table 7.8 it can be seen that a 90 × 450 mm beam of LC grade Whitewood is adequate to meet these requirements.
    From Table 7.8, $E_N I_Y = 216 \text{ kN.m}^2$.
    With restraint at each point load, lateral buckling takes the form of plan (c) in Fig. 9.1, and it is necessary to consider separately the parts of span between lateral restraints.

*Consider Part AB* (Part DC is the same). Loading case 8 in Table 9.3 applies, with $C_1 = 5.5$, $C_2 = 0.0$, and $L_b = 3$ m. From equation (9.8):

$$M_{buc} = \frac{5.5 \times 216}{5 \times 3} = 79 \text{ kN.m} > 18 \text{ kN.m}$$

Therefore part AB is adequate (because $C_2 = 0$, the part in brackets becomes unity).

*Consider Part BC.* Loading case 6 in Table 9.3 applies, with $C_1 = 3.14$, $C_2 = 0.0$ and $L_b = 6$ m. From equation (9.8):

$$M_{buc} = \frac{3.14 \times 216}{5 \times 6} = 22.6 \text{ kN.m} > 18 \text{ kN.m}$$

Therefore part BC is adequate.

If the beam is not restrained at the positions of applied load the form of buckling will be as plan (b) in Fig. 9.1 and $L_b = 12$ m. Loading case 5 in Table 9.3 applies with $C_1 = 3.27$ and $C_2 = 2.64$. From equation (9.8):

$$M_{buc} = \frac{3.27 \times 216}{5 \times 12} \left[ 1 - \frac{2.64 \times 0.45}{0.9 \times 12} \right] = 10.5 \text{ kN.m} < 18 \text{ kN.m}$$

This condition would not be stable against lateral buckling. A 115 × 450 mm glulam beam of LC grade gives:

$$M_{buc} = \frac{3.27 \times 451}{5 \times 12} \left[ 1 - \frac{2.64 \times 0.45}{0.9 \times 12} \right] = 21.9 \text{ kN.m} > 18 \text{ kN.m}$$

which is acceptable.

## 9.5   PARTIALLY RESTRAINED PLY WEB BOX BEAMS

### 9.5.1   Design method

This design method is a continuation of that outlined in § 9.3.

From Roark Table IX, the torsional constant $J$ for a Box beam as sketched in Fig. 9.5 may be taken as:

$$J = \frac{2b_0^2 h_a^2}{(b_0/h_f) + (h_a/t)}$$

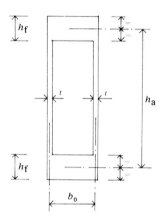

Fig. 9.5

For the range of Box beams in Table 8.5 (and this range may be regarded as typical) it can be calculated that $b_0/h_f$ is always less than 6% of $h_a/t$ and therefore can be neglected with little loss of accuracy. $J$ may then be approximated to:

$$J = 2th_a b_0^2$$

in which the flange contribution $b_0^2$ is at least 85% of the product $th_a b_0^2$. It seems reasonable therefore to adopt $G$ appropriate to flanges for deriving $GJ$.

Values of stability constants $N_1$ and $N_2$ are calculated and given in Table 9.4 for the range of Box beams with Finnish Birch-faced plywood webs and SS grade flanges in Table 8.5.

The $G$ value for Finnish Birch-faced plywood is $650 \, \text{N/mm}^2$.

$$N_1 = \sqrt{(EI_Y GJ)} \quad \text{(repeat of equation (9.2) with } \alpha \text{ taken as 1.0)}$$

$$N_2 = \sqrt{\left(\frac{EI_Y}{GJ}\right)} \quad \text{(repeat of equation (9.3))}$$

**Table 9.4** Stability constants $N_1$ and $N_2$ for selected ply web Box beams

| Box sections | $b_0$ (mm) | $h_a$ (mm) | $J$ (mm$^4 \times 10^6$) | $GI$ (kN.m$^2$) | $EI_Y$ (kN.m$^2$) | $N_1$ (kN.m$^2$) | $N_2$ | $E$ (N/mm$^2$) |
|---|---|---|---|---|---|---|---|---|
| 100.2 × 300 | 94.1 | 230 | 24.8 | 13.4 | 106 | 37.8 | 2.80 | 8680 |
| 100.2 × 400 | 94.1 | 330 | 35.6 | 19.3 | 119 | 48.0 | 2.48 | 8680 |
| 144.2 × 300 | 138.1 | 230 | 53.5 | 30.2 | 323 | 98.8 | 3.27 | 9030 |
| 144.2 × 400 | 138.1 | 330 | 76.8 | 43.3 | 350 | 123.1 | 2.84 | 9030 |
| 144.2 × 500 | 138.1 | 430 | 100.0 | 56.1 | 377 | 145.9 | 2.59 | 9030 |
| 144.2 × 600 | 138.1 | 530 | 123.3 | 69.5 | 404 | 167.7 | 2.41 | 9030 |
| 188.2 × 600 | 182.1 | 530 | 214.4 | 123.8 | 869 | 328.0 | 2.65 | 9240 |
| 188.2 × 700 | 182.1 | 630 | 254.9 | 147.2 | 917 | 367.4 | 2.50 | 9240 |
| 188.2 × 800 | 182.1 | 730 | 295.3 | 170.5 | 964 | 405.5 | 2.38 | 9240 |
| 188.2 × 900 | 182.1 | 830 | 335.8 | 193.9 | 1011 | 442.8 | 2.28 | 9240 |

$t = 6.1$ mm    $h_f = 70$ mm.

### 9.5.2  Design of a partially restrained ply web Box beam

A ply web Box beam of 9 m span is loaded on the top flange by point loads of 5 kN and 10 kN, as shown in Fig. 9.6. The medium term live load is 60% of the total loading. Determine a suitable section if the beam is restrained laterally at the positions of the applied loads and the beam is cambered to off-set dead load deflection.

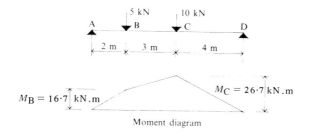

Fig. 9.6

The suggested procedure is to design a beam as explained in Chapter 8, assuming full restraint, and then to check for the actual partial lateral stability.

It can be shown that a 144.2 × 500 Box beam (Tables 8.5 and 9.4) is suitable for the loading if there is full lateral stability and the beam is cambered to off-set dead load deflection. However, in considering the actual case with partial restraint it is necessary to consider separately the parts of span between lateral restraints. The most critical part will govern the design. Lateral restraint occurs at B and C.

For the 144.2 × 500 Box beam, $N_1 = 145.9$ kN.m$^2$ and $N_2 = 2.59$.

*Part AB.* $L_b = 2$ m. $M_X = 16.7$ kN.m.
   Loading is case 8 of Table 9.3. $\therefore C_1 = 5.5$ and $C_2 = 0$.
   From equation (9.4):

$$M_{buc} = \frac{5.5 \times 145.9}{2.25 \times 2} = 178 \text{ kN.m} > 16.7 \text{ kN.m}$$

$\therefore$ Part AB is satisfactory.
(The part in the brackets of equation (9.4) becomes unity because $C_2 = 0$.)

*Part DC.* Similar to Part AB, except that $L_b = 4$ m and $M_X = 26.7$ kN.m.

$$M_{buc} = \frac{5.5 \times 145.9}{2.25 \times 4} = 89 \text{ kN.m} > 26.7 \text{ kN.m}$$

$\therefore$ Part DC is satisfactory.

*Part BC.* $L_b = 3$ m. $M_X = 16.7$ kN.m and 26.7 kN.m.
   The loading case is a condition between cases 6 and 7 in Table 9.3. $C_1$ can be calculated from the equation in §9.3. $\beta = 16.7/26.7 = 0.625$ (positive).

$$C_1 = 5.5 - 3.3\beta + 0.94\beta^2 = 5.5 - 3.3 \times 0.625 + 0.94 \times 0.625^2 = 3.80$$

$$M_{buc} = \frac{3.80 \times 145.9}{2.25 \times 3} = 82 \text{ kN.m} > 26.7 \text{ kN.m}$$

Therefore part BC is also satisfactory.
   This shows that BC is the most critical part of the span, but there is no lateral buckling with lateral restraints at B and C.

## 9.6    PARTIALLY RESTRAINED PLY WEB I BEAMS

### 9.6.1    Design method

I shaped sections are 'open' sections in contrast to 'closed' Box sections. The torsional rigidity of an open section is low compared to that of a Box section having similar flange and web areas and overall size, therefore the I section is more susceptible to lateral buckling than a Box beam. However, as I sections usually have a ceiling below them, the introduction of lateral restraints presents no aesthetic problems. The procedure for checking lateral stability is similar to that for Box beams, but some simplification is possible.
   Referring to Fig. 9.7 the torsional constant $J$ given by Roark (Table IX, Case 17) can be simplified, for ply web beams to:

$$J = \frac{2h_f^3}{3} \left[ B - 0.63h_f + \frac{t}{2} \right]$$

with no more than 3% error.
('$B$' is used to line through with the symbols used in Table 8.6.)

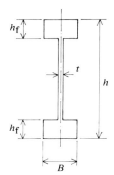

Fig. 9.7

The values for the stability constants $N_1$ and $N_2$ are calculated from equations (9.2) and (9.3) and are given in Table 9.5 for the range of I beams listed in Table 8.6. The flanges provide the major contribution to lateral stability, consequently it is the $G$ value for the flanges which is used in calculating $GJ$.

**Table 9.5   Stability constants $N_1$ and $N_2$ for selected ply web I beams**

| I section $B \times h$ (mm) | $E_{fN}$* | $J$ (mm$^4$ X 10$^6$) | $GJ$ (kN.m$^2$) | $EI_Y$* (kN.m$^2$) | $N_1$ (kN.m$^2$) | $N_2$ |
|---|---|---|---|---|---|---|
| 99.5 X 300 | 8680 | 13.98 | 7.58 | 99 | 27.4 | 3.61 |
| 99.5 X 400 | 8680 | 13.98 | 7.58 | 99 | 27.4 | 3.61 |
| 99.5 X 500 | 8680 | 13.98 | 7.58 | 99 | 27.4 | 3.61 |
| 99.5 X 600 | 8680 | 13.98 | 7.58 | 100 | 27.4 | 3.61 |
| 151.5 X 500 | 9240 | 25.87 | 14.94 | 375 | 74.8 | 5.01 |
| 151.5 X 600 | 9240 | 25.87 | 14.94 | 375 | 74.8 | 5.01 |
| 151.5 X 700 | 9240 | 25.87 | 14.94 | 375 | 74.8 | 5.01 |
| 151.5 X 800 | 9240 | 25.87 | 14.94 | 375 | 74.8 | 5.01 |

$G = E_{fN}/16$    $h_f = 70$    $t = 11.5$ mm.
* See Table 8.6.

### 9.6.2   Design of a partially restrained ply web I beam

A series of I beams spaced at 2 m centres is used to support a ceiling weighing 0.5 kN/m$^2$ attached to the lower flange. Beam span 15 m. There is no imposed loading and the beam is only laterally restrained at the ends.

$$\text{Total load} = 2 \times 0.5 \times 15 = 15 \text{ kN} \qquad \text{Shear} = 7.5 \text{ kN}$$

$$M_X = \frac{15 \times 15}{8} = 28.1 \text{ kN.m}$$

In this case there is no imposed loading and, because the deflection under dead loading could be cambered-out, theoretically there is no need to set a requirement on $EI_X$. If $EI_X$ is disregarded and a beam is chosen to match the bending and shear requirements only, a 151.5 X 600 mm ply web I beam (Table 8.6) is required. However, a check on the camber (or dead load deflection) shows a movement of around 53 mm (0.0035 X span).

If it is wished to limit the deflection to 0.003 X span, as a first approximation assume a ply web beam with an $EI_X$ value of 5.5 X $FL^2$ = 5.5 X 15 X $15^2$ = 18 560 kN.m².

Try a 151.5 X 700 mm I beam (Table 8.6).

$$\text{Total deflection} = \frac{5 \times 15 \times 15^3}{384 \times 19934} + \frac{1.0 \times 28.1}{5232} = 0.033 + 0.0054$$

$$= 0.0384 \text{ m} = 0.0026 \times \text{span}$$

Check lateral stability:

Lateral buckling will take the form of plan (b) of Fig. 9.1 and the loading condition is case 1, therefore $C_1 = 3.55$ and $C_2 = 1.41$.

For a 151.5 X 700 mm I beam with loading applied underneath, from equation (9.4):

$$M_{\text{buc}} = \frac{3.55 \times 74.8}{2.25 \times 15} \left[ 1 - \frac{1.41 \, (-0.35) \, 5.01}{15} \right]$$

$$= 9.16 \text{ kN.m} < 28.1 \text{ kN.m}$$

$M_{\text{buc}}$ is less than the $M_X$ therefore, either the compression flange must be restrained to give sufficient lateral stability, or a larger section must be used.

The type of design case just analysed can be encountered in the design of a ply web beam with no restraint to the tension flange, which becomes the compression flange when wind uplift exceeds the dead loading. For a detailed discussion, the designer may consult Bleich (1952). § 9.8 gives a further method of checking the condition where a U frame effect can be considered.

## 9.7   CONSIDERATION OF PARTIALLY RESTRAINED COMPRESSION FLANGES AS A HORIZONTAL COLUMN

With a ply web beam with the compression flange fully restrained, the full permissible bending stress can be used in calculations.

With ply web Box and I beams, the full permissible bending stress can be used providing the $I_X/I_Y$ ratio does not exceed the values given in Table 9.2 for the stated conditions of lateral stability. When these $I_X/I_Y$ values are exceeded, the design method shown in §§ 9.5 and 9.6 can be used. Alternatively, one design method is to assume that the top flange is a horizontal compression member taking a compression load equal to the bending moment divided by the effective depth (Fig. 9.8).

The effective length of this horizontal compression member for buckling

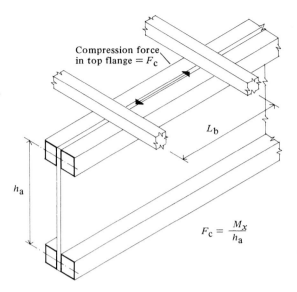

Fig. 9.8

in plan is a function of the distance $L_b$ between centres of lateral restraints. Usually the effective length is taken as $1.0 \times L_b$ although, if the restraints are fixed to the beam with a connection capable of giving some fixity on plan, or sheet materials are fixed to the restraints, the effective length may be reduced to $0.85L_b$ or $0.7L_b$.

The flange may be considered to be fully restrained in one direction by the web. The radius of gyration about the other axis may be calculated as $\sqrt{(I_{Yc}/A_c)}$, where $I_{Yc}$ is the inertia of the compression flange about the vertical axis and $A_c$ is the area of the compression flange.

It is quite normal to include the area of the web bounded by the timber flanges in calculations of $I_{Yc}$ and $A_c$ (Fig. 9.9), assuming this area for simplicity to be of the same material as the flanges.

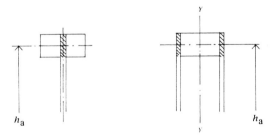

Fig. 9.9

## 9.8   U FRAME EFFECT FOR RESTRAINING COMPRESSION FLANGES

One occasionally encounters a design case where the compression flange is not laterally restrained, but beams are connected near the tension flange with moment connections to a cross member (as sketched in Fig. 9.10).

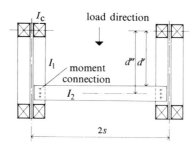

Fig. 9.10

A similar design case can occur when uplift occurs on a flat roof supported by ply web beams which have no restraints to the bottom flange (as sketched in Fig. 9.11). In either of these cases, the effective length of the compression flange may be calculated in a manner similar to that described for 'half-through bridges' in BS 5400:Part 3 by considering the 'U frame effect'. See Fig. 9.11 and the formulae below which use symbols modified from those used in BS 5400.

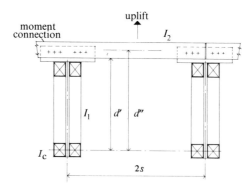

Fig. 9.11

$L_b$ = the effective length of the compression flange for lateral buckling.
$I_c$ = the maximum second moment of area of the compression flange about the vertical axis of the girder.
$s$ = half the distance between the girders.
$a$ = the distance between U frames along the length of the beam being designed.
$I_X$ = the second moment of area of one main beam.
$I_1$ = the second moment of area of a stiffener about the longitudinal plane of the beam.

$I_2$ = the second moment of area of a cross member in its plane of bending.

$$\delta = \frac{(d')^3}{3E_N I_1} + \frac{(d'')^2 s}{E_{Nb} I_2}$$

$E_N$ = the statistical minimum value appropriate to the number of pieces in the cross-section of one main beam.
$E_{Nb}$ = the statistical minimum value appropriate to the number of pieces in the cross-section of one cross beam.

If $\delta$ is determined and is found to be not greater than $a^3/40EI_X$, then the effective length $L_b = a$. Otherwise $L_b = 2.5 \times \sqrt[4]{(EI_X a\delta)}$, but $L_b$ must not be taken as being less than $a$.

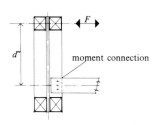

moment connection

Fig. 9.12

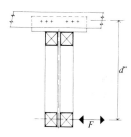

Fig. 9.13

If $F$ is the force causing buckling and hence the force causing moment on the moment connection (see Figs. 9.12 and 9.13), then:

$$F = \frac{1.4 \times 10^{-3} L_b}{\delta \left( \dfrac{\sigma_e}{\sigma_{c,a}} - 1.7 \right)}$$

where $\sigma_e$ = Euler critical stress in the flange = $\pi^2 E_{Nc}/(L_b/i)^2$. The $E_{Nc}$ value is the modified minimum value appropriate to the number of pieces in the compression flange.
$i$ = the radius of gyration.
$\sigma_{c,a}$ = the applied compression stress in the compression flange found by dividing $M_X$ by the effective depth of the beam and by the area of the compression flange.

If $\delta$ is less than $a^3/40EI_X$, the force $F$ is obtained by putting $\delta$ equal to $a^3/40EI_X$ and $L_b = a$. In calculating the permissible moment on the moment connection, the load-duration factor appropriate to the actual compression in the compression flange may be used.

# Chapter Ten
# Stress Skin Panels

## 10.1 INTRODUCTION

Plywood stress skin panels consist of plywood sheets attached to longitudinal timber members either by glue (usually glued/nailed joints) or mechanical means (usually nails or staples) to give a composite unit. With this construction it is possible to use smaller longitudinal members than those which would be required in a conventional joisting system or to extend the span of standard joist sizes. In addition, prefabricated panels can be used to reduce site work and speed erection. Most panels are sufficiently light to be erected by hand or with simple lifting gear. They have been used on floors, roofs and walls.

The maximum span of panels with simple joist longitudinal members is in the order of 7–9 m, but this can be extended if built-up beams are used as the longitudinal members. This latter case is not covered in this chapter, although the design principles are similar.

Stress skin panels can be either of double- or single-skin construction, the latter being sometimes referred to as 'stiffened panels'. The skins are usually either Canadian plywood or Finnish Birch-faced plywood. The longitudinal members are usually European Whitewood or Redwood.

If the full stress skin effect is required then it is essential to glue the plywood to the timber throughout the length of all member. Glue bonding is usually achieved with nails or staples. If nails or staples are used without adhesive, only a part of the full stress skin effect will be achieved no matter how close the spacing of the nails or staples. It is normal to surface the timber members on all four sides although it could be possible to regularize only the depth of the joists. The timber must be dried to the appropriate moisture content, particularly if in a double-skin construction.

In the designs in this chapter glued/nailed joints are assumed between the plywood and the timber joists, and the joists are surfaced on all four faces.

## 10.2 FORMS OF CONSTRUCTION

A basic form of double-skin panel is sketched in Fig. 10.1. Architectural details, insulation, ventilation, falls etc. can be added, and even plumbing or

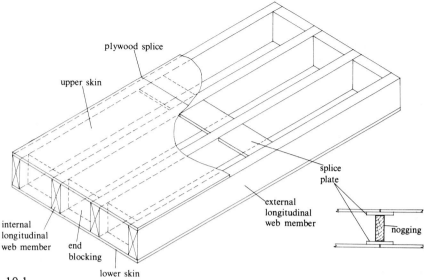

Fig. 10.1

electrics can be fitted in the factory. The edge joists are shown as the same thickness as the inner joists, but usually take half loading and could be thinner. Because of the size of available plywood sheets, panels are usually 1.2 m or 1.22 m wide, and usually have three or four joists in this width, designed as part of a load-sharing system. The designer should recognise that maximum economy will be achieved only if the panel width is matched to the available sizes of plywood sheets.

With Canadian plywood skins, the face grain usually runs parallel to the span of the panel; with Finnish plywood, at right angles to the direction of panel span.

The plywood splice plate shown in Fig. 10.1 can sometimes be replaced by a solid timber part-depth or full-depth nogging, providing a design check is carried out. Alternatively, it is occasionally possible to buy large plywood sheets suitably scarf jointed, or for the panel manufacturer to scarf standard sheets. If the joists are finger jointed, the joint must have adequate stength.

It is not usual to camber stress skin panels. End blocking is often provided at points of bearing.

Four of the many alternative forms are sketched in Fig. 10.2 and these can have modified edge details to prevent differential deflection.

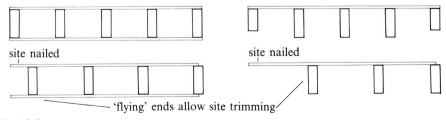

Fig. 10.2

## 10.3   SPECIAL DESIGN CONSIDERATIONS

### 10.3.1   Effect of shear lag on bending stresses and deflection

Tests on double-skin stress skin panels seem to have been carried out first in the US in the early 1930s, and the results showed that the interaction between the plywood skins and the timber webs requires a special design consideration. However, a design method was not developed until around 1940 when, because of the shortage of materials during the Second World War, an added interest was created in the high strength-to-weight ratio of plywood and timber, particularly in the aircraft industry. The method used a factor known as the 'basic spacing' which was based on the buckling characteristics of plywood loaded uniformly in compression. This method was used until reappraised in recent years by the Council of the Forest Industries of British Columbia from which a simplified design method was developed for use with Canadian plywood which is suitable for design office use.

This method takes account of the fact that the tension or compression stresses in the plywood skins under longitudinal bending result from a shear transfer from the web members into the plywood flanges, these stresses being a maximum at the junctions between web members and plywood and a minimum equidistant between webs. The stress distribution is sketched in Fig. 10.3. The variation is caused by shear deflection, and the variation from the elementary theory (i.e. that tension or compression stresses in the plywood are uniform across the panel) is usually called 'shear lag'.

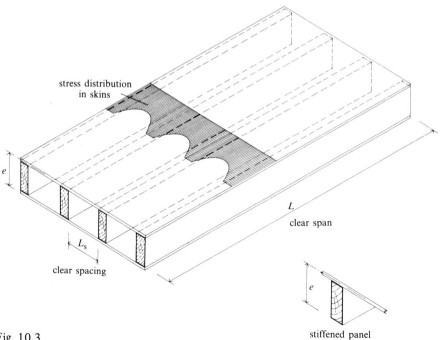

Fig. 10.3

A theoretical analysis undertaken by Foschi (1969) showed discrepancies between the 'shear lag' theory and the earlier 'basic spacing' theory which, while insignificant in many cases, showed that the earlier theory could result in unsafe designs for short span panels up to say 4 m span. A correction factor $K_c$ shown in graph form in Fig. 10.4 was derived, and when applied to bending stresses and deflections calculated from basic engineering formulae applied to the full section, this modifies the results and gives values consistent with those calculated by the 'shear lag' theory.

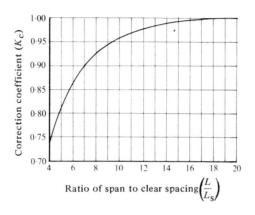

Fig. 10.4

Factor $K_c$ should be applied so as to increase the applied bending stress and the bending deflection. When applied to the bending deflection there is no need to have a separate calculation for shear deflection.

As far as the authors are aware these coefficients are available only for Douglas Fir plywood but they are used in this manual for designs in Douglas Fir-faced plywood. For panels using Finnish Birch-faced or other plywoods, the designer will have to use unmodified basic engineering formulae, unless coefficients become available from the relevant manufacturer's associations. Certainly separate calculations for shear deflection should be carried out if values for $K_c$ are not available.

## 10.3.2   Buckling of compression skin

The upper skin of a stress skin panel used as a floor or roof decking is largely in compression and is also taking a secondary moment from the imposed loading between the web members. Tests have shown that with normal proportions of plywood thicknesses and joist spacings, the plywood is unlikely to buckle but, for Douglas Fir plywood, coefficients have been evaluated from which the designer can check quickly whether buckling will occur. The buckling coefficients are given in Table 10.1. To avoid buckling of the compression skin, the uniformly distributed loading on the skin (not the total panel) must not exceed uniform loading $F_{cr}$ where:

**Table 10.1**  Buckling coefficient $K_{cr}$ for Douglas Fir-faced plywood with the face grain parallel to main direction of span

| Nominal thickness (unsanded) (mm) | Minimum thickness (mm) | $K_{cr}$ (kN.m) |
|:---:|:---:|:---:|
| 7.5 | 7.0 | 29.4 |
| 9.5 | 9.0 | 63.3 |
| 12.5 | 12.0 | 192 |
| 15.5 | 15.0 | 339 |
| 18.5 | 18.0 | 723 |

$$F_{cr} = \frac{eK_{cr}}{(L_s L)^2} \, \text{kN/m}^2 \tag{10.1}$$

where $e$ = the depth as illustrated in Fig. 10.3 to the centre of area of the plywood

$K_{cr}$ = the buckling coefficient in kN.m (Table 10.1)

$L_s$ = the clear spacing between webs (Fig. 10.3)

$L$ = the clear span of the panel (Fig. 10.3).

As far as the authors are aware, these coefficients are available only for Douglas Fir plywood, but in this manual they are taken as being equally applicable for Douglas Fir-faced plywood. For panels using Finnish Birch-faced or other plywoods, the designer may be able to obtain information from the relevant manufacturer's associations, or conduct a test or assume that buckling will not take place. Certainly it is unusual for buckling to take place with the proportions usual for roof and deck constructions of stress skin panels.

The CIB Structural Design Code gives a method which the authors read as being an alternative to an investigation of buckling instability of the plywood on the compression side. In this method a limit is placed on the width of plywood between joists which may be taken as part of the total section acting against bending of the whole panel (see Fig. 10.5). Where $L$ is the span of the panel (or, for a continuous panel, the distance between points of zero moment):

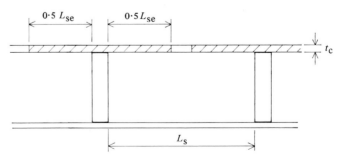

Fig. 10.5

$L_{se}/L$ should not exceed 0.1 for plywood with face grain perpendicular or parallel to the web members, and

$L_{se}$ should not exceed $20t_c$ for face grain parallel to web members or $25t_c$ for face grain perpendicular to web members.

For most cases it would make sense to limit $L_s$ to the calculated values of $L_{se}$, although this is not a requirement of the CIB Code method. However, it can be seen that the method makes no allowance for the actual stress in the plywood, hence it may lead to the assumption that buckling will take place even if the plywood skin is stressed to a low level (see the worked example in § 10.7).

## 10.4   SELECTING A TRIAL DESIGN CROSS-SECTION

The usual panel design criterion is deflection, and the inertia required can be provided by a multiplicity of variations of joist size and grade, plywood thickness and grade, single or double skins, therefore the design of a stress skin panel tends to be a case of trial and error. There are, however, a number of guide lines which can lead to a fairly accurate initial trial section.

For the initial trial design:

(a) The maximum span-to-overall-depth ratio should not exceed 35 for an imposed loading of 1.5 kN/m² and 40 for an imposed loading of 0.75 kN/m².

(b) With face grain parallel to the longitudinal members the compression skin should have at least the following minimum thicknesses: 9.0 or 12.0 mm for 0.75 kN/m² imposed loading and 12.0, 15.0 or 18.0 mm for 1.5 kN/m² imposed loading (but see §§ 5.7.2 and 5.7.3).

When the compression skin also carries loading between the web members, it is better to ensure the adequacy of this skin to carry this loading before proceeding to calculate the properties and stresses in the overall panel. The bending moment and deflection from transverse loading can be calculated from the formulae in Table 10.2 providing that the plywood skin is continuous across the full width of the panel.

The moment capacity and section rigidity per metre width for each of the plywood thicknesses likely to be used for the compression skin are given in Table 10.3 for Canadian Douglas Fir plywood and in Table 10.4 for Finnish Birch-faced plywood. In the case of Canadian Douglas Fir plywood the face grain is parallel to the supporting joists, and, in the case of Finnish Birch, perpendicular to the supporting joists.

(c) The tension skin is usually thinner than the compression skin, 6.5 or 9 mm sanded, good face quality if the soffit requires an architecturally improved finish, or unsanded if no special finish is required.

(d) Spacing of web members should not normally exceed 0.6 m for imposed loading of 0.75 kN/m², 0.4 m for imposed loading of 1.5 kN/m², or 0.3 m for imposed loading of more than 1.5 kN/m². Spacings of around 0.4 m are most common. Thickness of web members is normally 47 or 50 mm basic.

(e) The design criterion of the overall panel is usually deflection. From § 5.3.2 it can be seen that the $EI$ value of the trial section should be $4.34FL^2$ to limit bending deflection to $0.003 \times$ span.

(f) Site erection sequence and architectural or manufacturing requirements, rather than structural requirements, will often decide whether or not there is to be a lower skin.

**Table 10.2**

| Number of transverse spans over which plywood is continuous | Bending moment formulae* | Bending deflection formulae |
|---|---|---|
| 2 spans (over 3 joists) | $\dfrac{pL_s^2}{8}$ | $\dfrac{pL_s^4}{185\,EI}$ |
| 3 spans (over 4 joists) | $\dfrac{pL_s^2}{10}$ | $\dfrac{pL_s^4}{145\,EI}$ |
| 4 spans (over 5 joists) | $\dfrac{pL_s^2}{9.5}$ | $\dfrac{pL_s^4}{154\,EI}$ |

\* $p$ is the load per m²    $L_s$ is the clear spacing between webs.
Note: These formulae are based on the assumption that the designer will not insist on the centres of joists being taken in calculations but will accept the clear distance.

**Table 10.3   Properties of Canadian Douglas Fir plywood spanning between joists. Select Tight Face, Sheathing or Select Sheathing grades, unsanded**

Properties are based on the 'full area method'.

| Properties per 1 m width face grain parallel to joists | | Nominal thickness (mm) Canadian Douglas Fir plywood unsanded. Minimum thickness in brackets | | | | | |
|---|---|---|---|---|---|---|---|
| | | 7.5 (7.0) | 9.5 (9.0) | 12.5 (12.0) | 15.5 (15.0) | 18.5 (18.0) | 20.5 (20.0) |
| Permissible bending stress (N/mm²) | long term | 2.78 | 2.76 | 2.77 (3) | 3.41 | 3.58 (6) | 3.58 (6) |
| | medium | 3.48 | 3.45 | 3.46 (3) | 4.26 | 4.48 (6) | 4.48 (6) |
| Modulus (mm³ × 10⁶) | | 0.008 17 | 0.0135 | 0.0240 | 0.0375 | 0.0540 | 0.0667 |
| Moment capacity (kN.m) | long term | 0.022 7 | 0.0373 | 0.0665 | 0.1279 | 0.1933 | 0.2388 |
| | medium | 0.028 4 | 0.0466 | 0.0831 | 0.1598 | 0.2417 | 0.2985 |
| $E$ (bending) (N/mm²), face grain perp. | | 850 | 800 | 850 (3) | 2150 | 2500 (6) | 2650 (6) |
| $I$ (mm⁴ × 10⁶) | | 0.028 6 | 0.0607 | 0.144 | 0.281 | 0.486 | 0.667 |
| $EI$ (kN.m²) | | 0.024 3 | 0.0486 | 0.1224 | 0.6042 | 1.215 | 1.767 |

Where there are two or three layups for the same nominal thickness only the lowest value is given. The relevant layup is indicated in brackets.

**Table 10.4** Properties of Finnish Birch-faced plywood spanning between joists, sanded

Properties are based on the 'full area method'.

| Properties per 1 m width face grain perpendicular to joists | | Nominal thickness (mm) Finnish Birch-faced plywood Minimum thickness in brackets | | | | | |
|---|---|---|---|---|---|---|---|
| | | 6.5 (6.1) | 9 (8.8) | 12 (11.5) | 15 (13.9) | 18 (17.1) | 21 (19.9) |
| Permissible bending stress (N/mm²) | long term | 18.2 | 15.9 | 14.3 | 14.2 | 12.8 (13) | 12.4 |
| | medium | 22.75 | 19.88 | 17.88 | 17.75 | 16.0 (13) | 15.5 |
| Modulus (mm³ × 10⁶) | | 0.0062 | 0.0129 | 0.0220 | 0.0322 | 0.0487 | 0.0660 |
| Moment capacity (kN.m) | long term | 0.113 | 0.205 | 0.315 | 0.457 | 0.623 | 0.818 |
| | medium | 0.141 | 0.256 | 0.393 | 0.572 | 0.779 | 1.023 |
| $E$ (bending) (N/mm²), face grain par. | | 10 300 | 9500 | 8900 | 8550 (11) | 8050 | 7550 |
| $I$ (mm⁴ × 10⁶) | | 0.0189 | 0.0568 | 0.127 | 0.224 | 0.417 | 0.657 |
| $EI$ (kN.m²) | | 0.195 | 0.540 | 1.13 | 1.92 | 3.36 | 4.96 |

Where there are two layups for the same thickness only the lower value is given. The relevant layup is indicated in brackets.

## 10.5   PERMISSIBLE STRESSES

### 10.5.1   Joist webs

As there will usually be at least four members per panel or in adjacent panels supporting a common load, the joist webs may be considered as part of a load-sharing system. Permissible stresses for the webs are given in Table 10.5 for European Whitewood or Redwood of SS grade. The $EI$ capacity and $K_7$ depth factor of commonly used web members are given in Table 10.6.

**Table 10.5** Permissible grade stresses for SS grade European Whitewood or Redwood

| Stress type | SS grade stress (N/mm²) | Load-duration factor $K_3$ | | Load-sharing factor $K_8$ | Permissible stress* or $E_{mean}$ (N/mm²) |
|---|---|---|---|---|---|
| Bending | 7.5 | Long term | 1.00 | 1.1 | 8.25 |
| | | Medium | 1.25 | 1.1 | 10.31 |
| Shear | 0.82 | Long term | 1.00 | 1.1 | 0.902 |
| | | Medium | 1.25 | 1.1 | 1.13 |
| Modulus of elasticity | 10 500 | — | | — | 10 500 |

* Not increased by depth factor $K_7$.

**Table 10.6**

| Actual size of web members (mm) | 47 X 97 | 47 X 122 | 47 X 147 | 47 X 170 | 47 X 195 | 47 X 220 |
|---|---|---|---|---|---|---|
| $EI$ (kN.m$^2$) | 37.5 | 74.7 | 131 | 202 | 305 | 438 |
| Depth factor $K_7$ | 1.132 | 1.104 | 1.082 | 1.064 | 1.049 | 1.035 |

### 10.5.2   Canadian Douglas Fir plywood. Face grain parallel to span

Permissible stresses for Canadian plywood skins, face grain parallel to span, applicable to whole panel design are given in Table 10.7. Permissible stresses for the design of plywood taking loading transversely between web joists are dealt with separately in Table 10.3.

**Table 10.7   Permissible stresses for Canadian Douglas Fir plywood skins, face grain parallel to span. Select Tight Face, Sheathing or Select Sheathing grades, unsanded**

Properties based on 'full area method'.

| Stress type | Duration of load | Permissible stress in N/mm² for nominal thicknesses in mm Minimum thicknesses in brackets | | | | | |
|---|---|---|---|---|---|---|---|
| | | 7.5 (7.0) | 9.5 (9.0) | 12.5 (12.0) | 15.5 (15.0) | 18.5 (18.0) | 20.5 (20.0) |
| Tensile | Long term | 5.13 | 4.05 | 3.09 (4) | * | * | * |
| | Medium | 6.41 | 5.06 | 3.86 (4) | * | * | * |
| Compression | Long term | * | 6.31 | 4.82 (4) | 5.32 | 5.00 (5) | 5.04 (7) |
| | Medium | * | 7.89 | 6.03 (4) | 6.65 | 6.25 (5) | 6.30 (7) |
| Bearing on face | Long term | 2.16 | 2.16 | 2.16 | 2.16 | 2.16 | 2.16 |
| | Medium | 2.70 | 2.70 | 2.70 | 2.70 | 2.70 | 2.70 |
| Rolling shear† | Long term | 0.148 | 0.148 | 0.148 | 0.148 | 0.148 | 0.148 |
| | Medium | 0.185 | 0.185 | 0.185 | 0.185 | 0.185 | 0.185 |
| Modulus of elasticity in tension and compression | – | 10 500 | 8350 | 6350 (4) | 6900 | 6450 (5) | 6500 (7) |

Where there are two or three layups for the same nominal thickness only the lowest value is given. The relevant layup is indicated in brackets.
* Omitted for the reasons given in § 10.4.
† Already reduced by factors 0.5 ($K_{37}$) and 0.9 ($K_{70}$) (see Fig. 4.24) but not increased for load sharing.

### 10.5.3   Finnish Birch-faced plywood. Face grain perpendicular to span

Permissible stresses for Finnish Birch-faced plywood, face grain perpendicular to span, applicable to whole panel design are given in Table 10.8.

Permissible stresses for the plywood taking load transversely across web joists are dealt with separately in Table 10.4.

**Table 10.8**   **Permissible stresses for Finnish Birch-faced plywood skins, face grain perpendicular to span. Sanded**

| Stress type | Duration of load | Permissible stresses in N/mm² for nominal thicknesses in mm Minimum thicknesses in brackets | | | | | |
|---|---|---|---|---|---|---|---|
| | | 6.5 (6.1) | 9 (8.8) | 12 (11.5) | 15 (13.9) | 18 (17.1) | 21 (19.9) |
| Tensile | Long term | 3.51 | 7.07 | * | * | * | * |
| | Medium | 4.39 | 8.84 | * | * | * | * |
| Compression | Long term | * | 5.33 | 5.40 | 5.43 (9) | 5.30 | 5.20 |
| | Medium | * | 6.66 | 6.75 | 6.79 (9) | 6.63 | 6.50 |
| Bearing on face | Long term | 3.00 | 3.00 | 3.00 | 3.00 | 3.00 | 3.00 |
| | Medium | 3.75 | 3.75 | 3.75 | 3.75 | 3.75 | 3.75 |
| Rolling shear† | Long term | 0.20‡ | 0.27 | 0.27 | 0.27 | 0.27 | 0.27 |
| | Medium | 0.25‡ | 0.34 | 0.34 | 0.34 | 0.34 | 0.34 |
| Modulus of elasticity in tension and compression | | 4650 | 5300 | 5750 | 5000 (11) | 5100 | 5100 |

Where there are two layups for the same thickness only the lower value is given. The relevant layup is indicated in brackets.
* Omitted for the reason given in § 10.4.
† Already reduced by factors 0.5 ($K_{37}$) and 0.9 ($K_{70}$) (see Fig. 4.24) but not increased for loading sharing.
‡ If set by possible rolling shear in centre veneer (possibly over-conservative).

## 10.6   SELF-WEIGHT OF PANEL ELEMENTS

The designer can determine self-weight from the values given in Table 10.9 plus an allowance of around 10% for blocking, splice plates, glue and nails.

**Table 10.9**

| Canadian Douglas Fir plywood skins | | Finnish Birch-faced plywood skins | | Timber joists | |
|---|---|---|---|---|---|
| Nominal thickness (mm) | Weight (kN/m²) | Nominal thickness (mm) | Weight (kN/m²) | Size (mm) | Weight (kN/m) |
| 7.5 | 0.044 | 6.5 | 0.045 | 47 × 97 | 0.023 |
| 9.5 | 0.054 | 9 | 0.065 | 47 × 122 | 0.029 |
| 12.5 | 0.073 | 12 | 0.084 | 47 × 145 | 0.034 |
| 15.5 | 0.091 | 15 | 0.104 | 47 × 170 | 0.040 |
| 18.5 | 0.108 | 18 | 0.123 | 47 × 195 | 0.046 |
| 20.5 | 0.120 | 21 | 0.143 | 47 × 220 | 0.052 |

## 10.7   TYPICAL DESIGN. DOUBLE-SKIN PANEL

Consider the case of a double-skin panel spanning 4.8 m supporting a medium term imposed UDL of $1.5 \, \text{kN/m}^2$ and dead loading of $0.2 \, \text{kN/m}^2$ excluding self-weight. Panels are to be 1.2 m wide.

Estimate the trial section from the guiding principles discussed in § 10.4.

$$\text{Overall depth } (1.5 \, \text{kN/m}^2 \text{ imposed}) \simeq \frac{4.8 \times 10^3}{35} = 137 \, \text{mm}$$

Try cross-section arrangement shown in Fig. 10.6.
Top skin, assume 12.5 mm Canadian Douglas Fir plywood unsanded.
Lower skin, assume 9.5 mm Canadian Douglas Fir plywood unsanded.
Minimum joist depth $137 - 12.0 - 9.0 = 116.0 \, \text{mm}$.
Try SS grade Whitewood joists surfaced to $47 \times 122 \, \text{mm}$.

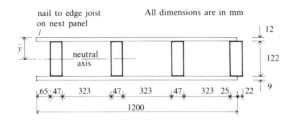

Fig. 10.6

*Top skin*
Check the top skin for transverse loading and buckling.

$$\text{The total load on the top skin} = 1.5 + 0.2 + \text{self-weight of top skin}$$
$$= 1.7 + 0.073 = 1.77 \, \text{kN/m}^2$$

$$\text{(See Table 10.2)} \qquad M = \frac{1.77 \times 0.323^2}{10} = 0.019 \, \text{kN.m}$$

which is less than the critical long term moment capacity of $0.0665 \, \text{kN.m}$ (from Table 10.3).

$$\text{Bending deflection} = \frac{pL_s^4}{145EI} = \frac{1.77 \times 0.323^4}{145 \times 0.1224} = 0.001 \, 09 \, \text{m} \, (0.0034 \, \text{span})$$

(*EI* from Table 10.3.).

Check if buckling of the compression skin will take place. Refer to § 10.3.2 and use the method given for Douglas Fir plywood.

$$e = 122 + 6.0 + 4.5 = 132.5 \, \text{mm} = 0.133 \, \text{m}$$

$$F_{cr} = \frac{eK_{cr}}{(L_s L)^2} = \frac{0.133 \times 192}{(0.323 \times 4.8)^2} = 10.62 \, \text{kN/m}^2$$

which is larger than the actual UDL, therefore buckling should not occur.

Check buckling using the method outlined in the CIB Structural Design Code (see Fig. 10.5). For face grain parallel to span:

$$\frac{L_{se}}{L} = \frac{323}{4800} = 0.067$$

which does not exceed 0.1, therefore one part of the method is satisfied.

$$\frac{L_s}{t_c} = \frac{323}{12} = 26.9$$

For face grain perpendicular to web members the CIB method gives a limit for $L_{se}/t_c$ of 25. Therefore, to satisfy this method, either the compression skin should be increased slightly in thickness or, in the overall design, $L_{se}$ should be taken as $25 \times 12.0 = 300$ mm.

After checking the strength of the top skin the self-weight of the trial panel can be calculated and shear, moment and stiffness characteristics checked.

Self-weight of the panel per m²:

$$\begin{aligned}
\text{Top skin} &= 0.073 \\
\text{Joists } 4 \times 0.029 \div 1.2 &= 0.097 \\
\text{Lower skin} &= 0.054 \\
\hline
&= 0.22 \,\text{kN/m}^2
\end{aligned}$$

$$\begin{aligned}
\text{Total loading per m}^2 &= 1.5 + 0.2 + 0.22 = 1.92 \,\text{kN/m}^2 \\
\text{Total loading} &= 1.92 \times 4.8 \times 1.2 = 11.0 \,\text{kN} \\
\text{Shear} &= 5.5 \,\text{kN}
\end{aligned}$$

$$M = \frac{11.0 \times 4.8}{8} = 6.6 \,\text{kN.m (on whole panel)}$$

*Deflection of panel*

The *EI* value required to limit bending deflection to 0.003 times span = $4.34 \times 11.0 \times 4.8^2 = 1100$ kN.m². To calculate the actual *EI* value, it is first necessary to locate the neutral axis which is distance $\bar{y}$ from top surface (see Fig. 10.6).

$$\bar{y} = \frac{\Sigma EAy}{\Sigma EA}$$

where $\Sigma EA$ = the product of the $E$ value and area of each element in the panel

$y$ = the distance of the centroid of each element from the top surface reference plane. (Any reference plane could be used, but the top and the bottom are most convenient.)

(*E* values and actual thicknesses are from Tables 10.5 and 10.7.)

**Table 10.10**

| Element | $E$ (N/mm$^2$) | $A$ (mm$^2$) | $y$ (mm) | $EA$ (N × 10$^6$) | $EAy$ (N.mm × 10$^6$) |
|---|---|---|---|---|---|
| Top skin | 6 350 | 12.0 × 1200 | 6.0 | 91.4 | 549 |
| Joists | 10 500 | 4 × 47 × 122 | 73.0 | 240.8 | 17 578 |
| Lower skin | 8 350 | 9.0 × 1200 | 138.5 | 90.2 | 12 490 |
| | | | | 422.4 | 30 617 |

From Table 10.10:

$$\bar{y} = \frac{30\,617}{422.4} = 72.48 \text{ mm}$$

The self $EI$ capacity of the top and bottom skins about their neutral axis is small and can be disregarded with little loss of accuracy. Therefore the bending rigidity of one panel $= EI_\text{webs} + \Sigma(EA)h_x^2$ where $h_x$ = distance from the neutral axis to the centroid of the element.

**Table 10.11**

| Element | $EA$ (N × 10$^6$) | $h_x$ (mm) | $(EA)h_x^2$ (kN.m) |
|---|---|---|---|
| Top skin | 91.4 | 66.47 | 404 |
| Joists | 240.8 | 0.53 | — |
| Lower skin | 90.2 | 66.03 | 393 |
| | | | 797 |

From Tables 10.11 and 10.6:

$$EI = (4 \times 74.7) + 797 = 1096 \text{ kN.m}^2$$

$$\text{Bending deflection} = \frac{5 \times 11.0 \times 4.8^3}{384 \times 1096} = 0.0144 \text{ m}$$

If the designer decides to adopt the CIB method of limiting buckling of the compression skin, then the $A$, $EA$ and $EAy$ values of the top skin (see Table 10.10) should be reduced accordingly (i.e. reduced in the proportion of 1131 mm to 1200 mm) which will lead to a small increase in the calculated value of bending deflection.

The shear deflection can be calculated by at least two methods. One would be to assume that only the four joists are effective in resisting shear deflection. This will give an over-estimate of shear deflection. Another method is to calculate the factor $K_c$ in Fig. 10.4 and increase deflection accordingly.

$$\frac{L}{L_s} = \frac{4800}{323} = 14.86 \quad \text{and therefore } K_c = 0.99$$

The bending deflection of 0.0144 m is increased to take account of shear deflection, to $0.0144/0.99 = 0.0145$ m (14.5 mm).
$0.003\,L = 0.0144$ m (14.4 mm).

*Bending stresses*
In the calculation of bending stresses the stresses in each element should be checked by Method 4 as discussed in § 8.4 on ply web beams.
The bending stress at any fibre of an element is:

$$\sigma_{m,a} = \frac{MyE}{EI}$$

where $M$ = the bending moment on the panel
$y$ = the distance from the neutral axis to the fibre under consideration
$E$ = the $E$ value of the element under consideration
$EI$ = the bending rigidity of the full panel.

Where applicable, and where the information is available, an allowance for the effect of shear lag (Figs 10.3 and 10.7) as an increase on the applied maximum stress in the plywood skins and joists should be made. (In this case $K_c = 0.99$.)

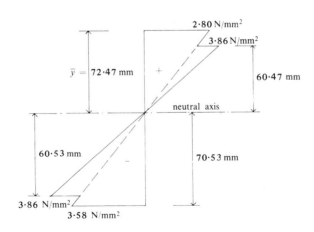

Fig. 10.7: *Bending stresses.*

At the extreme fibre of the top skin ($y = 72.47$ mm):

$$\sigma_{m,a} = \frac{6.6 \times 10^6 \times 72.47 \times 6350}{1096 \times 10^9 \times 0.99} = 2.80 \text{ N/mm}^2 < 6.03 \text{ N/mm}^2$$

(plywood mainly in compression)

At the extreme fibre of the lower skin ($y = 70.53$ mm):

$$\sigma_{m,a} = \frac{6.6 \times 10^6 \times 70.53 \times 8350}{1096 \times 10^9 \times 0.99} = 3.58 \text{ N/mm}^2 < 5.06 \text{ N/mm}^2$$

(plywood mainly in tension)

At the upper extreme fibre of the web joists ($y = 60.47$ mm):

$$\sigma_{m,a} = \frac{6.6 \times 10^6 \times 60.47 \times 10\,500}{1096 \times 10^9 \times 0.99} = 3.86 \text{ N/mm}^2 < 7.50 \times K_3 \times K_8 \times K_7$$

(see Tables 10.5 and 10.6) $= 11.08 \text{ N/mm}^2$

At the lower extreme fibre of the web joists ($y = 60.53$ mm):

$$\sigma_{m,a} = \frac{6.6 \times 10^6 \times 60.53 \times 10\,500}{1096 \times 10^9 \times 0.99} = 3.86 \text{ N/mm}^2 < 11.08 \text{ N/mm}^2$$

Hence all these bending stresses are acceptable.

*Horizontal and rolling shear stresses*
To calculate the horizontal and rolling shear stresses, first calculate the value of $ES = EAh_x$ taking into consideration those elements (or parts of elements) which occur either above or below the neutral axis.

By tabulation, consider parts above the neutral axis (see Table 10.12):

Table 10.12

| Element | $E$ (N/mm$^2$) | $A$ (mm$^2$) | $h_x$ (mm) | $EAh_x$ (N.mm $\times 10^9$) |
|---|---|---|---|---|
| Top skin | 6 350 | 12.0 $\times$ 1200 | 66.47 | 6.08 |
| Part web | 10 500 | 4 $\times$ 47 $\times$ 60.49 | 30.24 | 3.61 |
| | | | $\Sigma ES =$ | 9.69 |

Alternatively the same result is determined by considering parts below the neutral axis (see Table 10.13):

Table 10.13

| Element | $E$ (N/mm$^2$) | $A$ (mm$^2$) | $h_x$ (mm) | $EAh_x$ (N.mm $\times 10^9$) |
|---|---|---|---|---|
| Lower skin | 8 350 | 9.0 $\times$ 1200 | 66.03 | 5.95 |
| Part web | 10 500 | 4 $\times$ 47 $\times$ 60.51 | 30.76 | 3.74 |
| | | | $\Sigma ES =$ | 9.69 |

The horizontal shear stress at the neutral axis is:

$$\tau = \frac{F_v(ES)}{(EI)t} \qquad \text{where } t = \text{total thickness}$$

$$= \frac{5.5 \times 10^3 \times 9.69 \times 10^9}{1096 \times 10^9 \times 4 \times 47} = 0.26 \text{ N/mm}^2 < 1.13 \text{ N/mm}^2$$
$$\text{(see Table 10.5)}$$

which is acceptable.

At the junction of the web joists and the ply skins, the rolling shear stress is appropriate. For a known panel:

$$\tau_r = \frac{F_v(ES)}{(EI)t}$$

where $ES$ is the value for the skin only.

For a known panel arrangement and external shear $F_v$, $t$ and $EI$ are constant and the maximum rolling shear stress will occur at the junction where $ES$ (for skin only) is the larger. In this case it can be seen that the case at the top skin is more critical. Assuming that bonding pressure is achieved by nails, the $K_{37}$ factor of 0.5 and the reduction factor $K_{70}$ of 0.9 from Clause 47.4.2 of BS 5268 : Part 2 apply (§ 4.14 and Fig. 4.24).

Assuming that the shear in the panel is transferred uniformly between webs and skins, the rolling shear stress is over a contact width (see Fig. 10.6) of $(3 \times 47) + 25 = 166$ mm. Therefore:

$$\tau_{r,a} = \frac{5500 \times 6.08 \times 10^9}{1096 \times 10^9 \times 166} = 0.18 \text{ N/mm}^2$$

The permissible rolling shear stress from Table 8.8, increased for load sharing and duration of load, and modified by $K_{37}$ and $K_{70}$:

$$= 0.33 \times 1.10 \times 1.25 \times 0.5 \times 0.9 = 0.204 \text{ N/mm}^2$$

(or equals 0.185 N/mm² from Table 10.7 × load-sharing factor of 1.10).

## 10.8   SPLICE PLATES

When full-length plywood sheets are not available, it will be necessary to introduce splices in the plywood. Although the stress skin panel is in bending, the stress in the skin or skins approximates closer to pure compression or tension and the splice plates should be designed accordingly. Although splices may not occur at the position of maximum moment, it is convenient for the purpose of design to assume that they do so.

The longitudinal members also act as splices, and it is therefore only necessary to transfer through the plywood splices the force developed in the

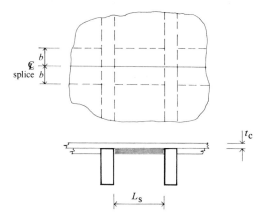

Fig. 10.8

area of the ply skin between longitudinal members. Thus the maximum force to be transferred will be expressed as:

$$\text{Force} = \sigma_{s,a} \times t_c \times L_s \quad \text{(Fig. 10.8)}$$

where $\sigma_{s,a}$ = the applied tensile or compression stress in the plywood skin
$t_c$ = skin thickness
$L_s$ = spacing between longitudinals.

This force must be resisted by the cross-sectional area of the splice plate. It is usual to make the splice plate of width $L_s$ (Fig. 10.8) and to fix it in the same direction as the skin thus, if the skin is stressed to only a small proportion of the permissible stress, some reduction in plywood splice thickness to a value less than $t_c$ may be acceptable. In practice, it is usual to specify the splice plate as the same thickness as the skin and no stress check is then required.

The force must also be transferred across the joint either by mechanical fasteners or glue. In the case of gluing, pressure may be achieved by clamps or nailing, the former being preferred, particularly for the lower skin if the soffit is of architectural quality.

The glue-line stress should not exceed the permissible rolling shear stress. In this case the $K_{37}$ factor of 0.5 is not applicable, and the $K_{70}$ factor of 0.9 will apply only if bonding pressure is achieved by nails or staples.

The half-length splice plate dimension $b$ (Fig. 10.8) is given by the expression:

$$b = \frac{\sigma_{s,a} t_c}{\tau_{r,adm}}$$

where $\tau_{r,adm}$ = the permissible rolling shear stress (not reduced by factors of 0.5 and 0.9).

For a tension skin splice plate 9.0 mm thick as in the example in § 10.7:

$$b = \frac{3.58 \times 9.0}{0.33 \times 1.25 \times 1.1} = 71 \text{ mm}$$

and the total splice plate length = $2b$ = 142 mm.

For a compression skin splice plate 12.0 mm thick, as in the example in § 10.7 (assuming bonding pressure by nailing):

$$b = \frac{2.80 \times 12.0}{0.33 \times 1.25 \times 1.1 \times 0.9} = 83 \text{ mm}$$

and the total splice plate length = $2b$ = 166 mm.

## 10.9 TYPICAL DESIGN TO COMPARE STRENGTH OF SINGLE-SKIN PANEL

To compare the strength added by the lower skin of a double-skin panel, calculate the properties of the panel detailed in § 10.7, but with the lower skin removed, and with loading and span as in § 10.7. The new section is shown in Fig. 10.9.

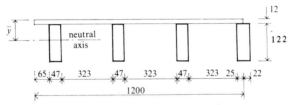

Fig. 10.9

The top skin will be adequate for transverse bending and deflection (as checked in § 10.7).

$$\text{Self-weight of the panel} = 0.073 + 0.097 = 0.17 \text{ kN/m}^2$$
$$\text{Total loading} = 4.8 \times 1.2(1.5 + 0.2 + 0.17) = 10.8 \text{ kN}$$
$$\text{Shear} = 5.4 \text{ kN}$$

$$M = \frac{10.8 \times 4.8}{8} = 6.48 \text{ kN.m (on whole panel)}$$

*Deflection*

**Table 10.14**

| Element | $E$ (N/mm²) | $A$ (mm²) | $y$ (mm) | $EA$ (N × 10⁶) | $EAy$ (N.mm × 10⁶) |
|---|---|---|---|---|---|
| Top skin | 6 350 | 12.0 × 1200 | 6.0 | 91.4 | 549 |
| Joists | 10 500 | 4 × 47 × 122 | 73.0 | 240.8 | 17 578 |
| | | | | 332.2 | 18 127 |

From Table 10.14:

$$\bar{y} = \frac{18\,127}{332.2} = 54.6\,\text{mm}$$

**Table 10.15**

| Element | $EA$ $(\text{N} \times 10^6)$ | $h_x$ (mm) | $(EA)h_x^2$ (kN.m) |
|---------|------------------------------|------------|--------------------|
| Top skin | 91.4 | 48.6 | 216 |
| Joists | 240.8 | 18.4 | 81 |
| | | | 297 |

From Tables 10.15 and 10.6:

$$EI = (4 \times 74.7) + 297 = 596\,\text{kN.m}^2$$

$$\text{Bending deflection} = \frac{5 \times 10.8 \times 4.8^3}{384 \times 596} = 0.026\,\text{m}$$

which is approximately 1.8 times that for the double-skin panel.

*Bending stresses* (see Fig. 10.10)
At the extreme fibre of the top skin ($y = 54.6$ mm):

$$\sigma_{m,a} = \frac{6.48 \times 10^6 \times 54.6 \times 6350}{596 \times 10^9 \times 0.99} = 3.81\,\text{N/mm}^2 < 6.03\,\text{N/mm}^2$$
$$\text{(acceptable)}$$

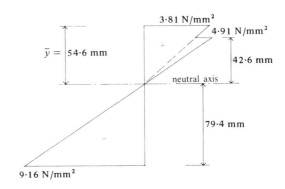

Fig. 10.10: *Bending stresses.*

At the upper extreme fibre of the web joists ($y = 42.6$ mm):

$$\sigma_{m,a} = \frac{6.48 \times 10^6 \times 42.6 \times 10\,500}{596 \times 10^9 \times 0.99} = 4.91\,\text{N/mm}^2 < 11.08\,\text{N/mm}^2$$
$$\text{(acceptable)}$$

At the lower extreme fibre of the web joists ($y = 79.4$ mm):

$$\sigma_{m,a} = \frac{6.48 \times 10^6 \times 79.4 \times 10\,500}{596 \times 10^9 \times 0.99} = 9.16\,\text{N/mm}^2 < 11.08\,\text{N/mm}^2$$

$$\text{(acceptable)}$$

*Horizontal and rolling shear stresses*

**Table 10.16**

| Element | $E$ (N/mm²) | $A$ (mm²) | $h_x$ (mm) | $EAh_x$ (N.mm × 10⁹) |
|---------|-------------|-----------|------------|---------------------|
| Top skin | 6 350 | 12.0 × 1200 | 48.6 | 4.44 |
| Part web | 10 500 | 4 × 47 × 45.8 | 21.3 | 1.79 |
| | | | $\Sigma ES =$ | 6.23 |

Alternatively:

$$\Sigma ES \text{ (for case below neutral axis)} = 10\,500 \times (4 \times 47 \times 79.4) \times 39.7$$
$$= 6.23 \times 10^9\,\text{N.mm}$$

The horizontal shear stress at the neutral axis is:

$$\tau = \frac{F_v(ES)}{(EI)t} = \frac{5.4 \times 10^3 \times 6.23 \times 10^9}{596 \times 10^9 \times 4 \times 47} = 0.30\,\text{N/mm}^2 < 1.13\,\text{N/mm}^2$$

$$\text{(acceptable)}$$

At the junction of the web joists and the ply skin, the rolling shear stress is appropriate (*ES* for skin only).

$$\tau_r = \frac{F_v(ES)}{(EI)t} = \frac{5400 \times 4.44 \times 10^9}{596 \times 10^9 \times 166} = 0.242\,\text{N/mm}^2 > 0.185 \times 1.1$$

$$= 0.204\,\text{N/mm}^2 \text{ (see Table 10.7)}$$

Therefore there is excessive rolling shear.

It can be seen that although there is a slight over-stress at the junction between ply skin and web, the main effect of removing the lower skin is that deflection is some 80% higher in this case than with the double-skin panel.

# Chapter Eleven
# Parallel-chord Lattice Beams

## 11.1 INTRODUCTION

For flat and mono-pitch roofs of 6–10 m span, ply web or glulam beams are popular, but there are cases in which parallel-chord lattice beams are the choice, either for functional or manufacturing reasons. This type of beam can be useful if plywood is not available or if the manufacturer is unable to glue, although it must be pointed out that glued plywood gussets are very acceptable in the construction of lattice beams, subject to the thickness of members and lack of distortion. Also a lattice beam is a natural choice (over whole or part of the span) when several fairly large diameter service pipes are to be accommodated in the depth of the beams. One possible disadvantage is that an economically designed lattice beam is deeper than either a glulam or ply web beam, usually having an effective depth around $\frac{1}{10}$–$\frac{1}{15}$ of the span. Lattice beams are normally placed at around 1.2–2 m centres using single or multiple chords.

## 11.2 CONSTRUCTION AND TYPES OF JOINT

Typical constructions consist of a top and bottom chord, often both of the same size, although with the latest published stresses of BS 5268:Part 2 there is a case for making the tension chord larger than the compression chord unless local bending on the compression chord is quite severe. Even if chords of different depths are used, the thickness is usually the same. Top and bottom chords will normally be jointed with a mechanical joint (Chapter 18), or a glued gusset or finger joint of adequate strength (Chapter 19). If a mechanical joint is used the effect of slip must be considered (§ 21.6.2).

The configuration of internal members is either of the Warren girder or N girder type (Fig. 11.1). Generally angle $\theta$ should not be much less than 30°. Because the internal members are so short, the slenderness ratio of compression members is unlikely to lead to any great reduction in strength and the choice of which internal configuration to use is much more dependent on manufacturing requirements or appearance than structural design. Warren girders are therefore more common than N girders. Internal members are

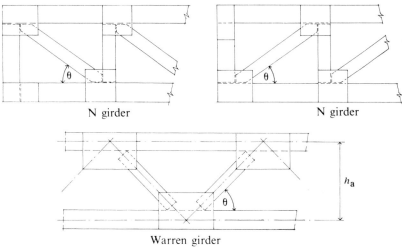

N girder                                    N girder

Warren girder

Fig. 11.1

usually the same thickness as the chords to facilitate jointing, and should be placed so that the centre lines of internals and chords intersect as closely as practical. For ease of manufacture, internal members are often the same size throughout the span.

The gusset plates can be of plywood (glued or nailed), conventional thin steel plates (nailed, screwed or even bolted) or proprietary punched metal-plate fasteners. Gusset plates normally occur on each side of the lattice, although one proprietary system covered by patent uses a single thin steel plate on the centre line connected by nails which are driven through the timber and the gusset. All the members of lattice girders are usually surfaced on all sides, no matter what gussets are used.

When internal members are of large sections of solid timber, consideration must be given to possible distortion of members when choosing the type of gusset to use. For example, if a web member is much over 50 mm thick and/or is very wide and may twist, it is unlikely that the manufacturer will be able to ensure a consistent glue line to plywood gussets. The use of bolted steel gussets is one alternative possibility when large size members are to be used.

Lattice girders can be constructed with a camber.

## 11.3   DESIGN METHOD

### 11.3.1   Calculation of axial load in members

If any UDL or point load occurs between node points on the top or bottom chords it is first necessary to transfer this loading to node points, usually either by treating each short span between nodes as being simply supported, or by taking the chord as a whole as being a continuous beam without sinking supports. Neither the size of the beam nor any likely economy seems to justify the use of a more complex method of analysis, unless computer facilities are available. The axial loads can then be calculated using the method of sections or other method, taking the effective depth as $h_a$ (Fig. 11.1) and assuming that all node points are pinned. Coefficients of axial loads are tabulated in Tables 21.17–21.25.

### 11.3.2    Design of compression chord

When the compression chord is supporting vertical loading between node points, a bending moment will be induced in the chord both at node points and between nodes. The various values of bending moment can be calculated by treating the whole chord as a continuous beam without sinking supports, or by considering each short length between nodes as fully fixed at any node where continuity occurs, and as pinned at the extreme ends. The chord is then designed as a strut taking combined bending and compression (§ 11.4).

In determining the permissible axial compression stress, the effective length for buckling in the plane of the beam is a function of the distance between node points. Clause 15.10 of BS 5268 : Part 2 suggests that the ratio should be between 0.85 and 1.0 times the distance between node points. As an alternative, when considering axial compression with the bending moment at a node point, the effective length for buckling in the plane of the beam can be taken as twice the distance from the node point to the furthest point of contraflexure A or B (Fig. 11.2), or as distance A plus B. Engineering judgement is required. The effective length for buckling out of the plane of the beam is a function of the distance between points of lateral restraint. Very often the cladding effectively restrains the chord fully about this plane.

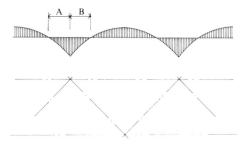

Fig. 11.2

In determining the permissible bending stress for bending due to vertical loading, only buckling out of the plane of the beam has to be considered. If the lateral restraint requirements of Table 4.2 are satisfied, the full permissible bending stress can be taken, otherwise one of the design methods outlined in Chapter 9 must be used to establish the reduced permissible stress.

### 11.3.3   Design of tension chord

The tension chord may be subjected to local bending between node points, in which case the bending moments are calculated as for the compression chord, and the chord is checked as a member taking combined tension and bending. The lower chord is normally in tension, but where, for example, the wind uplift on a flat roof is sufficient to put it into compression, the designer must check this design case, particularly in relation to the provision of lateral restraint. If beams are left exposed the usual wish is to omit any lateral restraints to the lower chord, but if it has to be designed for compression, one method of providing a fairly unobtrusive lateral restraint is to use wire fixed across the beams and tensioned to a fixing point at each end of the building. An alternative is to knee brace from a lateral restraint fixed at the level of the top chord.

### 11.3.4   Design of internal members

For internal members taking compression, the effective length is a function of the distance between node points. When the gussets give fixity in the plane of the beam, the effective length can be taken as about 0.85 and perhaps even as low as 0.7 times the distance between node points depending on the type of gusset and the rotational restraint which it gives at each end (see Table 19 of BS 5268:Part 2). Likewise, for buckling out of the line of the beam, a figure of between 1.0 and 0.7 times the distance between node points can usually be taken. A certain amount of engineering judgement is required in each case.

### 11.3.5   Load-duration/load-sharing factors for stress

The load-duration factor is always appropriate, but the load-sharing factor $K_8$ (or $K_{27}$ and $K_{29}$) should be applied only if load sharing can be seen to apply. On occasions, the designer may be able to justify using a load-sharing factor for one type of member in a design (e.g. the top chord) but not the internals or bottom chord (see §§ 20.5 and 2.5.1).

### 11.3.6   Deflection

The deflection of a lattice beam is caused by two effects which are the change of length due to strain in each member due to the axial load in it, and 'slip' at each connection. With glued gussets 'slip' can be disregarded.

The deflection caused by the first effect can be calculated using the strain energy formula $\Sigma FUL/AE$. (See worked examples in §§11.4 and 21.6.) Where there are two parallel members in cross-section, it seems reasonable to use a $K_9$ (or $K_{28}$) value of 1.14 for softwoods (applied to $E_{min}$) when determining the value for $E$. One could perhaps argue that because the deflection of the beam is caused by the strain in several members (two chords plus several internals) there is a case for using $E_{mean}$. The authors do not disagree with this, but are inclined to favour the use of $E_N$ for the chords and $E_{min}$ for the internals. However, if the engineer is satisfied that the beam is part of a load-sharing system, and particularly if beams are spaced at centres not in excess of 610 mm, a case for the use of $E_{mean}$ can be made.

The deflection caused by the second effect (slip) can be calculated using the strain energy formula $\Sigma U \Delta_s$, where $\Delta_s$ is the change in length between connections due to slip. With certain types of connections, this effect can lead to more deflection of the beam than the basic strain energy deflection. To calculate it, one has to know the load/slip characteristics of the connections or test the completed beam. See § 21.6.2 for notes and a worked example on slip deflection.

If a connection occurs between node points, the effect of slip in this connection must be taken into account in calculating the change in length of the member between the two adjacent node points.

## 11.4   DESIGN EXAMPLE

Determine the size of members required for the Warren girder lattice beam shown in Fig. 11.3. Assume that 40% of the loading is permanent dead loading applied through decking which restrains the beam laterally, and the remainder is medium term. The lattice beam is not considered to be part of

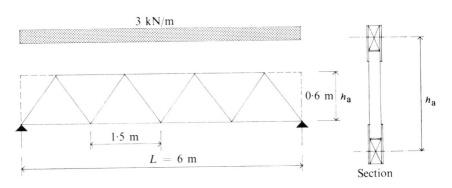

Fig. 11.3

a load-sharing system, SS grade Whitewood is to be used and the gussets are to be Finnish Birch-faced plywood glue/nailed each side of single members.

$$\text{Load per bay} = 4.5\,\text{kN} = F$$

From Fig. 27.10, maximum moment in top chord = $0.100 \times 4.5 \times 1.5$ = $0.675\,\text{kN.m}$. The axial load in members is derived from the coefficients in Table 21.25 with $h_a/L = 0.10 = a$.

Maximum axial compression in top chord $= 4.69F = 4.69 \times 4.5$
$$= 21.1 \text{ kN}$$
Maximum axial tension in bottom chord $= 4.69F = 21.1 \text{ kN}$
Maximum compression in diagonals $= 2.80F = 12.6 \text{ kN}$
Maximum tension in diagonals $= 1.60F = 7.2 \text{ kN}$

Because the members are glued it will be necessary to surface them (usually on all four sides). See Table 1.9 for surfacing reductions.

Before studying the detailed calculations which complete this example, the designer should be familiar with the design of compression members as covered in Chapters 14 and 15, and the design of ties as covered in Chapter 13.

*Design of top chord*
Try 47 $\times$ 145 mm Whitewood SS grade.

$$Z_X = 0.165 \times 10^6 \text{ mm}^3$$

$$\sigma_{m,a} = \frac{0.675 \times 10^6}{1.165 \times 10^6} = 4.09 \text{ N/mm}^2$$

$$\sigma_{m,adm} = \sigma_{m,g} \times K_3 \times K_7$$
$$= 7.5 \times 1.25 \times 1.083 = 10.1 \text{ N/mm}^2$$

$$\lambda_X = \frac{0.85 \times 1500 \sqrt{12}}{145} = 31$$

$$\lambda_Y = 0 \text{ (i.e. fully restrained by decking)}$$

$E/\sigma_c$ for determination of $K_{12} = E_{min}/(\sigma_{c,g} \times K_3) = 7000/(7.9 \times 1.25) = 709$.
From Table 14.2, with $\lambda = 31$, $K_{12} = 0.842$.

$$\sigma_{c,a} = \frac{21\,100}{47 \times 145} = 3.10 \text{ N/mm}^2$$

$$\sigma_{c,adm} = \sigma_{c,g} \times K_3 \times K_{12}$$
$$= 7.9 \times 1.25 \times 0.842 = 8.31 \text{ N/mm}^2$$

$$\sigma_e = \frac{\pi^2 E}{\lambda^2} = \frac{\pi^2 \times 7000}{31^2} = 72 \text{ N/mm}^2$$

Interaction formula:

$$\frac{4.09}{10.1 \left[ 1 - \dfrac{(1.5 \times 3.10 \times 1.25)}{72} \right]} + \frac{3.10}{8.31} = 0.44 + 0.37 = 0.81 < 1.0$$

which is acceptable.

*Design of bottom chord*
No local bending. SS grade Whitewood

$$\sigma_{t,g} = 4.5 \, \text{N/mm}^2$$
$$\text{Maximum tension} = 21.1 \, \text{kN}$$

Net area is gross area, since the joints are glued. From Table 13.2, select a 47 X 97 mm member. By proportion:

$$\bar{T} = 16.5 \times \frac{4.5}{3.2} = 23.2 \, \text{kN}$$

*Design of diagonals*
Design diagonals on the end compression member loading and use constant section throughout for simplicity in manufacture.

Length = 0.96 m. Axial compression = 12.6 kN medium term. SS grade Whitewood. Try 47 X 70 mm.

$$\lambda_X = \frac{0.9 \times 960 \sqrt{12}}{70} = 43$$

$$\lambda_Y = \frac{0.9 \times 960 \sqrt{12}}{47} = 64$$

With $E/\sigma_c = 709$ and $\lambda = 64$, $K_{12} = 0.597$.

$$\sigma_{c,a} = \frac{12\,600}{47 \times 70} = 3.83 \, \text{N/mm}^2$$

$$\sigma_{c,adm} = \sigma_{c,g} \times K_3 \times K_{12}$$
$$= 7.9 \times 1.25 \times 0.597 = 5.89 \, \text{N/mm}^2$$

which is acceptable.
   Use same section for tension diagonals. (If carrying out a design check, $K_{14}$ is applicable – see § 13.2.)

*Joints*
The glue joint at each end of the end diagonal has to take a medium term load of 12.6 kN. There is one gusset on each side of the member. The permissible shear stress on the glue line is the grade shear stress parallel to grain, or the rolling shear stress on the plywood, whichever is the lesser. In each loading case the permissible stress must be modified for duration of load ($K_3$) and for glue/nailing ($K_{70}$).
   For the timber (SS grade Whitewood), the permissible stress

$$= \text{grade shear parallel to grain stress (see Table 2.2)} \times K_3 \times K_{70}$$
$$= 0.82 \times 1.25 \times 0.9 = 0.92 \, \text{N/mm}^2$$

For the Finnish Birch-faced plywood, the permissible stress

= the rolling shear stress (see Table 8.7) $\times K_3 \times K_{70}$
= 0.60 $\times$ 1.25 $\times$ 0.9 = 0.67 N/mm²

(assuming the plywood to be at least 9 mm nominal thickness).

Length of glue line required on each of two gussets on a member 70 mm wide

$$= \frac{12\,600}{0.67 \times 2 \times 70} = 134\,\text{mm}$$

This can be provided, but from Fig. 11.4 one can see that it is not obvious what width of plywood can be taken in checking that the plywood is adequate for the compression it has to transmit from the diagonal into the chord. The design of any gusset is extremely complex, but from the sketch one can expect at least 100 mm of width at right angles to the diagonal to be effective on each gusset.

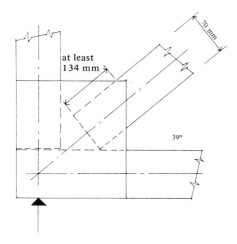

at least 134 mm

70 mm

39°

Fig. 11.4

∴ Compressive stress in nominal 9 mm thick Finnish Birch-faced plywood

$$= \frac{12\,400}{2 \times 8.8 \times 100} = 7.05\,\text{N/mm}^2$$

From Table 8.7 one can see that this actual stress compares with the permissible compressive stresses of 6.18 N/mm² (par) and 5.33 N/mm² (perp) times 1.25 for medium term loading, which shows a small over-stress if a width of 100 mm is retained, and may call for the use of a thicker gusset.

*Deflection*
The deflection of a truss including slip deflection is described in detail in § 21.6. This truss has glue joints, therefore slip deflection will not take place.

, Man'

...flection is calculated as in § 21.6.1, using the axial ...in Table 21.25, initially considering half of the truss. ...ue throughout. For the reason discussed in § 11.3.6, use $E_2$ ...rds (hence $K_9 = 1.14$), but $E_{min}$ for the internals.

| Member | P | U | L (m) | A (mm²) | E (N/mm²) | AE (kN) | $\dfrac{FUL}{AE}$ (m) |
|---|---|---|---|---|---|---|---|
| Top chord | ٦ ..٥ | +1.25 | 1.5 | 6815 | 7980 | 54 390 | 0.000 534 |
| | +21.1 | +2.5 | 0.75* | 6815 | 7980 | 54 390 | 0.000 727 |
| Bottom chord | −9.86 | −0.62 | 1.5 | 4559 | 7980 | 36 380 | 0.000 252 |
| | −21.1 | −1.67 | 1.5 | 4559 | 7980 | 36 380 | 0.001 453 |
| Internals | +2.80 | +0.80 | 0.96 | 3290 | 7000 | 23 030 | 0.000 093 |
| | −1.60 | −0.80 | 0.96 | 3290 | 7000 | 23 030 | 0.000 053 |
| | +1.60 | +0.80 | 0.96 | 3290 | 7000 | 23 030 | 0.000 053 |
| | 0 | −0.80 | 0.96 | 3290 | 7000 | 23 030 | 0 |
| | | | | | | | $\Sigma = 0.003\ 165$ |

* Half member length.

From Table 11.1:

$$\text{deflection of beam} = 2 \times 0.003\ 165 = 0.006\ 33\ \text{m} = 6\ \text{mm}$$

With a deflection as small as this, it hardly seems worthwhile to build in a camber against the dead loading only, but a manufacturer might wish to use a camber of 3–4 mm partly to eliminate the chance of a slight sag being built in. This beam has a large depth-to-span ratio, which is one reason why the deflection is so small. With shallow lattice beams, deflection can however, be critical, particularly if connections which can slip are used.

## Chapter Twelve

# Deflection. Practical and Special Considerations

## 12.1 DEFLECTION LIMITS

Any deflection limit which is set can be for functional reasons or purely for visual reasons. For example it has been found by experience that deflections below the horizontal not exceeding $0.003L$ are usually visually acceptable to anyone in a room below, unless there is a horizontal feature which makes the deflection much more obvious than it otherwise would be. With beams cambered to off-set dead load deflection, the deflection limit therefore applies to movement under imposed loading only.

BS 5268:Part 2 sets a limit 'for most general purposes' of 0.003 of the span. In addition, 'to avoid undue vibration' a further overall limit of 14 mm is set for domestic floor joists. It is likely that BS 5268:Part 3 will set a similar overall limit for trussed rafters.

For beams or decking spanning a short distance (say less than 2 m) it seems rather unnecessary to be too pedantic about observing a deflection limit of $0.003L$. At a span of 1.2 m this would give a limit of 3.6 mm. If the design permitted a deflection of 4.5 mm, this would be quite a large increase in percentage terms but hardly significant in real terms. It is important to realise that an over-deflection can take place without the beam necessarily approaching anywhere near to a stress–failure situation. This is a point often missed by architects, building inspectors etc. With short spans it is probably more important that the 'feel' of the beam is acceptable to any personnel who walk on it, and a simple test is often more satisfactory to determine suitability on short spans rather than deflection calculations.

In the design of beams for storage racking, it is usually acceptable to permit deflections in the order of 1/180 of the span if there is no access or only occasional access for personnel, and 1/240 of the span if there is general access for staff (not the general public). The design of roof decking in § 5.5 is based on a deflection limit of $L/240$.

If the designer finds himself in the position of having to decide a special deflection limit he should consider:

(a) the span,
(b) the type of structure and the usage,

(c) the possibility of damage to the ceiling or covering material,

(d) aesthetic requirements, particularly linked to any horizontal feature,

(e) the number of times and length of time when maximum deflection is likely to occur and whether there will be a camber,

(f) roof drainage,

(g) the effect on such items as partitions over or under the position of deflection,

(h) special items (see § 12.6),

(i) avoidance of secondary stresses at joints due to excessive rotation or distortion.

Only in exceptional circumstances is it normally necessary to tighten the deflection limit of 0.003 × span.

Deflection limits are also discussed in §§ 3.3, 3.14, and 4.15.1–4.15.4.

## 12.2   CAMBER

When a beam is prefabricated from several parts, as with glulam or ply web beams it is usually good practice to camber any simply supported beams (although it is usual for 'stock' glulam beams to be manufactured without camber). The usual object of providing a camber is to aim to have the beam deflect to a horizontal position under the action of the dead or permanent loading. In this way the beam has a horizontal soffit for the majority of the time, and deflects below the horizontal only when imposed load is added, and deflects to the maximum permitted position only in the possible extreme cases when the total imposed loading occurs, and then only for short periods of time.

With the variation in the weight of building materials and the $E$ value of timber, it is impossible for a designer to guarantee that a cambered beam will deflect exactly to the horizontal when the dead loading is in place. It should be possible, however, to design the beam to deflect to this position within the usual degree of building tolerance unless the builder departs from the design loadings to a large extent. It is well established that with a beam which is perfectly horizontal, there is an optical illusion that the beam is deflecting. (This is particularly so with a lattice beam.) Also, it is usually easier for a builder to pack down to obtain a level soffit from the centre of a beam which has a small residual camber, than to pack down from the ends of an under-cambered beam as this may interfere with an edge detail. The designer should usually therefore consider providing a camber slightly larger than the camber theoretically required.

It should be remembered that there is no such thing as a factor of safety on deflection. Double the load and the deflection doubles. If the agreed design loading is changed the beam deflection will be different from the calculated deflection. There is no way in which a beam can be designed to deflect to a required level, then deflect no more if further load is applied. If the designer is informed of an over-deflection having taken place on site and is satisfied that his calculations are correct, experience has shown that the first aspect to check on site is the weight of the applied loading. It may well be that this is excessive.

When a camber has been built-in with the object of bringing the beam to a horizontal position under the action of the permanent loading, the deflection limit (0.003 X span or other agreed limit) applies to deflection under the imposed loading only. If the designer has called for a camber slightly larger than the theoretical movement under permanent loading, it is not permissible to add the over-camber to the permitted deflection under imposed loading in checking the stiffness of the beam. However, if a designer opts for an 'under-camber' there is a case for checking the stiffness of the beam for the total loading minus the loading which would cause the beam to settle to the horizontal.

It is usual for a manufacturer to build-in a circular camber even though the beam may deflect in a parabolic manner. This is usually satisfactory for normal spans, but on spans of around 15 m or more, the designer would be wise to calculate several points on the ideal camber curve and instruct the manufacturer accordingly.

## 12.3 DEFLECTION DUE TO DEAD LOAD ONLY ON UNCAMBERED BEAMS

BS 5268:Part 2 does not lay down any criteria for limiting the deflection under dead loading only on uncambered beams (unless of course there is no imposed loading at all, in which case the limit automatically becomes 0.003 X span). With normal floor or roof beams, the dead loading is rarely more than 60% of the total loading, and therefore the deflected form under dead loading only, which is the position the beam will take up for the majority of its life, is approximately 0.002 X span or less. This is usually perfectly acceptable for the spans encountered with uncambered beams. If, however, for some reason it is impossible to camber a long-span beam, and the percentage of permanent loading to total loading is particularly high, the designer would be wise to discuss the possibility of tightening the deflection limit with the architect and/or the building user.

## 12.4 DEFLECTION DUE TO WIND UPLIFT ON ROOFS OR WIND ON WALLS

With the large wind uplifts on flat roofs required in designs to satisfy CP 3: Chapter V:Part 2:1972 it is quite possible to find that the residual uplift from the loading condition 'dead load + wind' exceeds the loading from 'dead + snow'. Because there is no load-duration factor for $E$ values, it is possible that the loading case of dead + wind uplift could be critical in setting the size of the beam from the point of view of upward deflection. When one considers that wind loading on beams is based on a five second gust which occurs very occasionally, if ever, and remembers that any deflection limit is purely an arbitrary limit, it seems rather unnecessary to apply the limit of 0.003 $\times$ span to this case, particularly bearing in mind that timber would not deflect to its full calculated amount in five seconds. The purpose of this handbook is certainly not to instruct a designer to disregard any part of any code of practice, but this is certainly a case where the designer could exercise some engineering discretion and discuss his opinions with the approving authority before the design is finalised.

Likewise it seems unnecessarily conservative to consider the full calculated deflection as occurring on a column in an external wall, particularly if a stud in a timber-framed house. At the very least it seems fair to claim that, if a beam is supposed to deflect no more than 80% of the calculated amount (see BS 5268 : Part 2 on testing) in the first 24 hours, a factor of 0.8 may be applied to the calculated deflection. By using this factor and a small amount of end restraint in the calculation of deflection of a stud, the calculated deflection is about one half of the 'full' simply-supported calculated deflection, which gives reasonable correlation with what available tests suggest happens in practice (see § 15.2).

## 12.5  DEFLECTION STAGES DUE TO SEQUENCE OF ERECTION

When an architect or building draughtsman draws a cross-section through a building with a flat roof and/or floor it will show beam soffits and the ceiling as straight horizontal lines, and will not usually consider at what stage the lines are horizontal. At the time of drawing the cross-section it may not even be known if the beams are to be cambered or not. On certain points, however, it is essential that someone coordinates the sequence of erection of items, either below or above the beams, bearing in mind that initial loading and further imposed loading will be applied to the beams in varying degrees throughout the lifetime of the structure. Some of these points are discussed here in general terms, and in slightly more detail in § 12.6, but the main object here is to make the designer aware of the type of situation which can occur, for which either the designer, the architect or the builder must have a solution.

If a non-load-bearing partition is placed under a beam before the total loading (including imposed) is applied to the beam, either an adequate gap and vertical sliding connection must be provided at the top of the partition or load will be transmitted to it.

If a builder adjusts the soffit of beams with packings etc. before all the dead loading is applied, the soffit will not be horizontal when the total dead loading has been applied.

If the gap between the top of a bottom run door and the underside of a roof beam is virtually nil once the dead loading is in place on the beam, the door will probably jam when snow falls on the roof. The designer must be alive to such potential troubles, some of which are discussed in § 12.6.

## 12.6 EXAMPLES OF CASES WHICH REQUIRE SPECIAL CONSIDER-ATION IN DEFLECTION/CAMBER CALCULATIONS

### 12.6.1 Beams over room with a change in width

Before the loading is applied to the beams as shown in Fig. 12.1, the undeflected form of cambered beams A and B is as indicated in Section *XX*. The erectors will have difficulty in fitting the secondary members between

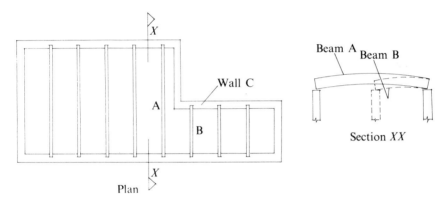

Fig. 12.1

the beams, particularly if the secondary members are continuous and particularly on the line of wall C. One method of easing the situation is to complete the roof to the left of beam A and to the right of beam B as far as is convenient, then infill between A and B. By the time the ceiling boards are to be fixed the beams will have deflected nearer to the horizontal.

### 12.6.2 Beams over room with tapered walls

The beams shown in plan in Fig. 12.2 would probably have the same cross-section, or at least the same depth, but, if cambered, each beam should be individually cambered to give a uniform soffit under permanent loading.

### 12.6.3 Edge condition with cambered beams

If beams B1, B2 etc. in the plan in Fig. 12.3 are built-up, they should prefer-ably be cambered. Beam A would be uncambered or could be a frame just sitting on the wall, rather than a beam. At the time of erecting the decking or secondary members between A and B1, the erectors would have to accept the difference in profile.

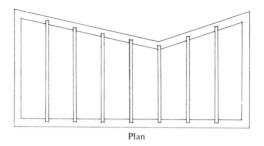

Fig. 12.2

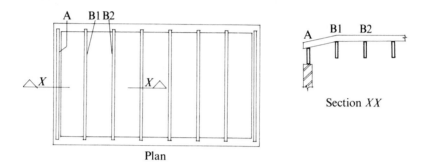

Fig. 12.3

### 12.6.4    Beams at clerestory glazing

If the members shown at B and C (Fig. 12.4) are designed in isolation for the roof loading which can occur on each, either the clerestory glazing can be subject to compression (which it may or may not be able to carry) or a

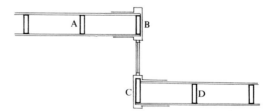

Fig. 12.4

gap may be induced at the top or bottom due to C deflecting more than B. Even accepting the many ways in which snow could lie on the roof it may be possible to design B and C as beams and link them to ensure equal deflection. However, another way is to manufacture B as a frame rather than a beam, and design beam C to be strong enough to take part of the upper roof through props at intervals to suit the glazing. Care must be taken with the flashing and the designer would have to give guidance on the sequence of erection.

### 12.6.5 Sliding doors

Comment regarding sliding doors has already been made in § 3.14.1. It is wise for the designer to ask for a copy of the manufacturing drawing of the doors to see whether sufficient tolerance is built into the top or bottom runners to ensure that the door will not jam under any condition of loading, whether the door is open or shut. Loading from doors should be considered as long term.

If the door is bottom run and occurs under a floor or roof which will deflect to different positions as the loading on it varies, it is usually wise to provide a beam or beams separate from the supporting structure, or secondary beams between the main beams to support the top door guide. If the door is top hung, then it is wise to hang it from one or two beams separate from the main support beams, so that deflection of the main beams will not cause the door to jam (Fig. 3.7), or from a secondary beam (Fig. 3.8).

In considering the clearance required in the runners for a top-hung folding door, the designer should realise that the difference in deflection in the support beam or beams between the cases with the door fully shut and partly open is not as great as one might imagine, due to the way the door folds. Compare the deflection calculations in Fig. 12.5.

### 12.6.6 Water or storage tanks or hoppers

The normal operation of a water-storage system is that replacement water is fed into the tank as water is drawn off, therefore the load to be taken by any supporting structure is usually constant. However, at the beginning and at stages in the life of the building, the tank will be empty. With small tanks this may have little effect on the structure, but with large tanks the designer should consider the effects of the load reduction. Although there is usually an overflow tube, it is advisable to assume that the tank can be filled to capacity. (Also see § 3.14.2.)

With storage tanks (say for oil) or hoppers (say for grain) the loading is not constant as the container regularly varies between being empty and being full. This variation must be considered in the design of the supporting structure. The designer must also take account of possible changes of moisture content on the weight of stored materials.

### 12.6.7 Cases in which beams should not normally be cambered

When a beam is continuous over two or more spans or rests on a more or less continuous support, it is normally either better or essential not to camber it. A few cases of non-cambered beams are sketched in Figs 12.6–12.10.

In Fig. 12.6, providing $a$ is less than $0.7L$, it is probably better not to camber the beam. If, however, $a$ increases as a percentage of $L$ it may be better to provide a beam cambered on span $L$ with no bearing taken initially on the internal wall. If an uncambered beam is provided when $b$ is a small

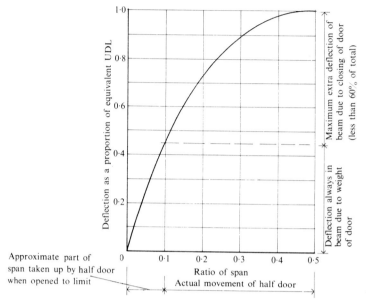

Note that shear deflection is omitted from this comparison exercise.

| Design case | 'Equivalent' UDL from $K_m$ factors (§§ 4.15.6 and 4.17) |
|---|---|
| $F$ (total weight of door) | 1.00$F$ |
| $0.1F$ $0.1F$ $0.4F$ $0.4F$ $0.4L$ $0.2L$ $0.4L$ | $(0.1 + 0.1)F \times 1.51 = 0.302F$ $(0.4 + 0.4)F \times 0.858 = 0.686F$ $\overline{\phantom{0000000000}0.988F}$ |
| $0.2F$ $0.2F$ $0.3F$ $0.3F$ $0.3L$ $0.4L$ $0.3L$ | $(0.2 + 0.2)F \times 1.267 = 0.5068F$ $(0.3 + 0.3)F \times 0.677 = 0.4062F$ $\overline{\phantom{0000000000}0.913F}$ |
| $0.3F$ $0.3F$ $0.2F$ $0.2F$ $0.2L$ $0.6L$ $0.2L$ | $(0.3 + 0.3)F \times 0.909 = 0.5454F$ $(0.2 + 0.2)F \times 0.467 = 0.1868F$ $\overline{\phantom{0000000000}0.7322F}$ |
| $0.4F$ $0.4F$ $0.1F$ $0.1F$ $0.1L$ $0.8L$ $0.1L$ | $(0.4 + 0.4)F \times 0.474 = 0.3792F$ $(0.1 + 0.1)F \times 0.238 = 0.0476F$ $\overline{\phantom{0000000000}0.4268F}$ |

Fig. 12.5

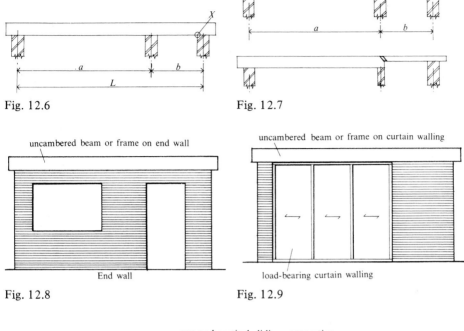

Fig. 12.6

Fig. 12.7

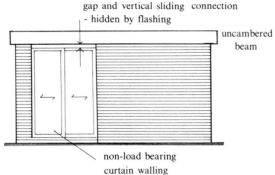

uncambered beam or frame on end wall

End wall

Fig. 12.8

uncambered beam or frame on curtain walling

load-bearing curtain walling

Fig. 12.9

gap and vertical sliding connection
- hidden by flashing

uncambered
beam

non-load bearing
curtain walling

Fig. 12.10

percentage of the span, the tendency is for the upward reaction at $X$ to lift the wall plate, with consequent cracking of plaster etc. The alternative arrangements shown in Fig. 12.7, in which the beams can be cambered, should be considered.

Special care is necessary in the design of purlins in cross-wall house construction (Fig. 12.11). With increase in roof pitch between 30° and 70°, the imposed loading reduces as a percentage of total loading, and under normal design conditions, with a beam cambered to off-set dead load deflection, the theoretical $EI$ capacity required at 70° roof slope to off-set live load deflection is nil. The effect of this approach may be to introduce excessive camber, because the $EI$ capacity obtained with a purlin designed to satisfy only bending and shear is probably exceptionally low. The effect of excessive

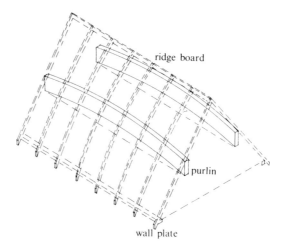

Fig. 12.11

camber is to change significantly the triangulation of the roof section at mid span of the purlin compared to the gable triangulation. The secondary continuous rafters have to be arched and sprung into place, which presents an unreasonable problem to the builder. This can be obviated by cambering the purlin only to a reasonable amount bearing erection in mind, and providing a section having an *EI* capacity appropriate to the total loading, or reduced only in proportion to the under-camber.

### 12.6.8   Combined deflection of trimmer and trimmed beam

As can be seen from Fig. 12.12 the deflection of a trimmer beam can have the visual effect of increasing the deflection of a trimmed beam. In certain cases, particularly if the trimmer is long span, it may be necessary to reduce the deflection of one or both to less than 0.003 times their span.

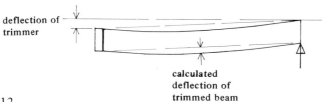

Fig. 12.12

### 12.6.9   Deflection of large overhangs

If a beam has a large overhang at one or both ends, it tends to deflect under uniform loading as sketched in Fig. 12.13.

The designer has two choices for the successful design of a beam with a cantilever:

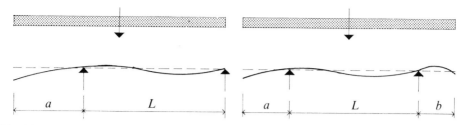

Fig. 12.13

(a) to decide against camber and limit the deflection all along the length to an acceptable amount.
(b) to camber for economy or functional reasons. In this case, by doing so, care must be taken not to cause an unacceptable situation at the end of the cantilever either before or after loading.

The method of choosing the best camber is to calculate and plot the deflected form under dead loading of an uncambered beam and then provide a camber shape as a 'mirror image' of this (Fig. 12.14). To suit manufacture, a modified camber curve may have to be accepted by the designer.

Figure 12.15 shows typical deflected forms of cambered beams with cantilever, before, during and after loading, which should assist the designer in visualising the conditions which can be encountered.

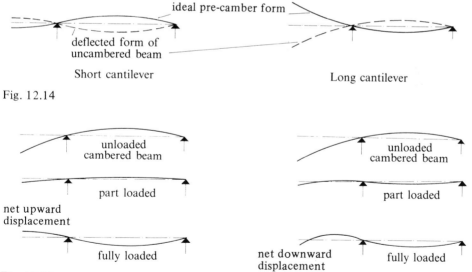

Fig. 12.14

Fig. 12.15

When the cantilever is more than about 2 m long it may not be advantageous to use a cambered beam.

BS 5268 : Part 2 : 1984 does not cover the case of deflection limits at the end of a cantilever. In many cases the designer will find that the only criterion is that the fascia at the end of the cantilever should be straight and the actual deflection is not important. In other cases a deflection of around 1/180 of the cantilever is found to be reasonable.

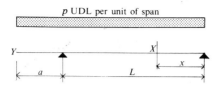

Fig. 12.16

Figures 12.16–12.20 give formulae and graphical representations for calculating deflections at the end of the cantilever and in the main span. In Fig. 12.16, $x$ is any distance between the supporters. (Also see Fig. 27.7.)

Bending deflection at $X = \dfrac{px}{24EIL}(L^4 - 2L^2 x^2 + Lx^3 - 2a^2 L^2 + 2a^2 x^2)$

Bending deflection at $Y = \dfrac{pa}{24EI}(4a^2 L - L^3 + 3a^3) = \dfrac{pL^4}{24EI}(K_{Y1})$

In evaluation of either of these formulae, a positive result indicates downward deflection and a negative result an upward one.

See Fig. 12.17 for values of $K_{Y1}$ for ratios of $a/L$.

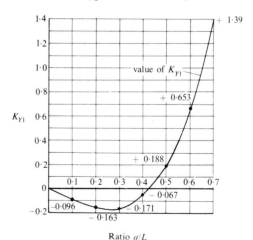

Fig. 12.17

To find the maximum deflection between the reactions, one way is to calculate deflection at two or three positions of $x$, calculate deflection at $Y$ and plot the deflected form of the beam.

In Fig. 12.18, $x$ is the distance to the point of maximum bending deflection between the supports:

Fig. 12.18

$$x = \frac{L}{\sqrt{3}}$$

$$\text{Bending deflection at } X = \frac{0.0642 FaL^2}{EI} \text{(upwards)} = \frac{FL^3}{EI}(K_{X1})$$

$$\text{Bending deflection at } Y = \frac{Fa^2}{3EI}(L + a) \text{(downwards)} = \frac{FL^3}{EI}(K_{Y2})$$

See Fig. 12.19 for values of $K_{X1}$ and Fig. 12.20 for values of $K_{Y2}$ for ratios $a/L$.

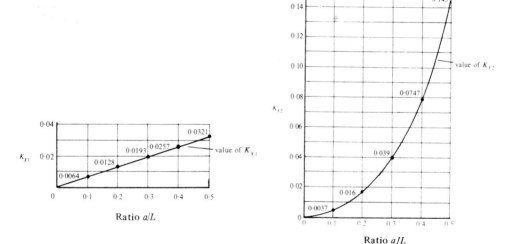

Fig. 12.19                     Fig. 12.20

## 12.7  EFFECT OF DEFLECTION ON END ROTATION OF BEAMS

In addition to vertical deflection, beams may displace horizontally as a result of bending stresses. The fibres in the compression zone shorten while the fibres in the tension zone lengthen. The resulting horizontal displacement ($\Delta_h$) at each end of a simple span beam can be calculated as:

$$\Delta_h = h_F \theta$$

where $h_F$ = height from the $XX$ axis to the point under consideration, being positive above and negative below the neutral axis

$\theta$ = end slope in radians. Values of $\theta$ are given in most standard textbooks on theory of structures.

$$\text{For a uniform load } F, \theta = \frac{FL^2}{24EI}$$

$$\text{For a central point load } F, \theta = \frac{FL^2}{16EI}$$

It may be assumed that end rotation is unaffected by shear deflection.

A positive value for $\Delta_h$ indicates a shortening of the beam and, conversely, a negative value indicates a lengthening of the beam.

To illustrate the method, calculate the horizontal end displacement of the upper and lower extremes of a 188.2 × 800 mm Box beam of the geometrical properties shown in Table 8.5 spanning 16.6 m and carrying a total uniform load of 20 kN.

For a uniform load:

$$\theta = \frac{FL^2}{24EI} \text{radians}$$

With an 800 mm deep beam:

$$h_F = \pm 0.4 \text{ m}$$

From Table 8.5:

$$EI = 32\ 845 \text{ kN.m}^2$$

$$\Delta_h = h_F \theta = \pm \frac{0.4 \times 20 \times 16.6^2}{24 \times 32\ 845} = \pm 0.0028 \text{ m}$$

i.e. 2.8 mm movement horizontally to the left and right of the neutral axis or 5.6 mm total horizontal displacement between top and bottom corners of the section.

The displacement at points between the neutral axis and the extreme fibres may be calculated proportionally.

The detailing of 'pin' joints, particularly between beam and post connections, must allow for this horizontal displacement. For beams of shallow depth the displacement may be sufficiently small to be accommodated within normal manufacturing tolerances and bolt-hole clearances. For deep sections the joint should be designed specifically to permit end rotation without developing secondary stresses. This is achieved by introducing positive clearances between faces of beam and column and by providing slotted holes for bolts etc.

# Chapter Thirteen
# Tension Members

## 13.1   AXIAL TENSILE LOADING

Until around 1970 the belief in the UK was that tensile strength parallel to the grain was one of the strongest properties of timber, and it was common to use the same grade stress for tension as for bending. Tables 3 and 4 of CP 112:Part 2:1971 presented stresses on this basis. Later research, including work at the Princes Risborough Laboratory, has shown that the tension strength for a particular grade is lower than the bending strength, and BS 5268:Part 2:1984 gives separate values for tension and bending stresses. For stress grades to BS 4978 the tension value is 60% of the bending value. For North American stress grades which can be graded only by visual methods the tension value is a higher percentage of the bending value as is the value for Strength Class SC1 (see notes in § 2.4.3).

The permissible tensile load in a member carrying axial loading only must be based on the permissible tensile stress and the effective cross-section of the member after provision for bolt holes, connector cut outs or other notches or cuts. In calculating the net area, it is not necessary to make a reduction for the wane permitted for the grade because wane is taken into account in setting the grade stress. Nor is it usual to consider stress concentrations as occurring at the net area, the adoption of the recommended permissible stresses and net area being considered to make provision for this aspect of design.

When assessing the effective cross-section of the member at a multiple-connector joint, all connectors within 0.75 connector diameters measured parallel to the grain should be considered as occurring at that cross-section.

With a member in tension there is no tendency to buckle, therefore the ratio of length to thickness is not critical, although some tension members, particularly in truss frameworks, may be subject to short term or very short term compression due to wind loading, and then the slenderness ratio $\lambda$ must be limited to 250.

The capacity of a member taking axial tension only may be determined by the strength at end connections or that of any joint along its length.

Particularly if using a low Strength Class in a design, it is important for the designer to realise and communicate to the supplier or manufacturer that

North American 'Light Framing', 'Utility' and 'Stud' grades, 'No. 3 Structural Light Framing' and 'No. 3 Joist and Plank' grades must not be used in designs to BS 5268 for members which are to carry any tension loading.

## 13.2  WIDTH FACTOR

For the first time in a UK code, BS 5268:Part 2 introduces a width factor $K_{14}$ for tension members. At a width of 72 mm or less, $K_{14}$ has the value of 1.17. For greater widths, $K_{14}$ has the value:

$$K_{14} = \left(\frac{300}{h}\right)^{0.11}$$

where $h$ = the greater dimension (of a rectangular section).

Therefore, for widths of between 72 and 300 mm of solid timber and glulam the value of $K_{14}$ applied to the grade tension stresses tabulated in BS 5268 for Strength Classes, timber graded to BS 4978, and North American 'Joist and Plank' grades is numerically equal to $K_7$ for which values are presented in Table 4.4 and Fig. 4.4. For widths greater than 300 mm the value of $K_{14}$ is also $(300/h)^{0.11}$.

For solid sections graded to the North American 'Structural Light Framing' grades (except No. 3 which must not be used in tension) the value of $K_{14}$ for sizes of 38 × 89 mm can be taken as 1.00. For values of $K_{14}$ for other sizes in 'Structural Light Framing' see Table 12 of BS 5268:Part 2.

From the comment made in § 13.1 it can be seen that there is no $K_{14}$ value for North American 'Light Framing', 'Utility' and 'Stud' grades, or 'No. 3 Structural Light Framing' or 'No. 3 Joist and Plank' grades.

## 13.3  COMBINED BENDING AND TENSILE LOADING

With a tension member of uniform cross-section also taking lateral loading (Fig. 13.1), the position of maximum stress occurs at the position of maximum bending moment. The sum of the tension and bending stress ratios must not exceed unity:

$$\frac{\sigma_{t,a,par}}{\sigma_{t,adm,par}} + \frac{\sigma_{m,a,par}}{\sigma_{m,adm,par}} \leqslant 1.0$$

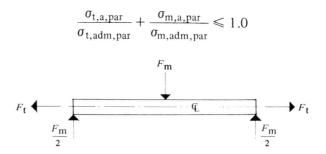

Fig. 13.1

with bending about one axis and:

$$\frac{\sigma_{t,a,par}}{\sigma_{t,adm,par}} + \frac{\sigma_{mX,a,par}}{\sigma_{mX,adm,par}} + \frac{\sigma_{mY,a,par}}{\sigma_{mY,adm,par}} \leqslant 1.0$$

with bending about both axes. The formulae can similarly be expressed in terms of the moment and tensile capacities:

$$\frac{M}{\overline{M}} + \frac{F_t}{\overline{T}} \leqslant 1.0 \qquad \text{or} \qquad \frac{M_X}{\overline{M}_X} + \frac{M_Y}{\overline{M}_Y} + \frac{F_t}{\overline{T}} \leqslant 1.0$$

where $F_t$ = the axial tension
$\overline{T}$ = the tensile capacity.

Values of $\overline{T}$ for several solid sections of GS grade Whitewood are given in Table 13.1.

In calculating $\sigma_{t,a}$ the designer is principally interested in the tension fibres, but $\sigma_{m,adm}$ should be determined from a consideration of the lateral restraint. The designer should also be alert to the possibility of the critical design case occurring on the compression edge if bending occurs with relatively low axial tension.

The eccentricity at an end connection or at a notch along the length of the tie causes a moment $F_t e$ which may be additive to the external moment at that point (see Fig. 13.2).

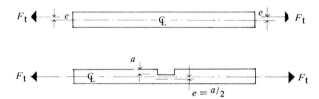

Fig. 13.2

## 13.4   TENSION CAPACITIES OF SOLID TIMBER SECTIONS CONTAIN-ING SPLIT-RING OR SHEAR-PLATE CONNECTORS

To simplify the design of tension members, particularly in trusses, tensile capacities for solid timbers with gross and net (or effective) areas are given in Table 13.1. The timber sizes are surfaced in accordance with the recommendations for constructional timber in Table 1.9. The effective area is calculated as the gross area less the projected area of the connector grooves (as required by Clause 45.2 of BS 5268:Part 2) and the projected area of the bolt hole not contained within the projected area of the connector grooves. Bolt holes are taken as bolt diameter plus 2 mm.

The table is prepared for GS grade European Whitewood (or Redwood) having a long term grade stress of 3.2 N/mm². Capacities for the same

**Table 13.1**  Tensile capacities for solid sections of GS grade Whitewood, long term loading for gross and net (effective) sections with deductions for split-ring or shear-plate connectors and their associated bolts

Net $\overline{T} = \sigma_{t,g} \times$ net area $\times K_{14} = 3.2 \times A_{Ne} \times K_{14}$ (long term $\times$ GS grade).
Values for other grades by proportion of $\sigma_{t,g}$.
Values for medium, short or very short term loading by proportion of $K_3$.

Net tensile capacity $\overline{T}$ (kN)

| | | Gross section | | Split-rings | | | | Shear-plates | | | |
|---|---|---|---|---|---|---|---|---|---|---|---|
| Actual section (mm) | $K_{14}$ | Area (mm²×10³) | $\overline{T}$ (kN) | 64 mm diameter | | 102 mm diameter | | 67 mm diameter | | 102 mm diameter | |
| 35 × 72 | 1.170 | 2.520 | 9.4 | — | — | — | — | + | + | + | + |
| 35 × 97 | 1.132 | 3.395 | 12.3 | 8.4 | 6.3 | — | — | + | + | + | + |
| 35 × 122 | 1.104 | 4.200 | 15.1 | 11.3 | 9.3 | — | — | + | + | + | + |
| 35 × 145 | 1.083 | 5.075 | 17.6 | 13.9 | 11.9 | 10.8 | — | + | + | + | + |
| 35 × 169 | 1.065 | 5.915 | 20.2 | 16.5 | 14.5 | 13.5 | 11.8 | + | + | + | + |
| 35 × 194 | 1.049 | 6.790 | 22.8 | 19.2 | 17.3 | 16.2 | 15.6 | + | + | + | + |
| 44 × 97 | 1.132 | 4.268 | 15.5 | 11.1 | 9.0 | — | — | 8.2 | — | — | — |
| 44 × 122 | 1.104 | 5.368 | 19.0 | 14.7 | 12.7 | — | — | 11.9 | — | — | — |
| 44 × 145 | 1.083 | 6.380 | 22.1 | 18.0 | 16.0 | 14.6 | 10.6 | 15.1 | — | 14.1 | 9.5 |
| 44 × 169 | 1.065 | 7.436 | 25.3 | 21.2 | 19.3 | 18.0 | 14.0 | 18.5 | — | 17.5 | 13.0 |
| 44 × 194 | 1.049 | 8.536 | 28.7 | 24.6 | 22.7 | 21.4 | 17.5 | 21.9 | 18.6 | 20.9 | 16.5 |
| 44 × 219 | 1.035 | 9.636 | 31.9 | 27.9 | 26.0 | 24.8 | 20.9 | 25.2 | 22.0 | 24.3 | 19.9 |
| 47 × 97 | 1.132 | 4.559 | 16.5 | 12.0 | 10.0 | — | — | 9.0 | — | — | — |
| 47 × 122 | 1.104 | 5.640 | 20.3 | 15.9 | 13.8 | — | — | 12.9 | — | — | — |
| 47 × 145 | 1.083 | 6.815 | 23.6 | 19.3 | 17.3 | 15.9 | 11.8 | 16.4 | — | 15.4 | 10.8 |
| 47 × 169 | 1.065 | 7.943 | 27.1 | 22.8 | 20.9 | 19.5 | 15.5 | 20.0 | — | 19.0 | 14.5 |
| 47 × 194 | 1.049 | 9.118 | 30.6 | 26.4 | 24.5 | 23.1 | 19.2 | 23.6 | 20.1 | 22.6 | 18.2 |
| 47 × 219 | 1.035 | 10.293 | 34.1 | 30.0 | 28.1 | 26.7 | 22.8 | 27.2 | 23.8 | 26.2 | 21.8 |

| | | | | | | | | | | | | | | | |
|---|---|---|---|---|---|---|---|---|---|---|---|---|---|---|---|
| 60 X 145 | 1.083 | 8.70 | 30.2 | 25.2 | 23.2 | — | — | 21.5 | 17.4 | 23.7 | 22.0 | — | — | 21.0 | 16.3 |
| 60 X 169 | 1.065 | 10.14 | 34.6 | 29.7 | 27.7 | — | — | 26.0 | 22.0 | 28.3 | 26.5 | — | — | 25.5 | 21.0 |
| 60 X 194 | 1.049 | 11.64 | 39.1 | 34.3 | 32.4 | 29.5 | 25.7 | 30.6 | 26.7 | 32.9 | 31.1 | 26.7 | 23.2 | 30.1 | 25.7 |
| 60 X 219 | 1.035 | 13.14 | 43.5 | 38.8 | 36.9 | 34.1 | 30.3 | 35.2 | 31.3 | 37.4 | 35.7 | 31.3 | 27.9 | 34.7 | 30.3 |
| 72 X 145 | 1.083 | 10.440 | 36.2 | 30.7 | 28.7 | — | — | 26.6 | 22.5 | 28.9 | 27.1 | — | — | 26.0 | 21.4 |
| 72 X 169 | 1.065 | 12.168 | 41.5 | 36.0 | 34.1 | — | — | 32.0 | 28.0 | 34.3 | 32.5 | — | — | 31.5 | 27.0 |
| 72 X 194 | 1.049 | 13.968 | 46.9 | 41.5 | 39.6 | 36.2 | 32.4 | 37.6 | 33.6 | 39.8 | 38.0 | 32.7 | 29.3 | 37.1 | 32.6 |
| 72 X 219 | 1.035 | 15.768 | 52.2 | 47.0 | 45.0 | 41.7 | 37.9 | 43.0 | 39.1 | 45.2 | 43.5 | 38.3 | 34.8 | 42.5 | 38.1 |

Connector arrangement (see Fig. 13.4)

| | | | $N_b$ | 1 | 1 | 2 | 2 | 1 | 1 | 1 | 1 | 2 | 2 | 1 | 1 |
|---|---|---|---|---|---|---|---|---|---|---|---|---|---|---|---|
| | | | $N_c$ | 1 | 2 | 1 | 2 | 2 | 1 | 2 | 1 | 2 | 1 | 1 | 2 |

— Not permitted owing to insufficient depth of section.

+ Not permitted owing to insufficient breadth of section.

sections in other grades or species can be calculated *pro rata* to the permissible grade tension stress (but see § 13.2 for the limit on certain sizes/stress grades of North American timber, and variation to $K_{14}$ values).

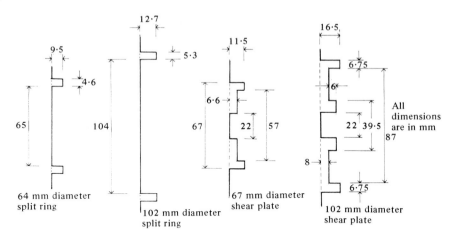

Fig. 13.3: *Dimensions of grooves.*

The dimensions of grooves which must be cut for split-ring and shear-plate connectors are shown in Fig. 13.3, and critical dimensions and areas are given in Table 13.2. From the values of $A_c$, $t_g$ and $d$ in this table, together with the values of $N_b$ and $N_c$ in Fig. 13.4 it is possible to evolve a general formula for the effective area for the more commonly occurring connector arrangements on which Table 13.1 is based. (The depth of recesses for shear-plate connectors given in BS 5268:Part 2 does not seem to link entirely with BS 1579:1960, but the differences are unlikely to be significant.)

**Table 13.2**

| | Connector | | | | Bolt dia. | Bolt hole dia. |
|---|---|---|---|---|---|---|
| Type | Nominal dia. (mm) | Overall dia. (mm) | Depth of groove $t_g$ (mm) | Projected area $A_c$ (mm²) | Bolt dia. (mm) | $d$ (mm) |
| Split-ring | 64 | 74.2 | 9.5 | 705 | 12 | 14 |
| | 102 | 114.6 | 12.7 | 1455 | 20 | 22 |
| Shear-plate | 67 | 67 | 11.5 | 770 | 20 | 22 |
| | 102 | 102.5 | 16.5 | 1690 | 20 | 22 |

The possible connector arrangements for solid timber of up to 219 mm depth are shown in Fig. 13.4. $N_b$ is the number of bolts and $N_c$ is the number of connectors on each bolt.

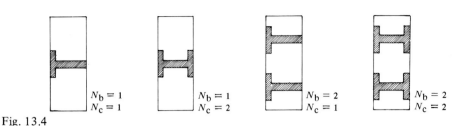

Fig. 13.4

For these arrangements the effective area $A_{Ne}$ is:

$$A_{Ne} = bh - N_b N_c A_c - N_b d(b - N_c t_g)$$

$$\therefore \text{ Tensile capacity } = \sigma_{t,adm}(\text{including } K_3 \text{ and } K_{14}) \times A_{Ne}$$

## 13.5   STEEL TIE BARS

Timber trusses and frames often incorporate steel tie bars as an alternative to timber tension members either to accommodate particularly high tensile loads or to 'lighten' or improve the appearance of the structure. Steel tie bars are manufactured in limited lengths and must be end jointed using couplers or turnbuckles on threaded ends if long lengths are required.

Table 13.3 gives technical data for steel tie bars as supplied by Reinforcement Steel Services (Sheffield). With larger diameter tie bars, erectors find it convenient if a turnbuckle with a tommy bar is used. Even on short tie bars a coupler or turnbuckle may be advantageous as a means of adjustment rather than relying on adjusting end nuts.

On tie bars over 8 m long, sag rods should be provided at 8 m centres or closer.

**Table 13.3 Tie bar data**

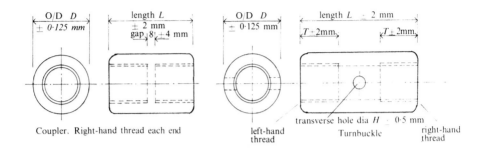

Coupler. Right-hand thread each end     left-hand thread    Turnbuckle    right-hand thread

| | Tie bar | | | Nut | | Couplers | | Turnbuckles | | |
|---|---|---|---|---|---|---|---|---|---|---|
| Material | Dia. (mm) | Maximum bar length (m) | Tensile working load (kN) | Length (mm) | Width across flats (mm) | Outside dia. (mm) | Overall length *L* (mm) | Outside dia. (mm) | Overall length *L* (mm) | Transverse hole diameter *H* (mm) |
| Cold | 20 | 18 | 110 | 25 | 42 | 40 | 55 | | | |
| worked | 25 | 18 | 170 | 32 | 47 | 50 | 75 | | Not | |
| high tensile | 32 | 18 | 275 | 40 | 56 | 60 | 100 | | available | |
| alloy steel* | 40 | 18 | 435 | 50 | 65 | 75 | 125 | | | |
| | 50 | 7.5 | 655 | 70 | 90 | 75 | 150 | | | |
| Mild steel | 50 | 12 | 200 | 47 | 70 | 75 | 110 | 75 | 170 | 15 |
| to RSS | 60 | 12 | 300 | 56 | 90 | 95 | 125 | 95 | 200 | 15 |
| Spec. 16M† | 75 | 12 | 450 | 70 | 106 | 110 | 155 | 110 | 230 | 15 |
| | 90 | 12 | 580 | 83 | 123 | 125 | 181 | 125 | 270 | 18 |
| | 100 | 12 | 730 | 96 | 141 | 140 | 205 | 140 | 310 | 18 |

* Tensile working load at Factor of Safety = 3.
† Tensile working load at 110 N/mm².

# Chapter Fourteen

# The Design of Columns. General Notes

## 14.1 RELATED CHAPTERS

This chapter deals in detail with the considerations necessary for the design of columns. As such it is a reference chapter. In addition there are several points in Chapter 4 on the general design of beams which also apply to columns. Three other chapters deal with the actual design of columns, each one being devoted to one main type:

Chapter 15    Columns of Solid Timber
Chapter 16    Columns of Simple Composites. Composites of two or three sections, rectangular or tee shapes, spaced columns
Chapter 17    Glulam Columns.

In addition, struts in triangulated frameworks are dealt with in Chapters 11, 20 and 21.

## 14.2 DESIGN CONSIDERATIONS

The principal considerations in the design of columns are:

axial stress
positional restraint of ends
directional restraint of ends (i.e. fixity)
lateral restraint along length
effective length and slenderness ratio
deflected form
bearing at bottom and top.

The relevant permissible stresses are computed by modifying the grade stresses by the factors from BS 5268:Part 2 which are discussed in this chapter.

As with beams BS 5268 gives stresses for 'dry exposure' conditions when the moisture content in service is not in excess of 18% and 'wet exposure' conditions when the moisture content is more than 18%. Normally a designer will deal with dry conditions and consequently all the values tabulated in this chapter are for moisture contents of 18% or less.

## 14.3   EFFECTIVE LENGTH

### 14.3.1   Introduction

The effective length is determined by:

the positional restraint at each end of the column (i.e. whether or not there is relative sway between the two ends)
the directional restraint at ends (i.e. whether or not there is fixity at either or both ends)
lateral restraint along the length.

Examples of effective lengths are sketched in Figs 14.1–14.3. It is emphasised, however, that there can be slight differences of opinion between engineers on deciding the effective length for actual cases.

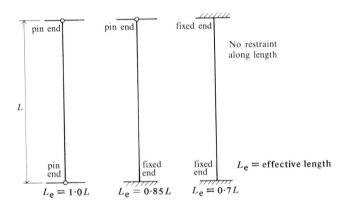

Fig. 14.1: *Effective lengths, XX axis. No sway.*

### 14.3.2   Effective lengths. No sway

Examples of effective lengths about the $XX$ axis are sketched in Fig. 14.1 for cases in which there is no relative sway between the two ends of the column. Examples about the $YY$ axis are sketched in Fig. 14.2.

### 14.3.3   Effective lengths. Sway possible

Examples of effective lengths about the $XX$ axis are sketched in Fig. 14.3 for cases in which sway is possible.

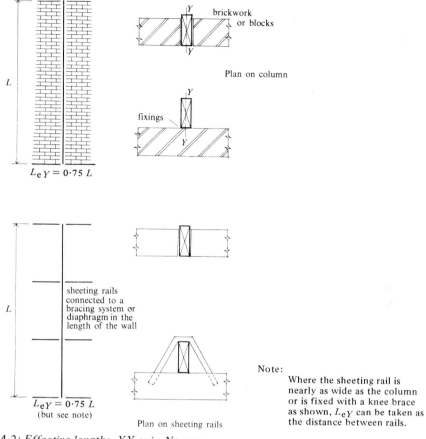

brickwork
or blocks

Plan on column

fixings

$L_{eY} = 0.75 L$

sheeting rails
connected to a
bracing system or
diaphragm in the
length of the wall

$L_{eY} = 0.75 L$
(but see note)

Plan on sheeting rails

Note:
Where the sheeting rail is
nearly as wide as the column
or is fixed with a knee brace
as shown, $L_{eY}$ can be taken as
the distance between rails.

Fig. 14.2: *Effective lengths, YY axis. No sway.*

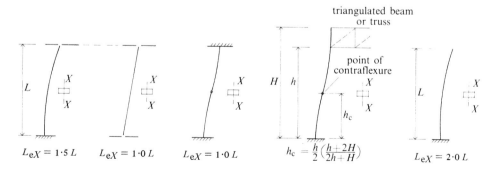

triangulated beam
or truss

point of
contraflexure

$L_{eX} = 1.5 L$     $L_{eX} = 1.0 L$     $L_{eX} = 1.0 L$     $h_c = \dfrac{h}{2}\left(\dfrac{h+2H}{2h+H}\right)$     $L_{eX} = 2.0 L$

$L_{eX} = 2 h_c$ for lower part of column
$L_{eX} = 2(h-h_c)$ for upper part of column

Fig. 14.3: *Effective lengths, XX axis. Sway possible.*

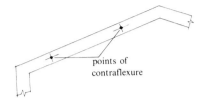

Fig. 14.4

### 14.3.4   Effective lengths. Points of contraflexure

In certain cases (e.g. in the frame sketched in Fig. 14.4), two points of contraflexure can occur in one member. As sketched, these are caused by bending about the $XX$ axis and, in considering the slenderness ratio of the member on the $XX$ axis against any axial loading, $L_{eX}$ can be taken as the distance between the points of contraflexure. (This is confirmed in Clause 15.2 of BS 5268:Part 2.) When considering buckling about the $YY$ axis due to bending about the $XX$ axis and axial loading, $L_{eY}$ is determined from the degree of lateral restraint.

   In the case of the propped cantilever with sway sketched in Fig. 14.5, $L_{eX}$ for axial loading can be taken as $2h_c$ when considering part AC of the column and $2(h - h_c)$ for part CB. The value of $L_{eY}$ for axial loading and for bending about the $XX$ axis is determined from the degree of lateral restraint.

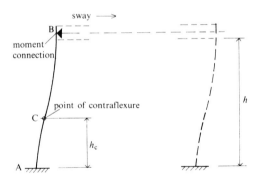

Fig. 14.5

### 14.4   SLENDERNESS RATIO. DURATION OF LOADING. PERMISSIBLE COMPRESSIVE STRESS

For any axis through the centre of gravity of a cross-section, the radius of gyration may be regarded as a measure of the dispersal of the area about that axis. The radius of gyration $i$ is defined as:

$$i = \sqrt{\frac{I}{A}}$$

where $I$ is the second moment of area

$A$ is the area.

For a member having the same effective length about both principal axes, buckling under axial compression occurs about the axis with the smaller radius of gyration, i.e. in the direction of the maximum slenderness ratio, where:

$$\text{slenderness ratio} = \frac{\text{effective length}}{\text{radius of gyration}} = \frac{L_e}{i} = \lambda$$

In many practical cases members have differing $L_e$ and $i$ values about each principal axis and hence differing slenderness ratios. If one denotes the principal directions of buckling by the suffixes $x$ and $y$, then the larger value of $(L_e/i)_x$ or $(L_e/i)_y$ will determine the permissible compression stress $\sigma_{c,adm,par}$.

Timber is used mainly in the form of rectangular sections and, because the relationship of $i$ to $b$ is constant $(i = b/\sqrt{12})$ the slenderness ratio can also be expressed as:

$$\lambda = \frac{L_e\sqrt{12}}{b}$$

In BS 5268:Part 2 the modification factor $(K_{12})$ for slenderness ratio is arrived at in a very different way to the $K_{18}$ and $K_{19}$ factors for slenderness ratio which occurred in CP 112:Part 2. Also, whereas $K_{18}$ and $K_{19}$ of CP 112 included an allowance for the duration of loading, $K_{12}$ of BS 5268 does not include for duration of loading, therefore $K_3$ applies to axial loading of columns. The formula for $K_{12}$ is given in Appendix C of BS 5268:Part 2 as:

$$K_{12} = \left\{\frac{1}{2} + \frac{(1+\eta)\pi^2 E}{2N\lambda^2 \sigma_c}\right\} - \left[\left\{\frac{1}{2} + \frac{(1+\eta)\pi^2 E}{2N\lambda^2 \sigma_c}\right\}^2 - \frac{\pi^2 E}{N\lambda^2 \sigma_c}\right]^{\frac{1}{2}}$$

where $\sigma_c$ = the compression parallel to the grain stress for the particular conditions of loading, exposure etc.

$E$ = the appropriate modulus of elasticity for the particular exposure condition:

| | |
|---|---|
| For a solid timber member acting alone<br>For a solid timber member acting in a<br>load-sharing system (such as a stud wall) | $E_{min}$ for the grade/<br>species or Strength<br>Class |
| For two or more solid timber members<br>connected together in parallel (by<br>mechanical fixings or adhesive) and<br>acting together (whether in a load-<br>sharing system or not) | $E_{min}$ for the grade/<br>species or Strength<br>Class $\times$ $K_9$ (or $K_{28}$<br>if $N = 5$ or more) |

For a horizontally laminated member $E_{\text{mean}}$ for SS grade of the species modified by $K_{20}$ or $K_{26}$ as relevant

$\lambda$ = the slenderness ratio ($L_e/i$).

$\eta$ = the eccentricity factor (taken as $0.005\lambda$).

$N = 1.5$ and takes account of the reduction factors used to derive grade compression stresses and moduli of elasticity.

For columns of solid timber, the value of $\sigma_c$ for entry into the formula for $K_{12}$ is the relevant grade stress modified only for moisture content (i.e. dry or wet exposure) and duration of loading (i.e. *not* for load sharing). For horizontally laminated members the value of $\sigma_c$ for entry into the formula for $K_{12}$ is the SS grade stress for the species modified by $K_{17}$ or $K_{23}$ as relevant (see Clause 18 of BS 5268:Part 2) and for duration of loading. It should be realised, however, that even though $\sigma_c$ as entered into the formula is modified for duration of loading, the permissible compressive stress for the column is derived from:

$$\sigma_{c,g,par} \times K_{12} \times K_3 \text{ (for duration of loading)} \times K_8 \text{ (or } K_{17} \text{ or } K_{23} \text{ or } K_{28})$$

Table 20 of BS 5268:Part 2 gives values of $K_{12}$ for rounded-off values of $E/\sigma_{c,par}$. However, to use Table 20, the designer had to interpolate in two directions which can be time-consuming. Therefore values of $K_{12}$ are given in Table 14.2 for the actual $E/\sigma_{c,par}$ values which a designer is most likely to encounter, as detailed in Table 14.1.

## 14.5   MAXIMUM SLENDERNESS RATIO

BS 5268:Part 2 states that the slenderness ratio $\lambda$ for a compression member should not exceed 180 unless it is a member normally subject to tension, or combined tension and bending arising from dead and imposed loads, but subject to a reversal of stress solely from the effects of wind, or a compression member such as wind bracing carrying self-weight and wind loads only, in which cases the slenderness ratio $\lambda$ can be as high as 250.

## 14.6   PERMISSIBLE AXIAL STRESSES AND LOAD SHARING

The permissible axial compressive stress is found by combining the grade stress for compression parallel to the grain with modification factors as outlined below.

For single solid members not part of a load-sharing system: $\sigma_{c,g,par} \times K_{12} \times K_3$ (for load duration). For two or more solid timber members (or a vertically glued laminated member) connected together in parallel (by mechanical fixings or adhesive) and acting together, but not part of a load-sharing system: $\sigma_{c,g,par} \times K_{12} \times K_{28} \times K_3$. (When compared to $K_8$, the value of 1.14 for $K_{28}$ for two members acting together against compression parallel to grain–see Table 24 of BS 5268:Part 2–seems somewhat illogical, although understandable for $E$ value. However, as drafted, $K_{28}$ applies to compression parallel to grain.)

**Table 14.1** $E/\sigma_c$ values for calculation of $K_{12}$.

| Species | Stress grade and factor | $E$ value | Long term loading | | Medium term loading | | Very short term loading | |
|---|---|---|---|---|---|---|---|---|
| | | | $\sigma_c$ | $E/\sigma_c$ | $\sigma_c$ | $E/\sigma_c$ | $\sigma_c$ | $E/\sigma_c$ |
| European | GS | 6000 | 6.8 | 882 | 8.5 | 706 | 11.9 | 504 |
| Redwood | SS | 7000 | 7.9 | 886 | 9.875 | 709 | 13.825 | 506 |
| or | M75 | 7000 | 8.7 | 805 | 10.875 | 644 | 15.225 | 460 |
| Whitewood | GS $K_9 = 1.14$ | 6840 | 6.8 | 1006 | 8.5 | 805 | 11.9 | 575 |
| | SS $K_9 = 1.14$ | 7980 | 7.9 | 1010 | 9.875 | 808 | 13.825 | 577 |
| | M75 $K_9 = 1.14$ | 7980 | 8.7 | 917 | 10.875 | 734 | 15.225 | 524 |
| | GS $K_9 = 1.21$ | 7260 | 6.8 | 1068 | 8.5 | 854 | 11.9 | 610 |
| | SS $K_9 = 1.21$ | 8470 | 7.9 | 1072 | 9.875 | 858 | 13.825 | 613 |
| | M75 $K_9 = 1.21$ | 8470 | 8.7 | 974 | 10.875 | 779 | 15.225 | 556 |
| | LB with $K_{17} = 1.04$ and $K_{20} = 0.9$ | 9450 | 8.216 | 1150 | 10.27 | 920 | 14.378 | 657 |
| | LC with $K_{17} = 0.92$ and $K_{20} = 0.8$ | 8400 | 7.268 | 1156 | 9.085 | 925 | 12.719 | 660 |
| Canadian | Various | 5500 | 6.8 | 809 | 8.5 | 647 | 11.9 | 462 |
| species | grade/species | 6000 | 4.6 | 1304 | 5.75 | 1043 | 8.05 | 745 |
| | combinations | | 6.1 | 984 | 7.625 | 787 | 10.675 | 562 |
| | | | 6.8 | 882 | 8.5 | 706 | 11.9 | 504 |
| | | | 6.9 | 870 | 8.625 | 696 | 12.075 | 497 |
| | | | 7.9 | 759 | 9.875 | 608 | 13.825 | 434 |
| | | 6500 | 7.9 | 823 | 9.875 | 658 | 13.825 | 470 |
| | | 7000 | 4.6 | 1522 | 5.75 | 1217 | 8.05 | 870 |
| | | | 6.1 | 1148 | 7.625 | 918 | 10.675 | 656 |
| | | | 6.9 | 1014 | 8.625 | 812 | 12.075 | 580 |
| | | | 7.7 | 909 | 9.625 | 727 | 13.475 | 519 |
| | | | 7.9 | 886 | 9.875 | 709 | 13.825 | 506 |
| | | | 8.5 | 824 | 10.625 | 659 | 14.875 | 471 |
| | | | 8.8 | 795 | 11.0 | 636 | 15.4 | 455 |
| | | 7500 | 7.9 | 949 | 9.875 | 759 | 13.825 | 542 |
| | | 8000 | 8.8 | 909 | 11.0 | 727 | 15.4 | 519 |
| | | | 9.3 | 860 | 11.625 | 688 | 16.275 | 492 |
| Various | SC2 | 5000 | 5.3 | 943 | 6.625 | 755 | 9.275 | 539 |
| | SC3 | 5800 | 6.8 | 853 | 8.5 | 682 | 11.9 | 487 |
| | SC4 | 6600 | 7.9 | 835 | 9.875 | 668 | 13.825 | 477 |
| | SC5 | 7100 | 8.7 | 816 | 10.875 | 653 | 15.225 | 466 |

$\sigma_c$ and $E$ values in N/mm$^2$.

**Table 14.2  Modification factor $K_{12}$ for compression members**

| $E/\sigma_c$ (see Table 14.1) | Values of slenderness ratio $\lambda$ ($L_e/i$) | | | | | | | | | | | | | | | | | | | |
|---|---|---|---|---|---|---|---|---|---|---|---|---|---|---|---|---|---|---|---|---|
| | <5 | 5 | 10 | 20 | 30 | 40 | 50 | 60 | 70 | 80 | 90 | 100 | 120 | 140 | 160 | 180 | 200 | 220 | 240 | 250 |
| | Equivalent $L_e/b$ for rectangular sections | | | | | | | | | | | | | | | | | | | |
| | <1.44 | 1.44 | 2.89 | 5.78 | 8.67 | 11.6 | 14.5 | 17.3 | 20.2 | 23.1 | 26.0 | 28.9 | 34.7 | 40.5 | 46.2 | 52.0 | 57.7 | 63.5 | 69.3 | 72.2 |
| 434 | 1.000 | 0.975 | 0.951 | 0.897 | 0.831 | 0.745 | 0.638 | 0.528 | 0.429 | 0.350 | 0.288 | 0.240 | 0.173 | 0.130 | 0.101 | 0.081 | 0.066 | 0.055 | 0.047 | 0.043 |
| 455 | 1.000 | 0.975 | 0.951 | 0.897 | 0.833 | 0.749 | 0.647 | 0.539 | 0.441 | 0.361 | 0.298 | 0.249 | 0.180 | 0.135 | 0.105 | 0.084 | 0.069 | 0.057 | 0.048 | 0.045 |
| 457 | 1.000 | 0.975 | 0.951 | 0.898 | 0.833 | 0.750 | 0.648 | 0.541 | 0.443 | 0.363 | 0.300 | 0.250 | 0.181 | 0.136 | 0.106 | 0.085 | 0.070 | 0.060 | 0.050 | 0.045 |
| 460 | 1.000 | 0.975 | 0.951 | 0.898 | 0.834 | 0.751 | 0.650 | 0.542 | 0.445 | 0.364 | 0.301 | 0.252 | 0.182 | 0.137 | 0.107 | 0.085 | 0.070 | 0.058 | 0.049 | 0.046 |
| 462 | 1.000 | 0.975 | 0.951 | 0.898 | 0.833 | 0.751 | 0.650 | 0.543 | 0.445 | 0.365 | 0.302 | 0.252 | 0.182 | 0.137 | 0.107 | 0.085 | 0.070 | 0.058 | 0.049 | 0.046 |
| 466 | 1.000 | 0.975 | 0.951 | 0.898 | 0.834 | 0.752 | 0.651 | 0.545 | 0.447 | 0.367 | 0.304 | 0.254 | 0.184 | 0.138 | 0.108 | 0.086 | 0.070 | 0.058 | 0.049 | 0.046 |
| 470 | 1.000 | 0.975 | 0.951 | 0.898 | 0.834 | 0.753 | 0.653 | 0.547 | 0.450 | 0.369 | 0.306 | 0.256 | 0.185 | 0.139 | 0.108 | 0.086 | 0.071 | 0.059 | 0.050 | 0.046 |
| 471 | 1.000 | 0.975 | 0.951 | 0.898 | 0.834 | 0.753 | 0.653 | 0.547 | 0.450 | 0.370 | 0.306 | 0.256 | 0.185 | 0.140 | 0.109 | 0.087 | 0.071 | 0.059 | 0.050 | 0.046 |
| 477 | 1.000 | 0.975 | 0.951 | 0.898 | 0.835 | 0.754 | 0.656 | 0.550 | 0.454 | 0.373 | 0.309 | 0.259 | 0.187 | 0.141 | 0.110 | 0.087 | 0.072 | 0.060 | 0.050 | 0.047 |
| 487 | 1.000 | 0.975 | 0.951 | 0.898 | 0.836 | 0.756 | 0.659 | 0.555 | 0.459 | 0.378 | 0.314 | 0.263 | 0.191 | 0.144 | 0.112 | 0.088 | 0.073 | 0.061 | 0.051 | 0.048 |
| 492 | 1.000 | 0.975 | 0.951 | 0.898 | 0.836 | 0.757 | 0.661 | 0.558 | 0.462 | 0.381 | 0.316 | 0.265 | 0.192 | 0.145 | 0.113 | 0.090 | 0.074 | 0.062 | 0.052 | 0.048 |
| 495 | 1.000 | 0.975 | 0.951 | 0.899 | 0.837 | 0.758 | 0.663 | 0.560 | 0.464 | 0.383 | 0.318 | 0.267 | 0.194 | 0.146 | 0.114 | 0.090 | 0.075 | 0.062 | 0.053 | 0.049 |
| 497 | 1.000 | 0.975 | 0.951 | 0.899 | 0.837 | 0.759 | 0.663 | 0.561 | 0.465 | 0.384 | 0.319 | 0.267 | 0.194 | 0.147 | 0.115 | 0.091 | 0.075 | 0.063 | 0.053 | 0.049 |
| 504 | 1.000 | 0.975 | 0.951 | 0.899 | 0.837 | 0.760 | 0.666 | 0.564 | 0.468 | 0.387 | 0.322 | 0.270 | 0.196 | 0.149 | 0.116 | 0.092 | 0.076 | 0.063 | 0.054 | 0.050 |
| 506 | 1.000 | 0.975 | 0.951 | 0.899 | 0.838 | 0.760 | 0.666 | 0.565 | 0.469 | 0.388 | 0.323 | 0.271 | 0.197 | 0.149 | 0.116 | 0.093 | 0.076 | 0.064 | 0.054 | 0.050 |
| 519 | 1.000 | 0.975 | 0.951 | 0.899 | 0.838 | 0.762 | 0.670 | 0.570 | 0.475 | 0.394 | 0.328 | 0.276 | 0.201 | 0.152 | 0.118 | 0.095 | 0.078 | 0.064 | 0.055 | 0.050 |
| 524 | 1.000 | 0.975 | 0.951 | 0.900 | 0.839 | 0.764 | 0.672 | 0.573 | 0.478 | 0.397 | 0.331 | 0.278 | 0.203 | 0.154 | 0.120 | 0.096 | 0.079 | 0.065 | 0.056 | 0.050 |
| 539 | 1.000 | 0.975 | 0.951 | 0.900 | 0.839 | 0.765 | 0.676 | 0.578 | 0.485 | 0.403 | 0.337 | 0.284 | 0.207 | 0.157 | 0.122 | 0.098 | 0.080 | 0.066 | 0.057 | 0.051 |
| 542 | 1.000 | 0.975 | 0.951 | 0.900 | 0.840 | 0.766 | 0.677 | 0.580 | 0.486 | 0.405 | 0.338 | 0.285 | 0.208 | 0.158 | 0.123 | 0.099 | 0.081 | 0.067 | 0.057 | 0.052 |
| 556 | 1.000 | 0.975 | 0.951 | 0.900 | 0.841 | 0.768 | 0.681 | 0.586 | 0.493 | 0.412 | 0.345 | 0.291 | 0.213 | 0.161 | 0.126 | 0.101 | 0.083 | 0.067 | 0.059 | 0.053 |
| 562 | 1.000 | 0.975 | 0.951 | 0.900 | 0.841 | 0.769 | 0.683 | 0.588 | 0.495 | 0.414 | 0.347 | 0.293 | 0.215 | 0.163 | 0.128 | 0.102 | 0.084 | 0.069 | 0.059 | 0.054 |
| 575 | 1.000 | 0.975 | 0.951 | 0.900 | 0.842 | 0.771 | 0.686 | 0.593 | 0.501 | 0.420 | 0.353 | 0.298 | 0.219 | 0.166 | 0.130 | 0.105 | 0.086 | 0.070 | 0.061 | 0.055 |
| 577 | 1.000 | 0.975 | 0.951 | 0.901 | 0.842 | 0.771 | 0.687 | 0.593 | 0.502 | 0.421 | 0.353 | 0.299 | 0.219 | 0.167 | 0.130 | 0.105 | 0.086 | 0.072 | 0.061 | 0.056 |
| 580 | 1.000 | 0.975 | 0.951 | 0.901 | 0.842 | 0.772 | 0.687 | 0.594 | 0.503 | 0.422 | 0.355 | 0.300 | 0.220 | 0.167 | 0.131 | 0.105 | 0.086 | 0.072 | 0.061 | 0.056 |
| 593 | 1.000 | 0.975 | 0.951 | 0.901 | 0.843 | 0.773 | 0.690 | 0.599 | 0.508 | 0.428 | 0.360 | 0.305 | 0.224 | 0.170 | 0.134 | 0.107 | 0.086 | 0.072 | 0.062 | 0.056 |
| 608 | 1.000 | 0.975 | 0.951 | 0.901 | 0.844 | 0.775 | 0.694 | 0.604 | 0.514 | 0.434 | 0.366 | 0.310 | 0.228 | 0.174 | 0.137 | 0.110 | 0.088 | 0.074 | 0.062 | 0.058 |
| 610 | 1.000 | 0.975 | 0.951 | 0.901 | 0.844 | 0.775 | 0.694 | 0.604 | 0.515 | 0.434 | 0.367 | 0.311 | 0.229 | 0.174 | 0.137 | 0.110 | 0.090 | 0.075 | 0.064 | 0.059 |
| 613 | 1.000 | 0.975 | 0.951 | 0.901 | 0.844 | 0.776 | 0.695 | 0.605 | 0.516 | 0.436 | 0.368 | 0.312 | 0.230 | 0.175 | 0.137 | 0.111 | 0.091 | 0.076 | 0.064 | 0.060 |
| 636 | 1.000 | 0.975 | 0.951 | 0.901 | 0.845 | 0.778 | 0.699 | 0.612 | 0.524 | 0.444 | 0.376 | 0.320 | 0.236 | 0.180 | 0.141 | 0.114 | 0.093 | 0.078 | 0.066 | 0.061 |
| 640 | 1.000 | 0.975 | 0.951 | 0.901 | 0.845 | 0.779 | 0.700 | 0.613 | 0.526 | 0.446 | 0.378 | 0.321 | 0.238 | 0.181 | 0.143 | 0.115 | | | | |
| 644 | 1.000 | 0.975 | 0.951 | 0.901 | 0.845 | 0.779 | 0.701 | 0.614 | 0.527 | 0.448 | 0.379 | 0.322 | 0.239 | 0.182 | 0.143 | 0.115 | | | | |
| 647 | 1.000 | 0.975 | 0.951 | 0.902 | 0.846 | 0.780 | 0.702 | 0.615 | 0.528 | 0.449 | 0.380 | 0.324 | 0.240 | 0.183 | 0.144 | 0.116 | 0.095 | 0.080 | 0.067 | 0.062 |

| | | | | | | | | | | | | | | | | | | | |
|---|---|---|---|---|---|---|---|---|---|---|---|---|---|---|---|---|---|---|---|---|
| 653 | 0.062 | 0.068 | 0.080 | 0.096 | 0.116 | 0.145 | 0.184 | 0.241 | 0.325 | 0.382 | 0.450 | 0.530 | 0.616 | 0.702 | 0.780 | 0.846 | 0.902 | 0.951 | 0.975 | 1.000 |
| 656 | 0.063 | 0.068 | 0.081 | 0.096 | 0.117 | 0.146 | 0.185 | 0.242 | 0.327 | 0.384 | 0.452 | 0.531 | 0.618 | 0.703 | 0.780 | 0.846 | 0.902 | 0.951 | 0.975 | 1.000 |
| 657 | 0.063 | 0.068 | 0.081 | 0.097 | 0.117 | 0.146 | 0.185 | 0.243 | 0.327 | 0.384 | 0.452 | 0.532 | 0.618 | 0.703 | 0.780 | 0.846 | 0.902 | 0.951 | 0.975 | 1.000 |
| 658 | 0.063 | 0.068 | 0.080 | 0.096 | 0.117 | 0.145 | 0.185 | 0.243 | 0.327 | 0.384 | 0.452 | 0.532 | 0.618 | 0.703 | 0.780 | 0.846 | 0.902 | 0.951 | 0.975 | 1.000 |
| 659 | 0.063 | 0.068 | 0.080 | 0.096 | 0.117 | 0.146 | 0.185 | 0.243 | 0.327 | 0.384 | 0.453 | 0.532 | 0.618 | 0.703 | 0.780 | 0.846 | 0.902 | 0.951 | 0.975 | 1.000 |
| 660 | 0.064 | 0.069 | 0.081 | 0.097 | 0.118 | 0.146 | 0.186 | 0.243 | 0.328 | 0.385 | 0.453 | 0.533 | 0.619 | 0.704 | 0.781 | 0.846 | 0.902 | 0.951 | 0.975 | 1.000 |
| 668 | 0.064 | 0.069 | 0.081 | 0.097 | 0.118 | 0.147 | 0.187 | 0.245 | 0.330 | 0.387 | 0.456 | 0.535 | 0.620 | 0.705 | 0.781 | 0.846 | 0.902 | 0.951 | 0.975 | 1.000 |
| 682 | 0.065 | 0.070 | 0.081 | 0.099 | 0.121 | 0.150 | 0.191 | 0.249 | 0.335 | 0.392 | 0.461 | 0.539 | 0.624 | 0.707 | 0.782 | 0.846 | 0.902 | 0.951 | 0.975 | 1.000 |
| 688 | 0.065 | 0.071 | 0.083 | 0.099 | 0.122 | 0.151 | 0.192 | 0.251 | 0.337 | 0.394 | 0.463 | 0.541 | 0.626 | 0.708 | 0.783 | 0.847 | 0.902 | 0.951 | 0.975 | 1.000 |
| 693 | 0.066 | 0.071 | 0.084 | 0.100 | 0.123 | 0.152 | 0.194 | 0.252 | 0.339 | 0.397 | 0.465 | 0.543 | 0.627 | 0.710 | 0.784 | 0.847 | 0.902 | 0.951 | 0.975 | 1.000 |
| 696 | 0.066 | 0.071 | 0.084 | 0.100 | 0.123 | 0.153 | 0.194 | 0.253 | 0.340 | 0.398 | 0.466 | 0.544 | 0.628 | 0.710 | 0.784 | 0.847 | 0.902 | 0.951 | 0.975 | 1.000 |
| 706 | | | | | 0.125 | 0.155 | 0.196 | 0.256 | 0.343 | 0.401 | 0.469 | 0.547 | 0.630 | 0.712 | 0.785 | 0.848 | 0.902 | 0.951 | 0.975 | 1.000 |
| 709 | | | | | 0.125 | 0.155 | 0.196 | 0.256 | 0.343 | 0.401 | 0.469 | 0.547 | 0.630 | 0.712 | 0.785 | 0.848 | 0.902 | 0.951 | 0.975 | 1.000 |
| 727 | | | | | 0.128 | 0.158 | 0.201 | 0.261 | 0.349 | 0.407 | 0.475 | 0.553 | 0.635 | 0.714 | 0.786 | 0.848 | 0.902 | 0.951 | 0.975 | 1.000 |
| 734 | 0.069 | 0.074 | 0.088 | 0.105 | 0.129 | 0.160 | 0.203 | 0.263 | 0.352 | 0.410 | 0.478 | 0.555 | 0.637 | 0.716 | 0.787 | 0.849 | 0.902 | 0.951 | 0.975 | 1.000 |
| 745 | | | | | 0.131 | 0.162 | 0.205 | 0.266 | 0.355 | 0.413 | 0.481 | 0.558 | 0.639 | 0.717 | 0.788 | 0.849 | 0.902 | 0.951 | 0.975 | 1.000 |
| 755 | 0.071 | 0.077 | 0.090 | 0.108 | 0.132 | 0.163 | 0.207 | 0.268 | 0.358 | 0.416 | 0.484 | 0.560 | 0.640 | 0.718 | 0.788 | 0.849 | 0.903 | 0.951 | 0.975 | 1.000 |
| 759 | 0.071 | 0.077 | 0.091 | 0.108 | 0.133 | 0.164 | 0.208 | 0.270 | 0.359 | 0.418 | 0.486 | 0.562 | 0.642 | 0.719 | 0.789 | 0.850 | 0.903 | 0.951 | 0.975 | 1.000 |
| 779 | | | | | 0.136 | 0.168 | 0.212 | 0.275 | 0.365 | 0.424 | 0.491 | 0.567 | 0.646 | 0.722 | 0.790 | 0.850 | 0.903 | 0.951 | 0.975 | 1.000 |
| 787 | | | | | 0.137 | 0.169 | 0.214 | 0.277 | 0.367 | 0.426 | 0.494 | 0.569 | 0.647 | 0.723 | 0.791 | 0.850 | 0.903 | 0.951 | 0.975 | 1.000 |
| 795 | 0.075 | 0.081 | 0.095 | 0.113 | 0.138 | 0.170 | 0.215 | 0.278 | 0.369 | 0.428 | 0.495 | 0.570 | 0.648 | 0.723 | 0.791 | 0.851 | 0.903 | 0.952 | 0.975 | 1.000 |
| 800 | | | | | 0.139 | 0.172 | 0.217 | 0.280 | 0.371 | 0.430 | 0.497 | 0.572 | 0.649 | 0.724 | 0.792 | 0.851 | 0.903 | 0.952 | 0.975 | 1.000 |
| 805 | | | | | 0.139 | 0.172 | 0.217 | 0.281 | 0.372 | 0.431 | 0.498 | 0.573 | 0.650 | 0.725 | 0.792 | 0.851 | 0.903 | 0.952 | 0.975 | 1.000 |
| 808 | | | | | 0.140 | 0.173 | 0.218 | 0.282 | 0.373 | 0.432 | 0.499 | 0.574 | 0.651 | 0.725 | 0.792 | 0.851 | 0.903 | 0.952 | 0.975 | 1.000 |
| 809 | 0.076 | 0.082 | 0.097 | 0.116 | 0.140 | 0.173 | 0.218 | 0.282 | 0.374 | 0.432 | 0.500 | 0.574 | 0.651 | 0.725 | 0.792 | 0.851 | 0.903 | 0.952 | 0.975 | 1.000 |
| 812 | | | | | 0.141 | 0.174 | 0.219 | 0.283 | 0.374 | 0.433 | 0.500 | 0.575 | 0.652 | 0.725 | 0.792 | 0.851 | 0.903 | 0.952 | 0.975 | 1.000 |
| 816 | | | | | 0.141 | 0.174 | 0.219 | 0.284 | 0.375 | 0.434 | 0.501 | 0.575 | 0.652 | 0.725 | 0.792 | 0.851 | 0.903 | 0.952 | 0.975 | 1.000 |
| 823 | 0.076 | 0.083 | 0.097 | 0.116 | 0.142 | 0.175 | 0.221 | 0.285 | 0.377 | 0.435 | 0.503 | 0.577 | 0.653 | 0.726 | 0.792 | 0.851 | 0.903 | 0.952 | 0.975 | 1.000 |
| 824 | 0.077 | 0.083 | 0.098 | 0.117 | 0.142 | 0.175 | 0.221 | 0.286 | 0.377 | 0.436 | 0.503 | 0.577 | 0.653 | 0.726 | 0.792 | 0.851 | 0.903 | 0.952 | 0.975 | 1.000 |
| 830 | 0.077 | 0.083 | 0.098 | 0.117 | 0.143 | 0.177 | 0.223 | 0.287 | 0.379 | 0.438 | 0.505 | 0.579 | 0.655 | 0.726 | 0.792 | 0.851 | 0.903 | 0.952 | 0.975 | 1.000 |
| 835 | | | | | 0.143 | 0.177 | 0.223 | 0.288 | 0.380 | 0.439 | 0.506 | 0.579 | 0.655 | 0.727 | 0.793 | 0.851 | 0.903 | 0.952 | 0.975 | 1.000 |
| 853 | 0.078 | 0.084 | 0.099 | 0.118 | 0.146 | 0.180 | 0.227 | 0.292 | 0.385 | 0.443 | 0.510 | 0.583 | 0.658 | 0.729 | 0.793 | 0.851 | 0.903 | 0.952 | 0.975 | 1.000 |
| 854 | 0.080 | 0.086 | 0.101 | 0.120 | 0.147 | 0.181 | 0.228 | 0.293 | 0.386 | 0.444 | 0.511 | 0.584 | 0.658 | 0.730 | 0.794 | 0.852 | 0.903 | 0.952 | 0.975 | 1.000 |
| 858 | | | | | 0.147 | 0.181 | 0.228 | 0.294 | 0.387 | 0.445 | 0.512 | 0.585 | 0.659 | 0.730 | 0.795 | 0.852 | 0.903 | 0.952 | 0.976 | 1.000 |
| 860 | | | | | 0.147 | 0.181 | 0.228 | 0.294 | 0.387 | 0.445 | 0.512 | 0.585 | 0.659 | 0.731 | 0.795 | 0.852 | 0.904 | 0.952 | 0.976 | 1.000 |
| 867 | 0.080 | 0.087 | 0.102 | 0.121 | 0.148 | 0.183 | 0.230 | 0.296 | 0.389 | 0.448 | 0.514 | 0.586 | 0.660 | 0.731 | 0.795 | 0.852 | 0.904 | 0.952 | 0.976 | 1.000 |
| 870 | | | | | 0.149 | 0.183 | 0.231 | 0.297 | 0.390 | 0.448 | 0.515 | 0.587 | 0.661 | 0.731 | 0.795 | 0.852 | 0.904 | 0.952 | 0.976 | 1.000 |
| 882 | | | | | 0.151 | 0.185 | 0.233 | 0.300 | 0.393 | 0.451 | 0.518 | 0.589 | 0.663 | 0.732 | 0.796 | 0.853 | 0.904 | 0.952 | 0.976 | 1.000 |
| 886 | 0.081 | 0.088 | 0.103 | 0.123 | 0.151 | 0.186 | 0.234 | 0.300 | 0.394 | 0.452 | 0.519 | 0.590 | 0.663 | 0.733 | 0.796 | 0.853 | 0.904 | 0.952 | 0.976 | 1.000 |
| 906 | | | | | 0.154 | 0.189 | 0.238 | 0.305 | 0.399 | 0.457 | 0.523 | 0.594 | 0.666 | 0.735 | 0.797 | 0.853 | 0.904 | 0.952 | 0.976 | 1.000 |

**Table 14.2** (*contd*)

| E/σc (see Table 14.1) | Values of slenderness ratio λ (Le/i) | | | | | | | | | | | | | | | | | | | | |
|---|---|---|---|---|---|---|---|---|---|---|---|---|---|---|---|---|---|---|---|---|---|
| | <5 | 5 | 10 | 20 | 30 | 40 | 50 | 60 | 70 | 80 | 90 | 100 | 120 | 140 | 160 | 180 | 200 | 220 | 240 | 250 |
| | Equivalent Le/b for rectangular sections | | | | | | | | | | | | | | | | | | | | |
| | <1.44 | 1.44 | 2.89 | 5.78 | 8.67 | 11.6 | 14.5 | 17.3 | 20.2 | 23.1 | 26.0 | 28.9 | 34.7 | 40.5 | 46.2 | 52.0 | 57.7 | 63.5 | 69.3 | 72.2 |
| 909 | 1.000 | 0.975 | 0.951 | 0.903 | 0.853 | 0.797 | 0.734 | 0.666 | 0.594 | 0.523 | 0.457 | 0.399 | 0.305 | 0.238 | 0.189 | 0.154 | 0.127 | 0.107 | 0.091 | 0.084 |
| 917 | 1.000 | 0.976 | 0.952 | 0.904 | 0.853 | 0.798 | 0.736 | 0.667 | 0.596 | 0.525 | 0.460 | 0.401 | 0.307 | 0.240 | 0.191 | 0.155 | | | | |
| 918 | 1.000 | 0.976 | 0.952 | 0.904 | 0.853 | 0.798 | 0.736 | 0.667 | 0.596 | 0.526 | 0.460 | 0.401 | 0.308 | 0.240 | 0.191 | 0.156 | | | | |
| 920 | 1.000 | 0.976 | 0.952 | 0.904 | 0.853 | 0.798 | 0.736 | 0.668 | 0.596 | 0.526 | 0.460 | 0.402 | 0.308 | 0.240 | 0.192 | 0.156 | | | | |
| 925 | 1.000 | 0.976 | 0.952 | 0.904 | 0.853 | 0.798 | 0.736 | 0.668 | 0.597 | 0.527 | 0.462 | 0.403 | 0.309· | 0.241 | 0.192 | 0.156 | | | | |
| 943 | 1.000 | 0.975 | 0.952 | 0.904 | 0.853 | 0.798 | 0.737 | 0.670 | 0.600 | 0.530 | 0.465 | 0.407 | 0.313 | 0.244 | 0.195 | 0.158 | 0.131 | 0.110 | 0.094 | 0.087 |
| 949 | 1.000 | 0.975 | 0.952 | 0.904 | 0.853 | 0.799 | 0.738 | 0.671 | 0.601 | 0.532 | 0.466 | 0.408 | 0.314 | 0.245 | 0.196 | 0.159 | 0.132 | 0.111 | 0.094 | 0.087 |
| 974 | 1.000 | 0.976 | 0.952 | 0.904 | 0.854 | 0.800 | 0.740 | 0.674 | 0.605 | 0.537 | 0.472 | 0.414 | 0.320 | 0.250 | 0.200 | 0.163 | | | | |
| 984 | 1.000 | 0.976 | 0.952 | 0.904 | 0.855 | 0.800 | 0.741 | 0.675 | 0.607 | 0.539 | 0.475 | 0.416 | 0.322 | 0.252 | 0.202 | 0.164 | | | | |
| 1006 | 1.000 | 0.976 | 0.952 | 0.904 | 0.855 | 0.801 | 0.742 | 0.678 | 0.610 | 0.543 | 0.479 | 0.421 | 0.326 | 0.256 | 0.205 | 0.167 | | | | |
| 1010 | 1.000 | 0.976 | 0.952 | 0.904 | 0.855 | 0.801 | 0.742 | 0.678 | 0.611 | 0.544 | 0.480 | 0.422 | 0.327 | 0.257 | 0.206 | 0.168 | | | | |
| 1014 | 1.000 | 0.976 | 0.952 | 0.904 | 0.855 | 0.802 | 0.743 | 0.679 | 0.611 | 0.544 | 0.481 | 0.423 | 0.328 | 0.258 | 0.206 | 0.168 | | | | |
| 1038 | 1.000 | 0.976 | 0.952 | 0.904 | 0.855 | 0.802 | 0.744 | 0.681 | 0.615 | 0.549 | 0.485 | 0.428 | 0.333 | 0.262 | 0.210 | 0.172 | | | | |
| 1043 | 1.000 | 0.976 | 0.952 | 0.905 | 0.855 | 0.803 | 0.745 | 0.682 | 0.616 | 0.549 | 0.486 | 0.429 | 0.333 | 0.263 | 0.211 | 0.172 | | | | |
| 1068 | 1.000 | 0.976 | 0.952 | 0.905 | 0.856 | 0.803 | 0.746 | 0.684 | 0.619 | 0.554 | 0.491 | 0.434 | 0.338 | 0.267 | 0.215 | 0.175 | | | | |
| 1072 | 1.000 | 0.976 | 0.952 | 0.905 | 0.856 | 0.803 | 0.746 | 0.684 | 0.619 | 0.554 | 0.492 | 0.434 | 0.339 | 0.268 | 0.215 | 0.176 | | | | |
| 1132 | 1.000 | 0.976 | 0.952 | 0.905 | 0.857 | 0.805 | 0.750 | 0.690 | 0.627 | 0.563 | 0.502 | 0.446 | 0.350 | 0.278 | 0.224 | 0.183 | | | | |
| 1148 | 1.000 | 0.976 | 0.952 | 0.905 | 0.857 | 0.806 | 0.750 | 0.691 | 0.629 | 0.566 | 0.505 | 0.448 | 0.353 | 0.280 | 0.226 | 0.185 | | | | |
| 1150 | 1.000 | 0.976 | 0.952 | 0.905 | 0.857 | 0.806 | 0.751 | 0.691 | 0.629 | 0.566 | 0.505 | 0.449 | 0.353 | 0.281 | 0.226 | 0.186 | | | | |
| 1156 | 1.000 | 0.976 | 0.952 | 0.905 | 0.857 | 0.806 | 0.751 | 0.692 | 0.629 | 0.567 | 0.506 | 0.450 | 0.354 | 0.282 | 0.227 | 0.186 | | | | |
| 1217 | 1.000 | 0.976 | 0.952 | 0.905 | 0.858 | 0.807 | 0.754 | 0.696 | 0.636 | 0.575 | 0.515 | 0.460 | 0.365 | 0.291 | 0.236 | 0.194 | | | | |
| 1304 | 1.000 | 0.976 | 0.952 | 0.905 | 0.858 | 0.809 | 0.757 | 0.702 | 0.644 | 0.585 | 0.527 | 0.473 | 0.378 | 0.304 | 0.247 | 0.204 | | | | |
| 1415 | 1.000 | 0.976 | 0.952 | 0.906 | 0.859 | 0.811 | 0.761 | 0.708 | 0.652 | 0.596 | 0.541 | 0.488 | 0.394 | 0.319 | 0.261 | 0.216 | | | | |
| 1522 | 1.000 | 0.976 | 0.952 | 0.906 | 0.860 | 0.813 | 0.764 | 0.713 | 0.659 | 0.605 | 0.552 | 0.500 | 0.408 | 0.333 | 0.273 | 0.227 | | | | |

For single solid members part of a load-sharing systems with members spaced at not more than 610 mm centres (as in a stud wall):

$$\sigma_{c,g,par} \times K_{12} \times K_3 \times K_8 \text{ (equal to 1.1)}$$

For two or more solid timber members acting together in parallel (by mechanical fixings or adhesive) and part of a load-sharing system with members spaced at not more than 610 mm centres, a certain amount of engineering judgement is called for but one presumes that either $K_8$ or $K_{28}$ should be used (certainly not both).

For a glulam member, horizontally laminated:

$$\sigma_{c,g,par} \text{ (for SS grade of the species)} \times K_{17} \text{ (or } K_{23}) \times K_{12} \times K_3$$

(For certain size/stress grade combinations of North American timber, a $K_{10}$ factor applies. However, these combinations are not common in the UK, therefore no other mention of $K_{10}$ is made in this manual. They appear in Table 12 of BS 5268:Part 2.)

## 14.7   COMBINED BENDING AND AXIAL LOADING

### 14.7.1   Introduction

A column is often subject to bending either about one or about both axes, and the combined effect of the bending and axial loading must be considered.

The designer must realise that, in considering two or more maximum stresses, they coincide at only one plane or one point in the section, and the combined stresses do not occur to the same extent over the whole section. This is an important consideration, because the plane or point being considered may have different permissible stresses to other parts of the section. See Fig. 14.6 for examples of how actual stress combinations can occur.

### 14.7.2   Interaction formula

A compression member subjected also to bending about one axis should be so proportioned that the interaction quantity (for parallel to grain stresses):

$$\frac{\sigma_{m,a}}{\sigma_{m,adm}\left(1 - \frac{1.5\,\sigma_{c,a}}{\sigma_e}K_{12}\right)} + \frac{\sigma_{c,a}}{\sigma_{c,adm}} \leqslant 1$$

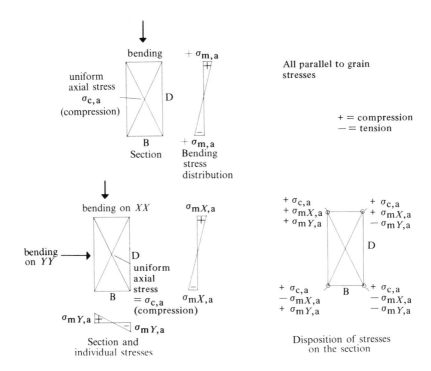

Fig. 14.6

where $\sigma_{m,a}$ = the applied flexural stress parallel to the grain
$\sigma_{m,adm}$ = the permissible flexural stress parallel to the grain
$\sigma_{c,a}$ = the applied compressive stress
$\sigma_{c,adm}$ = the permissible compressive stress
$\sigma_e$ = the Euler critical stress = $\dfrac{\pi^2 E}{(L_e/i)^2}$

utilising the appropriate minimum or modified minimum value of $E$ (i.e. $E_{min}$, $E_{min} \times K_9$ or $K_{28}$, or $E_{mean} \times K_{20}$ or $K_{26}$).

(Note that the 0.9 limitation for longer columns which applied in CP 112: Part 2 has been deleted from BS 5268:Part 2.)

Both fractions are influenced by slenderness, the compression part by buckling under axial loading and the bending part by lateral buckling due to flexure. As a column begins to buckle under axial loading a bending moment occurs related to the axial loading and the eccentricity of buckling. This is considered in determining the eccentricity factor $\eta$ (see formula in § 14.4). To take account of this, Clause 15 of BS 5268:Part 2 requires a compression member to be straight as erected (e.g. not more than about 0.003 of the length out of line).

The permissible bending stress is determined from a consideration of the degree of lateral restraint to the compression edge of the member acting as a beam. If the depth-to-breadth ratio of the member and the degree of restraint are within the limits of Table 17 of BS 5268: Part 2 (§ 4.6.1) the full grade bending stress may be used. If, however, the degree of restraint is less than that required by Table 17 it is necessary for the designer to see whether the reduced restraint will lead to a lower permissible stress:

$$= \frac{M_{\text{buc}}}{Z}$$

where $M_{\text{buc}}$ is determined in accordance with a method outlined in Chapter 9
　　$Z$ is the section modulus.

The designer may find it advantageous to express the interaction formula in terms of applied load and moment and sectional capacities. The interaction formula then takes the form:

$$\frac{M}{\overline{M} K_{\text{e}}} + \frac{F}{\overline{F}} \leqslant 1.0$$

where $M =$ the applied moment
　　$\overline{M} =$ the moment capacity
　　$F =$ the applied compressive load
　　$\overline{F} =$ the compressive capacity for the slenderness ratio
　　$K_{\text{e}} =$ Euler coefficient $= 1 - \dfrac{1.5\, \sigma_{\text{c,a}}\, K_{12}}{\sigma_{\text{e}}}$.

If a compression member has bending about both the $XX$ and $YY$ axes, the section should be so proportioned that (for parallel to grain stresses):

$$\frac{\sigma_{mX,a}}{\sigma_{mX,\text{adm}}\left(1 + \dfrac{1.5\,\sigma_{\text{c,a}} \times K_{12}}{\sigma_{\text{e}}}\right)} + \frac{\sigma_{mY,a}}{\sigma_{mY,\text{adm}}\left(1 + \dfrac{1.5\,\sigma_{\text{c,a}} \times K_{12}}{\sigma_{\text{e}}}\right)} + \frac{\sigma_{\text{c,a}}}{\sigma_{\text{c,adm}}} \leqslant 1.0$$

where $\sigma_{\text{e}} = \pi^2 E/\lambda_{\text{max}}^2$.

The assumption in writing the formula above with plus signs is that there is a point in the section where all three compressive stresses can occur simultaneously. If not, algebraic addition of compression and bending stress ratios applies and the designer is advised to check that the chosen point is where the combination of stress ratios is maximum. This need not necessarily be where the value of one of the applied stresses appears to dominate the case. In the majority of cases with columns of solid timber, the values of $\sigma_{mX,\text{adm}}$ and $\sigma_{mY,\text{adm}}$ will be the same but, in the case of glulam they are likely to differ (see Chapter 7).

## 14.8   EFFECTIVE AREA FOR COMPRESSION

For the purpose of calculating the actual compressive stress, open holes and notches must be deducted from the area but, if a bolt is inserted into a hole with only nominal clearance, no deduction is required.

Clause 15.6 of BS 5268:Part 2 requires the designer to take account of notches and holes but permits holes with a diameter not exceeding 0.25 × width of member to be positioned on the neutral axis within a distance of 0.25 and 0.4 of the length from the end or a support, without further calculation.

## 14.9   DEFLECTION AND SWAY OF COLUMNS

### 14.9.1   Deflection

When a column is subject to lateral loading over the whole or part of its height, the column deflects laterally. Little guidance is given in any code as to the limit of this deflection, and in the absence of any other ruling a limit of 0.003 × height is suggested. Where the lateral loading is by wind, it should be realised that this is a gust loading with the probability that the loading will be exceeded only once in 50 years. Composite action with claddings and partial end fixity will probably reduce the deflection somewhat, and a column will not deflect to its full calculated amounts during the short period of a gust. Therefore the deflection figure should be regarded as a guide rather than an absolute limit. The testing section of BS 5268:Part 2 requires that a beam should not deflect more than 0.8 times the calculated deflection in the first 24 hours of loading (Clause 62.1 of BS 5268:Part 2). Therefore, in calculating deflection due to wind loading it seems reasonable for a designer to introduce a factor of 0.8 in calculating the actual deflections (see Fig. 15.3).

### 14.9.2   Sway

When a building or frame can sway laterally, a limit should be set on the sway. Once more there is little guidance in codes, and a guide limit of 0.003 × height is suggested (Fig. 14.7).

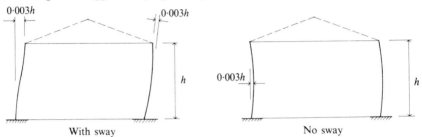

With sway                        No sway

Fig. 14.7

## 14.10 BEARING AT BASES

The permissible grade bearing stress at the base and top of columns is the value of $\sigma_{c,par}$ for $\lambda = 0$. However, if the column bears on a cross piece of timber (Fig. 14.8) the permissible grade bearing stress on the bearing area is $\sigma_{c,g,tra}$. If wane is excluded, the stress relevant to the full area can be taken (§ 4.10.1). The $K_4$ factor is also relevant (§ 4.10.1).

The detail at a column base often takes the form of a steel shoe bearing on concrete (Fig. 14.9). This type of base plate can be designed as a hinge or to give fixity.

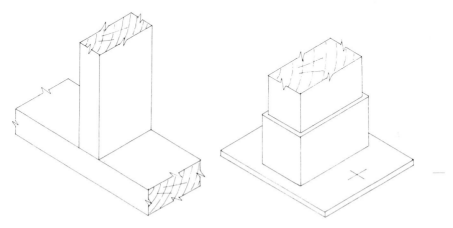

Fig. 14.8                         Fig. 14.9

If, for any reason (e.g. tolerance of manufacture), it is felt the fit of the timber to the steel plate or shoe will not be sufficiently accurate, then an epoxy resin–sand mix can be used to obtain tight fit or bearing. Resin glue has been used instead of epoxy resin.

If, for example in a humid service condition, it is felt that a means should be provided to prevent moisture being trapped between the timber and the steel shoe, one method is to make the inside of the steel shoe slightly over-size, and introduce an epoxy resin–sand mix as sketched in Fig. 14.10.

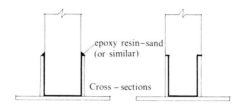

Fig. 14.10

Alternatively one can assume that moisture will intrude and allow it to drain or be ventilated away (Fig. 14.11). With this detail, the bearing area of timber to steel is considerably reduced, but usually sufficient area can be provided.

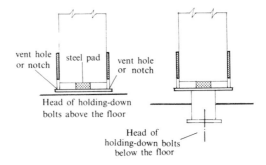

Fig. 14.11

## 14.11   BEARING AT AN ANGLE TO GRAIN

With the bearing surface inclined to the grain, the permissible compressive stress for the inclined surface is given by:

$$\sigma_{c,adm,\alpha} = \sigma_{c,adm,par} - (\sigma_{c,adm,par} - \sigma_{c,adm,tra}) \sin \alpha$$

where $\alpha$ = the angle between the load and the direction of grain.

Load-duration and load-sharing factors apply.

Where wane is excluded the value for $\sigma_{c,adm,tra}$ for solid timber and vertically glued laminated members can be based on 1.33 times the SS value. For horizontally laminated members the factors $K_{18}$ and $K_{24}$ have already been modified for the absence of wane. $\sigma_{c,adm,par}$ is not increased if there is no wane (§ 4.10.1).

(Also, see the notes in § 4.4.1 about inconsistency between $K_8$, $K_{27}$ and $K_{28}$ where two or more members act together, and where bearing is at an angle.)

# Chapter Fifteen
# Columns of Solid Timber

## 15.1   INTRODUCTION

The method of explaining the actual design of columns of solid timber in this manual is to give a typical design example. Chapter 14 details the various factors and aspects which must be taken into account and gives values of $K_{12}$ for $L_e/i$ (and $L_e/b$). Values of the grade compressive parallel to grain stress for the most common stress grade/species combinations and Strength Classes are given in Tables 2.2–2.5. These and the remainder are given in Tables 8–11 of BS 5268 : Part 2.

## 15.2   DESIGN EXAMPLE

The column sketched in Fig. 15.1 is supporting a medium term axial compression of 6 kN and a very short duration bending moment of 2.05 kN.m caused by a wind load of 4.00 kN on the $XX$ axis. The wind can either cause pressure or suction on the wall. Check that the chosen section is adequate. The top and bottom of the column are restrained in position but not fixed in direction. The column is restrained on its weak axis by rails at 1.4 m centres. There is no load sharing. Assume a 60 X 194 mm section of SS grade Whitewood.

*Medium term loading (no wind)*
For axial loading:

$$\lambda_X = \frac{4200 \times \sqrt{12}}{194} = 75 \qquad \text{(see § 14.4)}$$

$$\lambda_Y = \frac{1400 \times \sqrt{12}}{60} = 81$$

$$\frac{E}{\sigma_c} \text{ for derivation of } K_{12} = \frac{E_{min}}{\sigma_{c,g} \times K_3} = \frac{7000}{7.9 \times 1.25} = 709$$
(see § 14.4)

∴ From Table 14.2, with $\lambda = 81$, $K_{12} = 0.462$.

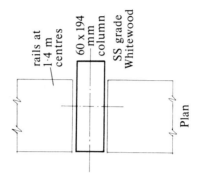

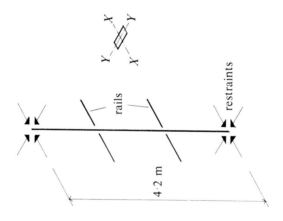

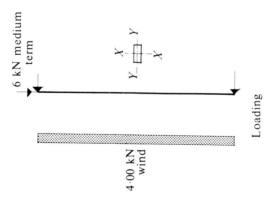

Fig. 15.1

$$\therefore \sigma_{c,adm} = \sigma_{c,g} \times K_{12} \times K_3 \quad \text{(see § 14.6)}$$
$$= 7.9 \times 0.462 \times 1.25 = 4.56 \, \text{N/mm}^2$$

$$\sigma_{c,a} = \frac{6000}{60 \times 194} = 0.52 \, \text{N/mm}^2 \quad \text{Acceptable}$$

*Very short term loading*

$$\frac{E}{\sigma_c} \text{ for derivation of } K_{12} = \frac{E_{min}}{\sigma_{c,g} \times K_3} = \frac{7000}{7.9 \times 1.75} = 506$$

$\therefore$ From Table 14.2, with $\lambda = 81$, $K_{12} = 0.381$.

$$\therefore \sigma_{c,adm} = \sigma_{c,g} \times K_{12} \times K_3$$
$$= 7.9 \times 0.381 \times 1.75 = 5.27 \, \text{N/mm}^2$$

$$\sigma_{c,a} = \frac{6000}{60 \times 194} = 0.52 \, \text{N/mm}^2$$

For bending about the *XX* axis, the column must be restrained about the *YY* axis.

$$\frac{h}{b} = \frac{194}{60} = 3.2$$

With the member held in line by rails at 1.4 m centres, the depth-to-breadth ratio may be as high as four to permit use of the full grade bending stress.

$$\sigma_{m,adm} = \sigma_{m,g} \times K_3 \times K_7$$
$$= 7.5 \times 1.75 \times 1.049 = 13.77 \, \text{N/mm}^2$$

$$\sigma_{m,a} = \frac{2.05 \times 10^6}{0.376 \times 10^6} = 5.45 \, \text{N/mm}^2$$

Euler critical stress $= \sigma_e = \pi^2 E/\lambda^2 = \pi^2 \, 7000/81^2 = 10.5 \, \text{N/mm}^2$.
Interaction formula:

$$\frac{5.45}{13.77 \left(1 - \dfrac{1.5 \times 0.52 \times 1.75}{10.5}\right)} + \frac{0.52}{5.27} = 0.55 < 1.0$$

which is satisfactory.

Check the deflection on the *XX* axis. With $E_{min} = 7000 \, \text{N/mm}^2$, $EI = 256 \, \text{kN.m}^2$:

$$\text{Bending deflection} = \frac{5 \times 4 \times 4.2^3}{384 \times 256} = 0.015 \, \text{m}$$

With $L/h = 4200/194 = 21.6$, from Fig. 4.26:

Total deflection (including shear deflection) $= 1.033 \times 0.015 = 0.0155 \, \text{m}$

If the designer considers the use of a 0.8 factor to be justified (§§ 14.9.1 and 4.5) the calculated deflection reduces to 0.0155 × 0.8 = 0.0124 m (= 0.003 × span).

Actual bearing stress at the base (Fig. 15.2) for a 60 × 194 mm section:

$$= \frac{6000}{60 \times 194} = 0.515 \text{ N/mm}^2$$

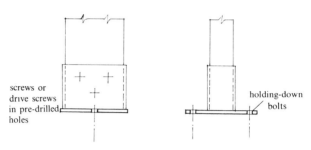

screws or drive screws in pre-drilled holes

holding-down bolts

Fig. 15.2

The permissible medium term bearing stress (parallel to grain):

$$= 7.9 \times 1.25 = 9.875 \text{ N/mm}^2$$

from which it can be seen that bearing with this type of detail is not likely to be critical on any column, particularly as the slenderness ratio increases, unless the end is severely reduced to suit a fixing detail.

Although, in calculating $\lambda_X$ the effective length was taken as 1.0 times the actual length, the actual end connections (even with a stud wall; particularly taking account of sheathing and lining) can usually justify $L_{eX}$ being taken as 0.85L. Certainly, the connection shown in Fig. 15.2 would more than justify a factor of 0.85 being taken.

This partial fixity must have an effect on the deflected form and hence the deflection of the column. By setting the distance between points of contraflexure as 0.85L the effect on bending deflection can be calculated (see Fig. 15.3) by working back from the BM diagram.

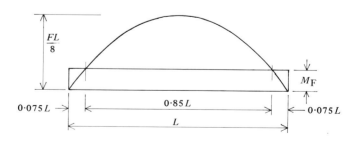

$\frac{FL}{8}$

$M_F$

0·075L      0·85L      0·075L

L

Fig. 15.3

From the properties of a parabola, the fixing moment $M_F$:

$$= \frac{FL}{8} \times \frac{0.075L \times 0.925L}{0.5L \times 0.5L} = 0.0347FL$$

where $F$ = the total UDL.

By the area moment method, the central bending deflection may be calculated as:

$$EI\, \delta_m = \frac{5FL^3}{384} - 0.0347FL \times \frac{L}{2} \times \frac{L}{4} = (0.013 - 0.0043)FL^3$$

$$= 0.0087FL^3$$

which is $\frac{2}{3}$ of the deflection for the simply supported case.

Therefore, if one takes account of this small amount of partial directional restraint and the 0.8 times factor, the actual bending deflection can be shown to be 66.6% × 0.8 = 53% of the 'full' calculated value. (This percentage actually agrees quite well with 'gust tests' on timber stud walls. With a stud wall $E_{mean}$ would be used in calculations.)

# Chapter Sixteen
# Columns of Simple Composites

## 16.1 INTRODUCTION

Simple composites are defined as components of more than one member which do not classify as glulam. The examples in this chapter are limited to components of two or three members.

Obviously it is possible to nail, screw or bolt two or more pieces of timber together to form a column and such a column will have considerably more strength than the sum of the strength of the two or more pieces acting alone. However, there is some doubt as to whether or not such mechanically jointed composites have the full composite strength. This depends on the strength, spacing and characteristics of the fastenings and the method of applying the loading; therefore this chapter deals only with glued composites other than those covered in the section on spaced columns. If a designer wishes to use a mechanically jointed composite (other than a spaced column for which design rules are detailed in BS 5268 and § 16.4) it is suggested that he or she reverts either to test or errs on the side of safety in calculating the permissible axial stresses. (See notes on composite action in § 6.2.) Typical glued composites of two or three members are sketched in Fig. 16.1.

When designing the columns shown in Fig. 16.1 see the notes in § 14.4 for guidance on the $E$ value to use in calculations, and § 14.6 for the modification factors which apply.

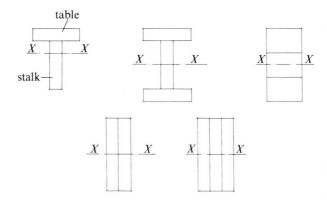

Fig. 16.1

## 16.2 COMBINED BENDING AND AXIAL LOADING. TEE SECTIONS

The method of combining bending and axial compression stresses is described in § 14.7 but there is one additional point to note with Tee sections. The section is not symmetrical about the $XX$ axis, and therefore bending about the $XX$ axis leads to a different stress in each of the extreme fibres, the maximum stress being at the end of the stalk. This maximum may be a tensile bending stress, in which case the critical compression case in the section will be the combination of axial compressive stress plus the bending compressive stress on the extreme fibre of the table. The axial compressive stress is uniform throughout the section.

The shape of a Tee section is such that with compressive bending stress on the table, the full permissible bending stress will be realised before any lateral instability takes place. For cases with the stalk of the Tee in compressive bending, a conservative estimate for lateral stability is to ensure that the proportion of the stalk member on its own complies with the requirements of § 4.6.1.

## 16.3 DESIGN EXAMPLE. TEE SECTION

A $_{47}T_{97}$ (Fig. 6.9) carries a long term UDL of 2.4 kN and a long term axial load of 12 kN on a span of 3 m as shown in Fig. 16.2. For axial loading, the

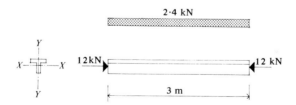

Fig. 16.2

effective length about the $YY$ axis is $0.75L$, and about the $XX$ axis is $1.0L$. Check the suitability of the section with SS grade Whitewood.

Data for the section from Fig. 6.9 is as follows:

$$A = 9.12 \times 10^3 \, \text{mm}^2$$
$$I_X = 16.2 \times 10^6 \, \text{mm}^4$$
$$Z_X \text{ min (stalk)} = 0.192 \times 10^6 \, \text{mm}^3$$
$$Z_X \text{ max (table)} = 0.273 \times 10^6 \, \text{mm}^3$$
$$i_X = 42.2 \, \text{mm}$$
$$i_Y = 22.0 \, \text{mm}$$

For the reason discussed in the example worked out in § 6.6.1 take $E_2$ in calculations. For SS Whitewood:

$$E_2 = 7000 \times (K_9 = 1.14) = 7980 \, \text{N/mm}^2$$

Do not increase permissible stress by $K_8$ (or $K_{27}$ or $K_{28}$) because only one member occurs at the extreme fibres which are in bending (or compression).

$$EI_X = 129 \text{ kN.m}^2$$

$$\text{Long term } M = \frac{2.4 \times 3}{8} = 0.90 \text{ kN.m}$$

$$\sigma_{m,a} \text{ (compression in table)} = \frac{0.90 \times 10^6}{0.273 \times 10^6} = 3.30 \text{ N/mm}^2$$

The shape of the section permits it to be considered as fully laterally restrained against bending on the $XX$ axis.

On balance the authors are inclined to take $K_7$ as relevant to the full depth of 144 mm, at which depth it equals 1.084, therefore long term permissible bending stress:

$$\sigma_{m,adm} = 7.5 \times 1.084 = 8.13 \text{ N/mm}^2$$

$$\sigma_{c,a} = \frac{12\,000}{9.12 \times 10^3} = 1.32 \text{ N/mm}^2$$

$$\lambda_X = \frac{3000 \times 1.0}{42.2} = 71$$

$$\lambda_Y = \frac{3000 \times 0.75}{22.0} = 102$$

$E/\sigma_c$ for the purpose of deriving $K_{12}$ for long term loading (§ 14.4)

$$= \frac{E_{min} \times K_9}{\sigma_c \times K_3} = \frac{7000 \times 1.14}{7.9 \times 1.00} = 1010$$

∴ From Table 14.2, with $\lambda = 102$, $K_{12} = 0.412$.

$$\therefore \sigma_{c,adm} = \sigma_{c,g} \times K_{12} \times K_3 \qquad (\text{§ 14.6})$$
$$= 7.9 \times 0.412 \times 1.0 = 3.25 \text{ N/mm}^2$$

Euler critical stress $= \sigma_e = \pi^2 7980/102^2 = 7.57 \text{ N/mm}^2$.

Interaction formula:

$$= \frac{3.30}{8.13 \left(1 - \frac{1.5 \times 1.32 \times 1.0}{7.57}\right)} + \frac{1.32}{3.25} = 0.96 < 1.0$$

which is satisfactory.

Bending deflection on the $XX$ axis:

$$\delta_m = \frac{5 \times 2.4 \times 3^3}{384 \times 129} = 0.0065 \text{ m}$$

$G = E/16 = 499 \text{ N/mm}^2$, $AG = 4550 \text{ kN}$ and $K_{\text{form}} = 2.0$ (§ 6.6.1).

$$\text{Shear deflection} = \frac{2M_0}{AG} = \frac{2 \times 0.9}{4550} = 0.0004 \text{ m}$$

Total deflection = 0.0069 = 0.0023 of span, which is satisfactory.

## 16.4   SPACED COLUMNS

Spaced columns are defined as two or more equal rectangular shafts spaced apart by end and intermediate blocking pieces suitably glued, bolted, screwed or otherwise adequately connected together. Figure 16.3 shows a typical assembly of a spaced column.

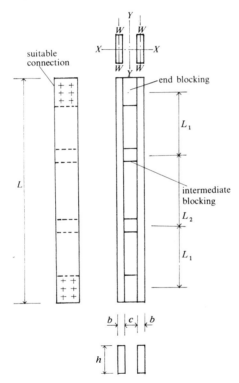

Fig. 16.3

$L$ is the overall length of the composite unit,

$L_1$ is the distance between the centroids of end and closest intermediate spacer,

$L_2$ is the distance between the centroids of intermediate spacers.

Under an axial compressive load it is possible for the unit to buckle about three axes, these being the $WW$, $XX$ and $YY$ axes shown in Fig. 16.3, $XX$ and

*YY* being related to the composite and *WW* to the individual sections. The lower axial capacity of the section determined by consideration of these three axes will govern the design.

The effective length about the *WW* axis is equal to the average centre-to-centre spacing of the blocking pieces. The axial loading capacity (of the total composite) is calculated as for a rectangular column (Chapter 15) whose section is that of one shaft multiplied by the number of shafts. The effective length about the *XX* axis is assessed in accordance with Chapter 15, $i_X$ being the same as that for the individual members. The axial loading capacity is that for a solid column with an area equal to the total area of the shafts. The effective length about the *YY* axis is first assessed in accordance with Chapter 15, but is then multiplied by a modification factor $K_{13}$ which provides for the method of connecting the blocking pieces to the shafts (see Table 16.1) and the ratio $c/b$ of spacing between shafts to shaft thickness (see Fig. 16.3).

**Table 16.1    Modification factor $K_{13}$**

| Method of connection | Ratio $c/b$ | | | |
|---|---|---|---|---|
| | 0 | 1 | 2 | 3 |
| Nailed | 1.8 | 2.6 | 3.1 | 3.5 |
| Screwed or bolted | 1.7 | 2.4 | 2.8 | 3.1 |
| Attached by connector | 1.4 | 1.8 | 2.2 | 2.4 |
| Glued | 1.1 | 1.1 | 1.3 | 1.4 |

The radius of gyration $i_Y$ is that for the cross-section of the built-up column. In the general case of two shafts:

$$i_Y = \frac{1}{\sqrt{12}} \left( 2b + \frac{5c}{3} \right)$$

In addition to these design requirements, BS 5268 sets further limitations on the geometry of the composite unit and on the method of fixing the spacer blocks to the shafts.

End blocks must be not less than 6*b* in length and suitably connected or glued to the shaft to transfer a shear force between the abutting faces of the packing and each adjacent shaft. The shear force acting at the abutting face of the blocking and one adjacent shaft is to be taken as:

$$\frac{1.3Ah\sigma_{c,a,par}}{Na}$$

where $A$ = the total cross-sectional area of the column
$h$ = the width (depth) of each shaft (see Fig. 16.3)
$N$ = the number of shafts
$a$ = the distance between centres of adjacent shafts (= $c + b$ in Fig. 16.3).

Although not made clear in BS 5268:Part 2, this shear acts parallel to the shafts.

The clear spacing between shafts should be not greater than $3b$.

The requirements for end blocking pieces do not apply to intermediate blocking pieces, which should be at least 230 mm long, and should be designed to transmit, between the abutting face of the blocking and one adjacent shaft, a shear force of half the corresponding shear force for the end packing.

If using glued packings, screws or bolts may be relied on to give adequate glue-line pressure. In this case there should be at least four screws or bolts per packing, spaced to provide uniform pressure over the area of the packing.

BS 5268 sets slenderness limitations on the individual shaft members. Where the length of the column does not exceed thirty times the thickness of the shaft, only one intermediate blocking need be provided. However, sufficient blockings should be provided to ensure that the greatest slenderness ratio $\lambda$ of the local portion of the individual shaft between packings is limited to seventy or 0.7 times the slenderness ratio of the whole column, whichever is the lesser. For the purpose of calculating the slenderness ratio of the local portion of an individual shaft, the effective length should be taken as the length between centroids of the groups of mechanical connectors or glue areas in adjacent packings.

With so many conditions to fill simultaneously, spaced-column design is one of trial and error. A design check on a spaced-column configuration for a given loading is detailed in § 16.5.

## 16.5   EXAMPLE OF SPACED-COLUMN DESIGN

Check that the spaced column shown in Fig. 16.4 complies with the design requirements of BS 5268. The column consists of two 35 × 120 mm White-wood SS grade members spaced 70 mm apart. All joints are glued.

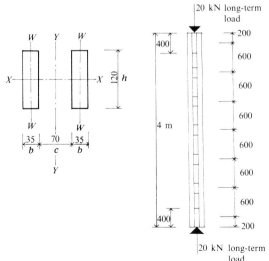

Fig. 16.4

Establish the maximum slenderness ratio about the $WW$, $XX$ and $YY$ axes with $c = 70$ and $b = 35$.

$c/b = 2$ which is less than the maximum permitted figure of 3.

$$\lambda_X = \frac{4000 \sqrt{12}}{120} = 115$$

For $c/b = 2$, $K_{13} = 1.3$ (Table 16.1).

$$\lambda_Y = \frac{L_e K_{13} \sqrt{12}}{\left(2b + \dfrac{5c}{3}\right)} = \frac{4000 \times 1.3 \sqrt{12}}{186.7} = 96.5$$

$\lambda_W$ must not exceed seventy or $0.7 \times$ the greater slenderness ratio of the whole column $= 0.7 \times 115 = 80.5$.

$$\lambda_W = \frac{600 \sqrt{12}}{35} = 59.4$$

which is satisfactory.

Although two members are connected together the authors are inclined to take an unmodified value of $E_{min}$ in determining the value of $K_{12}$ (see § 14.4), and $\sigma_c$ modified only for duration of loading.

$$\frac{E}{\sigma_c} = \frac{7000}{7.9 \times 1.0} = 886$$

∴ From Table 14.2, $K_{12} = 0.322$.

The authors are also inclined to disregard any increase in permissible stress due to load sharing (i.e. from $K_{28}$ factor).

$$\therefore \ \sigma_{c,adm,par} = 7.9 \times 1.0 \times 0.322 = 2.54 \, \text{N/mm}^2$$

$$\sigma_{c,a,par} = \frac{20\,000}{2 \times 35 \times 120} = 2.38 \, \text{N/mm}^2$$

which is acceptable.

Check compliance with geometrical limits and spacer requirements.

Permissible long term glue-line stress = permissible long term shear stress parallel to grain = $0.82 \, \text{N/mm}^2$.

Shear force acting on one face of an end blocking (§ 16.4)

$$= \frac{1.3 \times 2 \times 35 \times 120 \times 120 \times 2.38}{2 \times 105} = 14\,850 \, \text{N}$$

With 400 mm long end packing, glue-line stress on end pack parallel to grain

$$= \frac{14\,850}{120 \times 400} = 0.31 \, \text{N/mm}^2$$

which is acceptable.

Minimum length required for end pack = $6b$ = 6 × 35 = 210 mm, therefore 400 mm as sketched is adequate.

Minimum length of intermediate packs is 230 mm, therefore 250 mm as sketched is adequate. By simple mathematics, half the shear on the end packing must give a lower stress on the intermediate packings which are longer than half the end packings.

## 16.6   COMPRESSION MEMBERS IN TRIANGULATED FRAMEWORKS

### 16.6.1   Single compression members

Single compression members in triangulated frameworks should be designed using the same principles outlined in Chapter 15 for columns of solid timber.

For a continuous compression member the effective length may be taken as between 0.85 and 1.0 (according to the degree of end fixity) times the distance between node points for buckling in the plane of the framework, and times the actual distance between effective restraints for buckling perpendicular to the plane of the framework. Clause 15.10 of BS 5268: Part 2 defines some of the more usual types of restraint which may be considered as effective.

For a non-continuous compression member such as the internal web members of a truss, the effective length will depend upon the degree of end fixity. Most commonly the web members are assumed to be pin jointed at each end as they are frequently jointed at each end by a single connector unit. The effective length in this case is taken as the actual distance between the connector units at each end of the member. In the special case of web members held at each end by glued gusset plates (usually plywood) it may be assumed that a degree of fixity is developed and the effective length both in and out of the plane of the truss may be taken as 0.9 times the distance between the centre lines of the members connected.

### 16.6.2   Spaced compression members

Compression members in trusses frequently take the form of a twin member and the question arises as to when such an arrangement should be designed as a spaced column.

Consider a twin compression member which is restrained at the node points only. This would be designed as a spaced column adopting the principles given in § 16.4 except that the recommendations regarding the design of end packs will not apply (Clause 15.10(c) of BS 5268:Part 2).

The effective length for the calculation of $\lambda_X$ may be taken as between 0.85 and 1.0 (according to the degree of end fixity) times the distance between node points and for $\lambda_Y$ the effective length should be taken as the actual length between node points times $K_{13}$ (see Table 16.1) depending on the method of connection of intermediate packs. At least one intermediate pack must be provided in order to comply with Clause 15.8.2 of BS 5268: Part 2. The effective length for $\lambda_W$ will be the distance between centres of packs or between centres of packs and node points. $\lambda_W$ should not exceed 70 or $0.7\lambda_X$ or $0.7\lambda_Y$ whichever is the lesser. Although BS 5268 clearly indicates that end packings and adjacent shafts need not be designed for a transfer of shear force it appears that intermediate packs must be designed to transmit not less than half of the shear force calculated for the end packings. From § 16.4 the shear force would be:

$$F_v = \frac{0.65Fh}{Na}$$

where $F$ = total axial compression load in shafts
$\quad h$ = width of shaft
$\quad N$ = number of shafts
$\quad a$ = distance between centres of adjacent shafts
$\quad\quad = b + c$ (see Fig. 16.3).

This will frequently be found to be a very large force not easily developed by nails or screws, so that packings will invariably require to be bolted, connectored or glued.

If, instead of being restrained only at node points, there are further lateral restraints to the $YY$ and $WW$ axes from joisting etc. at intermediate points, then:

(a) $\lambda_X$ is calculated as above.
(b) for $\lambda_W$ the effective length is taken as the lesser of:
   (i)   distance between effective lateral restraints
   (ii)  the centres of intermediate packings, or
   (iii) the centres between a packing and node point.
(c) for $\lambda_Y$ the effective length is taken as the distance between effective restraints modified by $K_{13}$. (Consequently, with close spacings of lateral restraints, for example such as joisting at 600 mm centres, $\lambda_Y$ will be less than $\lambda_W$ and intermediate packings are unnecessary. In such a situation, in the authors' opinion, it would not be necessary to design the member as a spaced column.)

## Chapter Seventeen
# Glulam Columns

## 17.1   INTRODUCTION

A glulam section is one manufactured by gluing together at least four laminations with their grain essentially parallel. The notes in the sections listed below in Chapter 7, 'Glulam Beams', apply equally to columns.

§ 7.1   Introduction
§ 7.2   Timber Stress Grades for Glulam
§ 7.4   Appearance Grades for Glulam Members
§ 7.5   Joints in Laminations
§ 7.6   Choice of Glue for Glulam
§ 7.7   Preservative Treatment

## 17.2   TIMBER STRESS GRADES FOR GLULAM COLUMNS

BS 4978 describes three stress grades for softwood for particular use with horizontally laminated glulam. They are denoted LA, LB and LC. For vertically laminated glued members the stress grades for solid timber (e.g. GS, SS etc. to BS 4978) are those to be used in design and manufacture. For horizontally and vertically laminated members of hardwood, the HS grade from BS 5756 is to be used in design and manufacture. Only softwood is covered in this manual.

Grade LA is the strongest of the grades, but LC is not a particularly weak grade. It is possible to manufacture a glulam section using the same grade throughout or to mix the grades LA with LB, or LB with LC, as indicated in Fig. 17.1. No allowance is made in BS 5268 for a mixture of LA and LC.

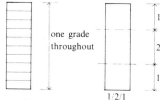

The higher grade must occupy at least the outer zones
Fig. 17.1          (i.e. 25% each of depth)

LB and LC grades are the more common in Redwood or Whitewood. The slope of grain restriction of LA is rather exacting and can lead to a high reject rate. Although the efficiency of finger joints in compression is high (see Table 19.2) it is still not easy to finger joint LA grade without sacrificing efficiency, and this is particularly so if the column is also taking bending (see § 7.5).

## 17.3   JOINTS IN LAMINATIONS

With structural glulam the end jointing of individual laminates is carried out almost certainly by finger jointing or scarf joints. If a butt joint is used, then the laminate in which it occurs has to be disregarded in the stress calculations, and this will usually make the member uneconomical.

In a column taking pure compression or a combination of compression and bending, either the joint must be strong enough to develop the full design strength of the laminations or the permissible strength must be reduced accordingly. Joint efficiencies and design are covered in § 19.7 and in § 7.5. Rather than repeat the notes here the designer is referred to these sections. Also, the worked example in § 17.7 includes a check on a finger joint in a member taking combined compression and bending.

## 17.4   PERMISSIBLE COMPRESSION STRESS FOR HORIZONTALLY LAMINATED GLULAM COLUMNS

The permissible compressive stress for a horizontally laminated glulam column is the product of the grade stress for SS grade of the same species as the laminations, modification factor $K_{17}$ or $K_{23}$ (for single grade or combined grade—see Tables 7.1 or 7.2), modification factor $K_{12}$ (see §§ 14.4 and 14.6, and Table 14.2), and load-duration factor $K_3$. (The designer will note that the values of $K_{17}$ and $K_{23}$ do not vary with the number of laminations, as was the case with factors $K_3$ and $K_8$ of CP 112:Part 2:1971.)

## 17.5   STANDARD COLUMN SIZES

Standard glulam sizes are discussed briefly in § 7.8. Those beam sizes listed in Table 7.7 with an $h/b$ ratio not exceeding four are suggested by the authors for use as columns. They are (in millimetres):

| | | | | |
|---|---|---|---|---|
| 65 × 180 | 65 × 225 | | | |
| 90 × 180 | 90 × 225 | 90 × 270 | 90 × 315 | 90 × 360 |
| 115 × 180 | 115 × 225 | 115 × 270 | 115 × 315 | 115 × 360 |
| 115 × 405 | 115 × 450 | | | |

| | | | | |
|---|---|---|---|---|
| 135 X 180 | 135 X 225 | 135 X 270 | 135 X 315 | 135 X 360 |
| 135 X 405 | 135 X 450 | 135 X 495 | 135 X 540 | |
| 160 X 180 | 160 X 225 | 160 X 270 | 160 X 315 | 160 X 360 |
| 160 X 405 | 160 X 450 | 160 X 495 | 160 X 540 | 160 X 585 |
| 160 X 630 | | | | |
| 185 X 180 | 185 X 225 | 185 X 270 | 185 X 315 | 185 X 360 |
| 185 X 405 | 185 X 450 | 185 X 495 | 185 X 540 | 185 X 585 |
| 185 X 630 | 185 X 675 | 185 X 720 | | |

## 17.6 DESIGN EXAMPLE OF COMBINED BENDING AND COM-PRESSION IN A GLULAM SECTION

Determine a suitable rafter section for the three-pin frame illustrated in Fig. 17.2, using LB grade Whitewood glulam. The frames are at 4 m centres,

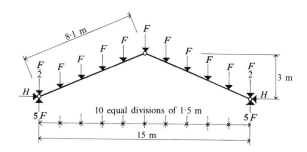

Fig. 17.2

supporting purlins (which may be assumed to restrain the frames laterally) at 1.5 m centres on true plan.

$$F_L \text{ (long term point load)} = 8 \text{ kN (each)}$$
$$F_M \text{ (medium term point load)} = 12.5 \text{ kN total (each)}$$
$$\theta = 21.8°$$
$$\text{Purlin spacing up the slope} = 1.62 \text{ m}$$

To determine the moments and forces, one must analyse the loaded frame in two parts, considering first local bending on the rafter, and secondly overall loading to calculate axial loading in the rafter (Fig. 17.3).

Consider the medium term loading condition:

The maximum rafter moment $M_M = 2F_M \times 3 - F_M \times 1.5 = 56.25 \text{ kN.m}$

The axial force in the rafter is $P = \dfrac{2.5F_M}{\sin \theta} = \dfrac{2.5 \times 12.5}{0.3714}$

$$= 84.1 \text{ kN}$$

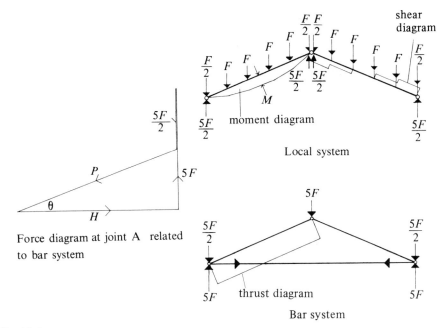

Fig. 17.3

The horizontal force is:
$$H = \frac{2.5F_M}{\tan\theta} = \frac{2.5 \times 12.5}{0.4} = 78.1 \text{ kN}$$

Effective lengths for the rafter members are 8.1 m about the $XX$ axis and 1.62 m about the $YY$ axis.

A trial section must be assumed. Experience shows that a fair first trial section is one having a moment capacity about 1.4 (i.e. 100/70) times the actual bending moment. In this case, because the purlins offer lateral restraint, choose a section with an $h/b$ ratio not exceeding four if visually acceptable. This will permit the adoption of the full grade bending stress (see Table 4.2).

Try a 135 × 540 mm LB section (twelve laminations).

$$\sigma_{c,a} = \frac{84\,100}{135 \times 540} = 1.15 \text{ N/mm}^2$$

$$\sigma_{m,a} = \frac{56.25 \times 10^6}{6.56 \times 10^6} = 8.58 \text{ N/mm}^2$$

$$\begin{aligned}
\sigma_{m,adm} &= 7.5 \times K_{15} \times K_3 \times K_7 \qquad (K_7 \text{ from Table 4.4}) \\
&= 7.5 \times 1.45 \times 1.25 \times 0.893 = 12.1 \text{ N/mm}^2
\end{aligned}$$

$$\lambda_X = \frac{8100\sqrt{12}}{540} = 52$$

$$\lambda_Y = \frac{1620\sqrt{12}}{135} = 42$$

$E$ for deriving $K_{12}$ is $E_{mean}$ for SS grade Whitewood $\times K_{20}$ (see § 14.4)

$$= 10\ 500 \times 0.90 = 9450\ \text{N/mm}^2$$

$\sigma$ for deriving $K_{12}$ (medium term loading) is the SS grade for the species $\times K_{17} \times K_3$

$$= 7.5 \times 1.04 \times 1.25 = 9.75\ \text{N/mm}^2$$

$$\frac{E}{\sigma_c} = \frac{9450}{9.75} = 969$$

$\therefore K_{12}$ at $\lambda = 52$, is 0.726 (see Table 14.2).

$$\therefore \sigma_{c,adm} = 7.5 \times K_{17} \times K_{12} \times K_3 \quad \text{(see § 14.6)}$$
$$= 7.5 \times 1.04 \times 0.726 \times 1.25 = 7.08\ \text{N/mm}^2$$

$$\sigma_e = \frac{\pi^2\, 9450}{52^2} = 34.5\ \text{N/mm}^2$$

Interaction formula:

$$\frac{8.58}{12.1 \left(1 - \dfrac{1.5 \times 1.15 \times 1.25}{34.5}\right)} + \frac{1.15}{7.08} = 0.92 < 1.0$$

which is satisfactory.

The interaction formula value of 0.92 is based on the assumption that the axial compression occurs at the neutral axis of the section. If the joints are carefully detailed so as to provide an off-centre application of the compressive force (i.e. below the neutral axis as in Fig. 17.4), then a moment equal to $Pe$ is artificially introduced into the system to relieve the moment due to lateral forces on the member.

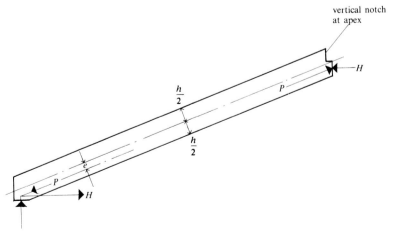

Fig. 17.4

A calculation for the required bearing area both vertically and horizontally and the consequent inclusion of adequate bearing plates determines the location of the intersection of the horizontal and vertical forces at the point of support. This leads to the determination of the eccentricity $e$ of the thrust $P$. The apex joint must then be accurately notched at the upper edge to ensure that the same relieving moment is introduced into the member.

In the example quoted, an eccentricity of approximately 135 mm could be accommodated, which would give a relieving moment of 11.3 kN.m and reduce the sum of the bending and compression ratios to 0.77. This in turn would justify checking whether a smaller section could be used.

In the design so far only medium term loading has been considered. A consideration of long term loading is also necessary.

$$F_L/F_M = 8/12.5 = 0.64$$

The long term moment $= M_L = 0.64 M_M = 0.64 \times 56.25 = 36 \text{ kN.m}$

The long term axial load $= P_L = 0.64 P_M = 0.64 \times 84.1 = 53.9 \text{ kN}$

$$\frac{E}{\sigma_c} = \frac{9450}{7.5 \times 1.04 \times 1.00} = 1212$$

$\therefore K_{12}$ at $\lambda = 52$, is 0.742.

$$\therefore \sigma_{c,\text{adm}} = 7.5 \times 1.04 \times 0.742 \times 1.00 = 5.79 \text{ N/mm}^2$$

$$\sigma_{c,a} = \frac{53\,900}{135 \times 540} = 0.74 \text{ N/mm}^2$$

$$\sigma_{m,a} = \frac{36 \times 10^6}{6.56 \times 10^6} = 5.49 \text{ N/mm}^2$$

$$\sigma_{m,\text{adm}} = 7.5 \times 1.45 \times 1.00 \times 0.89 = 9.68 \text{ N/mm}^2$$

$$\sigma_e = \frac{\pi^2 9450}{52^2} = 34.5 \text{ N/mm}^2$$

Interaction formula:

$$\frac{5.49}{9.68 \left(1 - \dfrac{1.5 \times 0.74 \times 1.0}{34.5}\right)} + \frac{0.74}{5.79} = 0.71 < 1.0$$

which is satisfactory.

## 17.7   CHECK ON STRENGTH OF A FINGER JOINT IN COMBINED BENDING AND COMPRESSION

As stated in § 7.5.1 there are two ways of checking in BS 5268 : Part 2 if a finger joint is acceptable to joint individual laminates. One is to provide a joint of a stated efficiency in bending (no matter how under-stressed the laminate may be, and even if it is stressed largely in compression), whilst the other is to design for the actual strength required for the actual loading.

Because LB laminates are used in the member being checked, reference to BS 5268: Part 2 and § 7.5.1 shows that a joint having 70% efficiency in bending may be used without any further design check. If, however, only the 12.5/3.0/0.5 joint (see Table 7.4) is available, which has an efficiency in bending of 65% and an efficiency in compression of 83%, it will be necessary to carry out a design check. One method is detailed below. (Note that it is not permitted to use a finger joint having an efficiency of less than 50% in bending.)

Check for the medium term loading.

The permissible bending stress in the member as limited by the joint being used is calculated as:

The SS grade stress for the species (7.5 N/mm²)
X the relevant load-duration factor $K_3$ (1.25 for medium term)
X the relevant moisture content factor $K_2$ (1.00 for dry)
X the modification factor for depth of *member*, $K_7$ (0.893 for 540 mm)*
X the ratio for efficiency of the joint in bending (0.65 in this case)
X factor $K_{30}$ (1.85)
(*See penultimate paragraph on page 180.)
= 7.50 X 1.25 X 1.00 X 0.893 X 0.65 X 1.85 = 10.07 N/mm²

The permissible compression stress in the member as limited by the joint being used is calculated as:

The SS grade stress for the species (7.9 N/mm²)
X the relevant load-duration factor $K_3$ (1.25 for medium term)
X the relevant moisture content factor $K_2$ (1.00 for dry)
X the ratio for efficiency of the joint in compression (0.83 in this case)
X factor $K_{32}$ (1.15)
= 7.90 X 1.25 X 1.00 X 0.83 X 1.15 = 9.43 N/mm²

BS 5268: Part 2 (Clause 20) is not specific in saying whether or not the strength of the joint in compression should be reduced (by $K_{12}$) to take account of the slenderness of the column. The authors are inclined to believe that no account need be taken of slenderness in establishing the compressive strength of the joint. Certainly it is not usual to do so with other joints in compression members (e.g. punched metal-plate fasteners if used to joint lengths in the rafter of a trussed rafter).

With medium term loading in the member being jointed:

$$\sigma_{c,a} = 1.15 \text{ N/mm}^2 \qquad \sigma_{m,a} = 8.58 \text{ N/mm}^2$$

Checking the combined effect of compression and bending on the joint:

$$\frac{1.15}{9.43} + \frac{8.58}{10.07} = 0.122 + 0.852$$
$$= 0.974 < 1.0$$

Therefore the joint with 65% efficiency in bending and 83% efficiency in compression is acceptable in this case.

# Chapter Eighteen

# Mechanical Joints

## 18.1 GENERAL

BS 5268 recommends basic loading for fasteners which fall into four categories related to the Stength Classes into which the species and stress grade combinations of the timber being used can be allocated. The level of these categories is indicated in Table 18.1.

Table 18.1

|  | Strength Classes | General description |
|---|---|---|
| (i) | SC1 and SC2 | Low stress grade softwoods |
| (ii) | SC3 and SC4 | The majority of softwoods |
| (iii) | SC5 | High stress grade softwoods and a few hardwoods |
| (iv) | SC6 to SC9 incl. | The majority of hardwoods |

The reader is referred to Tables 3–7 of BS 5268:Part 2 for the complete schedule of species/stress grade combinations and the Strength Classes which they satisfy.

The species most frequently used in timber engineering in the UK are European Whitewood and Redwood in GS and SS grades, and Spruce-Pine-Fir and Hem-Fir in GS, SS, No. 1 and No. 2 'Joist and Plank', and No. 1 and No. 2 'Structural Light Framing'. These fit into either SC3 or SC4 which, from Table 18.1, can be seen to be allocated the same fastener strengths. Even where the higher stress grades of these species are used (which would fall into SC5), BS 5268 requires the designer to use fastener loadings as tabulated for SC3 and SC4. Spruce-Pine-Fir and Hem-Fir of 'Construction' and 'Standard' Light Framing grades fall into SC2. (See Clause 38 of BS 5268:Part 2.)

It has long been acknowledged that stress concentrations occur when groups of fastenings are loaded eccentrically. Similarly, it is now generally accepted that stress concentrations occur even with concentric applications of load. Potter, reporting on work on nailed joints undertaken at Imperial College in 1969, showed that there is a distinct regression of apparent load per nail as the number of nails increases in line with the load. Similarly, from other sources there is evidence of a reduction in the apparent permissible load per bolt with an increasing number of bolts in line with the load. There is therefore ample evidence that stress concentrations do occur in multiple fastener joints, resulting in the introduction into BS 5268 of modification factors $K_{50}$ (nails), $K_{54}$ (screws), $K_{57}$ (bolts), $K_{61}$ (toothed-plate connectors), $K_{65}$ (split-ring connectors) and $K_{69}$ (shear-plate connectors).

The factors apply where a number of fasteners of the same size are arranged symmetrically in one or more lines parallel to the line of action of the load in an axially loaded member,

$$K_{50} = K_{54} = 1.0 \text{ for } n < 10$$

$$K_{50} = K_{54} = 0.9 \text{ for } n \geqslant 10$$

$$K_{57} = K_{61} = K_{65} = K_{69} = 1 - \frac{3(n-1)}{100} \text{ for } n < 10$$

$$K_{57} = K_{61} = K_{65} = K_{69} = 0.7 \text{ for } n \geqslant 10$$

where $n$ = the number of fasteners in each line (see Table 18.24 of this manual).

For all other cases where more than one fastener is used the factors above should be taken as 1.0.

The designer will note that permissible loadings for coach screws are not given in BS 5268:Part 2. The proposed loadings were obviously too low and were deleted, leaving manufacturers to justify loadings until such time as realistic values can be included in BS 5268.

BS 5268 tabulates basic loadings for nails, screws, bolts, toothed-plate connectors, split-ring connectors and shear-plate connectors which are applicable to long term loading duration. When other than long term loading is considered, the long term loading may be multiplied by a factor $K_{48}$ (nails), $K_{52}$ (screws), $K_{55}$ (bolts), $K_{58}$ (toothed-plate connectors), $K_{62}$ (split-ring connectors) or $K_{66}$ (shear-plate connectors).

The values of these modification factors are summarised in Table 18.2.

As far as practicable, fasteners should be arranged so that the line of force in a member passes through the centroid of the group. When this is not practicable, account should be taken of the secondary stresses induced through the full or partial rigidity of the joint and of the effect of rotation imposing higher loads on the fasteners furthest from the centroid of the group.

The loads specified for nails, screws and bolts apply to those which are not treated against corrosion. The loads specified for toothed-plate, split-ring and shear-plate connectors apply to those which are treated against corrosion. Some forms of anti-corrosion treatment may affect fastener performance,

**Table 18.2**

| Fastener | Modification factor | Duration of load | | |
|---|---|---|---|---|
| | | Long | Medium | Short or very short |
| Nails (tempered hardboard to timber) | $K_{48}$ | 1.0 | 1.25 | 1.62 |
| Nails (other than above) | $K_{48}$ | 1.0 | 1.12 | 1.25 |
| Screws<br>Toothed-plate connectors | $K_{52}$<br>$K_{58}$ | 1.0 | 1.12 | 1.25 |
| Bolts<br>Split-ring connectors<br>Shear-plate connectors | $K_{55}$<br>$K_{62}$<br>$K_{66}$ | 1.0 | 1.25 | 1.50 |

particularly when preservative or fire-retardant timber impregnation treatments are specified. When the designer is in doubt the manufacturer of the fastener should be consulted.

Because of the anisotropic nature of timber, the bearing stresses permitted in a direction parallel to the grain are higher than those permitted in a direction perpendicular to the grain. Similarly, with the exception of lateral loads on nails and screws, connector units carry maximum loading when loaded parallel to the grain. Permissible loads for intermediate angles of load to the grain from 0° (parallel) to 90° (perpendicular) are calculated using the Hankinson formula:

$$F_\alpha = \frac{F_0 F_{90}}{F_0 \sin^2 \alpha + F_{90} \cos^2 \alpha}$$

where $F_\alpha$ = value of the load at angle $\alpha$ to grain
$F_0$ = value of the load parallel to the grain
$F_{90}$ = value of the load perpendicular to the grain.

Several connector units remove part of the cross-sectional area of the timber and the designer must consider the effect of loading on the net section. The net area at a section is the full cross-sectional area of the timber less the total projected area of that portion of the connector within the member at the cross-section (e.g. the projected area of a split-ring and the projected area of the associated bolt not within the projected area of the split-ring). It is not usual to deduct the projected area of nails, screws, or the teeth of toothed-plate connectors in calculating the strength of a joint.

The correct location of a mechanical fastener with respect to the boundaries of a timber component is of utmost importance to the satisfactory performance of a joint and to the development of the design load. It cannot be emphasised too strongly that care in detailing (particularly of tension end distances) is all important to the performance of a joint.

The permissible load on a fastener is further influenced by moisture content. The tabulated basic values are for cases where the fastener is assembled into dry timber for use in a dry exposure situation. For other cases, the basic load is multiplied by the modification factor $K_{49}$ (nails), $K_{53}$ (screws), $K_{56}$ (bolts), $K_{59}$ (toothed-plate connectors), $K_{63}$ (split-ring connectors) or $K_{67}$ (shear-plate connectors).

A value of unity applies in all cases other than those listed in Table 18.3.

**Table 18.3    Modification factors for moisture content**

| Direction of load | Timber at time of jointing or fastener assembly | Exposure condition | Nails* $K_{49}$ | Screws $K_{53}$ | Bolts $K_{56}$ | Connectors | | |
|---|---|---|---|---|---|---|---|---|
| | | | | | | Tooth-plate $K_{59}$ | Split-ring $K_{63}$ | Shear-plate $K_{67}$ |
| Lateral | Dry | Wet | 1.0 | 0.7 | 0.7 | 0.7 | 0.7 | 0.7 |
| | Green | Wet | 0.7 | 0.7 | 0.7 | 0.7 | 0.7 | 0.7 |
| | | Dry | 0.7 | 0.7 | 0.4 | 0.8 | 0.8 | 0.8 |
| Withdrawal | Dry | Wet | 1.0 | 0.7 | | | | |
| | | Cyclic | 0.25 | 1.0 | | | | |
| | Green | Wet | 1.0 | 0.7 | | | | |
| | | Dry | 1.0 | 0.7 | | | | |
| | | Cyclic | 0.25 | 1.0 | | | | |

\* $K_{49} = 1.0$ for annular-ringed shank and helical-threaded shank nails under all exposure conditions.

## 18.2   NAILED JOINTS

### 18.2.1   Stock and special sizes. Ordinary round nails

The largest manufacturer of nails in the UK is GKN Distributors Ltd, whose list of stock and special ordinary round wire nails is given in Table 18.4.

The designer's attention is drawn to the stock sizes. Although there are 117 nails in this schedule only 22 are stock sizes. The remainder are specials and are subject to a cost premium and longer delivery period. While it is pertinent to state that manufacturers will generally make nails of almost any diameter and length providing that the quantities ordered will present a commercial proposition, to ensure that nails will be readily available only stock sizes should generally be specified.

**Table 18.4   Stock and special ordinary round wire nails (from GKN standard list)**

| Length (mm) | Diameter (mm) | | | | | | | | | | |
|---|---|---|---|---|---|---|---|---|---|---|---|
| | 8 | 6.7 | 6 | 5.6 | 5 | 4.5 | 4 | 3.75 | 3.35 | 3 | 2.65 |
| 150 | X | X | ● | | | | | | | | |
| 125 | X | X | X | ● | ● | | | | | | |
| 115 | X | X | X | X | ● | | | | | | |
| 100 | X | X | X | X | ● | ● | ● | | | | |
| 90 | X | X | X | X | X | X | ● | | | | |
| 75 | X | X | X | X | X | X | ● | ● | ● | | |
| 65 | X | X | X | X | X | X | X | X | ● | ● | ● |
| 60 | X | X | X | X | X | X | X | X | ● | ● | ● |
| 50 | X | X | X | X | X | X | X | X | ● | ● | ● |
| 45 | | | X | X | X | X | X | X | X | X | ● |
| 40 | | | X | X | X | X | X | X | X | X | ● |
| 30 | | | X | X | X | X | X | X | X | X | X |
| 25 | | | X | X | X | X | X | X | X | X | X |
| 20 | | | | | X | X | X | X | X | X | X |
| 15 | | | | | | X | X | X | X | X | X |

● = stock, X = special.
BS 1202 : Part 1 gives sizes up to 200 X 8 mm.

## 18.2.2   Lateral loads

The basic lateral load for a nail in single shear (Fig. 18.1) when inserted at right angles to the side grain of timber is given in Table 57 of BS 5268: Part 2. Values for Strength Classes SC3 and SC4 are extracted and given in Table 18.5. The tabulated values are applicable to load at any angle to the grain.

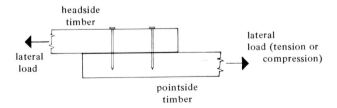

Fig. 18.1

For the basic load to apply the timber adjacent to the nail head (the head-side timber) and the timber receiving the point (the pointside timber) should be at least equal to the standard tabulated thickness. Greater thicknesses of timber do not permit loadings higher than the basic, although smaller thicknesses require a proportional reduction of the basic load. A nailed joint should normally contain at least two nails.

**Table 18.5**   **Basic lateral loads in single shear for ordinary round wire nails inserted at right angles to the side grain**

| Diameter of nail (mm) | Standard penetration of member | | Basic lateral load per nail for timbers in Strength Classes SC3 or SC4 (N) |
|---|---|---|---|
| | Headside (mm) | Pointside (mm) | |
| 2.65 | 19 | 25 | 180 |
| 3 | 22 | 29 | 230 |
| 3.35 | 25 | 32 | 285 |
| 3.75 | 29 | 38 | 355 |
| 4 | 32 | 44 | 405 |
| 4.5 | 38 | 51 | 515 |
| 5 | 44 | 57 | 635 |
| 5.6 | 51 | 67 | 795 |
| 6 | 57 | 76 | 910 |
| 6.7 | 64 | 89 | 1020 |
| 8 | 83 | 108 | 1130 |

The basic loads in Table 18.5 are based on tests performed at the Princes Risborough Laboratory. The strength of a joint is shown to depend on the number and diameter of nails, the specific gravity, moisture content and thickness of the timbers. For optimum loading in a two-member joint when the nail is in single shear, the headside member should be $\frac{3}{8}$ of the total joint thickness or nail length, whichever is the smaller.

No significant difference in maximum loading was found between tests carried out parallel and perpendicular to the grain.

The lateral strength of a nail depends on the bending strength of the nail and the local compression strength of the timber. Failure of the joint is accompanied by a high degree of bending of the nail and crushing of the timber. Typical modes of failure are shown in Fig. 18.2.

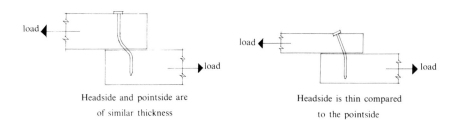

Headside and pointside are of similar thickness

Headside is thin compared to the pointside

Fig. 18.2

The lateral load on a nail driven into the end grain of timber should not exceed the recommended load for a similar nail driven into the side grain of timber multiplied by 0.7 (modification factor $K_{43}$).

*Example*

Calculate the dry medium term lateral load for two nails in single shear in SC3 timber as shown in Fig. 18.3 using 3.35 mm dia. × 65 mm ordinary round nails.

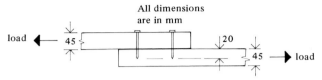

Fig. 18.3

The actual headside penetration is 45 mm and the basic headside penetration required is 25 mm, therefore the headside fixing permits the full basic load of 285 N. The pointside penetration is only 20 mm, whereas the basic pointside penetration required is 32 mm, therefore the basic load must be reduced proportionally, i.e. $\frac{20}{32} \times 285 = 178$ N.

Two nails give a basic load of $2 \times 178 = 356$ N.

The permissible load for a joint should be calculated as:

$$= \text{basic load} \times K_{48} \times K_{49} \times K_{50}$$
$$= 356 \times 1.12 \times 1.0 \times 1.0 = 399 \text{ N}$$

*Example*

Calculate the dry medium term load capacity of two 3.35 mm diameter × 50 mm long nails fixed into two pieces of SC3 timber as shown in Fig. 18.4. The headside penetrates side grain whereas the pointside penetrates end grain.

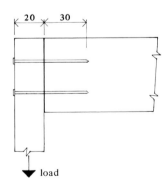

Fig. 18.4

The basic load headside in side grain after modifying for reduced penetration:

$$= \frac{20 \times 285}{25} = 228 \text{ N}$$

The basic load pointside after modifying for reduced penetration and for the point being in end grain ($K_{43}$):

$$= \frac{30 \times 285 \times 0.7}{32} = 187\,\text{N}$$

Two nails give a basic load of $2 \times 187 = 374\,\text{N}$.
∴ Permissible dry medium term load:

$$= \text{basic load} \times K_{48} \times K_{49} \times K_{50}$$
$$= 374 \times 1.12 \times 1.0 \times 1.0 = 419\,\text{N}$$

Clause 41.4.2 of BS 5268:Part 2 gives recommendations for determining the capacity of nails in double shear as shown in Fig. 18.5. The corresponding

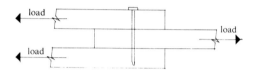

Fig. 18.5

basic loads are given in Table 18.6. For the basic load to apply, each of the members in the joint must have a thickness not less than $0.7\times$ the standard thickness for pointside timber shown in Table 18.5.

The slip under basic load depends on the density of the timber and the diameter of the nail. Larger diameter nails give larger slip and consequently it is preferable to use a greater number of smaller nails than a lesser number of larger diameter nails if slip is to be minimised. The design improvement

Table 18.6   **Basic lateral loads in double shear for ordinary round wire nails inserted at right angles to the side grain**

| Diameter of nail (mm) | Reduced standard thickness (see text) (mm) | Basic lateral load per nail for timbers in Strength Class SC3 or SC4 (N) |
|---|---|---|
| 2.65 | 17.5 | 324 |
| 3 | 20.3 | 414 |
| 3.35 | 22.4 | 513 |
| 3.75 | 26.6 | 639 |
| 4 | 30.8 | 729 |
| 4.5 | 35.7 | 745 |
| 5 | 39.9 | 1143 |
| 5.6 | 46.9 | 1431 |
| 6 | 53.2 | 1683 |
| 6.7 | 62.3 | 1836 |
| 8 | 75.6 | 2034 |

should be considered against any potential splitting which may occur if moisture content changes are likely.

Clause 41.4.4 permits the basic lateral load to be multiplied by modification factor $K_{44}$ (= 1.25) if square grooved or square twisted nails are specified. There has been a considerable amount of confusion about the type of nail to which this factor can be applied. Normally it has been assumed that it can be applied to any 'improved' nail. However, it has now been clarified that it is only for *square* nails that an improvement in lateral load can be claimed. Square grooved (untwisted) nails are more common in the Nordic area than in the UK. Square twisted nails have the twist over the whole length of the shank. Neither square grooved nor square twisted nails are illustrated in this manual. A nominal diameter of 0.75 times the diagonal should be assumed for these improved nails.

The following sizes of square twisted nails are usually available in stainless steel:

| | | |
|---|---|---|
| 32 × 3.75 mm | 38 × 3.75 mm | 75 × 3.75 mm |
| 75 × 4 mm | 90 × 4 mm | 100 × 4 mm |

BS 1202:Part 1 gives the following sizes (without defining the material):

| | | | |
|---|---|---|---|
| 65 × 3.35 mm | 50 × 3 mm | 50 × 2.65 mm | 40 × 2.36 mm |

If the headside timber in Table 18.5 is replaced by a metal plate and if the nail fits tightly into a hole in the plate then the headside portion of the nail will be constrained under load and the strength of the joint will be higher than a timber-to-timber joint. Where the metal plate is of sufficient strength to achieve this behaviour the code permits the basic lateral load to be multiplied by 1.25.

For nails used in green timber, or in timber under wet exposure conditions the basic loads should be multiplied by $K_{49} = 0.7$ (see Table 18.3).

Clause 41.6 of BS 5268:Part 2 introduces basic single shear lateral loads for nails in plywood-to-timber joints. The basic loads are related to nail diameter, plywood species and thickness, together with recommended lengths of nails. The recommendations of this clause of BS 5268:Part 2 are summarised in Table 18.7 for timber in Strength Classes SC3 and SC4.

For each nail diameter the upper load (i.e. smaller) value given in Table 18.7 is appropriate to plywoods of Group I comprising:

American Construction and Industrial plywood
Canadian Softwood plywood
Swedish Softwood plywood
Finnish Conifer plywood

**Table 18.7   Basic single shear lateral loads for round wire nails in a plywood-to-timber joint where the timber is in Strength Class SC3 or SC4**

| Nail diameter (mm) | Basic Load | | Minimum length (mm) of nail shall be the greater of: |
|---|---|---|---|
| | $t = 6$ mm (N) | 6 mm $< t <$ 20 mm (N) | |
| 2.65 | 145 | 145 | 40 or $(t + 25)$ |
| | 165 | 165 | |
| 3.00 | 190 | 190 | 45 or $(t + 29)$ |
| | 220 | 220 | |
| 3.35 | 230 | $224 + t$ | 45 or $(t + 32)$ |
| | 265 | 265 | |
| 3.75 | 260 | $254 + t$ | 55 or $(t + 38)$ |
| | 300 | $294 + t$ | |
| 4.00 | 270 | $258 + 2t$ | 60 or $(t + 44)$ |
| | 320 | $308 + 2t$ | |

$t$ = plywood thickness.

The lower (i.e. larger) load value in Table 18.7 is appropriate to plywoods of Group II comprising:

Finnish Birch-faced plywood
Finnish Birch plywood
British Hardwood plywood

Nails shall be fully embedded and have a length of at least the value calculated from Table 18.7.

The permissible load per nail = basic load $\times K_{48} \times K_{49} \times K_{50}$.

Similarly, Clause 41.7 of BS 5268 : Part 2 introduces basic single shear lateral loads for nails in tempered hardboard-to-timber joints which depend upon nail diameter, nominal tempered hardboard thickness and a minimum nail length. The recommendations of this clause of BS 5268 are summarised in Table 18.8 for timbers in Strength Classes SC3 or SC4.

For each nail diameter the upper load value relates to 3.2 mm nominal tempered hardboard thickness, and the lower value relates to thicknesses of 4.8, 6.4 and 8.0 mm.

The permissible load per nail = basic load $\times K_{48} \times K_{49} \times K_{50}$.

**Table 18.8** Basic single shear lateral load for round wire nails in a tempered hardboard-to-timber joint where timber is in Strength Class SC3 or SC4

| Nail diameter (mm) | Basic load (see text) (N) | Minimum length of nail required (mm) |
|---|---|---|
| 2.65 | $\dfrac{115}{130}$ | 40 |
| 3.00 | $\dfrac{140}{167}$ | 45 |
| 3.35 | $\dfrac{165}{205}$ | 45 |
| 3.75 | $\dfrac{187}{205}$ | 55 |
| 4.00 | $\dfrac{200}{205}$ | 60 |

## 18.2.3 Withdrawal loads

Ordinary round wire nails are relatively weak when loaded in withdrawal, and should be used only for relatively light loadings. Basic withdrawal loads as tabulated in Table 58 of BS 5268:Part 2 are based on tests carried out at the US Forest Products Laboratory. The formula for the average ultimate load in withdrawal in side grain (in green or dry timber) is given by the expression:

$$F_u = 47.6\rho^{2.5}d$$

where $F_u$ = the ultimate load per nail in N/mm of penetration
$\rho$ = the specific gravity taken as 0.358, 0.41, 0.505 and 0.61 for Strength Classes SC1 and 2, SC3 and 4, SC5, and SC6–9 respectively
$d$ = diameter of nail (mm).

The basic load is taken as $\frac{1}{6}$ of the ultimate and Table 58 of BS 5268: Part 2 is therefore derived from the equation:

$$F_{basic} = 7.93\rho^{2.5}d$$

The basic values in withdrawal apply to both green and dry timber, but when large changes in moisture content are anticipated after nailing, the basic values should be multiplied by 0.25 unless annular-ringed shank or helical-threaded shank nails are used. *No load in withdrawal should be carried by a nail driven into the end grain of timber.*

**Table 18.9**   **Stock sizes for annular-ringed shank nails (GKN standard list)**

| Length (mm) | Diameter (mm) | | | | |
|---|---|---|---|---|---|
| | 5 | 3.75 | 3.35 | 3 | 2.65 |
| 100 | ● | | | | |
| 75 | | ● | | | |
| 65 | | | ● | | |
| 60 | | | ● | | |
| 50 | | | ● | ● | ● |
| 45 | | | | | ● |
| 40 | | | | | ● |

BS 1202 : Part 1 gives sizes up to 200 × 8 mm.

Annular-ringed shank and helical-threaded shank nails have a greater resistance to withdrawal than ordinary round nails due to the effective surface roughening of the nail profile. Also their resistance to withdrawal is not affected by changes in moisture content. BS 5268 consequently permits the basic load for the 'threaded' part to be multiplied by a $K_{45}$ factor of 1.5 (Clause 41.4.4) for these particular forms of 'improved' nails and does not require a reduction in basic resistance even if there are likely to be subsequent changes in moisture content.

The range of available sizes for annular-ringed shank and helical-threaded shank nails is given in Tables 18.9 and 18.10 and the basic withdrawal values for both ordinary round, annular-ringed and helical-threaded shank nails are given in Table 18.11 for timber in Strength Classes SC3 and SC4.

**Table 18.10**   **More common sizes for helical-threaded shank nails**

| Length (mm) | Diameter (mm) | | | | | |
|---|---|---|---|---|---|---|
| | 6 | 5 | 3.75 | 3.35 | 3 | 2.65 |
| 150 | ● | | | | | |
| 100 | ● | ● | | | | |
| 75 | | | ● | | | |
| 65 | | | | ● | | |
| 60 | | | | ● | | |
| 50 | | | | ● | ● | ● |
| 45 | | | | | | ● |
| 40 | | | | | | ● |

The designer should check on availability before specifying.
BS 1202 : Part 1 gives sizes up to 200 × 8 mm.

**Table 18.11**  Basic resistance to withdrawal of ordinary round wire nails and annular-ringed shank and helical-threaded shank nails at right angles to grain

| Diameter (mm) | Ordinary round nails (N/mm) | Annular-ringed and helical-threaded nails (N/mm)* |
|---|---|---|
| 2.65 | 2.27 | 3.40 |
| 3 | 2.57 | 3.85 |
| 3.35 | 2.87 | 4.30 |
| 3.75 | 3.21 | 4.81 |
| 4 | 3.43 | 5.14 |
| 4.5 | 3.86 | 5.79 |
| 5 | 4.28 | 6.42 |
| 5.6 | 4.80 | |
| 6 | 5.14 | |
| 6.7 | 5.74 | |
| 8 | 6.85 | |

\* For 'threaded' length.

## 18.2.4  Spacing of nails

To avoid splitting of timber, nails should be positioned to give spacings, end distances and edge distances not less than those recommended in Table 56 of BS 5268:Part 2. All softwoods, except Douglas Fir may have values of spacing (but not edge distance) multiplied by 0.8. If a nail is driven into a glue-laminated section at right angles to the glue surface, then spacings (but not edge distances) may be multiplied by a further factor of 0.9 (note that this differs from CP 112:Part 2) because the glued surface provides a restraining action against cleavage of the face lamination. Although glue-laminated components have four or more laminations, the authors see no reason why the reduced spacing should not also apply to composites of two or three laminations suitably glued. The spacing, end distance and edge distance are functions of the diameter of the nail and are related to the direction of grain (but not to the direction of load) and are summarised in Fig. 18.6. To aid detailing, a schedule of spacings (rounded up to the next mm) is given in Table 18.12.

## 18.3  SCREW JOINTS

### 18.3.1  Stock and special sizes

Screws are used principally for fixings which require a resistance to with-drawal greater than that provided by either ordinary or improved nails. Although BS 5268 gives permissible lateral loadings for screws it should be recognised that nails of a similar diameter offer better lateral capacities and can be driven more economically. If, however, the depth of penetration is limited, some advantage may be gained by using screws.

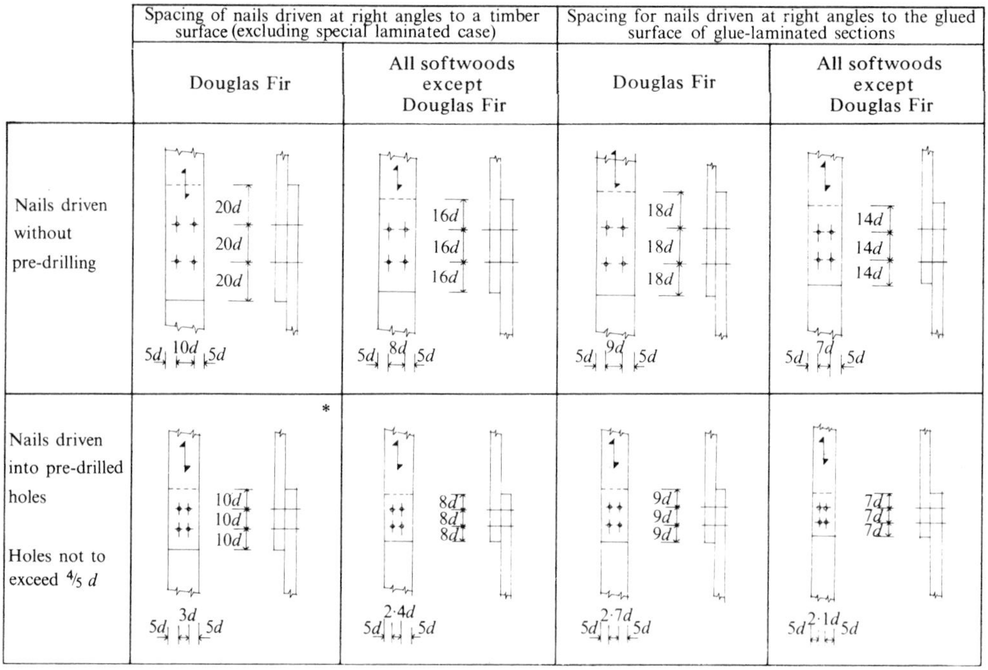

| | Spacing of nails driven at right angles to a timber surface (excluding special laminated case) | | Spacing for nails driven at right angles to the glued surface of glue-laminated sections | |
|---|---|---|---|---|
| | Douglas Fir | All softwoods except Douglas Fir | Douglas Fir | All softwoods except Douglas Fir |
| Nails driven without pre-drilling | 20d, 20d, 20d; 5d, 10d, 5d | 16d, 16d, 16d; 5d, 8d, 5d | 18d, 18d, 18d; 5d, 9d, 5d | 14d, 14d, 14d; 5d, 7d, 5d |
| Nails driven into pre-drilled holes. Holes not to exceed ⁴⁄₅ d | * 10d, 10d, 10d; 5d, 3d, 5d | 8d, 8d, 8d; 5d, 2·4d, 5d | 9d, 9d, 9d; 5d, 2·7d, 5d | 7d, 7d, 7d; 5d, 2·1d, 5d |

\* These spacings also apply to pre-drilled holes for screws in all timber species.
↕ = direction of grain.

Fig. 18.6: *Spacing of nails in timber-to-timber joints.*

**Table 18.12   Spacings of nails (mm)**

| Nail diameter* (mm) | Number of nail diameters | | | | | | | | | | | | | Nail number or gauge |
|---|---|---|---|---|---|---|---|---|---|---|---|---|---|---|
| | 2.1 | 2.4 | 2.7 | 3 | 5 | 7 | 8 | 9 | 10 | 14 | 16 | 18 | 20 | |
| 2.65 | 6 | 7 | 8 | 8 | 14 | 19 | 22 | 24 | 27 | 38 | 43 | 48 | 53 | 12 |
| 3 | 7 | 8 | 9 | 9 | 15 | 21 | 24 | 27 | 30 | 42 | 48 | 54 | 60 | 11 |
| 3.35 | 8 | 9 | 10 | 11 | 17 | 24 | 27 | 31 | 34 | 47 | 54 | 61 | 67 | 10 |
| 3.75 | 8 | 9 | 11 | 12 | 19 | 27 | 30 | 34 | 38 | 53 | 60 | 68 | 75 | 9 |
| 4 | 9 | 10 | 11 | 12 | 20 | 28 | 32 | 36 | 40 | 56 | 64 | 72 | 80 | 8 |
| 4.5 | 10 | 11 | 13 | 14 | 23 | 32 | 36 | 41 | 45 | 63 | 72 | 81 | 90 | 7 |
| 5 | 11 | 12 | 14 | 15 | 25 | 35 | 40 | 45 | 50 | 70 | 80 | 90 | 100 | 6 |
| 5.6 | 12 | 14 | 16 | 17 | 28 | 40 | 45 | 51 | 56 | 79 | 90 | 101 | 112 | 5 |
| 6 | 13 | 15 | 17 | 18 | 30 | 42 | 48 | 54 | 60 | 84 | 96 | 108 | 120 | 4 |
| 6.7 | 15 | 17 | 19 | 21 | 34 | 47 | 54 | 61 | 67 | 94 | 108 | 121 | 134 | 3 |
| 8 | 17 | 20 | 22 | 24 | 40 | 56 | 64 | 72 | 80 | 112 | 128 | 144 | 160 | — |

\* Nail diameter as per GKN literature.

Tables 63 and 64 of BS 5268:Part 2 give lateral and withdrawal values respectively for screws made from steel complying with BS 1210. Values are given for number (gauges) 6–20 (3.45 mm to 8.43 mm diameter). It is preferable to specify screw number (or diameter) followed by length, i.e. No. 6 X 38 mm to accord with industrial practice.

**Table 18.13   Preferred sizes for slotted countersunk head wood screws (GKN standard list)**

| Screw length (mm) | Screw number or gauge | | | | | | | |
|---|---|---|---|---|---|---|---|---|
| | 6 | 8 | 10 | 12 | 14 | 16 | 18 | 20 |
| 13 | ● | ○ | | | | | | |
| 16 | ● | ● | | | | | | |
| 19 | ● | ● | ○ | | | | | |
| 25 | ● | ● | ● | | | | | |
| 32 | ● | ● | ● | ● | ○ | ○ | | |
| 38 | ● | ● | ● | ● | ○ | ○ | | |
| 45 | ○ | ● | ● | ○ | ○ | | | |
| 50 | ● | ● | ● | ● | ○ | ○ | ○ | ○ |
| 57 | ● | ○ | ● | ● | ○ | | | |
| 63 | ● | ● | ● | ● | ○ | ○ | | |
| 75 | ● | ● | ● | ● | ○ | ○ | ○ | ○ |
| 89 | | | | ● | ○ | ○ | | |
| 100 | | | | ● | ○ | ○ | ○ | ○ |
| Diameter (mm) | 3.4 | 4.2 | 4.9 | 5.6 | 6.3 | 7.0 | 7.7 | 8.4 |

*(Screw diagram labelled "2/3 length (approx)" and "length")*

● = Also available Sherardised.

The most common form of wood screw is the slotted countersunk head wood screw available in the preferred sizes given in Table 18.13. Round headed screws may be used to advantage with metalwork as countersinking holes for the screw head is avoided. The range of round headed screws is, however, less comprehensive than that for countersunk screws. Preferred sizes are given in Table 18.14.

Steel Twinfast® Pozidriv® woodscrews have twin threads and a plain shank diameter less than the diameter of the thread. This eliminates the wedge action of the ordinary screw and reduces the danger of splitting. Pre-drilling can also be simplified to the drilling of a single hole. (See Table 18.20 for pre-drilling for ordinary screws.) Twinfast screws are threaded for the full length of the penetration depth (compared to an ordinary screw which is threaded for only approximately $\frac{2}{3}$ of the length under the head) giving at least 25% more withdrawal capacity. Preferred sizes are given in Table 18.15 from which it can be noted that the maximum length is less than that available with ordinary countersunk head wood screws.

For the reason given in § 18.1, basic loads for 'coach screws' are not given in BS 5268:Part 2. However, because they are still likely to be used in timber engineering, preferred sizes are given in Table 18.16.

**Table 18.14    Preferred sizes for round headed screws (GKN list)**

| Screw length (mm) | Screw number or gauge | | | | |
|---|---|---|---|---|---|
| | 6 | 8 | 10 | 12 | 14 |
| 13 | ○ | ○ | | | |
| 16 | ○ | ○ | ○ | | |
| 19 | ● | ● | ○ | | |
| 25 | ● | ● | ● | ○ | |
| 32 | ○ | ● | ● | | |
| 38 | ○ | ○ | ● | ○ | ○ |
| 45 | | ○ | ○ | | |
| 50 | | ○ | ○ | ○ | ○ |
| 63 | | | ○ | | |
| Diameter (mm) | 3.4 | 4.2 | 4.9 | 5.6 | 6.3 |

● = Also available Sherardised.

**Table 18.15    Preferred sizes for Twinfast ® Pozidriv ®**

| Screw length (mm) | Screw number or gauge | | | |
|---|---|---|---|---|
| | 6 | 8 | 10 | 12 |
| 13 | ● | ○ | | |
| 16 | ● | ● | | |
| 19 | ● | ● | ○ | |
| 25 | ● | ● | ● | ○ |
| 32 | ● | ● | ● | ○ |
| 38 | ● | ● | ● | ○ |
| 45 | | ● | ● | ○ |
| 50 | | ● | ○ | ○ |
| 57 | | | ○ | |
| 63 | | | ○ | ○ |
| Diameter (mm) | 3.4 | 4.2 | 4.9 | 5.6 |

○ = Also available bright zinc plated.

## 18.3.2    Lateral loads

Basic lateral loads for screws are given in Table 63 of BS 5268 : Part 2. Values for SC3 and SC4 are extracted and given in Table 18.17. Values are based on tests performed at the US Forest Products Laboratory.

The basic load capacity is given by the formula:

$$F_{basic} = Kd^2$$

where $F_{basic}$ = load (N)

   $d$ = diameter of screw (mm)

   $K$ = constant for timber species.

**Table 18.16**  Preferred sizes for coach screws (square heads)

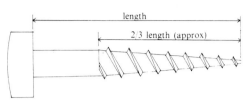

Other designs are also available. Check on availability before specifying.

| Screw length (mm) | Diameter (mm) | | | |
|---|---|---|---|---|
| | 6.5 | 8.0 | 10.0 | 12.0 |
| 25 | X | X | X | |
| 32 | X | X | X | |
| 38 | ⊗ | ⊗ | ⊗ | X |
| 45 | X | X | X | |
| 50 | ⊗ | ⊗ | ⊗ | X |
| 57 | X | X | X | |
| 63 | ⊗ | ⊗ | ⊗ | X |
| 75 | ⊗ | ⊗ | ⊗ | X |
| 89 | X | X | X | X |
| 100 | X | ⊗ | ⊗ | X |
| 115 | | X | X | X |
| 125 | X | X | ⊗ | X |
| 140 | | | X | |
| 150 | | X | X | X |

X = Popular sizes
O = Also available heavy bright zinc plated.

**Table 18.17**  Basic lateral loads for wood screws inserted at right angles to the grain for timber of Strength Class SC3 or SC4

| Screw number or gauge | Diameter of screw (mm) | Standard headside thickness (mm) | Standard pointside penetration (mm) | Basic load (N) |
|---|---|---|---|---|
| 6 | 3.45 | 12 | 24 | 220 |
| 8 | 4.17 | 15 | 29 | 325 |
| 10 | 4.88 | 17 | 34 | 445 |
| 12 | 5.59 | 20 | 39 | 580 |
| 14 | 6.30 | 22 | 44 | 740 |
| 16 | 7.01 | 25 | 49 | 915 |
| 18 | 7.72 | 27 | 54 | 1110 |
| 20 | 8.43 | 30 | 59 | 1320 |

For basic values the constant $K$ is equal to $14.5$, $18.6$, $22.8$ and $27.6$ for Strength Classes SC1 and 2, SC3 and 4, SC5, and SC6–9 respectively. For screws used in green timber, or in timber under wet exposure conditions, the basic loads should be multiplied by $K_{53} = 0.7$ (see Table 18.3).

If the headside thickness or the depth of penetration of the screw is less than the tabulated 'standard' value then the basic load is multiplied by the smaller of the two ratios:

(a) actual to standard headside thickness
(b) actual to standard pointside penetration.
(The actual pointside penetration should not be less than 0.6× standard pointside penetration.)

The lateral load on a screw driven into end grain should be taken as 0.7 times the recommended load for a similar screw driven into side grain. When a metal plate of adequate strength (generally 20 swg minimum) is screwed to a timber member, lateral loading values may be multiplied by 1.25 ($K_{46}$).

The capacities for coach screws are not given in BS 5268:Part 2:1984 but, in the absence of any better guide, the designer and approving authority may consider working from the US Forest Products Laboratory formula given above. It has been usual to take the 'standard penetration' for coach screws as approximately 7.5 times the diameter, consistent with Table 18.17, and to reduce the basic lateral load according to the method used for ordinary wood screws if the pointside penetration is less than the standard penetration. Current work of the Princes Risborough Laboratory may supersede this guidance.

Clause 42.6 of BS 5268:Part 2 introduces basic single shear lateral loads for screws in plywood-to-timber joints which depend upon screw diameter, plywood species and thickness, together with recommended minimum lengths of screws. The recommendations of this clause of BS 5268 are summarised in Table 18.18 for timbers in Strength Class SC3 and SC4.

**Table 18.18**  Basic single shear lateral loads for wood screws in a plywood-to-timber joint where the timber is in Strength Class SC3 or SC4

| Screw number or gauge | Diameter (mm) | Basic load (see text) | | Minimum length (mm) of screw shall be the greater of: |
|---|---|---|---|---|
| | | $t = 6$ mm (N) | 6 mm $< t <$ 20 mm (N) | |
| 8 | 4.17 | 225 / 300 | 225 / 300 | 38 or $(t + 29)$ |
| 10 | 4.88 | 305 / 360 | $269 + 6t$ / $329 + 6t$ | 44 or $(t + 34)$ |
| 12 | 5.59 | 365 / 485 | $341 + 4t$ / $455 + 5t$ | 51 or $(t + 39)$ |

$t$ = plywood thickness.

The permissible lateral load = basic load $\times K_{52} \times K_{53} \times K_{54}$.

For each screw diameter the upper value in Table 18.18 relates to plywood of Group I and the lower value relates to plywood of Group II as defined with Table 18.7.

### 18.3.3   Withdrawal loads

Basic withdrawal loads for screws are given in Table 64 of BS 5268: Part 2 and in Table 18.19 for timber in SC3 and SC4. Values are based on tests conducted in the US.

Table 18.19   Basic withdrawal loads for wood screws inserted at right angles to the grain in timber of Strength Class SC3 or SC4

| Screw number or gauge | Shank dia. (mm) | Basic withdrawal load per mm of pointside penetration of threaded part of screw (N) |
|---|---|---|
| 6 | 3.45 | 9.47 |
| 8 | 4.17 | 11.4 |
| 10 | 4.88 | 13.4 |
| 12 | 5.59 | 15.3 |
| 14 | 6.30 | 17.3 |
| 16 | 7.01 | 19.2 |
| 18 | 7.72 | 21.2 |
| 20 | 8.43 | 23.1 |

The average ultimate load in withdrawal for ordinary mild steel wood screws inserted into side grain of dry timber is given as:

$$F_u = 98\rho^2 d$$

where $F_u$ = the ultimate load per screw in N/mm penetration of thread. (*Note* – not full penetration of screw)
$\rho$ = the specific gravity of timber taken as 0.358, 0.41, 0.505 and 0.61 for Strength Classes SC1 and 2, SC3 and 4, SC5, and SC6–9 respectively
$d$ = the screw diameter (mm).

The basic withdrawal load is taken as $\frac{1}{6}$ of the ultimate value, therefore Table 64 of BS 5268 is derived from the equation:

$$F_{basic} = 16.3\rho^2 d \text{ (N/mm)}$$

For screws used in green timber, or in timber under wet exposure conditions, the basic loads should be multiplied by $K_{53} = 0.7$ (see Table 18.3). No withdrawal load or load in tension tending to cause withdrawal should be carried by a screw driven into end grain of timber.

The withdrawal capacities for coach screws are not given in BS 5268: Part 2:1984 but, in the absence of any better guide, the designer and approving authority may consider working from the formula above. Current work at the Princes Risborough Laboratory may supersede this guidance.

References to penetration of screws are only to that portion of penetration occupied by the threaded part of the screw, so that the permissible withdrawal load = basic load per mm × actual penetration of threaded portion × $K_{52}$ × $K_{53}$.

### 18.3.4   Spacing of screws

Clause 42.1 of BS 5268:Part 2 requires all screws to be driven with a screwdriver (or spanner in the case of coach screws) into pre-drilled holes. Spacings, end and edge distances shall be as for the requirements for pre-drilled holes in Douglas Fir in the lower first column of Fig. 18.6. The manufacturers' recommendations for sizes of pilot holes are given in Table 18.20. Clause 42.1 of BS 5268:Part 2 recommends that pilot holes to receive the screw thread should be between 0.4 and 0.6 times the diameter of the shank

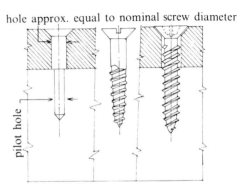

**Table 18.20   Pilot hole sizes. Drill sizes (mm)**

| Screw | | Ordinary screw | | | Pozidriv Twinfast | | |
|---|---|---|---|---|---|---|---|
| Number or gauge | Diameter (mm) | Pilot hole dia. (mm) | Drill size HW | Drill size SW | Pilot hole dia. (mm) | Drill size HW | Drill size SW |
| 6 | 3.4 | 4 | 2.1 | 1.5 | 4 | 2.3 | 1.60 |
| 8 | 4.2 | 6 | 2.5 | 1.8 | 4 | 2.95 | 1.95 |
| 10 | 4.9 | 6 | 2.75 | 2.15 | 5 | 3.15 | 2.25 |
| 12 | 5.6 | 8 | 3.15 | 2.5 | 6 | 3.6 | 2.60 |
| 14 | 6.3 | 9 | 3.6 | 2.75 | 6 | 4.2 | 2.95 |

HW = hardwood   SW = softwood

depending on density. Generally, the manufacturers' recommendations comply with this clause and coordinate the pilot hole size to suit drill sizes. Recommendations differ for hardwoods and softwoods (consequently both are listed) and for ordinary wood screws and Pozidriv Twinfast. To aid detailing, a schedule of spacings (rounded up to the next mm) is given in Table 18.21.

Table 18.21   **Spacings of screws (mm)**

| Screw diameter (mm) | Number of screw diameters | | | Screw number or gauge |
|---|---|---|---|---|
| | 3 | 5 | 10 | |
| 3.4 | 11 | 17 | 34 | 6 |
| 4.2 | 13 | 21 | 42 | 8 |
| 4.9 | 15 | 25 | 49 | 10 |
| 5.6 | 17 | 28 | 56 | 12 |
| 6.3 | 19 | 32 | 63 | 14 |
| 7.0 | 21 | 35 | 70 | 16 |
| 7.7 | 24 | 39 | 77 | 18 |
| 8.4 | 26 | 42 | 84 | 20 |

## 18.4   BOLT JOINTS

### 18.4.1   Stock and standard non-stock sizes

Bolts for timber engineering tend to have large $L/d$ ratios (where $L$ is the bolt length and $d$ the bolt diameter) compared to those used for structural steelwork connections, resulting from the need to joint thick sections of timber. Table 18.22 tabulates the stock and standard non-stock sizes catalogued by GKN Ltd.

For joints which will be permanently exposed to the weather or sited in a condition of high corrosion risk, the use of rust-proofing is essential. Also, when joints are in a non-hazard service condition but are to be of architectural merit (i.e. exposed to view) rust-proofing should be specified in order to avoid rust stains appearing on the timber surfaces.

If bolts are Sherardised or hot-dip galvanised, either the bolt threads must be run down or the nut tapped out before being treated, or the nut will not fit the thread. These disadvantages in the use of Sherardising and hot-dip galvanising favour the use of electro-galvanising, which is readily undertaken on the bolt sizes specified in Table 18.22, although the thickness of the coating is not as much as obtained by the two other methods.

### 18.4.2   Loads parallel and perpendicular to grain

Unlike nails and screws, the remaining forms of connectors discussed in this chapter have loading capacities which vary according to the direction of the applied load with respect to the direction of grain. For convenience of discussion, $F_0$ and $F_{90}$ represent loads parallel and perpendicular to grain respectively (as shown in Fig. 18.7).

Table 67 of BS 5268:Part 2 gives the basic values of $F_0$ and $F_{90}$ for one mild steel black bolt (to BS 4190) in single shear in a two-member joint in which the load acts perpendicular to the axis of the bolt. If a force $F$ acts

**Table 18.22   Stock and standard non-stock ISO metric bolts**

| Bolt length (mm) | ISO metric hex bolts grade 4.6 BS 4190 metric coarse thread | | | | | | | |
|---|---|---|---|---|---|---|---|---|
| | M8 | M10 | M12 | M16 | M20 | M24 | M30 | M36 |
| 35 | ✓ | ✓ | ✓ | | | | | |
| 40 | ✓ | ✓ | ✓ | | | | | |
| 45 | X | ✓ | ✓ | | | | | |
| 50 | ✓ | ✓ | ✓ | ✓ | ✓ | | | |
| 55 | X | X | X | ✓ | ✓ | | | |
| 60 | ✓ | ✓ | ✓ | ✓ | ✓ | | | |
| 65 | X | ✓ | ✓ | ✓ | ✓ | | | |
| 70 | ✓ | ✓ | ✓ | ✓ | ✓ | ✓ | | |
| 75 | X | ✓ | ✓ | ✓ | ✓ | ✓ | | |
| 80 | ✓ | ✓ | ✓ | ✓ | ✓ | ✓ | | |
| 90 | ✓ | ✓ | ✓ | ✓ | ✓ | ✓ | | |
| 100 | ✓ | ✓ | ✓ | ✓ | ✓ | ✓ | X | |
| 110 | X | X | ✓ | ✓ | ✓ | ✓ | X | X |
| 120 | ✓ | ✓ | ✓ | ✓ | ✓ | ✓ | X | X |
| 130 | X | X | ✓ | ✓ | ✓ | X | X | X |
| 140 | X | ✓ | ✓ | ✓ | ✓ | ✓ | X | X |
| 150 | X | X | ✓ | ✓ | ✓ | X | X | X |
| 160 | | | ✓ | ✓ | ✓ | ✓ | X | X |
| 170 | | | X | X | X | X | X | X |
| 180 | | | ✓ | ✓ | ✓ | ✓ | X | X |
| 190 | | | X | X | X | X | X | X |
| 200 | | | ✓ | ✓ | ✓ | ✓ | X | X |
| 220 | | | ✓ | ✓ | X | X | X | X |
| 240 | | | X | X | X | X | X | X |
| 260 | | | ✓ | ✓ | ✓ | X | X | X |
| 280 | | | X | X | X | X | X | X |
| 300 | | | ✓ | ✓ | ✓ | X | X | X |

✓ = preferred      X = non-preferred

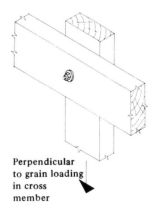

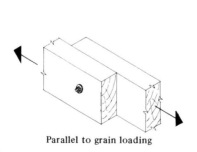

Parallel to grain loading

Perpendicular to grain loading in cross member

Fig. 18.7

at an angle $\theta$ to the axis of the bolt (Fig. 18.8) the component of load perpendicular to the load (i.e. $F \sin \theta$) must not be greater than the appropriate load given in Table 67 of BS 5268:Part 2. The component ($F \cos \theta$) acting axially on the bolt creates either tension or compression in the bolt and must be suitably resisted. (Permissible tension values for bolts are given in BS 449:Part 2.)

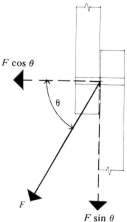

Fig. 18.8

When parallel members are of unequal thickness, the load for the thinner member should be used. Where members of unequal thickness are joined at an angle to the grain the basic load for each member must be determined and the smaller load used.

Table 18.23 gives values for $F_0$ and $F_{90}$ for bolts in single shear in timbers of Strength Class SC3 and SC4.

The derivation of basic loads for bolts is based principally on work carried out by Trayer at the US Forest Products Laboratory. Tests on bolts loaded parallel to the grain indicate that at small values of $L/d$ the bolt capacity is limited by the strength of the timber in compression parallel to the grain, whereas at larger values of $L/d$ the bending strength of the bolt influences the bolt capacity. $L$ is the thickness of the thinner timber in a single shear joint, or thickness of enclosed bearing in a multiple shear joint. Tests also show that over a certain value of $L/d$ the permissible load for a given bolt remains constant. Figure 18.9 shows the load/thickness curve for $F_0$ and $F_{90}$ for an M20 bolt in single shear in a two-member joint. The curve for $F_0$ shows a linear increase in capacity with increasing thickness, changing at $L/d = 2.75$ (approximately) to a distinct 'plateau' of constant load with increase in thickness. This plateau effect also commences at $L/d = 2.75$ for other diameters of bolt in single shear and at $L/d = 5.5$ for bolts in double shear. Work carried out by Trayer using mild steel side plates, applying load parallel to the grain, gave an average proportional limit some 25% higher than that for joints with timber side plates. BS 5268 recognises this increase by permitting a ($K_{46}$) 25% increase in load applied parallel to grain

**Table 18.23**   Values of $F_0$ and $F_{90}$ for bolts in single shear in timbers of Strength Class SC3 and SC4 (kN)

| Timber thickness (mm) | Load duration* | Bolt size | | | | | | | | | | | | | | | |
|---|---|---|---|---|---|---|---|---|---|---|---|---|---|---|---|---|---|
| | | M8 | | M10 | | M12 | | M16 | | M20 | | M24 | | M30 | | M36 | |
| | | $F_0$ | $F_{90}$ | $F_0$ | $F_{90}$ | $F_0$ | $F_{90}$ | $F_0$ | $F_{90}$ | $F_0$ | $F_{90}$ | $F_0$ | $F_{90}$ | $F_0$ | $F_{90}$ | $F_0$ | $F_{90}$ |
| 16 | L | 0.698 | 0.499 | 0.891 | 0.544 | 1.07 | 0.595 | 1.43 | 0.692 | 1.79 | 0.796 | 2.15 | 0.894 | | | | |
| | M | 0.872 | 0.623 | 1.11 | 0.680 | 1.33 | 0.743 | 1.78 | 0.865 | 2.23 | 0.995 | 2.68 | 1.11 | | | | |
| | S & VS | 1.047 | 0.748 | 1.33 | 0.816 | 1.60 | 0.892 | 2.14 | 1.03 | 2.68 | 1.19 | 3.22 | 1.34 | | | | |
| 19 | L | 0.771 | 0.593 | 1.04 | 0.646 | 1.27 | 0.706 | 1.70 | 0.822 | 2.13 | 0.945 | 2.55 | 1.06 | | | | |
| | M | 0.963 | 0.741 | 1.30 | 0.807 | 1.58 | 0.882 | 2.12 | 1.02 | 2.66 | 1.18 | 3.18 | 1.32 | | | | |
| | S & VS | 1.156 | 0.889 | 1.56 | 0.969 | 1.90 | 1.06 | 2.55 | 1.23 | 3.19 | 1.41 | 3.82 | 1.59 | | | | |
| 22 | L | 0.809 | 0.687 | 1.16 | 0.748 | 1.45 | 0.818 | 1.97 | 0.951 | 2.46 | 1.09 | 2.96 | 1.23 | | | | |
| | M | 1.011 | 0.858 | 1.45 | 0.935 | 1.81 | 1.02 | 2.46 | 1.18 | 3.07 | 1.36 | 3.70 | 1.53 | | | | |
| | S & VS | 1.21 | 1.03 | 1.74 | 1.12 | 2.17 | 1.22 | 2.95 | 1.42 | 3.69 | 1.63 | 4.44 | 1.84 | | | | |
| 25 | L | 0.819 | 0.780 | 1.24 | 0.850 | 1.61 | 0.929 | 2.23 | 1.08 | 2.80 | 1.24 | 3.36 | 1.40 | | | | |
| | M | 1.02 | 0.975 | 1.55 | 1.06 | 2.01 | 1.16 | 2.78 | 1.35 | 3.50 | 1.55 | 4.20 | 1.75 | | | | |
| | S & VS | 1.22 | 1.17 | 1.86 | 1.27 | 2.41 | 1.39 | 3.34 | 1.62 | 4.20 | 1.86 | 5.04 | 2.10 | | | | |
| 32 | L | 0.816 | 0.816 | 1.28 | 1.09 | 1.81 | 1.19 | 2.79 | 1.38 | 3.57 | 1.59 | 4.30 | 1.79 | | | | |
| | M | 1.02 | 1.02 | 1.60 | 1.36 | 2.26 | 1.48 | 3.48 | 1.72 | 4.46 | 1.98 | 5.37 | 2.23 | | | | |
| | S & VS | 1.22 | 1.22 | 1.92 | 1.63 | 2.71 | 1.78 | 4.18 | 2.07 | 5.35 | 2.38 | 6.45 | 2.68 | | | | |
| 36 | L | 0.816 | 0.816 | 1.28 | 1.21 | 1.83 | 1.34 | 3.00 | 1.56 | 3.97 | 1.79 | 4.84 | 2.01 | | | | |
| | M | 1.02 | 1.02 | 1.60 | 1.51 | 2.28 | 1.67 | 3.75 | 1.95 | 4.96 | 2.23 | 6.05 | 2.51 | | | | |
| | S & VS | 1.22 | 1.22 | 1.92 | 1.81 | 2.74 | 2.01 | 4.50 | 2.34 | 5.95 | 2.68 | 7.26 | 3.01 | | | | |
| 38 | L | 0.818 | 0.818 | 1.28 | 1.26 | 1.84 | 1.41 | 3.08 | 1.64 | 4.17 | 1.89 | 5.09 | 2.12 | | | | |
| | M | 1.02 | 1.02 | 1.60 | 1.57 | 2.30 | 1.76 | 3.85 | 2.05 | 5.21 | 2.36 | 6.36 | 2.65 | | | | |
| | S & VS | 1.22 | 1.22 | 1.92 | 1.89 | 2.76 | 2.11 | 4.62 | 2.46 | 6.25 | 2.83 | 7.63 | 3.18 | | | | |

| Size | Type | | | | | | | | | | | | | | | |
|---|---|---|---|---|---|---|---|---|---|---|---|---|---|---|---|---|
| 44 | L | 0.816 | 0.816 | 1.28 | 1.28 | 1.84 | 1.61 | 3.23 | 1.90 | 4.62 | 2.19 | 5.81 | 2.46 | | | | |
| | M | 1.02 | 1.02 | 1.60 | 1.60 | 2.30 | 2.01 | 4.03 | 2.37 | 5.77 | 2.73 | 7.26 | 3.07 | | | | |
| | S & VS | 1.22 | 1.22 | 1.92 | 1.92 | 2.76 | 2.41 | 4.84 | 2.85 | 6.93 | 3.28 | 8.71 | 3.69 | | | | |
| 47 | L | 0.816 | 0.816 | 1.28 | 1.28 | 1.84 | 1.69 | 3.26 | 2.03 | 4.79 | 2.34 | 6.17 | 2.62 | | | | |
| | M | 1.02 | 1.02 | 1.60 | 1.60 | 2.30 | 2.11 | 4.07 | 2.53 | 5.98 | 2.92 | 7.71 | 3.27 | | | | |
| | S & VS | 1.22 | 1.22 | 1.92 | 1.92 | 2.76 | 2.53 | 4.89 | 3.04 | 7.18 | 3.51 | 9.25 | 3.93 | | | | |
| 50 | L | 0.816 | 0.816 | 1.27 | 1.27 | 1.84 | 1.72 | 3.27 | 2.16 | 4.94 | 2.49 | 6.44 | 2.79 | | | | |
| | M | 1.02 | 1.02 | 1.58 | 1.58 | 2.30 | 2.15 | 4.08 | 2.70 | 6.17 | 3.11 | 8.05 | 3.48 | | | | |
| | S & VS | 1.22 | 1.22 | 1.90 | 1.90 | 2.76 | 2.58 | 4.90 | 3.24 | 7.41 | 3.73 | 9.66 | 4.18 | | | | |
| 63 | L | | | | | 1.84 | 1.68 | 3.27 | 2.63 | 5.12 | 3.14 | 7.21 | 3.52 | 10.1 | 4.09 | | |
| | M | | | | | 2.30 | 2.10 | 4.08 | 3.28 | 6.40 | 3.92 | 9.01 | 4.40 | 12.6 | 5.11 | | |
| | S & VS | | | | | 2.76 | 2.52 | 4.90 | 3.94 | 7.68 | 4.71 | 10.8 | 5.28 | 15.1 | 6.13 | | |
| 75 | L | | | | | | | 3.27 | 2.68 | 5.12 | 3.66 | 7.37 | 4.19 | 11.1 | 4.87 | | |
| | M | | | | | | | 4.08 | 3.35 | 6.40 | 4.57 | 9.21 | 5.23 | 13.8 | 6.08 | | |
| | S & VS | | | | | | | 4.90 | 4.02 | 7.68 | 5.49 | 11.0 | 6.28 | 16.6 | 7.30 | | |
| 100 | L | | | | | | | 3.27 | 2.52 | 5.10 | 3.79 | 7.36 | 5.18 | 11.5 | 6.50 | 16.4 | 7.48 |
| | M | | | | | | | 4.08 | 3.15 | 6.37 | 4.73 | 9.20 | 6.47 | 14.3 | 8.12 | 20.5 | 9.35 |
| | S & VS | | | | | | | 4.90 | 3.78 | 7.65 | 5.68 | 11.0 | 7.77 | 17.2 | 9.75 | 24.6 | 11.2 |
| 150 | L | | | | | | | | | | | 7.35 | 4.87 | 11.5 | 7.43 | 16.6 | 10.4 |
| | M | | | | | | | | | | | 9.18 | 6.08 | 14.3 | 9.28 | 20.7 | 13.0 |
| | S & VS | | | | | | | | | | | 11.0 | 7.30 | 17.2 | 11.1 | 24.9 | 15.6 |

*L = long    M = medium    S = short    VS = very short

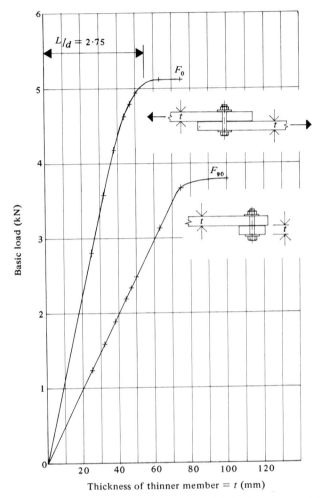

**Fig. 18.9:** *Typical load/thickness curve for an M20 bolt in timber of Strength Class SC3 and SC4.*

where a steel plate of adequate strength is joined to a timber member. However, no increase is applicable to values for loading perpendicular to the grain (see later notes).

Trayer's tests on bolted joints loaded perpendicular to the grain displayed similar dependence on the $L/d$ ratio, but in this instance the bolt diameter also influences the results. It was found that bolts of small diameter develop higher average stresses perpendicular to the grain, consistent with the timber's ability to resist concentrated loads in compression perpendicular to the grain as appropriate to bearing stresses (see BS 5268 : Part 2, Clause 14.2 and § 4.10 of this manual).

The curve for $F_{90}$ in Fig. 18.9 indicates a linear increase in load capacity up to a ratio of $L/d = 4.5$ (approximately). Trayer's experiments with mild steel side plates loading timber perpendicular to the grain gave no increase in the average proportional limit, and consequently BS 5268 does not permit increases for loading perpendicular to the grain (see Clause 43.5.2).

The basic loads for joints of more than two members may be taken as the sum of the basic loads for each shear plane, providing that any member having a shear plane on both sides has twice the tabulated thickness (Clause 43.4.2 of BS 5268:Part 2). To express this in another way, the capacity of a three-member joint, for example, as shown in Fig. 18.10, may be taken as twice the capacity of the two-member joint shown in the same figure.

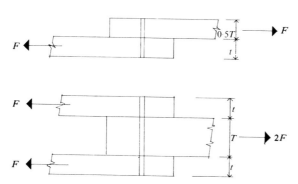

Fig. 18.10

Special precautions are required with bolted joints when the moisture content of the timber is above 18%. Firstly, if joints are to be in green timber or under wet exposure conditions, the basic loads should be multiplied by 0.7 ($K_{56}$). Secondly, if joints are prepared in timber which has not been dried (to about 18% or less) and subsequent drying-out is anticipated, the basic loads should be multiplied by 0.4 ($K_{56}$).

The basic load should be further modified by factor $K_{57}$ (see § 18.1) for the number of bolts ($n$) in a line parallel to the line of action of the load. Values of $K_{57}$ are given in Table 18.24.

**Table 18.24**  $K_{57}$ **values**

| $n$ | 1 | 2 | 3 | 4 | 5 | 6 | 7 | 8 | 9 | 10 | 11 or more |
|---|---|---|---|---|---|---|---|---|---|---|---|
| $K_{57}$ | 1.00 | 0.97 | 0.94 | 0.91 | 0.88 | 0.85 | 0.82 | 0.79 | 0.76 | 0.73 | 0.7 |

### 18.4.3  Loads at an angle to the grain

The basic load $F_\alpha$ acting in a direction at an angle $\alpha$ to the grain should be determined by the Hankinson formula:

$$F_\alpha = \frac{F_0 F_{90}}{F_0 \sin^2 \alpha + F_{90} \cos^2 \alpha}$$

which can be simplified using the function $\cos^2 \alpha + \sin^2 \alpha = 1$ to:

$$F_\alpha = \frac{F_0}{1 + [(F_0/F_{90}) - 1] \sin^2 \alpha}$$

**Table 18.25   Values of sin² α**

| α° | 0 | 2.5 | 5 | 7.5 |
|---|---|---|---|---|
| 0 | 0 | 0.001 90 | 0.007 60 | 0.0170 |
| 10 | 0.0302 | 0.046 8 | 0.067 0 | 0.0904 |
| 20 | 0.117 | 0.146 | 0.179 | 0.213 |
| 30 | 0.250 | 0.289 | 0.329 | 0.371 |
| 40 | 0.413 | 0.456 | 0.500 | 0.544 |
| 50 | 0.587 | 0.629 | 0.671 | 0.711 |
| 60 | 0.750 | 0.787 | 0.821 | 0.854 |
| 70 | 0.883 | 0.910 | 0.933 | 0.953 |
| 80 | 0.970 | 0.983 | 0.992 | 0.998 |
| 90 | 1.0 | | | |

Unfortunately, as can be seen from Table 18.23, the ratio $F_0/F_{90}$ varies, and for bolts no further simplification for $F_\alpha$ is possible. For the convenience of designers, values of $\sin^2 \alpha$ are listed in Table 18.25 for angles from $0°$ to $90°$ in $2.5°$ increments. In practice the joint capacity is estimated with sufficient accuracy if $\alpha$ is taken to the nearest $2.5°$ value.

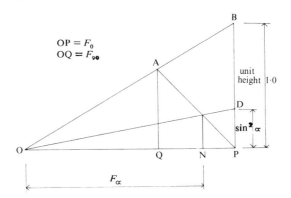

Fig. 18.11: *Construction of Scholten nomogram.*

The Hankinson formula can conveniently be resolved graphically by reference to Fig. 18.11 as follows:

Set out horizontally from an origin O a distance OP representing $F_0$, the load parallel to the grain, and from P to any convenient scale erect a vertical PB of unit height.

Along the line OP, set out from the origin a distance OQ representing $F_{90}$, the load perpendicular to grain, and from Q erect a vertical to meet OB at A.

On the line PB at a distance equal to $\sin^2 \alpha$, place joint D and join O to D.

Join A to P and from the intercept of AP on OD drop a vertical to meet OP at N.

It is now simple to show from similar triangles that the distance ON represents $F_\alpha$, the load at an angle $\alpha$ to grain. This graphical solution to

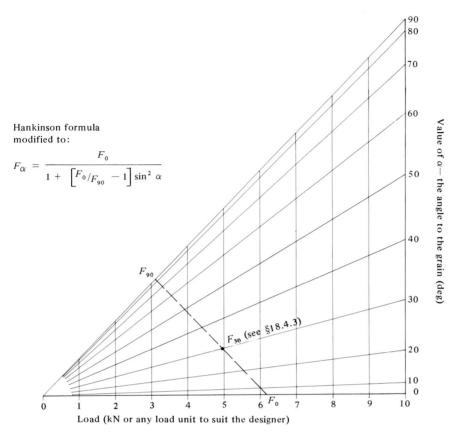

Fig. 18.12: *Scholten nomogram for the graphical solution of the Hankinson formula.*

Hankinson's formula is credited to Scholten and forms the basis of the Scholten nomogram drawn in Fig. 18.12.

As an example of using the Scholten nomogram, determine the medium term loading in the ceiling tie of the joint shown in Fig. 18.13 (timber SC3).

First consider the 50 mm thick timber side plates, where the bolt is loading the timber at $0°$ to the grain. From table 18.23, for a thickness of 50 mm, $F_0 = 6.17$ kN (medium term). There are two side plates, and therefore the capacity is 12.34 kN.

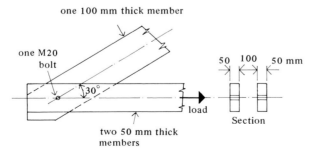

Fig. 18.13

The inner member, 100 mm total thickness, is loaded at 30° to grain (related to the load in the tie) and has a capacity equal to twice that calculated for a bolt in single shear through 50 mm thick timber. Therefore from Table 18.23, $F_0 = 6.17$ kN and $F_{90} = 3.11$ kN, from which $F_{30}$ can be determined.

From Fig. 18.12, $F_{30}$ is determined graphically as 4.9 kN (approximately) or mathematically from the Hankinson formula with $\sin^2\alpha = 0.25$ as:

$$F_{30} = \frac{6.17}{1.0 + [(6.17/3.11) - 1.0] \times 0.25} = 4.95\,\text{kN}$$

Therefore the capacity for the inner member is $2 \times 4.95 = 9.9$ kN, which for this joint sets the limit for the load in the tie.

### 18.4.4   Spacing of bolts

In general, bolt holes should be drilled as close as possible to the diameter of the specified bolt. Too great a hole tolerance will produce excessive joint slip under load which in turn will lead to additional deflections of a complete assembly, i.e. a truss or similar framed structure. On the other hand, a forced fit is not recommended as this is likely to cause splitting of the timber. In practice, holes are usually drilled up to 2 mm larger in diameter than the bolt, depending upon standard drill sizes.

A mild steel washer should be fitted under any head or nut which would otherwise be in direct contact with a timber surface. Minimum washer sizes (larger than those associated with structural steelwork) are given in Table 18.27. When appearance is important, round washers are preferred. They are easier to install than square washers, the latter requiring care in alignment if they are not to appear unsightly. The required washer diameters are large to avoid crushing of the fibres of the timber, and are consequently relatively thick in order to avoid cupping as the bolted joint is tightened. To avoid fibre crushing, care should be taken not to overtighten.

Spacing of bolts should be in accordance with Table 18.26, having due regard to the direction of load and grain. When a load is applied at an angle to the grain, the recommended spacings, both parallel and perpendicular, whichever are the larger, are applicable. No reductions in spacings are permitted even when the applied load is less than the bolt capacity for the joint.

To assist detailing, appropriate spacings are given in Table 18.27.

## 18.5   TOOTHED-PLATE CONNECTOR UNITS

### 18.5.1   General

Toothed-plate connectors (§1.12.7) are frequently used for small to medium span domestic trusses and for other assemblies, when loadings at joints are too high for simple bolts but too low to require the use of split-ring or

**Table 18.26   Minimum spacing, edge and end distances for bolts**

| Load parallel to grain | Load perpendicular to grain |
|---|---|

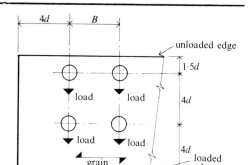

When using very large washers the designer should ensure that they do not foul each other.

Dimension $B$ should be not less than $3d$ for joints where the ratio $t/d$ is unity and $5d$ where the ratio is 3 or more. For values of the ratio $t/d$ between 1 and 3, the spacing should be calculated by linear interpolation from the equation:

$$B = (2 + d/t)\,d$$

where $t$ = thickness of timber
$d$ = bolt diameter.

**Table 18.27   Bolt spacings and washer dimensions**

| Bolt size | Spacings (mm) | | | | | Minimum washer size | |
|---|---|---|---|---|---|---|---|
| | $1.5d$ | $3d$ | $4d$ | $5d$ | $7d$ | Diameter (mm) | Thickness (mm) |
| M8 | 12 | 24 | 32 | 40 | 56 | 24 | 2 |
| M10 | 15 | 30 | 40 | 50 | 70 | 30 | 2.5 |
| M12 | 18 | 36 | 48 | 60 | 84 | 36 | 3 |
| M16 | 24 | 48 | 64 | 80 | 112 | 48 | 4 |
| M20 | 30 | 60 | 80 | 100 | 140 | 60 | 5 |
| M24 | 36 | 72 | 96 | 120 | 168 | 72 | 6 |
| M30 | 45 | 90 | 120 | 150 | 210 | 90 | 7.5 |
| M36 | 54 | 108 | 144 | 180 | 252 | 108 | 9 |

Also, note washer sizes in § 18.5.1.

shear-plate connectors. The only special equipment needed to form a toothed-plate joint is a special high tensile steel drawing stud to embed the teeth of the connectors into the face of the timbers at the joint.

A toothed-plate connector unit consists of one or other of the following:

(a) one double-sided connector with bolt in single shear (Fig. 18.14(a)).

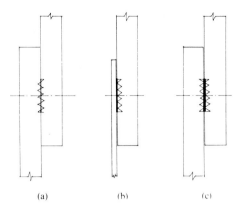

Fig. 18.14

(b) one single-sided connector with bolt in single shear, used with a steel
plate in a timber-to-metal joint (Fig. 18.14(b)).

(c) two single-sided connectors used back-to-back in the contact faces of a
timber-to-timber joint with bolt in single shear (Fig. 18.14(c)).

Not all sizes of tooth plates illustrated in manufacturers' literature have
been tested, and BS 5268 limits designers to single- and double-sided, round
and square tooth plates of 38 mm, 51 mm, 64 mm and 76 mm diameter
(or nominal side of square).

Where the nut or head of the bolt occurs on a timber surface a large
washer must be provided. Perhaps strangely the minimum sizes required by
BS 5268:Part 2 (as given below) do not link with the sizes required by
BS 5268 for bolts as shown below.

|  | Nominal size (mm) | Size of bolt | Washer diameter (mm) (or length of side) | Washer thickness (mm) |
|---|---|---|---|---|
| Round toothed-plate | 38 | M10 | 38 | 3 |
|  | 51 | M12 | 38 | 3 |
|  | 64 | M12 | 50 | 5 |
|  | 76 | M12 | 60 | 5 |
| Square toothed-plate | 38 | M10 | 38 | 3 |
|  | 51 | M12 | 50 | 5 |
|  | 64 | M12 | 60 | 5 |
|  | 76 | M12 | 75 | 5 |

### 18.5.2   Permissible loadings

Table 72 of BS 5268:Part 2 gives the dry basic loads permitted for one
toothed-plate connector unit for loads parallel to the grain $F_0$ and perpendicular to grain $F_{90}$. Note that loads for Strength Classes SC6–9 are not

tabulated. This is because the timbers are too dense for the connectors to be successfully embedded, there being risk that the teeth will deform. Table 18.28 summarises the values of $F_0$ and $F_{90}$ for timbers in Strength Classes SC3 and SC4. Note that a connector on both sides of a member and on the same bolt requires twice the thickness of timber required for a connector on one side only, if required to carry the same load.

The tabulated basic loads are based on tests carried out at the Princes Risborough Laboratory. As with most connectors, the improved loading capacity over that for bolts is achieved by a reduction in stress concentrations and an enlargement of the timber area made available to resist shear, without simultaneously removing too much timber from the joint. Load is transferred between members by bearing stresses developed between the teeth of the connector and the timber, and by bearing stresses developed between the bolt and the timber. Some 40% of the ultimate load is carried by the bolt, consequently it is important always to include a bolt, and to use the correct diameter. The triangular teeth of the plates are shaped in such a way that they tend to lock into the timber as the joint is formed and therefore tend to give a relatively stiff joint.

The basic load at an angle to the grain is calculated using the Hankinson formula or the Scholten nomogram (as described for bolts in §18.4.3). To aid the designer, the basic loads at an angle to the grain are presented in Table 18.29 for timbers in Strength Class SC3 and SC4, of 47 mm and 35 mm thicknesses as frequently encountered in truss design.

The permissible load for a toothed-plate connector is given by the formula:

$$\text{Permissible load} = \text{basic load} \times K_{58} \times K_{59} \times K_{60} \times K_{61}$$

where $K_{58}$ = the modification factor for duration of loading:

1.0 for long term loading
1.12 for medium term loading
1.25 for short term and very short term loading

$K_{59}$ = the modification factor for moisture content, taken as unity for dry conditions but reduced to 0.7 for joints in green timber or in timber which will be used in wet exposure conditions, or 0.8 for joints in green timber which will be under dry exposure conditions

$K_{60}$ = the lesser value of $K_C$, $K_D$ or $K_s$ which are the modification factors for sub-standard end distance, edge distance and spacing respectively

$K_{61}$ = the modification factor for a series of connectors in a single line parallel to the line of action of the load ($K_{61}$ is numerically equal to $K_{57}$ as given in Table 18.24).

**Table 18.28** Basic loads for one toothed-plate connector unit in softwood of Strength Class SC3 or SC4

| Nominal size of connector (mm) | Size of bolt | Thickness of members* | | Load parallel to grain | Load perpendicular to grain |
|---|---|---|---|---|---|
| | | Connector on one side only (mm) | Connector on both sides and on same bolt (mm) | $F_0$ (kN) | $F_{90}$ (kN) |
| 38 round or square | M10 | 16 | 32 | 2.17 | 1.50 |
| | | 19 | 38 | 2.38 | 1.60 |
| | | 22 | 44 | 2.54 | 1.70 |
| | | 25 | 50 | 2.65 | 1.80 |
| | | and over | and over | | |
| 51 round | M12 | 16 | 32 | 2.89 | 2.02 |
| | | 19 | 38 | 3.16 | 2.13 |
| | | 22 | 44 | 3.41 | 2.24 |
| | | 25 | 50 | 3.63 | 2.35 |
| | | 29 | – | 3.82 | 2.50 |
| | | – | 60 | 3.86 | 2.54 |
| | | – | 63 | 3.89 | 2.59 |
| | | 36 | 72 | 3.93 | 2.76 |
| | | 50 | 100 | 3.94 | 3.15 |
| | | and over | and over | | |
| 51 square | M12 | 16 | 32 | 3.20 | 2.32 |
| | | 19 | 38 | 3.47 | 2.44 |
| | | 22 | 44 | 3.72 | 2.55 |
| | | 25 | 50 | 3.94 | 2.66 |
| | | 29 | – | 4.13 | 2.81 |
| | | – | 60 | 4.17 | 2.84 |
| | | – | 63 | 4.20 | 2.90 |
| | | 36 | 72 | 4.24 | 3.07 |
| | | 50 | 100 | 4.25 | 3.45 |
| | | and over | and over | | |
| 64 round or square | M12 | 16 | 32 | 3.65 | 2.77 |
| | | 19 | 38 | 3.92 | 2.88 |
| | | 22 | 44 | 4.17 | 3.00 |
| | | 25 | 50 | 4.39 | 3.11 |
| | | 29 | – | 4.58 | 3.26 |
| | | – | 60 | 4.62 | 3.29 |
| | | – | 63 | 4.65 | 3.35 |
| | | 36 | 72 | 4.69 | 3.52 |
| | | 50 | 100 | 4.70 | 3.90 |
| | | and over | and over | | |

**Table 18.28** (*contd*)

| Nominal size of connector (mm) | Size of bolt | Thickness of members* | | Load parallel to grain | Load perpendicular to grain |
|---|---|---|---|---|---|
| | | Connector on one side only (mm) | Connector on both sides and on same bolt (mm) | $F_0$ (kN) | $F_{90}$ (kN) |
| 76 round | M12 | 16 | 32 | 4.22 | 3.34 |
| | | 19 | 38 | 4.49 | 3.45 |
| | | 22 | 44 | 4.74 | 3.57 |
| | | 25 | 50 | 4.95 | 3.68 |
| | | 29 | – | 5.15 | 3.83 |
| | | – | 60 | 5.19 | 3.86 |
| | | – | 63 | 5.22 | 3.92 |
| | | 36 | 72 | 5.26 | 4.09 |
| | | 50 | 100 | 5.27 | 4.47 |
| | | and over | and over | | |
| 76 square | M12 | 16 | 32 | 4.76 | 3.88 |
| | | 19 | 38 | 5.02 | 3.99 |
| | | 22 | 44 | 5.27 | 4.10 |
| | | 25 | 50 | 5.49 | 4.21 |
| | | 29 | – | 5.68 | 4.36 |
| | | – | 60 | 5.72 | 4.40 |
| | | – | 63 | 5.75 | 4.45 |
| | | 36 | 72 | 5.79 | 4.62 |
| | | 50 | 100 | 5.80 | 5.01 |
| | | and over· | and over | | |

* Actual thickness. Intermediate thicknesses may be obtained by linear interpolation.

## 18.5.3   Standard and sub-standard placement of connectors

To develop the full basic load (i.e. for $K_{60} = 1.0$), tooth plates must have the required standard end distance $C_{st}$, edge distance $D_{st}$, and spacing $S_{st}$, which can be expressed in terms of the tooth-plate diameter $d$. (In the case of a square connector $d$ is taken as the nominal side dimension plus 6 mm.)

End distance $C$ is the distance measured parallel to the grain from the centre of a connector to the square-cut end of the member (as shown in Fig. 18.15). If the end is splay cut, the end distance should be taken as the distance measured parallel to the grain taken from a point $d/4$ from the centre line of the connector as shown in Fig. 18.15.

The end distance is said to be loaded when the force on the connector has a component acting towards the end of the member, and is applicable to forces acting from $\alpha = 0°$ to $\alpha = 90°$ (Fig. 18.16). For a loaded end, $K_C = 1.0$ at $C_{st} = 0.5d + 64$ mm, and $K_C$ reduces for sub-standard end distances to a variable degree, depending on the type and size of connector. The

Table 18.29   Basic long term load (kN) for one toothed-plate connector unit loaded at an angle to grain

| Member thickness (mm) | Connector size | One face R* or S or two | Angle of load to grain (deg) | | | | | | | | | | | | | | | | | | |
|---|---|---|---|---|---|---|---|---|---|---|---|---|---|---|---|---|---|---|---|---|---|---|
| | | | 0 | 5 | 10 | 15 | 20 | 25 | 30 | 35 | 40 | 45 | 50 | 55 | 60 | 65 | 70 | 75 | 80 | 85 | 90 |
| 38 | 38 | R or S  1 | 2.65 | 2.64 | 2.61 | 2.56 | 2.51 | 2.44 | 2.37 | 2.29 | 2.21 | 2.14 | 2.07 | 2.01 | 1.95 | 1.90 | 1.87 | 1.83 | 1.81 | 1.80 | 1.80 |
| | | 2 | 2.59 | 2.58 | 2.55 | 2.50 | 2.45 | 2.38 | 2.31 | 2.23 | 2.16 | 2.08 | 2.02 | 1.95 | 1.90 | 1.85 | 1.81 | 1.78 | 1.76 | 1.75 | 1.75 |
| | 51 | R  1 | 3.94 | 3.93 | 3.90 | 3.86 | 3.81 | 3.74 | 3.67 | 3.59 | 3.52 | 3.44 | 3.37 | 3.30 | 3.24 | 3.18 | 3.14 | 3.10 | 3.08 | 3.06 | 3.06 |
| | | 2 | 3.52 | 3.50 | 3.46 | 3.39 | 3.31 | 3.21 | 3.10 | 2.99 | 2.88 | 2.77 | 2.67 | 2.58 | 2.50 | 2.44 | 2.38 | 2.34 | 2.31 | 2.29 | 2.29 |
| | | S  1 | 4.25 | 4.24 | 4.21 | 4.17 | 4.12 | 4.05 | 3.98 | 3.90 | 3.83 | 3.75 | 3.67 | 3.60 | 3.54 | 3.49 | 3.44 | 3.40 | 3.38 | 3.36 | 3.36 |
| | | 2 | 3.83 | 3.81 | 3.77 | 3.71 | 3.62 | 3.53 | 3.42 | 3.31 | 3.20 | 3.09 | 2.99 | 2.90 | 2.82 | 2.75 | 2.70 | 2.65 | 2.62 | 2.60 | 2.60 |
| 47 | 64 | R or S  1 | 4.70 | 4.69 | 4.66 | 4.62 | 4.57 | 4.51 | 4.44 | 4.36 | 4.28 | 4.20 | 4.13 | 4.06 | 3.99 | 3.94 | 3.89 | 3.85 | 3.83 | 3.81 | 3.81 |
| | | 2 | 4.28 | 4.26 | 4.22 | 4.16 | 4.08 | 3.99 | 3.88 | 3.77 | 3.66 | 3.56 | 3.46 | 3.36 | 3.28 | 3.21 | 3.15 | 3.10 | 3.07 | 3.05 | 3.05 |
| | 76 | R  1 | 5.27 | 5.26 | 5.23 | 5.19 | 5.14 | 5.08 | 5.01 | 4.93 | 4.86 | 4.78 | 4.70 | 4.63 | 4.57 | 4.51 | 4.46 | 4.43 | 4.40 | 4.38 | 4.38 |
| | | 2 | 4.84 | 4.82 | 4.79 | 4.73 | 4.65 | 4.56 | 4.46 | 4.35 | 4.24 | 4.14 | 4.04 | 3.94 | 3.86 | 3.79 | 3.72 | 3.68 | 3.64 | 3.62 | 3.62 |
| | | S  1 | 5.80 | 5.79 | 5.76 | 5.73 | 5.68 | 5.62 | 5.55 | 5.47 | 5.40 | 5.32 | 5.24 | 5.17 | 5.11 | 5.05 | 5.00 | 4.97 | 4.94 | 4.92 | 4.92 |
| | | 2 | 5.38 | 5.36 | 5.33 | 5.27 | 5.19 | 5.10 | 5.00 | 4.90 | 4.79 | 4.68 | 4.58 | 4.48 | 4.40 | 4.32 | 4.26 | 4.21 | 4.17 | 4.15 | 4.15 |

| | | | | | | | | | | | | | | | | | | | | | |
|---|---|---|---|---|---|---|---|---|---|---|---|---|---|---|---|---|---|---|---|---|---|
| 38 | R or S | 1 | 2.65 | 2.64 | 2.61 | 2.56 | 2.51 | 2.44 | 2.37 | 2.29 | 2.21 | 2.14 | 2.07 | 2.01 | 1.95 | 1.90 | 1.87 | 1.83 | 1.81 | 1.80 | 1.80 |
| | | 2 | 2.27 | 2.26 | 2.23 | 2.20 | 2.15 | 2.09 | 2.03 | 1.96 | 1.90 | 1.84 | 1.78 | 1.73 | 1.68 | 1.64 | 1.60 | 1.58 | 1.56 | 1.55 | 1.55 |
| 51 | R | 1 | 3.92 | 3.90 | 3.86 | 3.80 | 3.72 | 3.63 | 3.53 | 3.42 | 3.31 | 3.21 | 3.11 | 3.02 | 2.94 | 2.87 | 2.82 | 2.77 | 2.74 | 2.72 | 2.72 |
| | | 2 | 3.02 | 3.00 | 2.97 | 2.92 | 2.86 | 2.79 | 2.70 | 2.62 | 2.53 | 2.45 | 2.37 | 2.30 | 2.24 | 2.19 | 2.14 | 2.11 | 2.08 | 2.07 | 2.07 |
| | S | 1 | 4.23 | 4.21 | 4.18 | 4.12 | 4.04 | 3.95 | 3.84 | 3.74 | 3.63 | 3.53 | 3.43 | 3.34 | 3.26 | 3.19 | 3.13 | 3.08 | 3.05 | 3.03 | 3.03 |
| 35 | | 2 | 3.33 | 3.31 | 3.29 | 3.24 | 3.18 | 3.10 | 3.02 | 2.94 | 2.85 | 2.77 | 2.69 | 2.62 | 2.56 | 2.50 | 2.46 | 2.42 | 2.40 | 2.38 | 2.38 |
| 64 | R or S | 1 | 4.68 | 4.66 | 4.63 | 4.57 | 4.49 | 4.40 | 4.30 | 4.20 | 4.09 | 3.99 | 3.89 | 3.80 | 3.71 | 3.64 | 3.58 | 3.54 | 3.50 | 3.48 | 3.48 |
| | | 2 | 3.78 | 3.77 | 3.74 | 3.69 | 3.63 | 3.56 | 3.48 | 3.39 | 3.31 | 3.23 | 3.15 | 3.07 | 3.01 | 2.95 | 2.90 | 2.86 | 2.84 | 2.82 | 2.82 |
| 76 | R | 1 | 5.25 | 5.23 | 5.20 | 5.14 | 5.07 | 4.98 | 4.88 | 4.78 | 4.67 | 4.57 | 4.47 | 4.37 | 4.29 | 4.22 | 4.16 | 4.11 | 4.07 | 4.05 | 4.05 |
| | | 2 | 4.35 | 4.34 | 4.31 | 4.26 | 4.21 | 4.14 | 4.06 | 3.97 | 3.89 | 3.81 | 3.73 | 3.65 | 3.58 | 3.52 | 3.47 | 3.44 | 3.41 | 3.39 | 3.39 |
| | S | 1 | 5.78 | 5.76 | 5.73 | 5.68 | 5.60 | 5.52 | 5.42 | 5.32 | 5.21 | 5.11 | 5.00 | 4.91 | 4.83 | 4.75 | 4.69 | 4.64 | 4.60 | 4.58 | 4.58 |
| | | 2 | 4.89 | 4.88 | 4.85 | 4.81 | 4.75 | 4.68 | 4.60 | 4.52 | 4.44 | 4.35 | 4.27 | 4.20 | 4.13 | 4.07 | 4.02 | 3.98 | 3.95 | 3.93 | 3.93 |

* R = round    S = square

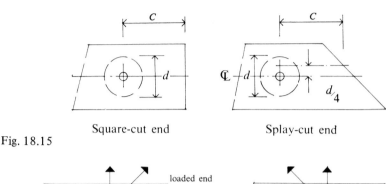

Square-cut end          Splay-cut end

Fig. 18.15

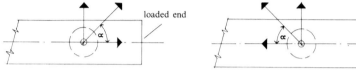

Fig. 18.16

minimum value of $K_C$ for sub-standard end distances varies between 0.85 and 0.5 and always occurs at $C_{min} = 0.5d + 13$ mm for any given case. There is no change in the values of $C_{st}$ and $C_{min}$ as the angle $\alpha$ varies from 0° to 90°, and for intermediate values of loaded end distance linear interpolation is permitted.

The end distance is said to be unloaded when the force on the connector has no component acting towards the end, and applies to forces acting from $\alpha = 90°$ to $\alpha = 0°$ (as shown in Fig. 18.16).

For an unloaded end, if the load is at $\alpha = 90°$ to grain:

$$C_{st} = 0.5d + 64 \text{ mm} \quad \text{and} \quad C_{min} = 0.5d + 13 \text{ mm}$$

as for the condition of a loaded end.

When $\alpha = 0°$:

$$C_{st} = C_{min} = 0.5d + 13 \text{ mm} \quad \text{and} \quad K_C = 1.0$$

For values between $\alpha = 0°$ and 90° the value of $C_{st}$ is determined by linear interpolation.

Edge distance $D$ is the distance from the edge of the member to the centre of the connector measured perpendicular to the edge (Fig. 18.17). If the edge is splay cut, the perpendicular distance from the centre of the connector to the splay edge should not be less than the edge distance $D$ (see Fig. 18.17).

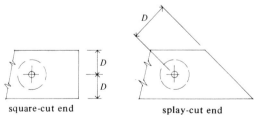

square-cut end          splay-cut end

Fig. 18.17

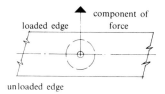

Fig. 18.18

The edge distance is said to be loaded or unloaded respectively according to whether or not the force on the connector has a component acting towards the edge of the member (Fig. 18.18).

In the case of tooth plates, the standard edge distance $D_{st}$ is also the minimum edge distance and applies to both the loaded and the unloaded edge, and $D_{st} = D_{min} = 0.5d + 6$ mm is applicable to all directions of load to grain. Therefore there are no sub-standard edge distances for tooth plates and $K_{60}$ is unaffected by edge distance.

Spacing $S$ is the distance between centres of adjacent connectors measured along a line joining their centres and known as the connector axis. Spacing can be parallel, perpendicular or at an angle to the grain as shown in Fig. 18.19.

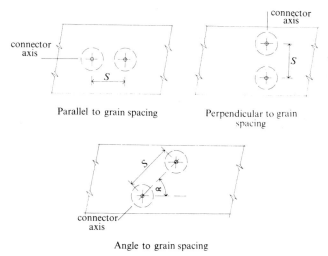

Fig. 18.19

The spacing of connectors is determined by the intersection of a diameter of an ellipse with its perimeter. The coordinates $A$ (parallel to grain) and $B$ (perpendicular to grain), as shown in Fig. 18.20, depend on the angle $\alpha$ of load to grain and on the size and shape of the tooth plate. The standard spacing $A$ parallel to grain and $B$ perpendicular to grain and the minimum spacing $S_{min}$ for loads parallel and perpendicular to the grain are expressed in terms of the connector diameter $d$ in Table 18.30. Standard spacings $A$ or $B$ permit $K_S = 1.0$ and minimum spacing $S_{min}$ permits $K_S = 0.75$ (where

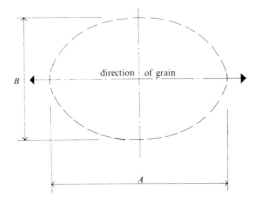

Fig. 18.20

Table 18.30    Spacing of toothed-plate connectors

| Angle of load to grain $\alpha$ | $A$ | $B$ | $S_{min}$ |
|---|---|---|---|
| $0°$ | $1.5d$ | $d + 13$ | $d + 13$ |
| $90°$ | $d + 13$ | $1.5d$ | $d + 13$ |

$A$ or $B$ coincide with $S_{min}$, $K_S$ is taken as 1.0). Intermediate values of $K_S$ are obtained by linear interpolation.

For angles of load to grain from $\alpha = 0°$ to $\alpha = 30°$, $A$ is greater than $B$, i.e. the major axis of the ellipse is parallel to the grain. At $\alpha = 30°$ the ellipse becomes a circle, and from $\alpha = 30°$ to $\alpha = 90°$ $A$ is less than $B$, i.e. the minor axis of the ellipse is parallel to the grain. These cases are illustrated in Fig. 18.21.

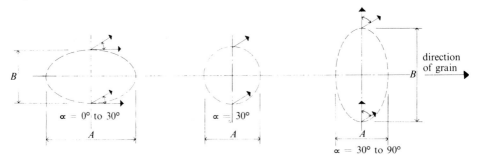

Fig. 18.21

When the load on a connector acts in a direction of $\alpha$ to the grain and the connector axis is at an angle $\theta$ to grain (Fig. 18.22), the standard spacing of connectors giving $K_S = 1.0$ may be determined from the equation:

$$S_{st} = \frac{AB}{\sqrt{(A^2 \sin^2 \theta + B^2 \cos^2 \theta)}} = \frac{A}{\sqrt{[1 + [(A/B)^2 - 1] \sin^2 \theta]}}$$

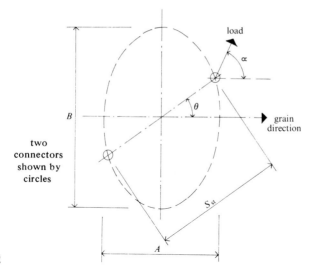

Fig. 18.22

$K_S = 0.75$ at $S_{min}$ and intermediate values of $K_S$ can be obtained by linear interpolation between $S_{st}$ and $S_{min}$:

$$K_S = 0.75 + 0.25 \left[ \frac{S - S_{min}}{S_{st} - S_{min}} \right]$$

where $S$ is the actual spacing.

Values of $A$, $B$ and $S_{min}$ for angle of load to grain $\alpha$ are given for tooth plates in Table 18.31.

**Table 18.31** Values for $A$, $B$ and $S_{min}$ for determination of $S_{st}$

| Size of toothed-plate connector (mm) | Round or square | Value of $A$ (mm) for angle of load to grain $\alpha$ (deg) | | | | | $S_{min}$ |
|---|---|---|---|---|---|---|---|
| | | $0°$ | $15°$ | $30°$ | $45°$ | $60-90°$ | |
| 38 | R | 57 | 57 | 54 | 54 | 51 | 51 |
| | S | 67 | 67 | 64 | 60 | 57 | 57 |
| 51 | R | 76 | 73 | 70 | 67 | 64 | 64 |
| | S | 86 | 83 | 79 | 76 | 70 | 70 |
| 64 | R | 95 | 92 | 86 | 83 | 76 | 76 |
| | S | 105 | 102 | 95 | 89 | 83 | 83 |
| 76 | R | 114 | 108 | 102 | 95 | 89 | 89 |
| | S | 124 | 117 | 111 | 102 | 95 | 95 |
| | | $90-60°$ | $45°$ | $30°$ | $15°$ | $0°$ | |

Value of $B$ (mm)
for angle of load to grain $\alpha$ (deg)

### 18.5.4   Example of determining $S_{st}$ and permissible connector load

Determine the value of $K_S$ for the arrangement of 64 mm diameter toothed-plate connectors shown in Fig. 18.23, given that $\theta = 50°$, $\alpha = 60°$ and $S = 80\,mm$. Determine the basic long term (dry) load per connector placed one side in 50 mm thick SC3 timber, assuming that edge and end distance requirements are satisfied.

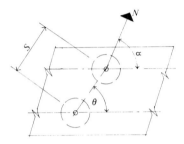

Fig. 18.23

From Table 18.31 with $\alpha = 60°$, $A = 76\,mm$, $B = 95\,mm$ and $S_{min} = 76\,mm$. The standard spacing at $\theta = 50°$ is:

$$S_{st} = \frac{76}{\sqrt{[1 + [(76/95)^2 - 1]\sin^2 50°]}} = \frac{76}{\sqrt{[1 - (0.36 \times 0.587)]}} = 86\,mm$$

therefore:

$$K_S = 0.75 + 0.25 \left[\frac{80 - 76}{86 - 76}\right] = 0.85$$

From Table 18.28, $F_0 = 4.70\,kN$ and $F_{90} = 3.90\,kN$.

$$F_{60} = \frac{F_0}{1 + [(F_0/F_{90}) - 1]\sin^2 60°} = \frac{4.70}{1 + (0.205)0.75} = 4.07\,kN$$

The basic long term load is:

$$F_{60} \times K_S = 4.07 \times 0.85 = 3.46\,kN$$
$$\text{(per connector)}$$

### 18.5.5   Charts for the standard placement of toothed-plate connectors

§§ 18.5.2 and 18.5.3 detail the background study into the determination of permissible loads on tooth plates and the formulae for establishing suitable placement of connectors with respect to one another and with respect to ends and edges of timber. To simplify the determination of loads and spacings for office use, Tables 18.32–18.39 give a graphical solution.

**Table 18.32**

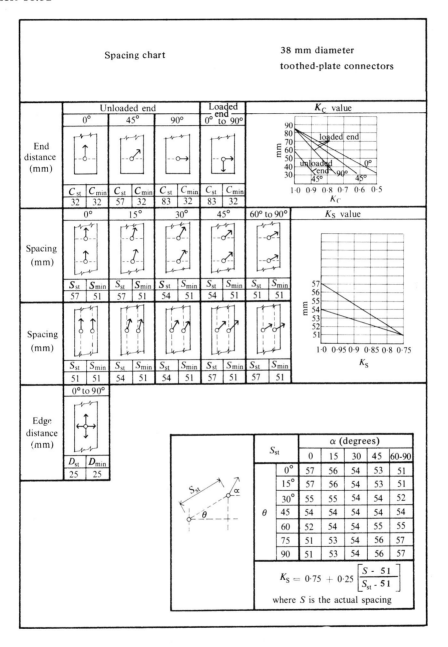

| Spacing chart | | 38 mm diameter toothed-plate connectors | |

| | Unloaded end | | | Loaded end 0° to 90° | $K_C$ value |
|---|---|---|---|---|---|
| **End distance (mm)** | 0° | 45° | 90° | | |
| | $C_{st}$ $C_{min}$ | $C_{st}$ $C_{min}$ | $C_{st}$ $C_{min}$ | $C_{st}$ $C_{min}$ | |
| | 32   32 | 57   32 | 83   32 | 83   32 | |

| | 0° | 15° | 30° | 45° | 60° to 90° | $K_S$ value |
|---|---|---|---|---|---|---|
| **Spacing (mm)** | $S_{st}$ $S_{min}$ | $S_{st}$ $S_{min}$ | $S_{st}$ $S_{min}$ | $S_{st}$ $S_{min}$ | $S_{st}$ $S_{min}$ | |
| | 57   51 | 57   51 | 54   51 | 54   51 | 51   51 | |

| **Spacing (mm)** | $S_{st}$ $S_{min}$ | $S_{st}$ $S_{min}$ | $S_{st}$ $S_{min}$ | $S_{st}$ $S_{min}$ | $S_{st}$ $S_{min}$ |
|---|---|---|---|---|---|
| | 51   51 | 54   51 | 54   51 | 57   51 | 57   51 |

| **Edge distance (mm)** | 0° to 90° |
|---|---|
| | $D_{st}$ $D_{min}$ |
| | 25   25 |

| $S_{st}$ | | $\alpha$ (degrees) | | | | |
|---|---|---|---|---|---|---|
| | | 0 | 15 | 30 | 45 | 60-90 |
| | 0° | 57 | 56 | 54 | 53 | 51 |
| | 15° | 57 | 56 | 54 | 53 | 51 |
| | 30° | 55 | 55 | 54 | 54 | 52 |
| $\theta$ | 45 | 54 | 54 | 54 | 54 | 54 |
| | 60 | 52 | 54 | 54 | 55 | 55 |
| | 75 | 51 | 53 | 54 | 56 | 57 |
| | 90 | 51 | 53 | 54 | 56 | 57 |

$$K_S = 0.75 + 0.25\left[\frac{S - 51}{S_{st} - 51}\right]$$

where $S$ is the actual spacing

**Table 18.33**

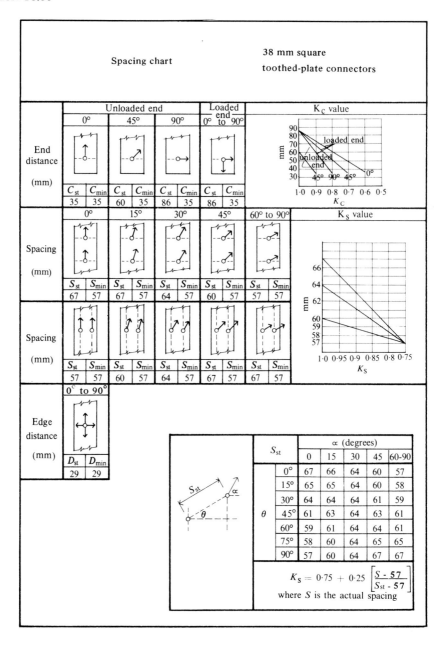

**Table 18.34**

Spacing chart — 51 mm diameter toothed-plate connectors

| End distance (mm) | Unloaded end | | | Loaded end 0° to 90° | $K_C$ value |
|---|---|---|---|---|---|
| | 0° | 45° | 90° | 0° to 90° | |
| | $C_{st}$ / $C_{min}$ | $C_{st}$ / $C_{min}$ | $C_{st}$ / $C_{min}$ | $C_{st}$ / $C_{min}$ | graph: loaded end, unloaded end |
| | 38 / 38 | 63 / 38 | 89 / 38 | 89 / 38 | $K_C$ axis: 1·0 0·9 0·8 0·7 0·6 0·5 |

| Spacing (mm) | 0° | 15° | 30° | 45° | 60° to 90° | $K_S$ value |
|---|---|---|---|---|---|---|
| | $S_{st}$ / $S_{min}$ | $S_{st}$ / $S_{min}$ | $S_{st}$ / $S_{min}$ | $S_{st}$ / $S_{min}$ | $S_{st}$ / $S_{min}$ | graph mm: 76 74 72 70 68 66 64 |
| | 76 / 64 | 73 / 64 | 70 / 64 | 67 / 64 | 64 / 64 | $K_S$ axis: 1·0 0·95 0·9 0·85 0·8 0·75 |

| Spacing (mm) | 0° | 15° | 30° | 45° | 60° to 90° |
|---|---|---|---|---|---|
| | $S_{st}$ / $S_{min}$ | $S_{st}$ / $S_{min}$ | $S_{st}$ / $S_{min}$ | $S_{st}$ / $S_{min}$ | $S_{st}$ / $S_{min}$ |
| | 64 / 64 | 67 / 64 | 70 / 64 | 73 / 64 | 76 / 64 |

| Edge distance (mm) | 0° to 90° |
|---|---|
| | $D_{st}$ / $D_{min}$ |
| | 32 / 32 |

| $S_{st}$ | α (degrees) | | | | |
|---|---|---|---|---|---|
| | 0 | 15 | 30 | 45 | 60-90 |
| θ 0° | 76 | 73 | 70 | 67 | 64 |
| 15° | 75 | 73 | 70 | 67 | 65 |
| 30° | 72 | 71 | 70 | 68 | 66 |
| 45° | 69 | 70 | 70 | 70 | 69 |
| 60° | 66 | 68 | 70 | 71 | 72 |
| 75° | 65 | 67 | 70 | 73 | 75 |
| 90° | 64 | 67 | 70 | 73 | 76 |

$$K_S = 0.75 + 0.25\left[\frac{S - 64}{S_{st} - 64}\right]$$

where $S$ is the actual spacing

**Table 18.35**

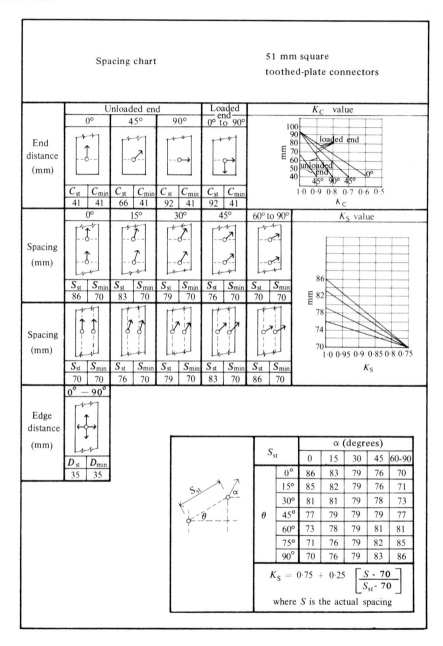

| | Spacing chart | | | | 51 mm square toothed-plate connectors | |

**Table 18.36**

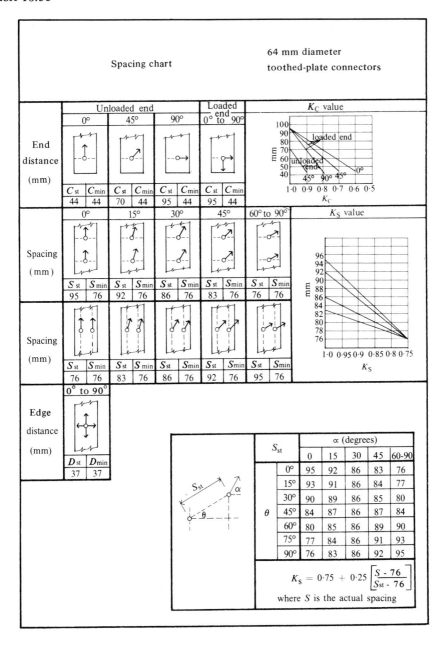

**Table 18.37**

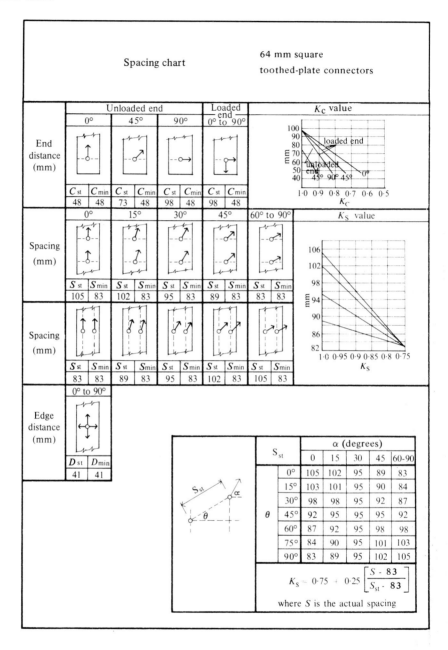

**Table 18.38**

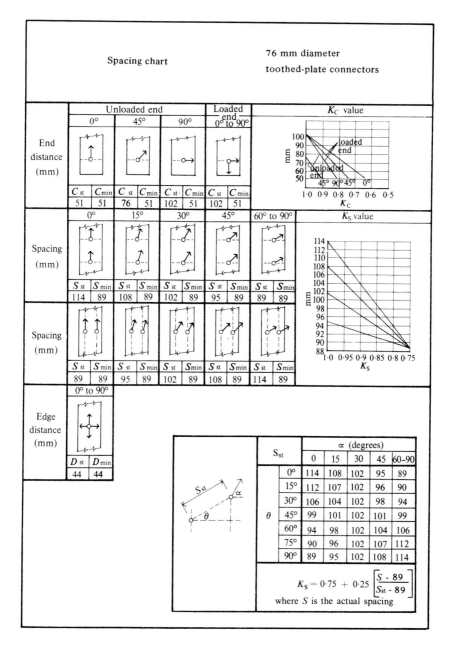

**Table 18.39**

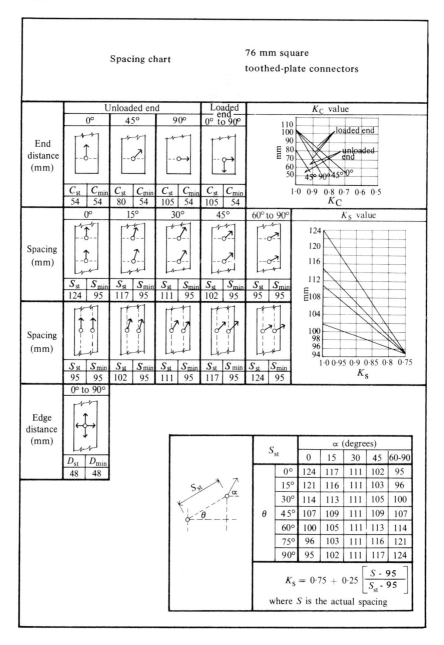

| | Unloaded end | | | | | | Loaded end | | $K_C$ value |
|---|---|---|---|---|---|---|---|---|---|
| | 0° | | 45° | | 90° | | 0° to 90° | | |
| End distance (mm) | $C_{st}$ | $C_{min}$ | $C_{st}$ | $C_{min}$ | $C_{st}$ | $C_{min}$ | $C_{st}$ | $C_{min}$ | |
| | 54 | 54 | 80 | 54 | 105 | 54 | 105 | 54 | |

Spacing chart — 76 mm square toothed-plate connectors

| $S_{st}$ | | α (degrees) | | | | |
|---|---|---|---|---|---|---|
| | | 0 | 15 | 30 | 45 | 60-90 |
| θ | 0° | 124 | 117 | 111 | 102 | 95 |
| | 15° | 121 | 116 | 111 | 103 | 96 |
| | 30° | 114 | 113 | 111 | 105 | 100 |
| | 45° | 107 | 109 | 111 | 109 | 107 |
| | 60° | 100 | 105 | 111 | 113 | 114 |
| | 75° | 96 | 103 | 111 | 116 | 121 |
| | 90° | 95 | 102 | 111 | 117 | 124 |

$$K_S = 0.75 + 0.25 \left[ \frac{S - 95}{S_{st} - 95} \right]$$

where $S$ is the actual spacing

Spacing (mm) — 0°, 15°, 30°, 45°, 60° to 90°
$S_{st}$ / $S_{min}$: 124 / 95, 117 / 95, 111 / 95, 102 / 95, 95 / 95

Spacing (mm):
$S_{st}$ / $S_{min}$: 95 / 95, 102 / 95, 111 / 95, 117 / 95, 124 / 95

Edge distance (mm) — 0° to 90°
$D_{st}$ / $D_{min}$: 48 / 48

## 18.6   SPLIT-RING AND SHEAR-PLATE CONNECTORS

### 18.6.1   General

General details for split rings and shear plates are shown in §§ 1.12.8 and 1.12.9. These are regarded generally as the most efficient ring-type connectors available, and are commonly used for most joints which require a high degree of load transference. Split rings are available in two sizes–64 mm and 104 mm nominal internal diameter–and consist of a circular band of steel, either (a) with parallel sides, or (b) with bevelled sides. The dimensions are given in Fig. 18.24.

| Nominal size | Shape | Dimensions (mm) | | | |
|---|---|---|---|---|---|
| | | A | B | C | D |
| 64 mm diameter split ring | Parallel side | 63.5 | 4.1 | – | 19 |
| | Bevelled side | 63.5 | 4.1 | 3.1 | 19 |
| 104 mm diameter split ring | Bevelled side | 101.6 | 4.9 | 3.4 | 25.4 |

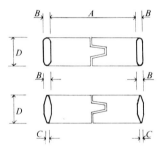

Fig. 18.24

Split rings are relatively flexible and are able to adjust their shape to off-set distortion which may occur in the grooves during any subsequent drying-out. This type of connector is therefore particularly suitable for joints where there may be some change of moisture content after assembly.

Shear plates are also available in two sizes–67 mm and 104 mm nominal outside diameter. 67 mm diameter shear plates are made from pressed steel, whereas 104 mm diameter shear plates are made from malleable cast iron. Each varies significantly in detail as illustrated in Fig. 18.25.

A split-ring unit consists of one split ring with its bolt in single shear (Fig. 18.26). The 64 mm diameter split ring requires a 12 mm diameter bolt, and the 104 mm diameter split ring requires a 20 mm diameter bolt. The 20 mm diameter bolt should also be used if 64 mm and 104 mm diameter split rings occur simultaneously at a joint concentric about a single bolt.

A shear-plate unit consists of either (a) one shear-plate connector with its bolt in single shear, used with a steel side plate in a timber-to-metal joint,

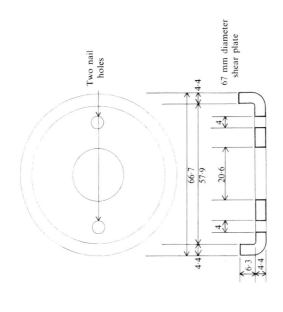

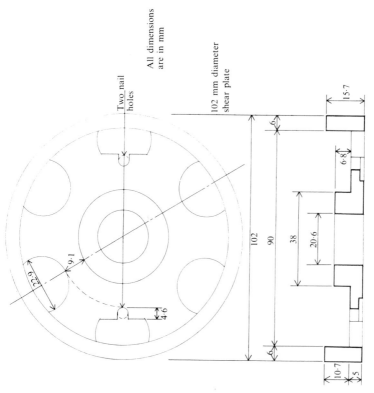

Fig. 18.25: *Details of shear plates.*

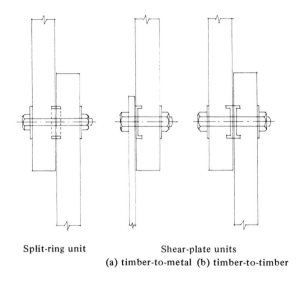

Split-ring unit       Shear-plate units
(a) timber-to-metal (b) timber-to-timber

Fig. 18.26

or (b) two shear plates used back to back in a timber-to-timber joint. Although this latter unit requires two shear plates to develop a load only slightly in excess of that provided by a single split ring, it can be justified where demountable joints for easier site assembly are required (i.e. split rings require to be drawn into a jointing groove under pressure, whereas shear plates are, or can be, a sliding fit followed by a bolt insertion).

### 18.6.2 Permissible loadings

The load-carrying actions of split rings and shear plates differ considerably. The split-ring connectors are embedded in a pre-drilled groove (see Table 75 of BS 5268:Part 2), the groove being slightly greater in diameter than the ring. The tongue-and-groove slot in the connector (see Fig. 18.24) permits the split ring to expand into the pre-formed groove. The depth of the split ring is shared equally between two mating faces of timber at a joint. The capacity of the unit is the combined values of the connector taking approximately 75% of the total load and the associated bolt taking approximately 25% of the total load. To function to its optimum value, therefore, the bolt (of the correct size) should not be omitted.

If a split ring is subjected to load, having been sprung into position in the pre-formed groove (as shown in Fig. 18.27), compression stresses occur where the outside of the connector bears against the side of the groove and where the inside of the connector bears against the internal annulus of the joint. The bolt also bears against the side of the central hole. These compression forces under the rings must be resisted by shear forces in the timber at the end of the member, and consequently, because of timber's low resistance to horizontal shear stresses, the end distance greatly influences the force which can be transferred. Failure of a joint is usually accompanied by

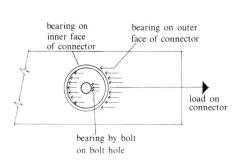

Fig. 18.27

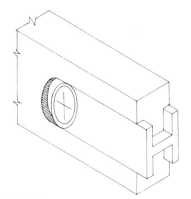

Fig. 18.28: *Typical split-ring mode of failure.*

a shearing action and the displacement of an I section piece of timber from under the bolt and ring, as sketched in Fig. 18.28.

The action of a shear plate is to transfer peripheral bearing stresses from the timber via the disc of the shear plate to the central hole in the plate. Load transfer is completed by bearing stress from the plate on to the central bolt and shear resistance in the bolt. The 67 mm diameter pressed steel shear plate is limited by bearing at the central hole, whereas the 102 mm diameter malleable cast iron shear plate with an increased centre hub thickness has a permissible load limited by the shear capacity of the 20 mm diameter bolt.

Tables 79 and 83 of BS 5268:Part 2 give the basic loads for one split-ring unit and one shear-plate unit respectively. Tables 18.40 and 18.41 give these values, whilst Table 18.42 gives limiting values for the permissive loads on one shear plate.

**Table 18.40    Basic loads for one split-ring connector unit**

| Split-ring diameter (mm) | Bolt size | Connector on one side only (mm) | Connectors on both sides and on same bolt (mm) | $\overline{F}_0$ Load parallel to grain SC1 and SC2 | SC3 and SC4 | SC5 | SC6– SC9 | $\overline{F}_{90}$ Load perpendicular to grain SC1 and SC2 | SC3 and SC4 | SC5 | SC6– SC9 |
|---|---|---|---|---|---|---|---|---|---|---|---|
| 64 | M12 | 22 | 32 | 5.23 | 6.00 | 7.39 | 8.92 | 3.66 | 4.20 | 5.17 | 6.25 |
| | | 25 | 40 | 6.32 | 7.25 | 8.92 | 10.8 | 4.42 | 5.08 | 6.25 | 7.55 |
| | | 29 | 50 | 7.68 | 8.81 | 10.8 | 13.1 | 5.38 | 6.17 | 7.59 | 9.18 |
| | | or over | or over | | | | | | | | |
| 102 | M20 | 29 | 41 | 10.1 | 11.5 | 14.2 | 17.2 | 7.04 | 8.08 | 9.94 | 12.0 |
| | | 32 | 50 | 11.5 | 13.1 | 16.2 | 19.5 | 8.01 | 9.19 | 11.3 | 13.7 |
| | | 36 | 63 | 13.5 | 15.4 | 19.0 | 23.0 | 9.42 | 10.8 | 13.3 | 16.1 |
| | | 40 | 72 | 14.3 | 16.4 | 20.1 | 24.3 | 9.98 | 11.4 | 14.1 | 17.0 |
| | | 41 | 75 | 14.4 | 16.5 | 20.3 | 24.5 | 10.1 | 11.5 | 14.2 | 17.2 |
| | | or over | or over | | | | | | | | |

\* Actual thickness. Intermediate thicknesses may be obtained by linear interpolation.

**Table 18.41** Basic loads* for one shear-plate connector unit (also see Table 18.42)

| Shear-plate diameter (mm) | Bolt size | Thickness of members† Connector on one side only (mm) | Thickness of members† Connectors on both sides and on same bolt (mm) | $\bar{F}_0$ Load parallel to grain SC1 and SC2 | $\bar{F}_0$ SC3 and SC4 | $\bar{F}_0$ SC5 | $\bar{F}_0$ SC6– SC9 | $\bar{F}_{90}$ Load perpendicular to grain SC1 and SC2 | $\bar{F}_{90}$ SC3 and SC4 | $\bar{F}_{90}$ SC5 | $\bar{F}_{90}$ SC6– SC9 |
|---|---|---|---|---|---|---|---|---|---|---|---|
| 67 | M20 | – | 41 | 6.45 | 7.39 | 9.10 | 11.0 | 4.51 | 5.18 | 6.37 | 7.70 |
| | | – | 50 | 7.72 | 8.85 | 10.9 | 11.9 | 5.40 | 6.20 | 7.63 | 9.22 |
| | | – | 63 | 8.21 | 9.42 | 11.6 | 11.9 | 5.75 | 6.59 | 8.11 | 9.80 |
| | | 41 and over | 72 and over | 8.32 | 9.54 | 11.7 | 11.9 | 5.82 | 6.68 | 8.22 | 9.93 |
| 102 | M20 | – | 44 | 8.32 | 9.55 | 11.7 | 14.2 | 5.83 | 6.68 | 8.22 | 9.94 |
| | | – | 50 | 9.18 | 10.5 | 12.9 | 15.6 | 6.42 | 7.37 | 9.07 | 11.0 |
| | | – | 63 | 10.5 | 12.1 | 14.5 | 18.0 | 7.38 | 8.47 | 10.4 | 12.6 |
| | | 41 | – | 11.7 | 13.4 | 16.5 | 19.9 | 8.17 | 9.37 | 11.5 | 13.9 |
| | | – | 75 | 11.9 | 13.6 | 16.8 | 20.3 | 8.33 | 9.55 | 11.8 | 14.2 |
| | | 44 and over | 97 and over | 12.5 | 14.4 | 17.7 | 21.4 | 8.79 | 10.1 | 12.4 | 15.0 |

* The tabulated values apply for timber-to-timber and steel-to-timber joints, except that for steel-to-timber joints made with 102 mm shear plates, the parallel-to-grain loads may be increased by:

3% for SC3 and SC4 timber
11% for SC5 timber and
16% for SC6–SC9 timber.

† Actual thickness. Intermediate thicknesses may be obtained by linear interpolation.

**Table 18.42** Limiting values for permissible loads on one shear-plate connector unit‡

| Shear-plate diameter (mm) | Bolt size | All loading except wind (kN) | All loading including wind (kN) |
|---|---|---|---|
| 67 | M20 | 12.9 | 17.2 |
| 102 | M20 | 22.1 | 29.5 |

‡ These values may cause a reduction in a value obtained from Table 18.41.

In addition to the design check on the adequacy of the connectors, the designer must also check the net area of the timber at the connection. BS 5268 : Part 2 makes it clear that it is the projected area of split-ring and shear-plate connectors which must be deducted in arriving at the net area (see Tables 13.1 and 13.2), plus of course the area of the bolt hole which falls outside the projected area of the connector. To aid the designer, particularly in the design of truss joints, values for loading parallel to the grain in SC3 and SC4 timbers surfaced in accordance with the recommendations for constructional timber (§ 1.6.3) are given in Table 18.43.

The tabulated basic loads are based on tests carried out by Scholten at the US Forest Products Laboratory, and it has been shown that the capacity of a

**Table 18.43**   **Basic connector capacities ($\overline{F}_0$) in surfaced timber of Strength Classes SC3 and SC4**

| Type of connector | Diameter (mm) | Thickness of member (mm) | Connector on one side of member | | | Connector on each side of member | | |
|---|---|---|---|---|---|---|---|---|
| | | | long term load | medium term load | short or very short term load | long term load | medium term load | short or very short term load |
| Split ring | 64 | 35 | 8.81 | 11.0 | 13.2 | 6.47 | 8.09 | 9.70 |
| | | 47 | 8.81 | 11.0 | 13.2 | 8.34 | 10.4 | 12.5 |
| | | 60 | 8.81 | 11.0 | 13.2 | 8.81 | 11.0 | 13.2 |
| | | 72 | 8.81 | 11.0 | 13.2 | 8.81 | 11.0 | 13.2 |
| | 104 | 35 | 14.8 | 18.5 | 22.2 | X | X | X |
| | | 47 | 16.5 | 20.6 | 24.7 | 12.5 | 15.7 | 18.8 |
| | | 60 | 16.5 | 20.6 | 24.7 | 14.8 | 18.5 | 22.3 |
| | | 72 | 16.5 | 20.6 | 24.7 | 16.4 | 20.5 | 24.6 |
| Shear plate | 67 | 35 | X | X | X | X | X | X |
| | | 47 | 9.54 | 11.9 | 14.3 | 8.36 | 10.4 | 12.5 |
| | | 60 | 9.54 | 11.9 | 14.3 | 9.29 | 11.6 | 13.9 |
| | | 72 | 9.54 | 11.9 | 14.3 | 9.54 | 11.9 | 14.3 |
| | 104 | 35 | X | X | X | X | X | X |
| | | 47 | 14.4 | 18.0 | 21.6 | 10.0 | 12.5 | 15.0 |
| | | 60 | 14.4 | 18.0 | 21.6 | 11.7 | 14.6 | 17.6 |
| | | 72 | 14.4 | 18.0 | 21.6 | 13.5 | 16.9 | 20.3 |

Connector capacity $\overline{F}_0$ (kN)

X = not permitted.

connector is dependent on the size and type of the connector, the species (principally the density) of timber, the direction of load with respect to grain, and the thickness of the timber.

The basic load at an angle to the grain is calculated using the Hankinson formula or the Scholten nomogram described in § 18.4.3. It has been shown that:

$$F_\alpha = \frac{F_0}{1 + [(F_0/F_{90}) - 1]\sin^2\alpha}$$

A consideration of the ratio $F_0/F_{90}$ for the values tabulated for $F_0$ and $F_{90}$ in Tables 18.40 and 18.41 gives a standard value of 1.428 for cases other than 64 mm diameter shear plates in Strength Classes SC6–SC9. Hence (for other than that case):

$$F_\alpha = K_\alpha F_0 \text{ where } K_\alpha = \frac{1}{1 + 0.428\sin^2\alpha}$$

The formula should not be used for 67 mm diameter shear plates in SC6 to SC9 timbers (where the formula would give a maximum under-estimate of 20% at $\alpha = 90°$) or where advantage is taken of the increase in loads parallel to grain permitted in the footnote to Table 18.41. Values of $K_\alpha$ are given in Table 18.44.

**Table 18.44**

| $\alpha$ | $K_\alpha$ |
|---|---|
| 0 | 1.0 |
| 5 | 0.997 |
| 10 | 0.987 |
| 15 | 0.972 |
| 20 | 0.952 |
| 25 | 0.929 |
| 30 | 0.903 |
| 35 | 0.877 |
| 40 | 0.850 |
| 45 | 0.824 |
| 50 | 0.799 |
| 55 | 0.777 |
| 60 | 0.757 |
| 65 | 0.740 |
| 70 | 0.726 |
| 75 | 0.715 |
| 80 | 0.707 |
| 85 | 0.702 |
| 90 | 0.700 |

The permissible load for a split-ring connector unit is given by the formula:

$$\text{Permissible load} = \text{basic load} \times K_{62} \times K_{63} \times K_{64} \times K_{65}$$

and, for a shear-plate connector unit:

$$\text{Permissible load} = \text{basic load} \times K_{66} \times K_{67} \times K_{68} \times K_{69}$$

where $K_{62} = K_{66}$ = the modification factor for duation of load, being:

1.0 for permanent loading
1.25 for medium term loading
1.5 for short term or very short term loading

$K_{63} = K_{67}$ = the modification factor for moisture content, being unity for dry timber to be used in dry exposure conditions, and as listed in Table 18.3 for other cases

$K_{64} = K_{68}$ = the lesser value of $K_C$, $K_D$ and $K_S$ which are the respective modification factors for sub-standard end distance, edge distance and spacing

$K_{65} = K_{69}$ = the modification factor for the number of connectors in each line (see Table 18.24).

### 18.6.3　Standard and sub-standard placement of connectors

To develop the full basic load split rings and shear plates must have the required standard end distance $C_{st}$, edge distance $D_{st}$ and spacing $S_{st}$, which can be expressed in terms of the nominal connector diameter 'd'. In calculating end distance, edge distance, and spacing (all as defined in § 18.5.3), 'd' for a 67 mm diameter shear plate should be taken as 64 mm. Note that BS 5268:Part 2 gives identical values for end and edge distances, and spacings, in Tables 76, 77, 78, 80 and 81.

With loaded end distance (Fig. 18.16), $K_C = 1.0$ at $C_{st} = d + 76$ mm, and the minimum value of $K_C$ for sub-standard end distance is $K_C = 0.625$ at $C_{min} = 0.5 \times C_{st} = 0.5d + 38$ mm. There is no change in the values of $C_{st}$ and $C_{min}$ as the angle $\alpha$ of load to grain varies from $0°$ to $90°$, and for intermediate values of loaded end distance linear interpolation of $K_C$ is permitted.

When the end distance is unloaded at $\alpha = 0°$ (see Fig. 18.16), $K_C = 1.0$ at $C_{st} = d + 38$ mm and $K_C = 0.625$ at $C_{min} = 0.5d + 32$ mm with intermediate values of $K_C$ determined by linear interpolation. When the load acts perpendicular to grain ($\alpha = 90°$), the end distance is to be regarded as loaded; hence $K_C = 1.0$ at $C_{st} = d + 76$ mm and $K_C = 0.625$ at $C_{min} = 0.5d + 38$ mm. When the load acts at an intermediate angle to grain, $C_{st}$ and $C_{min}$ may be reduced linearly (i.e. at any angle $\alpha$) hence:

$$C_{st} = d + 0.422\alpha + 38 \quad \text{and} \quad C_{min} = 0.5d + \alpha/15 + 32$$
$$(\alpha \text{ in degrees})$$

The minimum edge distance is $D_{min} = 44$ mm for 64 mm diameter split rings and 67 mm diameter shear plates, and $D_{min} = 70$ mm for 102 mm diameter split rings and shear plates. For an unloaded edge distance $D_{st} = D_{min}$ and $K_D = 1.0$ (see Fig. 18.18).

For loaded edge distance, $D_{st} = D_{min}$ at $\alpha = 0°$, whereas at $\alpha = 45°$ to $90°$, $D_{st} = 70$ mm for 64 mm diameter split rings and 67 mm diameter shear plates and $D_{st} = 95$ mm for 102 mm diameter split rings and shear plates. For intermediate angles to grain between $\alpha = 0°$ and $45°$, there is a linear variation in $D_{st}$. There is a reduction in $K_D$ for sub-standard loaded edge distance.

For 64 mm diameter split rings and 67 mm diameter shear plates:

$$K_D = 1 - 0.17 \left[ \frac{\alpha}{45} - \frac{(D-44)}{26} \right]$$

where $\alpha$ = angle of load to grain between $0°$ and $45°$
　　$D$ = actual edge distance (mm).

Note that where $\alpha$ is greater than $45°$, $\alpha/45$ is taken as unity. $D$ is in the range of 44 mm to 70 mm and $K_D$ is not greater than 1.0.

For 102 mm diameter split rings and shear plates:

$$K_D = 1 - 0.17 \left[ \frac{\alpha}{45} - \frac{(D-70)}{25} \right]$$

where $\alpha$ = angle of load to grain between 0° and 45°

$D$ = actual edge distance (mm).

Note that where $\alpha$ is greater than 45°, $\alpha/45$ is taken as unity. $D$ is in the range of 70 mm to 95 mm and $K_D$ is not greater than 1.0.

The spacing for split-ring and shear-plate connectors is determined in a similar way to that for toothed-plate connectors. Referring to Fig. 18.20, the appropriate standard spacings $A$ and $B$ parallel and perpendicular to the grain respectively, and the minimum spacing $S_{min}$ for loads parallel and perpendicular to the grain, can be expressed in terms of the nominal connector diameter as given in Table 18.45.

Table 18.45  Spacing for split rings and shear plates

| Angle of load to grain $\alpha$ (degrees) | $A$ | $B$ | $S_{min}$ |
|---|---|---|---|
| 0 | $1.5d + 76$ | $d + 25$ | $d + 25$ |
| 90 | $d + 25$ | $1.5d + 12$ | $d + 25$ |

Table 18.46  Values of $A$, $B$ and $S_{min}$ for the determination of $S_{st}$

| Type and size of connector | Angle of load to grain (degrees) | $A$ (mm) | $B$ (mm) | $S_{min}$ (mm) |
|---|---|---|---|---|
| 64 mm | 0 | 171 | 89 | |
| split ring | 15 | 152 | 95 | |
| 67 mm | 30 | 130 | 98 | 89 |
| shear plate | 45 | 108 | 105 | |
| | 60–90 | 89 | 108 | |
| 102 mm | 0 | 229 | 127 | |
| split ring | 15 | 203 | 137 | |
| 102 mm | 30 | 178 | 146 | 127 |
| shear plate | 45 | 152 | 156 | |
| | 60–90 | 127 | 165 | |

Values of $A$, $B$ and $S_{min}$ for angles $\alpha$ of load to grain are given in Table 18.46. When the load on a connector acts at $\alpha$ to grain (see Fig. 18.22) and the connector axis is at an angle $\theta$ to grain, the standard spacing of connectors, giving $K_S = 1.0$, is determined as for tooth plates:

$$S_{st} = \frac{A}{\sqrt{[1 + [(A/B)^2 - 1] \sin^2 \theta]}}$$

similarly $K_S = 0.75$ at $S_{min}$ and intermediate values of $K_S$ are given as:

$$K_S = 0.75 + 0.25 \left[ \frac{S - S_{min}}{S_{st} - S_{min}} \right]$$

where $S$ is the actual spacing.

When $K_S$ is known, this formula may be transposed to:

$$S = \left[ \frac{(K_S - 0.75)(S_{st} - S_{min})}{0.25} \right] + S_{min}$$

### 18.6.4   Example of determining $S_{st}$, $K_S$ and permissible connector load

Determine the value of $K_S$ and the permissible long term load for each connector arranged as in Fig. 18.29. Connectors are 64 mm diameter split rings placed in two faces of a 47 mm thick SC3 timber. $\theta = 30°$, $\alpha = 15°$ and $S = 100$ mm. Assume that edge and end distance requirements are satisfied.

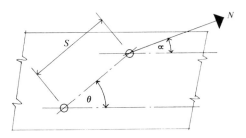

Fig. 18.29

From Table 18.46 with $\alpha = 15°$, $A = 152$ mm, $B = 95$ mm and $S_{min} = 89$ mm:

$$S_{st} = \frac{152}{\sqrt{[1 + [(152/95)^2 - 1] \sin^2 30°]}} = 129 \text{ mm}$$

$$\therefore K_S = 0.75 + 0.25 \left[ \frac{100 - 89}{129 - 89} \right] = 0.819$$

From Table 18.43, $\bar{F}_0 = 8.34$ kN
From Table 18.44, $K_\alpha = 0.972$

Permissible load (per connector) $= \bar{F}_0 \times K_\alpha \times K_S$
$$= 8.34 \times 0.972 \times 0.819 = 6.64 \text{ kN}$$

### 18.6.5   Charts for the standard placement of split-ring and shear-plate connectors

§§ 18.6.2 and 18.6.3 detail the background study into the determination of permissible loads for shear plates and split rings, and formulae for establishing suitable placements of connectors with respect to one another and with respect to ends and edges of timber. To simplify the determination of loads and spacings for office use, a graphical solution is given in Tables 18.47 and 18.48 and Figs 18.30 and 18.31.

**Table 18.47**

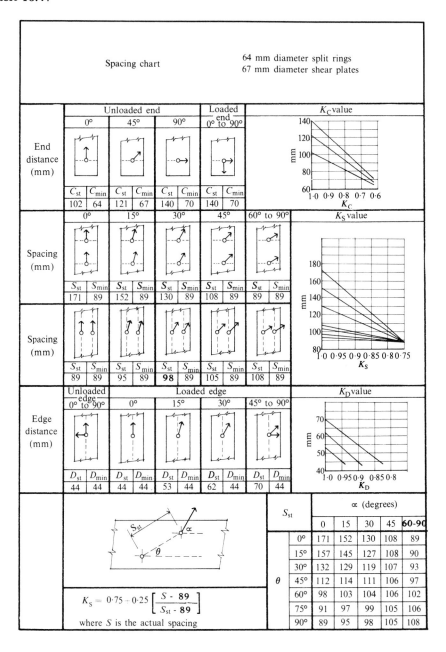

**Table 18.48**

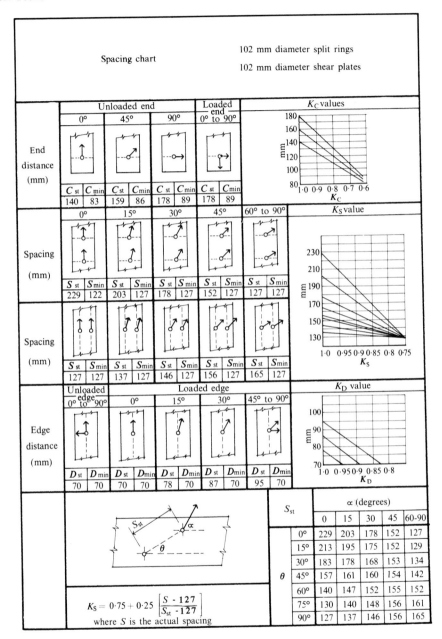

| | Unloaded end | | | Loaded end 0° to 90° | $K_C$ values |
|---|---|---|---|---|---|
| | 0° | 45° | 90° | | |

**End distance (mm)**

| $C_{st}$ | $C_{min}$ | $C_{st}$ | $C_{min}$ | $C_{st}$ | $C_{min}$ | $C_{st}$ | $C_{min}$ |
|---|---|---|---|---|---|---|---|
| 140 | 83 | 159 | 86 | 178 | 89 | 178 | 89 |

**Spacing (mm)** — 0°, 15°, 30°, 45°, 60° to 90° — $K_S$ value

| $S_{st}$ | $S_{min}$ | $S_{st}$ | $S_{min}$ | $S_{st}$ | $S_{min}$ | $S_{st}$ | $S_{min}$ | $S_{st}$ | $S_{min}$ |
|---|---|---|---|---|---|---|---|---|---|
| 229 | 122 | 203 | 127 | 178 | 127 | 152 | 127 | 127 | 127 |

**Spacing (mm)**

| $S_{st}$ | $S_{min}$ | $S_{st}$ | $S_{min}$ | $S_{st}$ | $S_{min}$ | $S_{st}$ | $S_{min}$ | $S_{st}$ | $S_{min}$ |
|---|---|---|---|---|---|---|---|---|---|
| 127 | 127 | 137 | 127 | 146 | 127 | 156 | 127 | 165 | 127 |

| Unloaded edge 0° to 90° | Loaded edge | | | | $K_D$ value |
|---|---|---|---|---|---|
| | 0° | 15° | 30° | 45° to 90° | |

**Edge distance (mm)**

| $D_{st}$ | $D_{min}$ | $D_{st}$ | $D_{min}$ | $D_{st}$ | $D_{min}$ | $D_{st}$ | $D_{min}$ | $D_{st}$ | $D_{min}$ |
|---|---|---|---|---|---|---|---|---|---|
| 70 | 70 | 70 | 70 | 78 | 70 | 87 | 70 | 95 | 70 |

$$K_S = 0.75 + 0.25 \left[ \frac{S - 127}{S_{st} - 127} \right]$$

where $S$ is the actual spacing

| $S_{st}$ | | $\alpha$ (degrees) | | | | |
|---|---|---|---|---|---|---|
| | | 0 | 15 | 30 | 45 | 60-90 |
| | 0° | 229 | 203 | 178 | 152 | 127 |
| | 15° | 213 | 195 | 175 | 152 | 129 |
| | 30° | 183 | 178 | 168 | 153 | 134 |
| $\theta$ | 45° | 157 | 161 | 160 | 154 | 142 |
| | 60° | 140 | 147 | 152 | 155 | 152 |
| | 75° | 130 | 140 | 148 | 156 | 161 |
| | 90° | 127 | 137 | 146 | 156 | 165 |

102 mm diameter split rings

102 mm diameter shear plates

Spacing chart

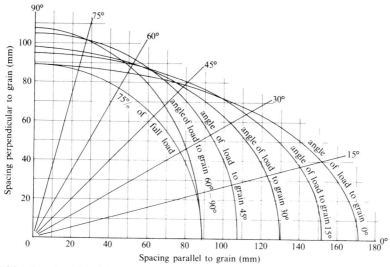

Fig. 18.30: *Spacing for 64 mm diameter split rings and 67 mm diameter shear plates.*

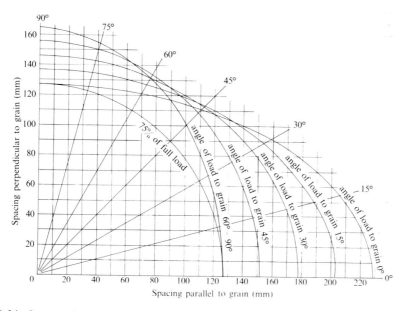

Fig. 18.31: *Spacing for 102 mm diameter split rings and shear plates.*

## 18.7 STEEL GUSSETS

Although the tendency in timber engineering constructions is to use connections and gussets specially made for timber, there are cases, particularly in truss design, where the neatest answer is to use steel gussets drilled for nails, screws or bolts, or even to use them with the specialist connectors. To limit slip, the diameter of the holes should be as close as possible to the diameter of the fastener. Gussets can be painted or hot-dip galvanised, depending on service conditions. Gussets are usually of Grade 43 steel, with a permissible bearing stress of 250 N/mm² (Clause 50 of BS 449).

## 18.8   PUNCHED METAL-PLATE FASTENERS

Punched metal-plate fasteners are usually fabricated from 18–20 gauge thick coil steel (the Code of Practice on trussed rafter roofs sets a minimum thickness of 0.91 mm over the protective coating). Stainless steel plates, and thicker plates (e.g. 14 gauge) are also available, if not so common. Integral nails or teeth are pressed-out at right angles to give spikes at close centres, which will subsequently be pressed into the timber by specialist machinery. Plain steel plates are hot-dip galvanised or otherwise zinc coated.

To date, the largest use of punched metal-plate fasteners has been as gusset plates in trussed rafters (Chapter 20) where all the members are in line and the plates are fixed on each side. Many manufacturers of proprietary metal-plate fasteners have obtained an Agrément certificate for their plates. These certificates present general information and also details of tension and shear strength of the plates. In certain cases, plates are called upon to take a bending moment (e.g. in the rafter of a trussed rafter). This can be justified by test or an investigation of the forces acting on individual spikes and abutting faces at joints.

Although metal-plate fasteners can resist bending moments, the engineer should exercise caution in placing plates in situations where they will take moment. Although plates have been used to splice simple joists, the engineer must realise that they do not give a fully rigid joint, and additional deflection will take place due to deformation and slip at the joint. (See § 21.6.2 for notes on slip.)

If using metal-plate fasteners as gussets in a framework which is not fully triangulated, the engineer is advised to revert to prototype testing to confirm the design.

## 18.9   H CLIPS

Aluminium H clips approximately 12 mm long can be obtained for fitting between plywood sheets, and in certain cases can take the place of a nogging or tongued and grooved detail (see Fig. 18.32). In theory these clips can be obtained for all thicknesses of plywood, but unless a large order is placed they are normally only available for thicknesses up to 12.5 mm. Clips are usually placed at 150–300 mm centres.

When the final finishing is shingles or tile hanging on battens, H clips may

Fig. 18.32

well be adequate to prevent differential movement of the plywood, although it should be realised that there is a narrow gap between all sheets. However, when the plywood is acting as a base for asphalt, felt or any such waterproof membrane, the authors advise against the use of H clips. During laying or at any subsequent time, the weight of a man acting between clips, or even the differential movement which can occur in service, is likely to tear the felt or cause a leak.

H clips should not be used if the junction on which they occur is on a slope. During erection or subsequently, slight vibration can cause the clips to slide down the slope.

## 18.10   MILD STEEL HANGERS

Proprietary or specially designed mild steel hangers are produced by a number of manufacturers for fixing timber joists to brickwork, concrete, steel, timber etc. (Fig. 18.33) and for fixing ply web or glulam beams. Hangers can be painted, galvanised, Sherardised or otherwise zinc coated. The thickness of steel used in manufacture and the specification varies, as consequently does the safe load on the hanger. If the engineer carries out a special design he should consider the bearing stress between the incoming beam and the hanger base, the strength of the welding, the tension in the steel at the top bend, and the bearing of the hanger at the top on the supporting material. In addition, with larger solid joists and ply web and glulam beams, the engineer must consider the possibility of the eccentricity of application of loading causing the trimmer to rotate. The sketch in Fig. 18.34 indicates the tendency of an incoming beam to rotate a trimmer. This can obviously crack a ceiling, look unsightly, or be dangerous.

The easiest solution is sketched in Fig. 18.35 with screw or drive screw fixing into the trimmer, nails into the side of the beam, and nails through the top of the hanger into the trimmer. Although most hangers of this type have a small 'tail' bent into them, the tolerances are such that the designer can not rely on there being any bearing between it and the back of the trimmer. Its only effective purpose is as a safety device during assembly to prevent the hanger slipping off the trimmer. The tension in the screws A and the shear in the nails (or drive screws) B can be taken as $F_r e/h_e$, where $F_r$ is the reaction from the incoming beam.

When a large beam is supported by a hanger with the top located between bricks or blocks, the engineer must consider the possibility of rotation as

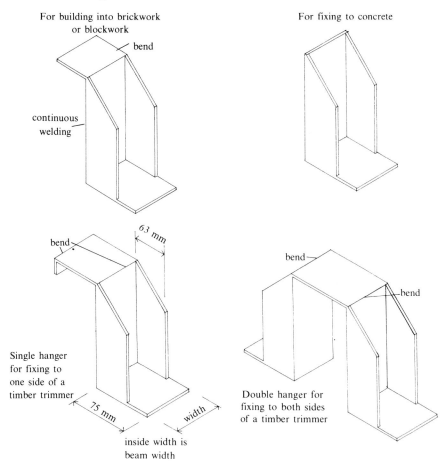

Fig. 18.33: *Typical hanger shapes–approximate dimensions.*

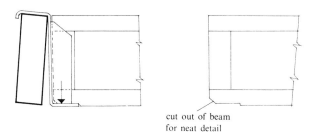

Fig. 18.34

sketched in Fig. 18.36 due to the hanger not acting as a shear connection and the top being withdrawn by tension. This can be resisted by fixings through the back of the hanger or through side lugs near the top of the hanger (Fig. 18.37). The tension on these fixings can be taken as $F_r e/h_e$. The resisting moment is completed by the bearing of the lower vertical part of the hanger against the wall. Hangers of the type described for ply web and glulam beams are often fabricated of 10 gauge (3 mm) steel.

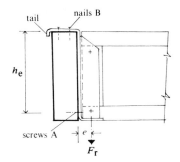

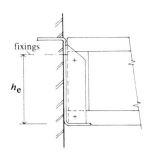

Screws A can be
fixed through side
lugs to ease erection.

Fig. 18.35

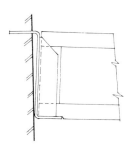

Fig. 18.36

Fig. 18.37

Small proprietary hangers for fixing small beams of solid timber may
support the beams more by nail shear and tension in the steel than bearing
on the base, as with the larger hangers sketched in this section.

Also see BS 6178:Part 1:1982, 'Specification for joist hangers for
building into masonry walls of domestic dwellings'.

## 8.11   HOLDING-DOWN BOLTS AND STRAPS

Although timber components can often be adequately fixed down by nailing
or clipping to wallplates, there are many cases where it is necessary to use
bolts through steel brackets into a padstone or concrete (Fig. 18.38). These

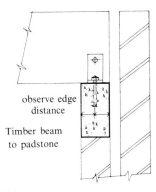

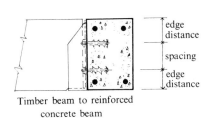

Timber beam to reinforced
concrete beam

Fig. 18.38

bolts can be ordinary black bolts with washers, rag bolts or one of the proprietary expanding type. Any of these give a satisfactory fixing if fitted correctly.

Figure 18.39 shows sketches of two of the more common expanding type of Rawlbolts manufactured by the Rawlplug Co. Ltd. Rawlbolts are available both in 'loose bolt' and 'bolt projecting' types. The loose bolt type is suitable

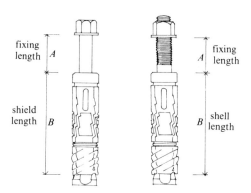

fixing length *A*

shield length *B*

fixing length *A*

shell length *B*

Fig. 18.39

for insertion after the component has been positioned. The bolt can be removed from the shield at any time if required. The bolt projecting type is positioned before the component is put into place and is the type more commonly used with timber components. Load capacity is achieved by the tightening of the bolt, causing the expansion into the sides of the pre-drilled hole in the shield (loose bolt type) or the shell (bolt projecting type). In detailing the connection, adequate edge distances and bolt spacings must be provided to prevent cracking of the padstone or concrete.

One very common method of holding down timber components such as wallplates, ends of beams, stud walls etc. is the use of steel straps anchored into concrete or masonry, with holes pre-drilled for fixings into the timber. Such straps can be fairly long but thin, allowing tolerances in construction to be taken up.

# Chapter Nineteen

# Glue Joints, Including Finger Joints

## 19.1 INTRODUCTION

Structural adhesives are available which are suitable for either interior or exterior conditions and which give rigid permanent glue joints in timber providing the correct quality control is exercised. The history of gluing timber goes back many centuries, and modern synthetic adhesives have been used successfully for structural applications since the beginning of this century. Finger jointing is also well proven, having been developed and used successfully since the Second World War.

Glue joints should be designed so that the adhesive is stressed in shear, with little if any secondary stresses to cause tension on the adhesive. Adhesives such as epoxy resin with a sand additive can be used as a filler in a compression joint, either between timber and timber, or timber and steel or concrete, but the comments in this chapter mainly refer to joints in which the adhesive is completely or mainly in shear (Fig. 19.1).

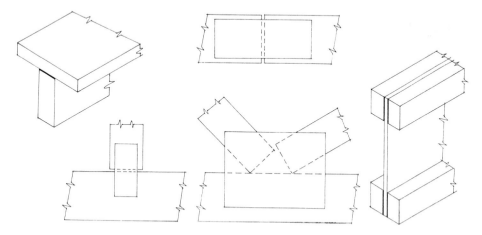

Fig. 19.1: *Typical glue joints.*

The object in design and manufacture is to make the glue line as thin as possible, compatible with good 'wetting' of both surfaces. 'Gap-filling' glues are available–in fact these are the types usually used in timber engineering–

but the designer is advised to regard this phrase as an indication of good quality and not as encouraging gaps to be built-in. Such adhesives still retain adequate strength while filling localised gaps, for example those due to slight imperfections in machining. The gap usually quoted as the maximum acceptable between two mating surfaces is 1.3 mm, which is the test gap for glues required to satisfy BS 1204 : Part 1 : 1979.

The adhesive can usually be applied to one or both surfaces, the same total amount being used whichever way is adopted. The spreading rate specified by the adhesive manufacturer should be used unless sufficient testing is carried out to prove that less is adequate for a particular detail. Often, 'squeeze-out' can be used as an indication of whether or not sufficient adhesive has been applied.

The glue joint must be held in close contact during curing. The pressure can be applied by clamps which are removed after curing, or by nails or other fasteners which are left in the final assembly. Even if using nails for bonding pressure there are advantages in holding the assembly with clamps until the nailing is carried out.

If a glue joint is tested to destruction, failure should normally take place in the timber or plywood close to the glue face, not in the thickness of the adhesive, and the permissible design stresses used in the design of glue joints are those appropriate to the timber, not the adhesive.

A correctly designed structural glue joint may be expected not to slip.

## 19.2   TYPES OF ADHESIVE USED IN TIMBER ENGINEERING

### 19.2.1   General

The adhesives used in the assembly of structural components should normally satisfy BS 1204 : Part 1 : 1979, 'Synthetic resin adhesives for wood. Gap-filling adhesives', or BS 1444 : 1970, 'Cold-setting casein adhesive powders for wood'. (It is possible that BS 1444 will be withdrawn although objections have been lodged as it is referenced in a few British Standards and Codes.)

The choice of adhesive is determined by the conditions the component will encounter in service, the method of manufacture and perhaps the conditions during delivery and construction. BS 1204 (Parts 1 and 2) details a system for classifying adhesives by their known or required durability, and it is normally acceptable for a designer to specify the classification type required to be used (§ 19.2.2).

Current formulations of PVA or PVAC adhesives should not be used for structural use.

### 19.2.2 Classification for durability

The BS 1204 classifications are as follows:

*Type WBP: Weather proof and boil proof*
Adhesives of the type which, by systematic tests and by their records in service over many years, have been proved to make joints which are highly resistant to weather, micro-organisms, cold and boiling water, steam and dry heat.

*Type BR: Boil resistant*
Joints made with these adhesives have good resistance to weather and to the test for resistance to boiling water, but fail under the very prolonged exposure to weather that type WBP adhesives will withstand. The joints will withstand cold water for many years and are highly resistant to attack by micro-organisms.

*Type MR: Moisture resistant and moderately weather resistant*
Joints made with these adhesives will survive full exposure to weather for only a few years. They will withstand cold water for a long period and hot water for a limited time, but fail under the test for resistance to boiling water. They are resistant to attack by micro-organisms.

*Type INT. Interior*
Joints made with these adhesives are resistant to cold water but are not required to withstand attack by micro-organisms.

### 19.2.3 Exposure category

At the time in 1970 of a world shortage of resorcinol (one of the main chemicals in WBP adhesives) an *ad hoc* meeting of staff of the Princes Risborough Laboratory, adhesive manufacturers, users and the Ministry of Technology was convened to discuss this problem, and guidance on exposure categories and conditions was made and published in Timberlab Paper No. 35, 1970. Further developments led to the publication in 1975 of *BRE Digest*, No. 175, 'Choice of glues for wood'. Table 1 of that digest has been modified over the years and the generally accepted current (1984) version is presented in Table 19.1. It is considered that, of glued components only glued laminated members should be considered for use in the 'Exterior/high hazard category'. The adhesive types necessary to match these exposure categories are discussed in § 19.2.4.

### 19.2.4 Adhesives used and their classification

From the many adhesives which are available, the few types discussed below are those most used in timber engineering. Generally the WBP types are the most expensive, INT types the cheapest.

**Table 19.1   Choice of adhesives for different exposure categories**

| Exposure category | Typical exposure conditions | Adhesive type (see text for explanation of abbreviations) | BS reference |
|---|---|---|---|
| Exterior high hazard | Full exposure to the weather (e.g. marine structures, exterior structures where the glue line is exposed to the elements). (See note in §19.2.3 about use only of glued laminated members.) | RF <br> PF <br> PF/RF | Type WBP <br> BS 1204 : Part 1 |
| Exterior low hazard | Protected from sun and rain (e.g. inside roofs of open sheds and porches). Temporary structures such as concrete formwork. | RF <br> PF <br> PF/RF <br> MF/UF* <br> Other modified UF* <br> U/F* | Type WBP <br> BS 1204 : Part <br> Type BR <br> BS 1204 : Part 1 <br> Type MR. BS 1204 : Part 1 |
| Interior high hazard | Building with warm and damp conditions where a moisture content of 18% is exceeded and where the glue-line temperature can exceed 50°C (e.g. laundries and unventilated roof spaces). Chemically polluted atmospheres (e.g. chemical works, dye works and swimming pools). External single-leaf walls with protective cladding. | RF <br> PF <br> PF/RF | Type WBP <br> BS 1204 : Part 1 |
| Interior low hazard | Heated and ventilated buildings where the moisture content of the timber will not exceed 18% and where the temperature of the glue line will remain below 50°C (e.g. interiors of houses, halls, churches and other buildings and the Inner leaf of cavity walls). | RF <br> PF <br> PF/RF <br> MF/UF* <br> Other modified UF* <br> UF* <br> Casein | Type WBP <br> BS 1204 : Part 1 <br><br> Type BR. BS 1204 : Part 1 <br> Type MR. BS 1204 : Part 1 <br> BS 1444 : 1970 |

* The glue should always be a particular formulation is suitable for the actual service conditions and for the intended life of the structure.

*Resorcinol-formaldehyde (RF) and Phenol-formaldehyde (PF)*
Adhesives of these types can be expected to satisfy the WBP requirements of BS 1204:Part 1 and rate exterior/high hazard exposure category. Such types have been used successfully in the construction of timber components (particularly glulam) in chemically polluted atmospheres, dyeworks and swimming-pool superstructures. One can also encounter phenol/resorcinol-formaldehyde (PF/RF) which has similar characteristics in use.

*Urea-formaldehyde (UF)*
Only those which satisfy the BR or MR requirements of BS 1204:Part 1 should be considered for use in timber engineering in the relevant exposure categories indicated in Table 19.1. Assurance should be obtained from the adhesive manufacturer before using a particular formulation in any particular exposure category, particularly if the temperature in service is likely to exceed 50°C or the moisture content to exceed 18% for other than short periods.

*Melamine/urea-formaldehyde (MF/UF) or other modified UF*
Only those which satisfy the BR requirements of BS 1204:Part 1 should be considered for the categories exterior/low hazard and interior/low hazard as indicated in Table 19.1.

*Casein*
Casein is not a synthetic adhesive, being manufactured from milk, and should be used only in interior/low hazard conditions. The type specified for timber engineering should conform to BS 1444 and preferably should contain a fungicide and insecticide. Although casein is not a synthetic adhesive the 1970 edition of BS 1444 amended the method of test to align it with the test described in BS 1204:Part 2, Appendices B, C and D and one does see casein referred to as satisfying the INT (interior) requirements of BS 1204:Part 2.

*Epoxy resin*
Epoxy resins are waterproof and unaffected by acids, alkalis or solvents, but are usually used as adhesives for non-porous materials. However, they do find uses in timber engineering for special applications, for example as a filler between timber and steel or concrete at a compression joint, or to seal a timber-to-steel bearing bracket in an environment such as a swimming pool. Epoxy resins are expensive and when using them for a compression joint it is usual to bulk them up with sand.

## 19.3    QUALITY CONTROL REQUIREMENTS. GENERAL GLUE JOINTS

### 19.3.1    Glue mixing and spreading

The weights or volumes of the parts of the mix must be accurately measured and mixed in accordance with the adhesive manufacturer's instructions. The mixing must be carried out in clean containers. With factory gluing there should usually be a separate room or area set aside for mixing. The separate parts should be stored in the correct temperature conditions, and mixing should be carried out with the air above the minimum stated temperature. Normal gel tests and tests for rate of spread of adhesive must be carried out and records kept. The rate of spread must be in accordance with the adhesive manufacturer's instructions unless testing is carried out to determine that less is adequate for a particular joint.

### 19.3.2    'Open' and 'closed' storage times

Once the adhesive is mixed, spreading must be carried out within the stated 'open storage' time, and once the two mating surfaces are brought together any adjustment in alignment or clamping must take place within the stated 'closed storage' time.

### 19.3.3    Temperature

The air temperature during the period of glue storage, glue mixing, spreading and curing must be as stated by the glue manufacturer. The temperature of the timber must be above freezing. If any timber to be glued has become frozen on the surface it must be stored inside until the surface thaws. During spreading, and during curing, the joint should not be subjected to cold draughts, particularly if the glue area is small.

### 19.3.4    Curing period

Unless an accelerated method of curing is used, such as radio-frequency curing, the joint should be stored in a suitable temperature and humidity for the curing period (which varies with the temperature). Although the initial curing may take place within a matter of hours, there are certain glues, even WBP glues, which for seven days must not be placed where rain could affect them. Until the end of that period the glue strength could be reduced by chemicals near the edges being 'washed-out'.

Accelerated curing can be completed in seconds. The methods of achieving this are expensive in themselves and may require the timber to be kilned to around 12% moisture content, but may save overall by suiting a particular flow-line technique. Some manufacturing processes utilise a part-accelerated curing technique followed by natural curing. Even with accelerated curing, full cure may still take several days. During the curing period care will be required in handling glued members.

If radio-frequency (R/F) curing is used it is usually necessary to exclude any metal from the area of the glue joint. Metal can cause serious 'shorting'.

### 19.3.5   Moisture content

At the time of gluing solid timber to solid timber, the moisture content of the timber must not exceed the limit stated either by the glue manufacturer or in BS 5268:Parts 2 and 3, BS 6446 on glued structural components, BS 5291 on structural finger joints, or BS 4169 on glued-laminated members. This figure is usually around 20% or less. A certain amount of evidence exists that satisfactory glue joints can be made at moisture contents up to around 24–28% if care is taken, but probably not on a mass-production basis, and the joint may suffer during subsequent drying.

The two pieces or surfaces being glued must be at approximately the same moisture content, usually within 5–6% (less for glulam to BS 4169) and preferably within a few per cent of the equilibrium moisture content of the component in service. See § 24.8.1 and the relevant Standard for the actual levels laid down for various situations.

If accelerated methods of curing are to be used it may be necessary to dry the timber to around 12% before gluing takes place.

When gluing solid timber to plywood, the plywood can be at a much lower moisture content (around 9–12%) than the timber. The difference in moisture content between the plywood and the timber should not exceed 10% (Table 87 of BS 5268:Part 2).

Water-based glues such as casein add to the moisture content of the timber, and if several layers are being glued (as with site construction of a shell roof of several layers of boarding) the designer should instruct the site staff not to seal both sides until the extra moisture has been able to 'breathe' away. This may take several weeks and may require site checks with moisture meters during this period.

### 19.3.6   Machining surfaces to be glued. Site gluing

The surfaces of solid timber to be glued must be machined and the gluing must be carried out within a prescribed period of machining unless special precautions are taken to ensure that the surface stays suitable for gluing. Various time limits are given in Standards. These vary from 24–72 hours for unpreserved timber, down to 12 hours maximum for finger jointing preserved timber. However, factory conditions etc. can have a significant influence and generally the time should be kept to a minimum for solid timber. If the timber is left too long, the cut-through cells tend to close over, a condition sometimes called 'case hardening', and if the glue joint is then produced it will have less strength than otherwise. This is one reason why site gluing is not to be encouraged unless carried out under strict supervision with surfaces perhaps dressed just prior to gluing, or if the adhesive is mainly to provide added stiffness rather than to take stress.

The machined surfaces must be kept clean and free from any dust etc. It is essential to carry out the gluing operations in a clean part of the factory.

When gluing plywood, experience has shown that the time limit does not apply, but the surface of the plywood must be clean. Any paper stickers which coincide with the glue area must be removed. Any previously dried adhesive must be cleaned off before being reglued or the piece must be rejected.

### 19.3.7    Pressure during curing

The requirement for most adhesives used in timber engineering is that the mating surfaces must be held together in close contact during curing rather than that there must be pressure. In practice this means that pressure is applied but (except for finger joints) the pressure which actually occurs on the glue line is rarely measured. The pressure is applied by clamps or pads which are removed after the curing period, or by nails or staples (occasionally screws), which are left in place after curing although not considered in the design to add to the shear strength of the glue joint. Except for finger jointing (see § 19.6.6), pressure is usually considered to be adequate when 'squeeze-out' of the glue occurs. If clamps are used to hold several components together, it is necessary to tighten them occasionally during the closed assembly period.

Even when the bonding/cramping pressure is achieved by nails there are advantages in assembling units with clamps before and during nailing. Guidance on the spacing of nails is given in BS 5268 : Part 2 and BS 6446 on structural glued assemblies.

When curing at high temperatures the viscosity of the adhesive decreases and the glue-line pressure must be sufficient to prevent adhesive 'run-out'.

### 19.3.8    Quality control tests to destruction

The manufacturer must have a sampling system for testing joints made from standard production glue mixes, usually at least two or more from any shift or major glue mix, and several more if large productions are involved. The tests should be to destruction and records should be kept. It is useful for the manufacturer to plot these results on graphs with degrees of strength on the horizontal axis and number of tests on the vertical axis, allocating one square to each test. If the correct quality control is being exercised, the shape of the plot on the graph will gradually build up to the normal Gaussian shape typical of timber strength (Fig. 19.2). If any result falls below a level set by the designer or by quality control in Standards, (such as BS 5291), the components involved should be inspected and perhaps rejected, the cause ascertained and corrective action taken.

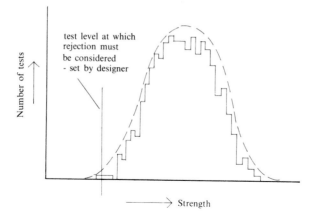

Fig. 19.2

### 19.3.9   Appearance

Particularly if using a WBP adhesive, the 'squeeze-out' which occurs with most glue joints will affect the appearance of the component. When appearance is important, as with some glulam, for example, machining after gluing is desirable or essential. Wiping will not usually remove WBP, BR or MR adhesive, and may even spread it around. After machining glulam or finger jointed timber, the thin appearance of the edge of the glue lines is generally considered neat and acceptable visually, particularly with finger joints if a type with no gap at the tips is being used.

### 19.3.10   Compatibility with preservatives or fire retardants

When gluing preserved timber or preserving glued timber, care must be taken to ensure that there is compatibility. This is particularly the case with preservatives containing water-repellent waxes or additives such as resins. It is normally possible to glue preserved timber, but extra care has to be taken. There are cases, however, in which gluing is not possible. For example it is not considered possible to glue timber which has already been treated with a fire retardant containing ammonia or inorganic salts, although it is possible to treat with such a formulation providing a suitable period (usually seven days) has elapsed after gluing and the adhesive has achieved full cure.

When encountering for the first time a combination of glue and preservative, the designer should check their compatibility with the manufacturer of each. When gluing preservative-treated plywood, one would not normally wish to machine or sand the surfaces before gluing, but brushing to remove surplus salts is desirable.

Technical Note No. 31 of the Princes Risborough Laboratory gives guidance on gluing preservative-treated timber although later experience may modify some of the guidance given in that Note.

Currently there is disagreement on whether or not timber pre-treated at the sawmill by the boron diffusion process can be glued. The treatment process is not practised to any extent. Few mills offer it but, if the designer is called upon to use timber treated in this way, it would be prudent to obtain the latest information on whether or not it can be glued.

### 19.3.11   Temperature and moisture content in service

If the temperature in service is likely to be high (particularly if over 50°C) and other than a WBP adhesive is to be used, the designer should check the suitability of the adhesive with the manufacturer (see Table 19.1). Likewise if the moisture content in service is likely to exceed 18% for other than short periods the designer should check with the adhesive manufacturer.

## 19.4   THE STRENGTH OF A GLUE JOINT

### 19.4.1   Permissible stresses

The permissible stresses of a glue joint are determined by the strength of the timber or plywood face to which the glue is adhering, and not to the strength within the glue line itself.

The permissible shearing stress parallel to the grain of solid timber is the appropriate stress of the timber for shear parallel to the grain (Tables 2.2–2.5). Load-duration and load-sharing factors apply. The permissible shearing stress perpendicular to the grain of solid timber is one-third the value parallel to the grain and is referred to as 'rolling shear' (see § 4.14 and § 8.8.2).

In the event of a glued joint being designed so that one face of solid timber is loaded at an angle to the direction of the grain, the permissible shear stress is determined by applying the following formula from Clause 47.4.2 of BS 5268 : Part 2:

$$\tau_\alpha = \tau_{\text{adm,par}} (1 - 0.67 \sin\alpha)$$

where $\tau_{\text{adm,par}}$ = the permissible shear parallel to the grain stress for the timber

$\alpha$ = the angle between the direction of the load and the longitudinal axis of the piece of timber.

When considering the face of plywood, the permissible shear stress is given the name permissible 'rolling shear' stress, described in detail, for example, in § 8.8.2 for ply web beams. Even when the shear stress on the face of the plywood is parallel to the face grain, the permissible rolling shear stress should be taken in calculations, because the perpendicular veneer next to the face is so close to the surface that one cannot be certain that full dispersal of a face stress could occur before rolling shear starts in the perpendicular veneer. Therefore the formula given above does not apply to glue lines on plywood faces. Rolling shear stresses for two plywoods are given in Tables 8.7 and 8.8.

### 19.4.2 Reduction in permissible stresses for stress concentrations

In the special case of the flange-to-web connection of a ply web beam and the connection of plywood (or other board) to the outermost joist of a glued stress skin panel, Clause 30 of BS 5268 requires that the permissible shear stress at the glue line be multiplied by the $K_{37}$ factor of 0.5. This is an arbitrary factor to take account of likely stress concentrations, but it is pertinent to add that some manufacturers of mass-produced proprietary ply web beams who have tested their beams to destruction have found this type of joint to have factors of safety above ten and have considered themselves justified therefore in disregarding this clause of BS 5268 in their particular case. Normally, however, a designer would be expected to work to the clause. (See Fig. 4.24.)

### 19.4.3 Glued/nailed joints

BS 5268:Part 2 gives certain clauses relating to glued/nailed joints. When gluing plywood to timber the permissible glue-line stresses should be multiplied by 0.90 $(K_{70})$ if assembly is by nailing (see §4.14.1). When assembling a ply web I beam with glue/nailing, because the plywood is fixed and clamped between two pieces of timber there seems less reason to apply the 0.9 reduction factor than in the design of a Box beam, however, a strict interpretation of Clause 47.4.2 of BS 5268:Part 2 requires the 0.9 factor to be applied.

The maximum spacing of nails required to give bonding pressure is detailed in BS 5268:Part 2 (Clause 47.3) and in standard BS 6446 on glued structural components. They are not repeated here (see §4.14.1).

In nail-pressure gluing, the nails are not considered to add to the strength or stiffness of the glue line. Screws, improved wire nails and power-driven fastenings such as staples may be used if proved to be capable of applying pressure to the glue line at least equal to the nailing procedures described in BS 5268 and BS 6446.

## 19.5 STRUCTURAL FINGER JOINTS

### 19.5.1 Types

Structural finger joints are generally considered to be of two basic types which are the longer finger joints which are deliberately made with a small gap at the tips to ensure contact on the sloping sides, and the more recent but well-established short joints which have no measurable gap at the tips. Both types are sketched in Fig. 19.3.

For maximum strength a finger joint should have its sloping surfaces as close as possible to the longitudinal direction of the timber, and have as small a tip width as possible commensurate with it being possible to cut the

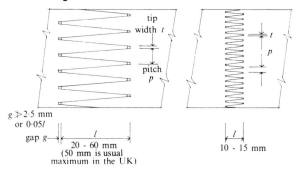

Fig. 19.3: *Types of structural finger joint.*

joint. The length of the glue line per unit width of member also affects the strength. Tests show that a correctly made short joint can have a strength as high as most of the longer joints. This is partly due to the increased pressure at which the short joint can be assembled without causing the timber to split. The short joint is suitable for joinery as well as structural use, whereas the gap at the tips of the longer structural joints makes them unsuitable for most joinery uses.

### 19.5.2    Maximum size and length of finger jointed timber

There is no theoretical limit to the size of timber which can be finger jointed. Size is normally limited only by the capacity of the machine which has been installed. Finger jointing in softwood sizes of 75 × 200 mm or 50 × 300 mm is quite common, and finger jointing of certain larger glulam sections has been carried out in Germany and Sweden, including its use at the angle of portal frame knee braces.

The only limit to the length being finger jointed is that of handling in the factory, on transport and on site. Lengths of 12 m are very common and do not represent the maximum by any means.

### 19.5.3    Appearance and wane

It is possible to leave a piece of finger jointed timber in the 'as sawn' condition without planing. In this case the adhesive which has squeezed-out is very obvious, particularly if WBP, and it is possible for the two pieces at a joint to be off-set due to tolerances and the lining-up of the fingers (Fig. 19.4).

This has little effect on strength but, when appearance is important, the timber should be surfaced after finger jointing. The finger joint then usually has a very neat appearance, particularly if a short joint is being used. Wiping off the glue squeezed-out instead of machining is unlikely to improve appearance and may make it worse by spreading the glue over a larger area.

Wane in the length of a finger joint and within a short distance of a finger joint acts as a stress-raiser and must be limited. BS 5291 permits wane to

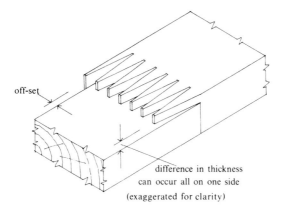

off-set

difference in thickness
can occur all on one side
(exaggerated for clarity)

Fig. 19.4

occur on one or two corners but gives limits. Within the finger length and within 75 mm of the roots of the fingers, if the efficiency rating of the joint is equal to or less than 60%, the sum of the dimensions of the wane should not exceed 10% of the width plus thickness of the piece (which represents a maximum of about 0.6% of the area). If the efficiency is in excess of 60%, the sum of the dimensions of the wane should not exceed 5% of the width plus thickness of the piece.

### 19.5.4   Types of adhesive

Although it is possible to finger joint with an adhesive having only interior classification, it is normal for a manufacturer to use a WBP adhesive or perhaps a BR or MR type, even if using a different type elsewhere in the assembly of the component. The adhesive must be adequate for the service conditions (see Table 19.1). BS 5291 emphasises the use of WBP adhesives for structural finger joints.

### 19.5.5   Minimum centres of joints

There is no reason to suppose that the existence of several finger joints at close centres reduces the strength of a component (indeed some joinery components deliberately introduce finger joints at close centres to reduce the possibility of distortion). However, even if only to avoid the possibility of adverse reaction from site staff or occupants, it is wise to place some limit on the number of joints in a structural member, therefore the distance between the centres of any two finger joints in a piece of timber or single lamination should not usually be less than 1 metre. This is stated in BS 5291.

### 19.5.6   Species mix

Most of the experience in the UK of structural finger jointing has been with European Whitewood or European Redwood. Even though there are known cases where these two species have been finger jointed together successfully,

BS 5291 is quite clear in stating that species should not be mixed at a joint. Therefore, particularly if a designer is using North American timber (such as the SPF combination), this point must be remembered.

## 19.6   QUALITY CONTROL REQUIREMENTS. STRUCTURAL FINGER JOINTS

### 19.6.1   General

The quality control requirements listed in §§ 19.3.1–19.3.7 apply equally to finger joints, and there are also additional requirements. These are detailed in BS 5291 on finger joints in structural softwood and in §§ 19.6.2–19.6.7.

### 19.6.2   Machining

The cutters which actually cut the fingers must be kept sharp, and, particularly with the longer fingers, the backing blocks must be adequate to prevent 'spelching' (i.e. part of the individual fingers being torn off by the cutters).

When appearance is important it is essential to machine the completed piece to eliminate any off-sets and to remove the glue which will squeeze-out from the joint.

The amount of cup must be limited on pieces to be finger jointed, or splitting is likely to generate from the tips of the fingers, resulting from face pressure applied during assembly of the joint.

### 19.6.3   Moisture content

The shape and method of assembling finger joints is such an excellent example of a glue joint that it is usually quite acceptable to work to the upper limit of moisture content in the timber relevant to the glue being used. The moisture content of the timber at assembly should not exceed 20% and should be within a few percent of the average equilibrium moisture content expected in service. The moisture content of the pieces to be jointed should not differ by more than 6%.

### 19.6.4   Knots

Knots must be limited both in and close to the fingers. BS 5291 limits the dimension of knots within the finger length to half the pitch of the fingers or 5 mm whichever is the lesser. Outside the length of the fingers no knot shall be closer to the root of the fingers than three times its maximum dimension $d$ measured parallel to the grain (see Fig. 19.5), although knots with a dimension $d$ of 5 mm or less can be disregarded. In trimming the end of a piece to be finger jointed, disturbed grain should be removed as well as over-size knots.

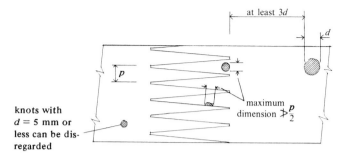

Fig. 19.5: *Maximum permitted size of knots.*

### 19.6.5 Wane

Wane within the length of fingers and close to fingers should be limited. The limits of BS 5291 are detailed in § 19.5.3.

### 19.6.6 End pressure and fissures

The end pressure during assembly of the finger joints must be sufficient for maximum strength to be developed, but not too great or splitting will occur from the fingers. With short joints, the pressure is in the order of 5.0–15.0 N/mm², reducing to around 1.5–5.0 N/mm² with the longest joints.
    BS 5291 details the limits on fissures which may occur at a finger joint.

### 19.6.7 Quality control tests to destruction

It is necessary for a manufacturer of structural finger joints to have access to a test machine to test specimens of standard production joints. The test is by four-point loading (Fig. 19.6), and the manufacturer can consider plotting the results as indicated in Fig. 19.2. BS 5291 details the test and result requirements. If a correctly made joint is found to have low strength one of the first checks to make is on the density of the timber.

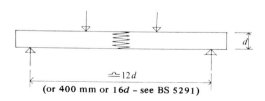

Fig. 19.6

### 19.6.8 Restrictions on position of joints

At the time of finger jointing, the position of the finger joints in the final component or assembly is not known. It is extremely important, therefore, to know whether finger joints can occur at any position either along a span or at a joint, and normally this is the case.

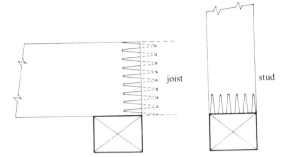

Fig. 19.7

If a finger joint is actually cut through at the end of a joist (see Fig. 19.7), this is satisfactory and may even be an improvement on a normal cut end.

See §§ 19.7.3 on the use of finger joints in load-sharing and non-load-sharing systems.

At the time of writing this edition of the manual, the situation on the extent to which finger joints should be permitted under punched metal-plate fasteners is being reviewed, following analysis of tests carried out in the UK and Sweden. The conclusions as far as design of trussed rafters is concerned will appear in BS 5268:Part 3 (which will be the rewrite of CP 112: Part 3). CP 112:Part 3 stated that there are certain positions at which a finger joint should not occur under a punched metal-plate, and therefore implied that there is no strength left at such a position. It is now accepted that, although a punched metal-plate with teeth coinciding in a certain way with a finger joint can reduce the strength of the finger joint, the residual strength is still significant. It is therefore expected that BS 5268:Part 3 will give reduction factors for bending and tension for joint efficiencies and leave the designer to establish whether or not the residual strength is adequate for the maximum combined stresses which can occur at the various node points where a finger joint may occur.

From the known results of the tests it looks as though a normal punched metal-plate as used in a trussed rafter could reduce the bending and tension strength (but not the compression strength) of a 15 mm long finger joint in 35 mm thick material by about 20% (and less in thicker material). Other finger joints in 35 mm thick material might have to be considered as being subject to a somewhat larger reduction (perhaps 30%). Even then, and even without taking compression into account, it would be possible to use a joint of 75% bending efficiency in 35 mm thick, M50 material (75% × 0.7 = 52.5%) without having to make a check on the actual stresses encountered at the joint. In practice, trussed rafters made with M75 material are often well under-stressed therefore, in many cases, it should be possible for even M75 material to be finger jointed with a joint of 75% bending efficiency, providing a design check is carried out for the relevant node points. Finger joints are certainly satisfactory under glued/nailed or nailed plywood gussets without reducing the quoted efficiencies.

In glulam members, it is advisable to stagger joints in adjacent laminations, This is a 'belt and braces' restriction, however, as no consideration is taken in the design of joints in laminations of the strengthening effect of adjacent laminations acting as a lap joint (Fig. 7.12). BS 4169 on glulam calls for 'excessive grouping of finger joints in adjacent laminations' to be avoided.

### 19.6.9 Machine stress grading of finger jointed timber

Feeding finger jointed timber through the current designs of stress grading machines does not give a complete method of checking the strength of the finger joint because the machines measure stiffness not strength. It does, however, give a method of proof testing the joints in bending which could be considered to be better than nothing. However, at present the accepted method of checking production joints is a visual examination of knot size and position, plus a check on any other clauses which may be required by standards, plus exercise of the quality control and tests to destruction detailed in §§ 19.3 and 19.6.

### 19.6.10 Timber finger jointed abroad

Sweden and Finland have control organisations which authorise manufacturers to produce finger jointed structural timber to Nordic document NKB 13:Chapter IX and police them at regular intervals. The quality control regulations are sufficiently acceptable for the foreword of BS 5291 to state that joints made by Nordic manufacturers authorised under NKB 13 may be assumed to have the same strength as joints of the same profile produced to BS 5291 providing the requirements for small knots given in BS 5291 are satisfied. The authors know of no other countries exporting significant quantities of finger jointed timber to the UK.

## 19.7 THE STRENGTH AND DESIGN OF FINGER JOINTS

### 19.7.1 General

The general philosophy in BS 5268:Part 2 on the use of finger joints in a structural situation is either to require a finger joint to have a certain efficiency in bending when using timber of a certain stress grade without carrying out a design check on the actual stress combinations, or to design for the actual stress combinations and provide a joint accordingly. Design examples are given in § 7.5 for combined bending and tension and, in § 17.7 for combined bending and compression, using both methods.

Also see § 19.6.8 on the use of a finger joint under a punched metal-plate fastener and § 19.7.3 on the use of finger joints in a load-sharing system and a non-load-sharing system.

Finger joints are not considered to affect the $E$ value of a piece of timber or of a built-up member.

### 19.7.2   Joint efficiencies in bending, tension and compression

The joint efficiency in bending is determined by test, and in the UK is expressed as a percentage of the bending strength of unjointed defect-free timber of the same cross-section and species. The proving test is a four-point bending test with the finger joint occurring in the central part. The method of establishing the bending efficiency of a new profile, or an established profile being produced on a new production line is detailed in BS 5291.

The joint efficiency in tension can be established by test but is usually taken in the UK as having the same efficiency value as in bending (even though the grade stress of timber in tension is considerably less than the grade stress in bending).

Tests tend to show that the joint efficiency in compression is 100% or close to 100%. However, in the UK, the efficiency is taken as:

$$\frac{p-t}{t} \times 100\%$$

where $p$ = the pitch of the fingers
and    $t$ = the width of the tip.

Normally no finger joint with an efficiency in bending of less than 50% should be used structurally. This is a requirement of BS 5268 : Part 2.

Guide efficiency values for well-established joint profiles are given in Table 19.2 with the more common profiles indicated by an asterisk. See the various design examples in this manual on how to apply efficiency ratings.

**Table 19.2**

| Finger profiles | | | Efficiency rating in bending and tension (per cent) | Efficiency rating in compression (per cent) |
|---|---|---|---|---|
| Length $l$(mm) | Pitch $p$(mm) | Tip width $t$(mm) | | |
| 55 | 12.5 | 1.5 | 75 | 88 |
| 50* | 12.0 | 2.0 | 75 | 83 |
| 40 | 9.0 | 1.0 | 65 | 89 |
| 32* | 6.2 | 0.5 | 75 | 92 |
| 30 | 6.5 | 1.5 | 55 | 77 |
| 30 | 11.0 | 2.7 | 50 | 75 |
| 20* | 6.2 | 1.0 | 65 | 84 |
| 15* | 3.8 | 0.5 | 75 | 87 |
| 12.5* | 4.0 | 0.7 | 65 | 82 |
| 12.5* | 3.0 | 0.5 | 65 | 83 |
| 10.0 | 3.7 | 0.6 | 65 | 84 |
| 10.0 | 3.8 | 0.6 | 65 | 84 |
| 7.5 | 2.5 | 0.2 | 65 | 92 |

* Profiles more likely to be available.

As stated above, when using a finger joint, one can either use one of a stated efficiency in bending related to various stress grades, or design for the actual stresses or stress combinations encountered in a particular design. Table 19.3 gives joint efficiency ratings in bending which, if matched for a particular stress grade, may be used without any further design check.

**Table 19.3**

| BS 4978 stress grades | NLGA and NGRDL stress grades | | | Finger joint efficiency in bending % (see § 19.7.2) |
|---|---|---|---|---|
| | Joist and Plank | Structural Light Framing | Light Framing Utility and Stud | |
| LA | | | | see § 7.2 |
| LB | | | | 70 |
| LC | | | | 55 |
| M75 | | | | 75 |
| SS or MSS | | | | 60 |
| M50 | | | | 50 |
| GS or MGS | | | | 50 |
| | Select | Select | | 65 |
| | No. 1, No. 2, No. 3 | No. 1, No. 2, No. 3 | Construction Standard Utility Stud | 50 |

When it is necessary to carry out a design check on glulam for the actual stresses rather than simply use the efficiency figures given in Table 19.3, see modification factors $K_{30}$, $K_{31}$ and $K_{32}$ of BS 5268:Part 2 and § 7.5 for combined bending and tension, and § 17.7 for combined bending and compression.

In certain cases it may be necessary to check if a finger joint is strong enough in solid timber, rather than simply refer to Table 19.3. This may occur for example if a manufacturer wishes to use M75 material which is in stock, yet has only a 65% efficient joint (in bending), but knows that the component will not be stressed to the full M75 level, or will be stressed partly in compression as well as in bending. To enable the designer to check if this is possible it is necessary to know the efficiency of each stress grade in bending, tension and compression. Because of the elimination of 'basic' stresses from BS 5268:Part 2 it is not possible to quote grade stress as a percentage of basic stress therefore, in Table 19.4, the authors present suggested grade efficiencies which are calculated taking the grade stresses in bending, tension and compression from Table 2.2 for M75 as being at the 75% efficiency level for bending, tension and compression. The designer may evolve another system. Figures are not given for Canadian SPF mix for the reason given in § 19.5.6; and are not given for Douglas Fir-Larch because

**Table 19.4**

| Species | Stress grade | Stress grade efficiency % (see § 19.7.2) | | |
|---------|-------------|---------|---------|------------|
| | | Bending | Tension | Compression |
| European Whitewood | M75 | 75 | 75 | 75 |
| or Redwood | SS or MSS | 56 | 56 | 68 |
| | M50 | 50 | 50 | 63 |
| | GS or MGS | 40 | 40 | 59 |
| Canadian | M75 | 75 | 75 | 75 |
| Hem-Fir | SS or MSS | 56 | 56 | 63 |
| | M50 | 50 | 50 | 62 |
| | GS or MGS | 40 | 40 | 57 |
| | J and P | | | |
| | Select | 60 | 77 | 71 |
| | No. 1 | 42 | 54 | 64 |
| | No. 2 | 42 | 54 | 56 |
| | S.L.Fmg.* | | | |
| | Select | 60 | 78 | 71 |
| | No. 1 | 42 | 54 | 64 |
| | No. 2 | 42 | 54 | 56 |
| | L.Fmg.* | | | |
| | Const. | 35 | – | 49 |
| | Standard | 26 | – | 37 |

\* Bending and tension values from Table 2.4 reduced by $K_7 = 1.143$ to give comparison.

BS 5268:Part 2:1984 does not quote values for M75 (or M50) for this species.

Therefore, using Tables 2.2 and 19.4, make the following checks on the suitability of a finger joint (assuming other requirements, see § 19.7.3, are met):

(a) If a piece of European Whitewood (immaterial of grade) is subject to a long term bending stress of 8.3 N/mm² check if a finger joint of 65% efficiency is adequate.

From Table 2.2 it can be seen that the long term grade stress for M75 (i.e. 75% efficiency) is 10.0 N/mm², therefore 8.3 N/mm² requires a joint efficiency of

$$75 \times \frac{8.3}{10.0} = 62.25\%$$

from which it can be seen that the available joint is adequate.

(b) If a piece of SS European Whitewood has a long term bending stress of 7.3 N/mm² (as against the grade stress of 7.5 N/mm²) check if a finger joint of 50% efficiency is adequate.

From Table 19.4 it can be seen that 7.5 N/mm² represents 56% efficiency in bending, therefore 7.3 N/mm² represents 54.5% efficiency and therefore a joint of 50% efficiency is not sufficient.

(c) Check if the 15/3.8/0.5 mm finger joint (see Table 19.2) is adequate in European Whitewood for a bending stress of 5.2 N/mm² and a compression stress of 4.0 N/mm², both long term.

From Table 19.2 it can be seen that the joint has an efficiency in bending of 75% (which represents a long term grade stress of 10.0 N/mm²) and an efficiency in compression of 87% which represents a long term grade stress of 10.1 N/mm².

$$\frac{5.2}{10.0} + \frac{4.0}{10.1} = 0.52 + 0.40 = 0.92 < 1.0$$

which shows the joint to be adequate.

## 19.7.3  Load-sharing/non-load-sharing systems

Clause 47.4.1 of BS 5268:Part 2 puts a limitation on certain uses of finger joints by stating that they should not be used if failure of any one joint would lead to 'collapse of the system'. (This may be considered to be similar to progressive collapse.) When finger jointing is used in a load-sharing system such as four or more members acting together and spaced at not more than 610 mm centres (Clause 13 of BS 5268;Part 2), such as rafters, joists, trusses or wall studs, with adequate provision for lateral distribution of loads, there is no restriction on the use of finger jointing (but see § 19.6.8).

Generally, there is also no restriction on the use of finger jointing in a glulam member of four or more laminates.

The restriction is quite clear when one considers a member such as a trimmer beam consisting of one piece of solid timber and supporting one or more floor joists. In this case, if a finger joint in the trimmer beam fails, the load can not normally be transmitted laterally (unless the designer can prove that secondary members would support the floor joist or joists), and therefore it would not be permissible to use finger jointing in the trimmer beam. (Because of the reliability of finger jointing and the relative ease with which one is likely to be able to detect a faulty joint if it should occur in a trimmer, before it is erected, it is worth noting that not all engineers agree with this clause in BS 5268, but it stands at present.)

Generally, it is unlikely that an engineer would be called on to satisfy Clause 47.4.1 if finger jointing occurs in a member in pure compression.

Before one can use finger jointing in laterally interconnected members which fall outside the limits of Clause 13 (i.e. 610 mm spacings), or in members built-up in such a way that bending or tension stresses are resisted by at least two pieces of timber interconnected so as to be equally strained, it is necessary to carry out a design check as required by Clause 47.4.1. This is best illustrated by taking examples.

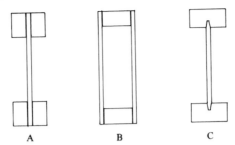

Fig. 19.8
A                    B                    C

Assume that ply web beams such as the one shown at A in Fig. 19.8 are spaced at 1.2 m centres and that the flanges are stressed to 95% of the full permissible stress and are finger jointed. Assume initially that the secondary members between the beams do not transmit load laterally, therefore, for the purpose of checking whether or not finger jointing is permitted, the beam will have to be regarded as a principal member.

Assume that one finger joint in one of the two bottom flange pieces fails. Disregarding the splice plate effect of the web, assume that one of the flange pieces has to act by itself (in combined tension and bending – see §8.4.2). Assume that the beam was designed for medium term loading (∴ $K_3 = 1.25$) with the permanent dead loading equal to 40% of the total loading.

Clause 47.4.1 requires the check on the remaining finger jointed member to be carried out assuming $K_3 = 1.00$ (even if it was 1.25 or 1.75 in the basic design), but permits the design load to be taken as dead loading plus one-third of the live loading, and the remaining structures to be considered as having permissible stress values of twice the normal permissible stress value (but with $K_3 = 1.00$).

Therefore, by proportion (and disregarding the web effect) it is relatively easy to see that the ratio of actual reduced stresses compared to the 'new' permissible stresses on the bottom flange equals the product of the following ratios:

$$= \frac{\text{new reduced loading}}{\text{previous design loading}} \times \frac{\text{two flange pieces}}{\text{one flange piece}}$$

$$\times \frac{\text{original actual stress as a percentage of permissible (incl. } K_3)}{\text{twice original permissible stress but with } K_3 = 1.00}$$

$$= \frac{40\% + (\frac{1}{3} \times 60\%)}{100\%} \times \frac{2}{1} \times \frac{95\% \times 1.25}{200\% \times 1.00}$$

$$= 0.6 \times 2 \times (0.475 \times 1.25) = 0.7125 < 1.0$$

Therefore, in this case, the beam would satisfy Clause 47.4.1 (which does not call for a further deflection check). In an actual case, the designer could use actual values rather than proportions but it can be seen by studying the ratios above that, even if all the loading is long term, finger jointing is normally permitted if two or more members act together, and are equally strained.

The really serious nature of the clause as far as ply web beams are concerned becomes apparent when beams of the type shown at B and C in Fig. 19.8 having only a one-piece flange on the bending tension side are considered. In the case of the Box beam the easiest way of satisfying the clause is to manufacture the flanges with two pieces glued together (see Table 8.5), although the plywood webs act as gussets and perhaps this fact would satisfy many designers.

Consider beams of the type shown at C in Fig. 19.8 at 1.2 m centres spanning 8 m, and supporting a dead load of 6.0 kN (0.625 kN/m² made up of 0.425 at upper level and 0.200 at ceiling level) and a medium term imposed load on the top of 7.2 kN (0.75 kN/m²). This equals a total load of 13.2 kN. If one finger joint in the bottom flange of one beam fails, although the remaining beam has the form of a Tee beam it is safer to assume that all the load from that beam must be transferred to the beams on each side (or, in the case of the end beam, to the gable wall on one side and the adjacent beam on the other side). Assume noggings top and bottom at 0.6 m centres, continuous across at least three beams and with joints staggered so that at least one nogging (either top or bottom) forms a continuous span between the adjacent beams every 0.6 m centres (see Fig. 19.9).

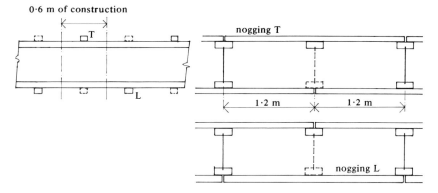

Fig. 19.9

From Clause 47.4.1 it can be seen that two requirements must be satisfied:

(a) The beams on each side of the beam with the (assumed) finger joint failure must be designed for a dead load of 6 × 1.5 = 9.0 kN and an imposed load of 7.2 × 1.5 × $\frac{1}{3}$ = 3.6 kN, with $K_3$ = 1.00. Therefore, instead of a total medium term design load of 13.2 kN, the beam must support a load, assumed to be long term, of 12.6 kN. However, the permissible stresses may be doubled. Therefore, assuming the beam was fully stressed for the normal design (as against being limited by deflection) the new stress ratio becomes:

$$\frac{13.2}{12.6} \times \frac{1}{2} = 0.52 < 1.0$$

Therefore, in this case one part of Clause 47.4.1 is satisfied.

(b) It is still necessary to check if the load from the 'failed' beam can be transferred to the beams on each side via the noggings as detailed above and shown in Fig. 19.9.

For the purposes of Clause 47.4.1 the imposed loading applied at the top level is $\frac{1}{3} \times 0.75 = 0.25\,\text{kN/m}^2$, the dead loading applied at the top level is $0.425\,\text{kN/m}^2$, and the dead loading applied at the lower level is $0.2\,\text{kN/m}^2$. Because both rows of noggings would deflect the same amount (both being fixed to the same beam) a certain amount of load sharing would take place. However, for simplicity, assume that the noggings carry the loading as calculated below for a 0.6 m run of construction (along the beam) and disregard any assistance from the cantilever effect of noggings on to the 'failed' beam.

Nogging T (see Fig. 19.9) carries a UDL from loading at the top level

$$= (0.25 + 0.425)\,0.6 \times 2.4 = 0.972\,\text{kN total}$$

and a central point load from the lower level transmitted via the beam

$$= 0.2 \times 0.6 \times 2.4 = 0.288\,\text{kN}$$

Nogging L (see Fig. 19.9) carries a UDL from loading at the lower level

$$= 0.2 \times 0.6 \times 2.4 = 0.288\,\text{kN total}$$

and a central point load from the upper level transmitted via the beam

$$= (0.25 + 0.425)\,0.6 \times 1.2 = 0.486\,\text{kN}$$

$$M \text{ on nogging T} = \frac{0.972 \times 2.4}{8} + \frac{0.288 \times 2.4}{4} = 0.464\,\text{kN.m}$$

$$M \text{ on nogging L} = \frac{0.288 \times 2.4}{8} + \frac{0.486 \times 2.4}{4} = 0.378\,\text{kN.m}$$

Try a 60 $\times$ 50 mm section, 60 mm vertical, SS grade Whitewood. Check nogging T.

$$Z = 30 \times 10^3\,\text{mm}^3$$

$$\therefore \sigma_{m,a} = \frac{0.464 \times 10^6}{30 \times 10^3} = 15.47\,\text{N/mm}^2$$

The permissible bending stress with $K_3 = 1.0$, $K_7 = 1.17$, and allowing the factor of 2 from Clause 47.4.1 is:

$$7.5 \times 1.0 \times 1.17 \times 2 = 17.55 \, \text{N/mm}^2$$

which satisfies Clause 47.4.1.

$K_8 = 1.1$ has not been used in the calculation of permissible stress but could perhaps be justified. Clause 47.4.1 does not call for a deflection check. (Assuming $E_{\text{mean}}$, the calculated deflection would be in the order of 30 mm. If such a deflection occurred, without collapse, it is likely that the building owner would investigate and repair, which is one intention of Clause 47.4.1.)

In addition to the checks above, the fixings of the noggings to the beams would have to be checked.

As far as nogging T is concerned the end reaction is in bearing and equals:

$$(0.972 + 0.288)\,0.5 = 0.63 \, \text{kN}$$

With a bearing area of say 50 $\times$ 50 mm, the bearing stress is very small. The point load of 0.288 kN (i.e. 288 N) is hung from the beam and could be taken by two nails. From Table 18.11 it can be seen that two 3.35 mm diameter nails with a penetration of 30 mm give a total resistance to withdrawal (allowing permissible loading to be doubled)

$$= 2 \times 2 \times 30 \times 2.87 = 344 \, \text{N}$$

As far as nogging L is concerned the end reaction is:

$$0.5\,(0.288 + 0.486) = 0.387 \, \text{kN} \, (387 \, \text{N})$$

With the load-duration factor = 1.00, pre-drilling to allow two nails in the thickness of 50 mm, and a penetration into the beam of say 35 mm it should be possible to take this load, particularly bearing in mind that the permissible stress may be increased by a factor of 2 (Clause 47.4.1).

Therefore, with care and added cost, it should be possible to satisfy Clause 47.4.1 on many occasions even if using beams as indicated at C in Fig. 19.8.

## 19.8  SCARF JOINTS

### 19.8.1  Types

Types of scarf joint are sketched in Fig. 19.10. One can regard a scarf joint as an earlier and simple form of finger joint. The amount of timber wasted on the overlap is much greater than a finger joint, as slopes vary from 1 in 6

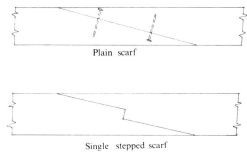

Fig. 19.10: *Types of scarf joint.*

to 1 in 12 and therefore as much as twelve times the thickness can be wasted at an overlap.

BS 5268: Part 2 permits scarf joints to be used only in glulam.

### 19.8.2   Quality control

BS 5268 does not go into detail on the quality control required when using a scarf joint, but the quality control requirements for any glue joint should be exercised, and in addition great care must be taken in cutting, lining-up and assembling the joint. Assembly is often by nailing to hold the joint during curing of the glue. Nails must be recessed to avoid damaging the cutters at any subsequent machining operation.

### 19.8.3   Strength of scarf joints

BS 5268: Part 2 gives efficiency ratings for plain scarf joints used in glulam as shown in Fig. 7.11. It seems somewhat illogical to permit 100% in compression when one compares the permitted efficiencies of finger joints, but perhaps it is the finger joint figures which are too conservative. See the notes in §7.5.1.

When deciding whether or not a scarf joint is adequate for use in glulam one can either provide for an efficiency in bending of 70% or 55% for LB and LC laminates respectively, or work through the design check outlined in §7.5.1 for bending or combined bending and tension, or in §17.7 for compression or combined bending and compression.

# Chapter Twenty
# Trussed Rafters

## 20.1 INTRODUCTION

The term 'trussed rafter' rather than 'truss' is used in relation to lightweight prefabricated timber trusses designed to be placed at close centres (usually 600 mm) on spans normally not exceeding around 9–11 m for 35 mm thick members and 12–15 m for 47 mm thick members to support a tiled or slated roof. Trussed rafters can be mono-pitch or have unsymmetrical shapes, but by far the majority used are symmetrical duo-pitch with 'Fan' or 'Fink' type configurations as shown in Fig. 20.1. In recent years experience has

Fink type            Fan type

Fig. 20.1

shown that Fink trussed rafters are cheaper to produce than Fan configurations supporting the same loading and therefore Fink trussed rafters are more common. Standardisation of designs and/or design procedures has reached a stage where, particularly if proprietary punched metal-plate gussets are used, little if any structural design is required unless the designer is actually working for a manufacturer of trussed rafters, which is why trussed rafters are separated in this manual from Chapter 21 on trusses. The Code of Practice covers most points required by a designer.

On a straightforward duo-pitch roof, the only additional structural timbers normally required are diagonal bracings as indicated in Fig. 20.3 at the rafter and ceiling tie levels and on the line of the internal strut members, lateral ties near the apex and node points along the ceiling tie, plus framing under any water tanks (unless supported on walls), tiling battens to the rafters, plus any framing required around openings for chimneys or access openings at ceiling tie levels (see Fig. 20.4).

A trussed rafter roof in a house uses around 30% less timber than a roof constructed traditionally, and can be erected in a fraction of the time it takes to construct a traditional timber roof. The construction is therefore very economical and accounts for the majority of pitched house roofs built in the UK, and is also being used on other building types of similar span (e.g. recreation halls) and even in industrial buildings. Many manufacturers specialise in trussed rafters and most have produced standard designs of duo-pitch and other shapes for 100 mm increments of span and $2\frac{1}{2}°$ increments of slope between 15° and 35° to suit a tiled roof, and lower slopes for other roof claddings. With a tiled roof a $22\frac{1}{2}°$ slope tends to give the most economical overall design, but slope is usually dictated by architectural considerations.

CP 112:Part 3 was produced in 1973 on 'Trussed rafters for roofs for dwellings', but since then so much more accent has been given to the need to provide bracing in trussed rafter roofs, and trussed rafters are being used extensively in other than roofs of dwellings, that the rewrite of CP 112: Part 3 will cover bracing etc. and will be called 'Trussed rafter roofs'. Although still a Code of Practice it will be re-numbered BS 5268:Part 3. At the time of rewriting this manual the general form and details which BS 5268:Part 3 will take are well known but there may still be changes before the document is printed. Therefore, although the notes in this chapter are as up to date as possible the designer should certainly refer to BS 5268: Part 3 which is anticipated to be available in 1984.

Whereas CP 112:Part 3 gave the finished size of members and the sawn size from which they should be processed, BS 5268:Part 3 will give only the finished sizes. It is usual to obtain finished sizes by surfacing or skimming the timber from timber sawn perhaps 1 or 2 mm thicker than the finished size. The vast majority of trussed rafters have been manufactured from timber having a finished thickness of 35 mm and, from the UK experience of some 30 million trussed rafters in service, this thickness has proved to be adequate for most applications (e.g. such as tiled roofs on battens) providing reasonable care is taken with handling and erection of the rafters, and of course that the timber is properly dried before assembly.

The most common ceiling is 12.7 mm thick plasterboard (with or without a plaster skim) for use with trussed rafters at 600 mm centres.

Standard trussed rafters manufactured and used in the UK are invariably designed for use at 600 mm centres. However, costing by some builders has shown that in certain cases it is more economical to use a wider spacing, even though this requires heavier trusses and secondary members. In Nordic designs, trussed rafters are usually spaced at 1.2 m centres.

Rafters, ceiling ties and internal members are placed in the same vertical plane with gusset plates on each side. The majority of trussed rafters are made with gussets of one of the proprietary punched metal-plate fasteners, although many have gussets of nailed plywood. Gussets of glued/nailed plywood or proprietary metal plates fixed by ordinary nailing are also used. (American C–D plywood is excluded by BS 5268:Part 3 for use as gussets.)

Trussed rafters are usually erected on a timber wall plate (usually 75–

100 mm basic width) which may be on a timber or brick wall and must be fixed down to it. It is common to order trussed rafters with a span slightly more than that required (say 25–50 mm), thus permitting tolerance on the positioning of the wall plate (Fig. 20.2). BS 5268:Part 3 permits distance $x$ as shown in Fig. 20.2 to be up to 50 mm, or:

$$\frac{\text{depth of ceiling tie}}{3 \times \text{sine of truss slope}}$$

whichever is the greater.

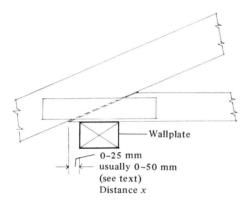

Fig. 20.2

It is considered preferable to fix trussed rafters to wallplates by proprietary purpose-made holding-down fixings but two skew nails are often used. BS 5268:Part 3 specifies that they should be a minimum of 4.5 mm diameter × 100 mm long galvanised round wire nails. Obviously, at that size, great care is required not to split the timber in the trussed rafter. Particularly if using a lightweight cladding, a low slope, or if the roof is in an area of high wind loading, care must be taken to ensure that the fixings of the rafters to the wallplate are adequate to resist uplift. BS 5268:Part 3 requires that the forces causing uplift should be multiplied by 1.4 (which seems somewhat on the high side) in calculating restraint against uplift.

## 20.2   SERVICE CONDITIONS AND PRESERVATION

BS 5268:Part 3 states that trussed rafters should be used only in roofs where the moisture content of the timber in service will not exceed 18% for any significant period. Recommendations are given on ventilation to minimise any risk of condensation. Requirements are given for galvanised and other finishes for metal-plate fasteners. Stainless steel fasteners are available at (considerable) extra cost.

BS 5268:Part 3 points out that it is not normally necessary to preserve trussed rafters (particularly as they are not usually used where there is any 'aggressive chemical pollution') except in those areas specified in the Building Regulations for England and Wales which are subject to infestation by the house longhorn beetle. Where a decision is taken to use preservation, the clear recommendation in BS 5268:Part 3 is to use an organic-solvent preservative rather than a water-borne CCA process. Organic-solvent preservatives do not increase the moisture content of the timber whereas water-borne preservatives increase moisture content considerably. Also, because the truss manufacturer is rarely given time to allow timber to air dry after preservation, the cost of kilning after a water-borne process would have to be incurred and, even then, the manufacturer would have to wait about seven days to allow the salts to 'fix' in the timber before re-processing the timber and assembling the trussed rafters. Also, in service, there is the added risk (no matter how small in a properly ventilated roof) that a CCA preservative could increase the possibility of corrosion of the metal fasteners if the moisture content of the timber increased.

## 20.3   BRACING

Experience in recent years has shown that trussed rafters are not always erected sufficiently vertical and this is one reason why BS 5268:Part 3 will put so much emphasis on bracing of roofs. In BS 5268:Part 3 bracing is shown for duo-pitch roofs in considerable detail, based largely on recommendations made earlier by the Truss Plate Association. The bracing design is largely empirical although on the lines of good engineering practice. The designer is referred to Part 3 for the full details which are shown in outline form in Fig. 20.3 for a straightforward roof, and are intended to apply mainly to domestic and similar roofs.

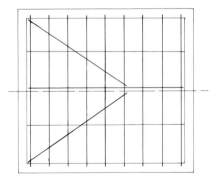

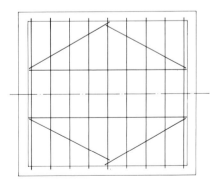

Outline of timber bracing at rafter level.
On occasions, on a long narrow roof, the designer may choose to call for four braces rather than two in each roof unit. The final wording in Part 3 may require diagonal bracing to be in contact with each rafter.

Outline of timber bracing at ceiling level.
This layout is used if the distance between the centres of the cross-walls does not exceed 1.2 X trussed rafter span. Where the distance does exceed 1.2 X trussed rafter span, four diagonal braces are used each side in a W formation.

Fig. 20.3

As can be seen from Fig. 20.3 the diagonal bracing under the rafters plus the ridge member at apex level, acts against longitudinal movement/loads, whilst the diagonal bracing and binders at ceiling level act against any lateral movement/loads (parallel to the trussed rafters). The intention is that both these bracing systems should occur in every roof unit even if a row of terraced houses is involved. The minimum standard bracing size is given as 25 × 100 mm and there is no stated requirement that the timber must be stress graded. If individual lengths of bracing are not long enough, pieces can be lapped side by side, the lap occurring across at least two rafters.

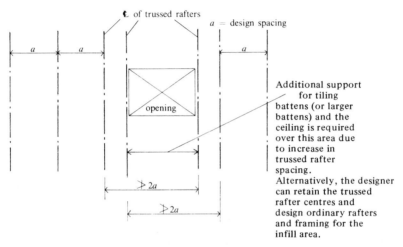

Fig. 20.4

Generally two 3.35 mm diameter × 75 mm long round wire nails are used at each connection (but refer to BS 5268:Part 3).

In designing the layout of a roof, the designer must ensure that the rafter diagonal bracing can be fixed without fouling the internal members of intermediate trussed rafters. Ceiling ties must be positioned to avoid supports for water tanks.

In addition, for spans of 5 m and over, Part 3 calls for longitudinal diagonal bracing to occur at each end of each roof unit (usually across two spacings) in the line of the internal struts. One can see the theoretical sense in tying the two bracing systems together, but obviously it is difficult to fix.

Where trussed rafters occur with masonry cross-walls, steel brackets and timber blockings are required at 2 m centres at rafter level and ceiling tie level, at gable and party walls. Full details are given in Part 3.

## 20.4 STANDARD DESIGNS

By testing, it has been found that a trussed rafter designed by conventional means will not fail where expected (in the lower part of the rafter) but usually in the rafter near the apex at external loading somewhat in excess of the anticipated failure loading. To maximise the economy in use of trussed rafters, the results of hundreds of tests have been pooled and for symmetrical Fink trusses and mono-pitch profiles the span tables thus derived are printed in BS 5268:Part 3 (for spans up to 12 m using stress grades M50, SS/MSS, and M75). (The previous 'Composite' grade is no longer covered.) Tables for GS may be added. In addition many manufacturers of trussed rafters or metal-plate manufacturers have their own standard designs.

The limits which apply when using the span tables from BS 5268:Part 3 are detailed at length in Part 3 and are not repeated here. (Even if producing a special design which could justify longer spans for the same members as tabulated, Part 3 advises that the quoted spans should not be exceeded.) Therefore when an engineer is involved with a contract using standard designs, his or her function often becomes one of checking that the grading and size of the timber used complies with the specification, and that the quality control of manufacture meets the requirements set out in BS 5268: Part 3. This Code of Practice was written after many years of experience in the manufacture and use of trussed rafters and, because much of this practical experience has been written into the Code, the authors have little to add, and refer the reader to it. Although the primary object of the Code was to link quality control requirements to standard designs, it is worth emphasising that if special designs are prepared the quality control requirements still apply.

The extent to which finger joints can be permitted under punched metal-plate fasteners is still to be finalised (see § 19.6.8).

Although a method for supporting water tanks is sketched in BS 5268: Part 3, it is preferable to support tanks from walls rather than the trussed rafters.

## 20.5 SPECIAL DESIGNS

In certain cases it may not be possible or desirable for the designer to rely on the standard design tables in BS 5268:Part 3, in which case calculations will have to be produced. Even when this is the case, if the trussed rafters are to be manufactured using gussets of proprietary metal-plate fasteners or nailed plywood, the engineer is advised to have the calculations produced by one of the design groups to which most manufacturers have access. These design groups are familiar with the load-slip-rotation characteristics of particular joints, are closely involved with their costing and can usually prepare a more economical design than a designer who does not have this intimate knowledge. If however, a designer is faced with producing his own calculations, there are a few alternative ways which can be adopted, each of which would satisfy the design Clause 15 of BS 5268 : Part 3.

The most simple design method is to assume that all joints are pinned for the purpose of calculating axial forces in members, having first transposed all loading which occurs between node points to the node points. Any local bending on the rafter or ceiling tie is calculated usually by treating each as a continuous beam without sinking supports. The members are then designed for tension, compression, combined bending and tension, or combined bending and compression for the load combinations detailed in Clause 14 of the Code.

Design assumptions for trussed rafters are detailed in BS 5268 : Part 3, but some are worth highlighting and discussion here concerns fully triangulated configurations.

The axial forces in members can be taken as for a pin-jointed framework. (Even with computer analysis taking account of deformation of the truss, this assumption has been found to hold true.)

Bending moments can be determined by any method which the designer can justify. However, a 'simplified analysis' has been incorporated into the Code which assumes that rafters and ceiling joists are continuous throughout their length, but that all free ends are pin-jointed. With this method a redistribution of moments is permitted in that moments over node points may be reduced by 10% whilst moments mid way between nodes are increased accordingly. (Note, not necessarily by 10% unless there are three or more bays.) This method can be quite useful but it is worthwhile realising that, for example for a Fink truss, a 10% redistribution is far less than may occur, particularly if joint slip is taken into account, and takes no account of varying slopes. Even when joint slip is disregarded, the redistribution effect can be significant at a slope of $22\frac{1}{2}°$, whilst at 35° it is considerably less.

When considering axial compression in the rafter, the useful point is made that the effective length on the $XX$ axis of the rafter is different if one considers the position at the node, to the mid span position. For mid-bay checks $L_{eX}$ may be taken as 0.8 of the bay length and, for the node point may be taken as 0.2 of the sum of the adjacent bay lengths. If the positions of points of contraflexure are calculated they may be used in determining the effective length.

When determining the permissible bending stress for the rafter the effect of tiling battens and diagonal bracing may be considered to give full restraint. However, when determining the permissible axial compression stress for the rafter a more conservative assumption is required, even where tiling battens occur, with $L_{eY}$ (in metres) being taken as:

$$0.01 b L_{bay} K^{\frac{1}{4}}$$

where $b$ = the thickness of the member (mm)

$L_{bay}$ = the length of the rafter bay along the slope (m)

$K$ = equals 1 or the ratio of the distance between points of increased lateral stability along the slope divided by $L_{bay}$ whichever is the greater.

Points of 'increased lateral stability' are the eaves, ridge and any point braced by a lateral member which itself is braced by diagonal members as required in the Code.

The effective length of compression internal members should be taken as 0.9 times the distance between node points. No internal member should have a width less than 60 mm. Limits of slenderness are as for BS 5268 : Part 2 (see §14.5). Note the form of the interaction formula in Clause 15.3 of Part 3.

Even where trussed rafters spaced at 600 mm centres are part of a load-sharing system, Clause 15.3 of Part 3 is quite clear in stating that the $K_8$ factor of 1.1 should not be applied to the permissible stresses in the ceiling tie (unless binders or boarding can provide lateral distribution of load). However, one wonders if this is over-conservative. If a ceiling tie starts to 'give', presumably the load would begin to be relieved from the truss (not just the ceiling tie) hence load transference would be achieved (see §4.4.1 and §2.5.1). Part 3 may permit some re-distribution of the concentrated load on the ceiling tie.

Unless a trussed rafter or rafters acts as a principal member one would expect to design for deflection using $E_{mean}$. Where two or more trussed rafters are connected together to act as a principal member, factors numerically equal to $K_8$ for stress, and $K_9$ (multiplied to $E_{min}$) are applicable.

BS 5268 : Part 3 gives a deflection method for calculating the effect of joint slip.

Arbitrary deflection limits of 12 mm for spans of 12.0 m, and 15 mm for spans of 15.0 m are given for trussed rafters which are not cambered in addition to the usual limit of 0.003 × span. Most manufacturers prefer not to build-in a camber, but camber is permitted.

Where a plated butt joint is used in a rafter or ceiling tie BS 5268 : Part 3 details certain arbitrary rules which must be met.

Whereas CP 112 : Part 3 excluded wane from the upper edge of rafters and the lower edge of ceiling ties, BS 5268 : Part 3 permits wane up to the limits of SS grade of BS 4978 (i.e. not more than one quarter on any face), but does not permit wane to occur within the area of any jointing device (Clause 22.1 of Part 3). Also, wane which reduces the effective thickness of a member to less than 35 mm should not occur on the top edges of rafter members, the bottom edges of ceiling ties, or within bearing areas. Therefore it appears as though, for example, 47 mm thick members could have wane on edges of as much as 12 mm (or, strictly speaking, 11.75 mm to match SS grade).

## 20.6 QUALITY CONTROL AND CARE IN HANDLING AND ERECTION

The quality control required with mechanical or glue joints is generally as explained in Chapters 18 and 19 respectively, with certain special points applicable to trussed rafters as set out below and others contained in BS 5268:Part 3 which are not repeated here.

With mechanical joints, the moisture content at time of manufacture should not exceed 22%, and should be within a few per cent of the equivalent moisture content in service.

Only WBP adhesive is permitted by BS 5268:Part 3 to fix glued or glue/nailed plywood gussets.

Metal-plate fasteners should be evenly embedded to ensure full penetration of all teeth, bursts or nails without damage to the timber or fastener. Within the area of a punched metal-plate fastener the average gap between any two adjacent members should not exceed 1.5 mm unless specifically allowed for. The difference in thickness between adjacent members at a joint should not exceed 1 mm where punched metal-plate fasteners or nailed plywood gussets are used. For joints with glued plywood gussets the difference in thickness should not exceed 0.5 mm.

If metal-plate fasteners are used to joint the rafter or ceiling tie between node points, normally they should not be placed at the centre but usually between 0.1 and 0.25 of the distance between nodes.

Handling, storage and the accuracy of erection can be critical to the subsequent performance of trussed rafters and all concerned should ensure that care is taken during delivery, site storage, handling and erection. Storage should be arranged so that no unit is deformed or distorted. Covers during storage should protect from rain, snow and sun whilst allowing ventilation. Erection should be arranged to ensure that units are subjected to rain, snow, sun and wind for the minimum possible time.

## 20.7 TILING BATTENS

The size of tiling battens should be in accordance with BS 5534:Part 1 including AMD 3554. The table of sizes is given in Table 20.1. The grade should generally be as GS of BS 4978, although there is no specific requirement to stress grade or mark the timber. Wane up to one-third is permitted on one arris and it is better if this is not the arris against which the tile lip rests. Strangely enough BS 5534 does not limit the species required to satisfy the table of sizes. European Whitewood or Redwood are commonly used.

Table 20.1    Batten sizes for pitched roofs and vertical work

| Examples of slating or tiling | Basic sizes* | | | |
| | 450 mm span | | 600 mm span | |
| | width (mm) | depth (mm) | width (mm) | depth (mm) |
|---|---|---|---|---|
| *Slates* | | | | |
| Sized | 38 | 19 | 38 | 25 |
| Random | 50 | 25 | 50 | 25 |
| Asbestos-cement | 38 | 19 | 38 | 25 |
| Concrete | 38 | 19 | 38 | 25 |
| *Clay and concrete tiles* | | | | |
| Plain: pitched roofs | 32 | 19 | 32 | 25 |
|       vertical work | 38 | 19 | 38 | 25 |
| Single lap | 38 | 22 | 38 | 25 |

\* All the sizes in this table are subject to production/resawing allowances of

   *either*: width: $+0, -2$ mm

   *or*:    depth: $+0, -1$ mm.

In addition the sizes are subject to a minus tolerance of 0.5 mm in width and depth.

Where larger sizes of battens are used the designer has to revert to normal design methods. Variations may also be required for certain designs to satisfy the Building Standards (Scotland) Regulations 1981.

To minimise splitting, the maximum nail diameter should not exceed one-tenth of the batten width for nails fixed on the centre line of the batten.

## 20.8    COEFFICIENTS OF AXIAL LOADING

### 20.8.1    Introduction

When trussed rafters have to be designed rather than proved by test, it is necessary to calculate the axial loading in the members due to the external loading. To save design time, coefficients are tabulated in Tables 20.2–20.5 for cases of Fink, Fan and mono-pitch configurations, on the assumption that loading is by point loads at node points and that all node points are pinned.

## 20.8.2 Fink Trusses

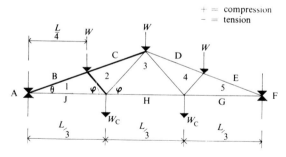

+ = compression
− = tension

Although *F* is generally used in BS 5268 and this manual to represent load, *W* is retained in Figs 20.5–20.8 and in Tables 20.2–20.8 where there is no chance of confusion.

Fig. 20.5

**Table 20.2**

| Slope $\theta$ | Axial loading in members due to point loads $W$ and $W_C$ (Fig. 20.5) | | | | | | Angle $\varphi$ |
|---|---|---|---|---|---|---|---|
| | B1 | C2 | H3 | J1 | 12 | 23 | |
| 15° | +5.796W | +4.830W | −3.733W | −5.600W | +1.197W | −1.197W | 38°47′ |
| | +3.864W_C | +3.864W_C | −2.488W_C | −3.732W_C | 0 | −1.595W_C | |
| 17½° | +4.988W | +4.157W | −3.171W | −4.757W | +1.092W | −1.092W | 43°24′ |
| | +3.326W_C | +3.326W_C | −2.115W_C | −3.172W_C | 0 | −1.455W_C | |
| 20° | +4.386W | +3.655W | −2.747W | −4.121W | +1.018W | −1.018W | 47°31′ |
| | +2.924W_C | +2.924W_C | −1.832W_C | −2.748W_C | 0 | −1.356W_C | |
| 22½° | +3.920W | +3.267W | −2.414W | −3.621W | +0.963W | −0.963W | 51°10′ |
| | +2.613W_C | +2.613W_C | −1.611W_C | −2.414W_C | 0 | −1.301W_C | |
| 25° | +3.549W | +2.957W | −2.145W | −3.217W | +0.922W | −0.922W | 54°26′ |
| | +2.366W_C | +2.366W_C | −1.428W_C | −2.145W_C | 0 | −1.230W_C | |
| 27½° | +3.249W | +2.708W | −1.921W | −2.881W | +0.891W | −0.891W | 57°22′ |
| | +2.166W_C | +2.166W_C | −1.283W_C | −1.921W_C | 0 | −1.187W_C | |
| 30° | +3.000W | +2.500W | −1.732W | −2.598W | +0.866W | −0.866W | 60°00′ |
| | +2.000W_C | +2.000W_C | −1.155W_C | −1.732W_C | 0 | −1.155W_C | |
| 32½° | +2.792W | +2.327W | −1.569W | −2.354W | +0.846W | −0.846W | 62°23′ |
| | +2.000W_C | +1.861W_C | −1.048W_C | −1.570W_C | 0 | −1.128W_C | |
| 35° | +2.615W | +2.179W | −1.428W | −2.142W | +0.831W | −0.831W | 64°33′ |
| | +1.743W_C | +1.743W_C | −0.948W_C | −1.428W_C | 0 | −1.109W_C | |

Axial loading in members due to unit load at apex

| | | | | | |
|---|---|---|---|---|---|
| +0.5 cosec $\theta$ | +0.5 cosec $\theta$ | −0.5 cotan $\theta$ | −0.5 cotan $\theta$ | 0 | 0 |

Axial loading in members due to half a unit load at each ceiling node point

| | | | | | |
|---|---|---|---|---|---|
| +0.5 cosec $\theta$ | +0.5 cosec $\theta$ | −0.5 (cotan $\theta$ − cotan $\varphi$) | −0.5 cotan $\theta$ | 0 | −0.5 cosec $\varphi$ |

**Table 20.3**

| Slope $\theta$ | Axial forces in members due to point loads $W$ and $W_C$ (Fig. 20.6) | | | | | | | | Angle $\varphi$ |
|---|---|---|---|---|---|---|---|---|---|
| | B1 | C2 | D3 | K4 | L1 | 12 | 23 | 34 | |
| 15° | $+9.660W$ | $+7.728W$ | $+7.728W$ | $-5.599W$ | $-9.332W$ | $+1.932W$ | $+1.000W$ | $-2.395W$ | 38° 47' |
| | $+3.864W_C$ | $+3.864W_C$ | $+3.864W_C$ | $-2.489W_C$ | $-3.732W_C$ | $0$ | $0$ | $-1.283W_C$ | |
| 17½° | $+8.314W$ | $+6.651W$ | $+6.651W$ | $-4.757W$ | $-7.929W$ | $+1.663W$ | $+1.000W$ | $-2.183W$ | 43° 24' |
| | $+3.326W_C$ | $+3.326W_C$ | $+3.326W_C$ | $-2.114W_C$ | $-3.172W_C$ | $0$ | $0$ | $-1.376W_C$ | |
| 20° | $+7.310W$ | $+5.848W$ | $+5.848W$ | $-4.121W$ | $-6.868W$ | $+1.462W$ | $+1.000W$ | $-2.035W$ | 47° 31' |
| | $+2.924W_C$ | $+2.924W_C$ | $+2.924W_C$ | $-1.832W_C$ | $-2.748W_C$ | $0$ | $0$ | $-1.480W_C$ | |
| 22½° | $+6.533W$ | $+5.226W$ | $+5.226W$ | $-3.621W$ | $-6.036W$ | $+1.307W$ | $+1.000W$ | $-1.926W$ | 51° 10' |
| | $+2.613W_C$ | $+2.613W_C$ | $+2.613W_C$ | $-1.611W_C$ | $-2.414W_C$ | $0$ | $0$ | $-1.595W_C$ | |
| 25° | $+5.916W$ | $+4.733W$ | $+4.733W$ | $-3.217W$ | $-5.361W$ | $+1.183W$ | $+1.000W$ | $-1.844W$ | 54° 26' |
| | $+2.366W_C$ | $+2.366W_C$ | $+2.366W_C$ | $-1.428W_C$ | $-2.145W_C$ | $0$ | $0$ | $-1.719W_C$ | |
| 27½° | $+5.415W$ | $+4.332W$ | $+4.332W$ | $-2.881W$ | $-4.802W$ | $+1.083W$ | $+1.000W$ | $-1.781W$ | 57° 22' |
| | $+2.166W_C$ | $+2.166W_C$ | $+2.166W_C$ | $-1.282W_C$ | $-1.921W_C$ | $0$ | $0$ | $-1.854W_C$ | |
| 30° | $+5.000W$ | $+4.000W$ | $+4.000W$ | $-2.598W$ | $-4.330W$ | $+1.000W$ | $+1.000W$ | $-1.732W$ | 60° |
| | $+2.000W_C$ | $+2.000W_C$ | $+2.000W_C$ | $-1.155W_C$ | $-1.732W_C$ | $0$ | $0$ | $-2.000W_C$ | |
| 32½° | $+4.653W$ | $+3.722W$ | $+3.722W$ | $-2.354W$ | $-3.924W$ | $+0.931W$ | $+1.000W$ | $-1.693W$ | 62° 23' |
| | $+1.861W_C$ | $+1.861W_C$ | $+1.861W_C$ | $-1.047W_C$ | $-1.570W_C$ | $0$ | $0$ | $-2.157W_C$ | |
| 35° | $+4.358W$ | $+3.487W$ | $+3.487W$ | $-2.142W$ | $-3.570W$ | $+0.872W$ | $+1.000W$ | $-1.663W$ | 64° 33' |
| | $+1.743W_C$ | $+1.743W_C$ | $+1.743W_C$ | $-0.948W_C$ | $-1.428W_C$ | $0$ | $0$ | $-2.319W_C$ | |
| Axial loading in members due to unit load at apex | | | | | | | | | |
| | $+0.5\,\mathrm{cosec}\,\theta$ | $+0.5\,\mathrm{cosec}\,\theta$ | $+0.5\,\mathrm{cosec}\,\theta$ | $-0.5\,\mathrm{cosec}\,\theta$ | $-0.5\,\mathrm{cotan}\,\theta$ | $0$ | $0$ | $0$ | $0$ |
| Axial loading in members due to half a unit load at each ceiling node point | | | | | | | | | |
| | $+0.5\,\mathrm{cosec}\,\theta$ | $+0.5\,\mathrm{cosec}\,\theta$ | $+0.5\,\mathrm{cosec}\,\theta$ | $-0.5\,(\mathrm{cotan}\,\theta - \mathrm{cotan}\,\varphi)$ | $-0.5\,\mathrm{cotan}\,\theta$ | $0$ | $0$ | $-0.5\,\mathrm{cosec}\,\varphi$ | $0$ |

## 20.8.3 Fan trusses

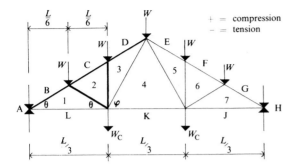

Fig. 20.6

Although *F* is generally used in BS 5268 and this manual to represent load, *W* is retained in Figs 20.5–20.8 and in Tables 20.2–20.8 where there is no chance of confusion.

## 20.8.4 Mono-pitch trusses

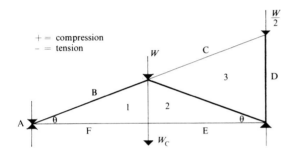

Fig. 20.7

Athough *F* is generally used in BS 5268 and this manual to represent load, *W* is retained in Figs 20.5–20.8 and in Tables 20.2–20.8 where there is no chance of confusion.

### Table 20.4

| Slope $\theta$ | Axial loading in members due to point loads $W$ and $W_C$ (Fig. 20.7)* | | | | | | |
|---|---|---|---|---|---|---|---|
| | B1 | C3 | D3 | E2 | F1 | 12 | 23 |
| $15°$ | $+1.932W$ | 0 | $+0.500W$ | $-1.866W$ | $-1.866W$ | 0 | $+1.932W$ |
| | $+1.932W_C$ | 0 | 0 | $-1.866W_C$ | $-1.866W_C$ | $-1.000W_C$ | $+1.932W_C$ |
| $17\frac{1}{2}°$ | $+1.663W$ | 0 | $+0.500W$ | $-1.586W$ | $-1.586W$ | 0 | $+1.663W$ |
| | $+1.663W_C$ | 0 | 0 | $-1.586W_C$ | $-1.586W_C$ | $-1.000W_C$ | $+1.663W_C$ |
| $20°$ | $+1.462W$ | 0 | $+0.500W$ | $-1.374W$ | $-1.374W$ | 0 | $+1.462W$ |
| | $+1.462W_C$ | 0 | 0 | $-1.374W_C$ | $-1.374W_C$ | $-1.000W_C$ | $+1.462W_C$ |
| $22\frac{1}{2}°$ | $+1.307W$ | 0 | $+0.500W$ | $-1.207W$ | $-1.207W$ | 0 | $+1.307W$ |
| | $+1.307W_C$ | 0 | 0 | $-1.207W_C$ | $-1.207W_C$ | $-1.000W_C$ | $+1.307W_C$ |

* The coefficients can also be used to calculate axial loading from a unit load placed at the centre node point of the rafter or ceiling joist.

**Table 20.5**

| Slope $\theta$ | | Axial forces in members due to point loads $W$ and $W_C$ (Fig. 20.8)* | | | | | | | | | | |
|---|---|---|---|---|---|---|---|---|---|---|---|---|
| | | B1 | C2 | D5 | E5 | F4 | G3 | H1 | 12 | 23 | 34 | 45 |
| 15° | | $+3.864W$ | $+3.864W$ | 0 | $+0.500W$ | $+1.864W$ | $+1.864W$ | $-3.732W$ | $+1.000W$ | $-2.681W$ | 0 | $+2.115W$ |
| | | $+3.864W_C$ | $+3.864W_C$ | 0 | 0 | $+1.864W_C$ | $+1.864W_C$ | $-3.732W_C$ | 0 | $-2.681W_C$ | $-1.000W_C$ | $+2.115W_C$ |
| $17\frac{1}{2}°$ | | $+3.326W$ | $+3.326W$ | 0 | $+0.500W$ | $+1.590W$ | $+1.590W$ | $-3.172W$ | $+1.000W$ | $-2.341W$ | 0 | $+1.878W$ |
| | | $+3.326W_C$ | $+3.326W_C$ | 0 | 0 | $+1.590W_C$ | $+1.590W_C$ | $-3.172W_C$ | 0 | $-2.341W_C$ | $-1.000W_C$ | $+1.878W_C$ |
| 20° | | $+2.924W$ | $+2.924W$ | 0 | $+0.500W$ | $+1.379W$ | $+1.379W$ | $-2.748W$ | $+1.000W$ | $-2.091W$ | 0 | $+1.703W$ |
| | | $+2.924W_C$ | $+2.924W_C$ | 0 | 0 | $+1.379W_C$ | $+1.379W_C$ | $-2.748W_C$ | 0 | $-2.091W_C$ | $-1.000W_C$ | $+1.703W_C$ |
| $22\frac{1}{2}°$ | | $+2.613W$ | $+2.613W$ | 0 | $+0.500W$ | $+1.210W$ | $+1.210W$ | $-2.414W$ | $+1.000W$ | $-1.900W$ | 0 | $+1.569W$ |
| | | $+2.613W_C$ | $+2.613W_C$ | 0 | 0 | $+1.210W_C$ | $+1.210W_C$ | $-2.414W_C$ | 0 | $-1.900W_C$ | $-1.000W_C$ | $+1.569W_C$ |

* If the coefficients for axial loading from loads $W$ and $W_C$ are halved, this will give the axial loading from half a unit load placed at each rafter node point and ceiling joist node point respectively.

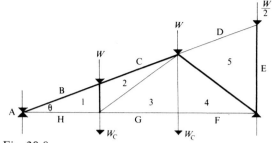

Fig. 20.8

Although $F$ is generally used in BS 5268 and this manual to represent load, $W$ is retained in Figs 20.5–20.8 and in Tables 20.2–20.8 where there is no chance of confusion.

# Chapter Twenty-one
# Trusses

## 21.1 INTRODUCTION

The choice of truss profile with timber construction is probably far greater than with any other structural material. Possibly this is due to the long tradition in the use of timber for this element of structure, or possibly because of the relative ease with which unusual truss shapes can be fabricated and assembled in timber. Several of the profiles regarded as traditional are still specified for architectural reasons, and the engineer needs to be familiar with both modern and traditional forms of truss design.

The structural function of a truss is to support and transfer loads from the points of application (usually purlins) to the points of support as efficiently and as economically as possible. The efficiency depends on the choice of a suitable profile consistent with the architectural requirements and compatible with the loading conditions. Typical 'idealised' truss profiles for three loading conditions are sketched in Fig. 21.1.

With a symmetrical system of loading (particularly important in the second case in Fig. 21.1, which is a four-pin frame and therefore unstable) in each idealised case, the transfer of loading is achieved without internal web members, because the chord profile matches the bending moment of the simple span condition. Unfortunately it is seldom possible to use a profile omitting internal members, because unbalanced loading conditions can nearly always occur from snow, wind or permanent loadings. Unbalanced conditions can also occur due to manufacturing and erection tolerances (see also § 3.12), nevertheless, the engineer should try to use a truss profile closely related to the idealised profile (the moment diagram), adding a web system capable of accommodating unbalanced loading. In this way the loading in the internal members and the connections is minimised, with consequent design simplicity and economy.

Undoubtedly the engineer will encounter cases in which the required architectural profile is at conflict with the preferred structural profile, therefore high stresses may be introduced into the web system and the connections. Economy must then be achieved by adopting the most suitable structural arrangement of internal members in which it is necessary to create

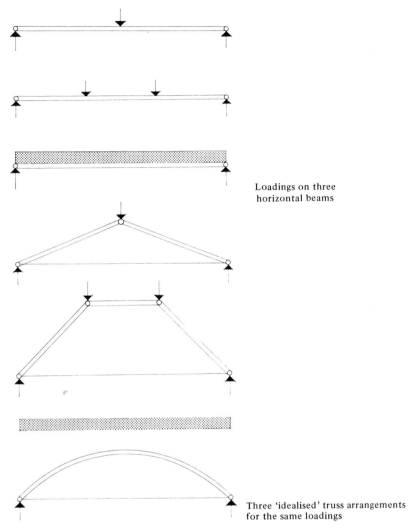

Loadings on three
horizontal beams

Three 'idealised' truss arrangements
for the same loadings

Fig. 21.1

an economical balance between materials and workmanship. The configuration of internal members should give lengths between node points on the
rafters and ceiling ties such as to reduce the numbers of joints. Joints should
be kept to a sensible minimum because the workmanship for each is expensive, and also the joint slip at each (except with glued joints) generally adds to
the overall deflection of the truss (§ 21.6.2). On the other hand, the slenderness ratio of the compression chords and the internal struts must not be
excessive, local bending on the chords must not be too large, and the angle
between internal diagonals and the chords must not be too small.

The engineer is usually influenced by architectural considerations, the
type and length of roof material, support conditions, span and economy, and
probably chooses from three basic truss types: pitched (mono- or duo-pitch),
parallel-chord or bowstring trusses (Fig. 21.2).

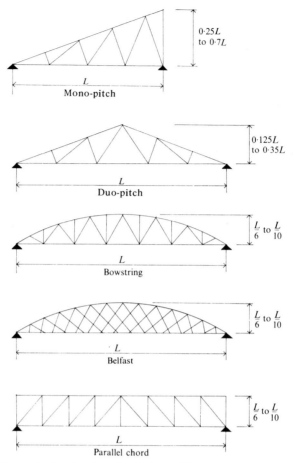

Fig. 21.2: *General span-to-depth ratios for basic profiles.*

The most common form both for domestic and industrial uses is the pitched truss. The shape fits the usual moment diagram reasonably well and is compatible with traditional roofing materials, such as tiles for domestic uses and corrugated sheeting for industrial applications. A portion of the applied loading is transferred directly through the top chord members to the points of support, while the web members transfer loads of relatively small to medium magnitude, and the joints can usually be designed to develop these loads with little difficulty. Mono-pitch trusses are suitable generally for spans only up to around 9 m. Above this span the vertical height is usually too large for architectural reasons, even if the truss slope is reduced below that suitable for tiling. Duo-pitch domestic trusses span up to around 12 m with duo-pitch industrial trusses spanning up to around 15 m, above which span they become difficult to transport unless fabricated in parts (Fig. 21.8).

For large-span industrial uses, bowstring trusses (Fig. 21.2) can be very economical. These may be regarded as the current alternative to the traditional all-nailed 'Belfast truss' (Fig. 21.2). With uniform loading and no large concentrated loads the arched top chord profile supports almost all of the applied loading, and spans in excess of 30 m are not uncommon. A parabolic profile is the most efficient theoretical choice to support uniform loading, but practical manufacturing considerations usually make it more convenient or necessary to adopt a circular profile for the top chord member. The top chord member is usually laminated (not necessarily with four or more members), using either clamp pressure or nail pressure for assembly. The curvature may be introduced while laminating (laminations of course visible on the sides), or alternatively, the chord may be fabricated straight and then bent to the required curvature. The designer must be aware of the method of manufacture, or will be unable to allow for the correct curvature stresses. The least radial stresses occur if the curvature is introduced during laminating. Bending after laminating usually leads to a flattening of the curve close to the heel joints, due to a lack of leverage at the ends during assembly. This flattening, if not recognised, can lead to a change of tangency at the heel joint, resulting in incorrect dimensioning and placing of the critical tension connectors in the tie member. Also, bending leads to difficulties in that the member tries to straighten during assembly and service. Bending after laminating is not normally used.

With mono-pitch and duo-pitch trusses, secondary bending on the chord members should be avoided where possible, by placing purlins at the node points, whereas with bowstring trusses the purlins may be placed between the node points deliberately to create a secondary moment to off-set the moment caused by the product of the axial tangential loading and the eccentricity of the chord.

When the manufacture and erection of bowstring trusses is a matter of some urgency, the shape has the disadvantage of requiring curved sheeting. Curved sheeting is usually a non-standard or non-stock item, which may have a very long delivery period. When this is the case, the delay can be overcome by adding to the bowstring as shown in Fig. 21.3 to give it a mansard profile, and revert to the use of flat sheets. In producing this artificial profile, adequate lateral restraint must still be provided to the curved compression chord, which is no longer restrained directly by the purlins.

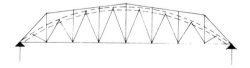

Fig. 21.3: *Bowstring converted to mansard.*

Parallel-chord trusses are frequently specified as an alternative to ply web or glulam beam on long spans where the beams may be uneconomical. The loads in the web members are frequently very large, which causes some difficulty in providing adequate joints. The choice of web configuration is between the Howe (diagonals in compression, Table 21.17), the Pratt (diagonals in tension, Table 21.20) and the Warren type (diagonals in alternate compression and tension, Table 21.23).

When a parallel-chord truss is joined to a timber or steel column with connections at both the top and bottom chords (Fig. 21.4), this gives fixity in a building subject to sway (Figs 23.5–23.8). A Pratt truss would be favoured.

When it is required to minimise the height of the perimeter wall, the Howe truss is favoured (Fig. 21.5).

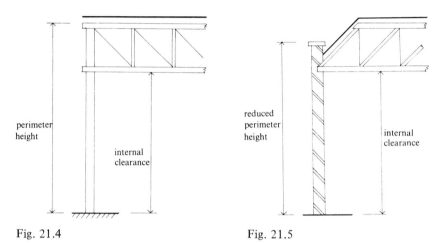

Fig. 21.4                                 Fig. 21.5

As an indication of the difference in magnitude of forces in internal members dictated by the choice of truss profile, coefficients are presented in Fig. 21.6 for the three basic types at a typical span-to-depth ratio.

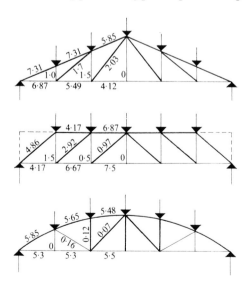

Fig. 21.6

Each of the three basic profiles may have raised bottom chords to give extra central clearance. This can be particularly useful in storage buildings with central access. The guide span-to-central-depth ratios given in Fig. 21.2 should be maintained for maximum economy in design. Modified pitched, parallel-chord and bowstring trusses are sketched in Fig. 21.7, plus some traditional configurations. In practice, the collar beam is used only for small spans with steep slopes. If using a raised chord truss (and particularly if using the collar beam type), the designer should consider the possibility of thrusts occurring at the support points due to deflection of the framework.

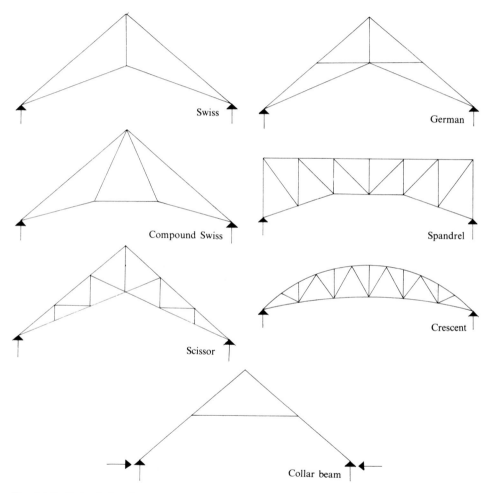

Fig. 21.7: *Raised chord trusses.*

The larger the span the more necessary it may be to use the smaller span-to-depth ratios. Although shallower depth trusses may be preferred aesthetically, they tend to have large deflections which may create secondary stresses. Deflection can be minimised by:

    (a) using lower stress grades and hence larger size members,
    (b) keeping the number of joints and mechanically jointed splices to a
        minimum, and
    (c) using fastenings with low slip characteristics.

The use of lower-grade material may be regarded initially as an uneconomical proposition as this makes larger sizes necessary. However, in many cases the spacings and edge distances for connector groups call for the use of timber sections larger than those required to satisfy the axial loadings and moments. This may invalidate the apparent benefits of a higher stress grade. From the values in Chapter 18 it can be seen that connector capacities do not increase to the same extent as the efficiency increase of higher stress grades, and the required minimum spacings and edge distances do not reduce *pro rata* to the increase in grade stress. These points encourage the use of lower stress grades although of course the strength of the net timber section is less compared to the same size of a higher stress grade.

To obtain an indication of the likely effect on the deflection of a truss by increasing the stress grade from GS grade Whitewood to SS grade Whitewood, compare the *AE* values of the tension members, assuming members fully stressed. The ratios show that the strain in the SS members leads to approximately 20% more deflection of the truss than if using GS members. Providing the slenderness ratio of the compression members is quite small, the *AE* ratio in compression is similar to that in tension, and it can be assumed that a truss designed to use the minimum amount of SS material deflects approximately 20% more than a truss designed for the minimum amount of GS material, providing that there is no slip at joints. However, if the same type of mechanical connector is used in each truss, the slip effect is constant for each, which reduces the difference in deflection from 20% to approximately 10%.

Trusses (not trussed rafters) used on spans similar to domestic applications are usually spaced at around 1.5–2.0 m centres, perhaps with solid purlins supporting intermediate rafters. For industrial uses where lighter-weight roof specifications can be expected, trusses are spaced at 5–6 m centres with solid or composite purlins at 0.8–1.8 m spacings supporting corrugated sheeting. Economy usually results if truss spacing increases with truss span.

Transportation is frequently a limiting factor with deep or long-span trusses. Trusses deeper than 3 m, or longer than 20 m require special attention. The transport problem can usually be overcome by a partial or complete breakdown of the truss. For example, bowstring trusses may have the main members spliced at mid span, and in many cases the entire assembly can be carried out on site, although it is preferable to carry out an initial assembly in the works to ensure correct fit, then break down for transport. Pitched trusses, especially those of large span, can be fabricated in two halves and linked together on site, with a loose centre tie perhaps with an optional

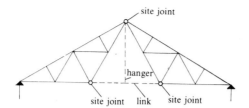

Fig. 21.8

hanger (Fig. 21.8). § 27.3 gives guidance on the maximum size of components which can be accommodated on normal lorries without special transport arrangements having to be made.

## 21.2   LOADING ON TRUSSES

Figure 21.9 shows a typical four-panel truss supporting a UDL of $4F$. It is common and accepted practice to assume equal point loads at each node point from this UDL in calculating the axial loading in the truss members. If, however, each rafter is isolated, as shown in Fig. 21.10 and treated as a

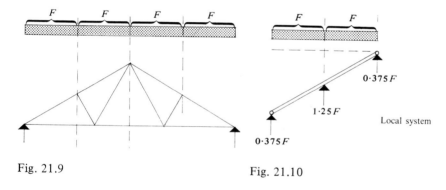

Fig. 21.9                                    Fig. 21.10

member continuous over two bays without sinking supports (and disregarding shear deflection of the rafter), this has the effect of reducing the overall moment on the truss by reducing the central point load (Fig. 21.11). Whether or not the designer takes account of this in normal truss design, it should be

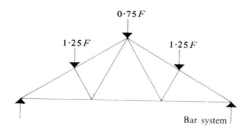

Fig. 21.11

realised that, if secondary rafters are used parallel to trusses and are supported by a mid point purlin between trusses (ridge beam at the top), the point load transferred to the truss will be related to $1.25F$ (Fig. 21.11) rather than $1.0F$.

With a truss having six panels, the point loads on the node points intermediate on the rafters become $1.10F$ and the apex load equals $0.8F$. With trusses of more than six bays it is usually acceptable to disregard any such increase in the intermediate node point loads.

## 21.3   TYPES OF MEMBERS AND JOINTS

### 21.3.1   Mono-chord and duo-chord trusses

It is usual for the mono-chord truss to use chords and internal members in one plane with mild steel gusset plates or perhaps even thick (say 14 gauge) punched metal-plate fasteners on each side at all connections (Fig. 21.12). The design is similar to a steelwork truss, and has the advantage that all connectors are loaded parallel to grain. With lightly loaded members, bolts at each end are adequate. Single-sided tooth plates are usually used for medium loads, with shear-plate connectors required for heavily loaded members. This type of truss is extremely economical. The gussets placed on the outside may not be visually acceptable for certain applications, in which case the double-chord truss (Fig. 21.13) can be considered.

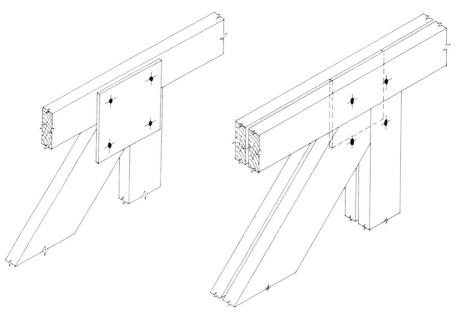

Fig. 21.12                          Fig. 21.13

A duo-chord truss as shown in Fig. 21.13 uses double members for the chord and internal members placed in the same two planes with mild steel gussets between the members. Washers must be placed under the heads and nuts of bolts, circular washers being used if appearance is important.

Large-span trusses may use laminated members (two or more laminations, clamp/glued or glue/nailed) to give sections with a larger area and/or improved appearance.

### 21.3.2   Rod-and-block assemblies

Rod-and-block assemblies (Fig. 21.14) are not used to any great extent for truss design in the UK. This system uses steel rods for the internal tension members, and as such may be the choice for buildings where a traditional

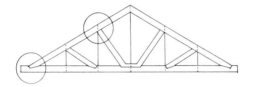

Fig. 21.14

appearance is desirable. Because the steel rods will not take compression, this type of construction is not permitted when the wind loading creates reversal of stress in the internal members.

The tension in the tie rods is transferred into the top chord by bearing at an angle $\theta$ to the grain (Fig. 21.15). A large round or square washer may be necessary to develop the load, and advantage should be taken of the $K_3$ and $K_4$ coefficients of BS 5268 (§§ 4.5, 4.10.1) and the formula for modifying bearing stresses at an angle to grain (§14.11) to keep the bearing area to a minimum and remove as little material from the top chord as possible. The

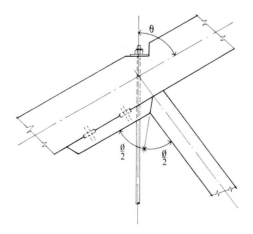

Fig. 21.15

tension is transferred into the bottom chord by bearing at right angles to the grain. If the bolt head is recessed into the bottom chord, the sides of the notches should be chamfered and the chord designed on net area (see § 4.12).

The compression in the timber diagonal has components of load perpendicular to and parallel to the slope of the top chord. The component perpendicular to the slope is taken in bearing, while the component parallel to the slope is taken first in bearing by the bearing block and is then transferred in shear (either by connectors or a glue line) into the top chord (Fig. 21.15). Maximum bearing pressure is developed by ensuring that the bearing line between the internal strut and the bearing block bisects the angle $\phi$ between them, giving the same angle of load to grain for both the strut and the bearing block.

The heel joint of a rod-and-block truss requires careful detailing. It is a stepped (or bridle) joint and can be weakened by eccentricities if the loads do not intercept at one point. The detail is sketched in Fig. 21.16.

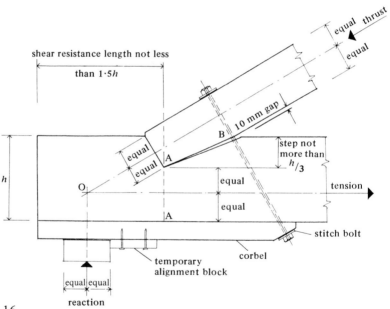

Fig. 21.16

To ensure that the joint is correctly designed the following conditions should be met:

(a) the bearing between the compression chord and the ceiling tie should be perpendicular to the line of the compression chord, with a gap between the mitre angle (line AB) of approximately 10 mm at the extreme end of the mitre, to prevent bending or thrust being applied to the ceiling tie along this line. The compression load should align with the centre of the bearing area.

(b) The lines of the thrust, tension and reaction should coincide at one point O so that the tension force is concentric with the net section AA below the step.

(c) Adequate length of tie must be provided beyond the step joint, to give adequate shear resistance to resist the component of thrust from the top chord parallel to the grain (numerically equal to the tension in the ceiling tie). An adequate depth of member must be provided at section AA to resist the shear from the vertical component of thrust from the top chord (numerically equal to the end reaction).

(d) A corbel should be added underneath the tie member to provide a bearing for the stitch bolt. The corbel should be attached to the tie to transfer a shear component from the stitch bolt should shearing commence beyond the step joint. The stitch bolt is introduced because the structure would otherwise be totally dependent on the shear resistance of the timber beyond the step. Its inclusion also locates the chord member.

(e) A bearing pad should be fixed underneath the ceiling tie, to ensure that the reaction occurs directly below point O.

When the required detail means that it is not possible for the line of the thrust, tension and reaction to coincide (Fig. 21.17), the net section AA must be designed for the moment $M_L$, where:

$$M_L = Ve_V - Te_T$$

The design of the chord must also take into account the moment $Pe_P$ resulting from the eccentric end bearing.

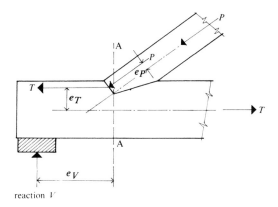

Fig. 21.17

## 21.3.3  Multi-member assemblies

Most timber trusses are of the multi-member type with the chords and internal members interlaced, the connections being made with split-rings or tooth-plates. The Timber Research and Development Association have produced a comprehensive range of standard designs for domestic and industrial use, and certain tile manufcturers also issue design guides for domestic trusses using multi-member assemblies.

With a short-span or lightly loaded truss, or a truss with lightly loaded internals (bowstring or Warren), it is possible to use a single solid member for the internal compression members with twin member chords, but only if an eccentricity is accepted at joints. (See Fig. 21.18(a).) It is more usual to use twin members for internal struts and single members for internal ties leading to the type of joint assembly sketched at (b) and (c) in Fig. 21.18.

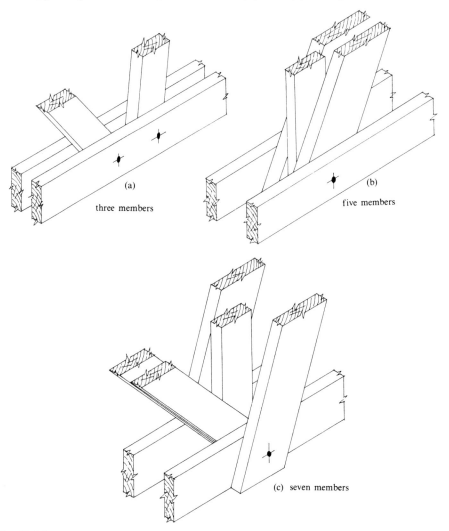

(a)
three members

(b)
five members

(c) seven members

Fig. 21.18

When there is an eccentricity at a joint or joints, the effect of eccentricity and secondary moments should be considered (Fig. 21.19). Unless an external load is applied at the connection, the transverse components of the com-

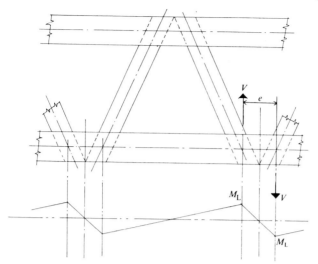

Fig. 21.19: *Secondary moment in chord due to eccentric joint.*

pression and tension members are the same, and if equal to $V$, give a local moment of $V.e$, which is shared (usually equally) to each side, giving $M_L = V.e/2$.

The seven-member arrangement sketched at (c) in Fig. 21.18 may occur in large-span parallel-chord trusses, particularly those involving cantilevers, and in eight-panel Fink trusses (Table 21.10) at the centre of the rafter. The direction and magnitude of the load as it affects each part of the connection is determined by considering each interface. This is best illustrated by the example given below.

Consider the members 1, 4 and 12 of the eight-panel Howe truss, for which coefficients are given in Table 21.19, assuming the angle between diagonal and vertical (and horizontal) to be 45°. Assume a point load of $W$ at each upper-panel node point, and no loading on the bottom chord. The equilibrium diagram for the upper, outermost joint is shown in Fig. 21.20.

Members 1 and 4 are in compression and are twin members. Member 12 is a single section. Symmetry is obviously desirable in the assembly of the truss,

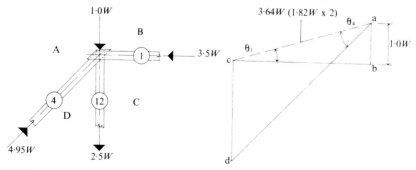

Fig. 21.20

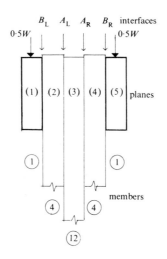

Fig. 21.21

therefore member 12 is placed on the centre line. There is then a choice for the positioning of member 1 and member 4 which will influence the joint design and economy. This is illustrated in Fig. 21.21 if one considers the two parts of member 1 to be placed at the outer positions (1) and (5) and the two parts of member 4 to be placed in positions (2) and (4). The joint has to be considered on each interface, and at this stage the designer must decide how the vertical external loading is to be applied. Assume it is applied through brackets or by direct bearing on the top chord (i.e. members 1) and analyse the loading in two parts.

*Loads to the left of interface $B_L$* (Fig. 21.21)
The $0.5W$ external load is applied at the top of the half of member 1 to the left of interface $B_L$. Because of this external load, the load in the connector at interface $B_L$ cannot be taken direct from the vector diagram in Fig. 21.20 except by joining points c and a. The loads in this connector to the left of interface $B_L$ are $0.5W$ vertical and $1.75W$ horizontal (i.e. half $3.5W$) giving $1.82W$ at an angle of $16°$ to member 1 and $29°$ to member 4. (See Fig. 21.22.)

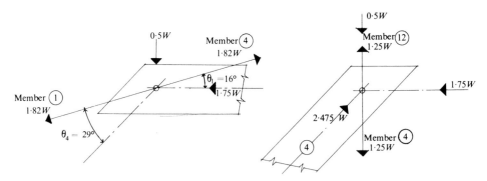

Fig. 21.22: *Loads to left of interface $B_L$.*     Fig. 21.23: *Loads to left of interface $A_L$.*

If either single- or double-sided connector units are used, the part in member 1 acts as an angle of 16° and the part in member 4 at an angle of 29° to grain. As part of the loading is taken by the connecting bolt, whichever type of connector is used, it is important to use the correct diameter bolt.

*Loads to the left of interface $A_L$ (Fig. 21.21)*
The connector unit at this interface is being acted upon by the 0.5W and 1.75W loads plus 2.475W from member 4 which, from Fig. 21.20, can be seen to lead to a resultant of $\frac{1}{2} \times 2.5W = 1.25W$ acting vertically. The connector acts at an angle of 45° to the grain in member 4 and parallel to the grain in member 12. (See Fig. 21.23.) An easier method of looking at this interface would be to cut the truss on its vertical axis and see how half the load in member 12 affects this interface.

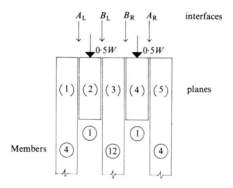

Fig. 21.24

The other method of positioning members 1 and 4 is shown in cross-section in Fig. 21.24. With this arrangement the two halves of member 4 are fixed on the outside of the chord with the external point load 1.0W applied in two halves to the top of the chord 1. The load to the left of interface $A_L$ (Fig. 21.25) is simply 2.475W, applied parallel to member 4 and at 45° to member 1.

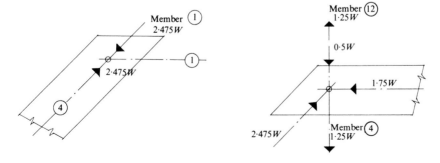

Fig. 21.25: *Loads to left of interface $A_L$.*    Fig. 21.26: *Loads to left of interface $B_L$.*

The loads to the left of interface $B_L$ are shown in Fig. 21.26, and result in a load of $1.25W$ parallel to the grain in member 12 and at right angles to member 1. This answer could be derived in an easier way by considering the truss cut on its vertical axis and calculating how half the load in member 12 affects this interface.

At each interface, the number and size of connectors required to develop the load at an interface is dictated by the maximum angle of load to grain and can be expressed in general terms as:

$$\text{Number of connectors} = \frac{\text{load}}{\text{connector capacity}} = \frac{\text{load}}{\overline{F}_0 K_\alpha}$$

where $\overline{F}_0$ = the capacity of one connector parallel to the grain ($\alpha = 0°$)

$K_\alpha$ = the coefficient for angle of load to grain given in Table 18.44 for split-ring and shear-plate connectors. For the sake of discussion below, the values derived in §§ 18.4.3 and 18.5.2 for bolts and toothed-plate connectors at an angle to the grain may be considered as similar to $\overline{F}_0 K_\alpha$.

One can now compare the connector requirements for the joint assembled in the two ways sketched in Figs 21.21 and 21.23 to determine which assembly is likely to offer most advantages. This comparison is shown in tabular form in Table 21.1 where $W$ is the applied load at node points and $\overline{F}_0$ is the capacity parallel to grain required of the connectors.

**Table 21.1**

| Interface | With the chords at the outer positions (Fig. 21.21) | | | | With the chords at the inner positions (Fig. 21.24) | | | |
|---|---|---|---|---|---|---|---|---|
| | Load | Maximum angle to grain | $K_\alpha$ | Required number of connectors | Load | Maximum angle to grain | $K_\alpha$ | Required number of connectors |
| A | $1.25W$ | $45°$ | 0.824 | $1.52W/\overline{F}_0$ | $2.475W$ | $45°$ | 0.824 | $3.00W/\overline{F}_0$ |
| B | $1.82W$ | $29°$ | 0.908 | $2.00W/\overline{F}_0$ | $1.25W$ | $90°$ | 0.700 | $1.78W/\overline{F}_0$ |
| | | | | $3.52W/\overline{F}_0$ | | | | $4.78W/\overline{F}_0$ |

The summation in Table 21.1 shows that with the chords in the inner positions, $(4.78 - 3.52) \times 100/3.52 = 36\%$ more connector capacity is required. The designer will not wish to work through the various combinations before deciding on the optimum arrangement, and a simple rule to follow, which can be seen from a quick examination of Table 21.1, is to assemble members so that the angle between adjacent members is kept to a minimum.

In a multi-member joint, it is inevitable that certain members will be loaded at an angle to the grain and that members such as members 4 in

Fig. 21.21 and 1 in Fig. 21.23 will receive a load from a different angle on each face. The entire joint assembly is in equilibrium when looked at in elevation, the components perpendicular to the grain on each face being of equal magnitude but acting in opposite directions. This leads to a cleavage action within the thickness of the member which is illustrated in Fig. 21.27.

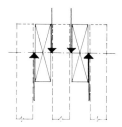

Fig. 21.27

Although this action is recognised, no design method has been evolved to calculate the effect through the thickness of the timber, and the best guidance one can give to the designer is to limit the effect. The cleaving tendencies of the alternative joint assemblies (Figs 21.21 and 21.23) are seen in Fig. 21.28 by examining the magnitude of the components perpendicular to the grain.

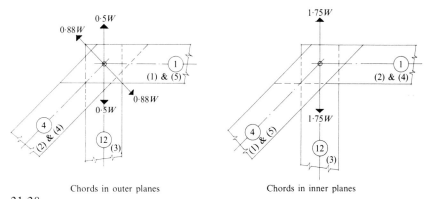

Chords in outer planes · Chords in inner planes

Fig. 21.28

With the chords in the inner planes, two adjacent faces are at 90°, which is bad from the point of view of efficiency of connectors (Table 21.1), and also the cleavage load of $1.75W$ perpendicular to the grain of members 1 is much larger than either of the cleavage loads with the chords in the outer planes. The arrangement with the chords in the outer planes is therefore preferable from these aspects and also tends to have better appearance.

Figure 21.22 illustrates how the effect of an external load being applied at a node point throws the line of action of load on the connectors off the axes of members. When no external load occurs, the direction of the connector loadings at a joint is in line with the member axes. Consider, for example, a lower chord joint in which members 8, 9, 12 and 5 meet in the Howe truss illustrated in Table 21.19. The loads in members and the vector diagram are shown in Fig. 21.29.

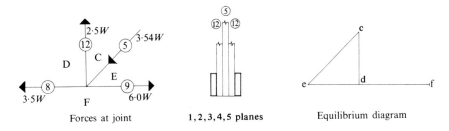

Fig. 21.29

The chord members 8 and 9 are continuous through the joint and in the outer planes (1) and (5). Member 12 is in planes (2) and (4), with member 5 in the centre. The difference in load in members 9 and 8 is shared equally to each half of member 12 loading the connector with $1.25W$ parallel to members 8–9 and at 90° to grain in members 12. By splitting the truss on the centre line, it can be seen that half the load in member 5 (i.e. $1.77W$) acts parallel to member 5 and at 45° to grain in member 12.

## 21.4   DESIGN OF A PARALLEL-CHORD TRUSS

### 21.4.1   Loading

Determine the member sizes and principal joint details for the 12 m span parallel-chord truss shown in Fig. 21.30, which is carrying a medium term UDL of 6 kN/m applied to the top chord by joisting at 0.6 m centres. Try to use GS grade Whitewood. The top and bottom chords are twin members in

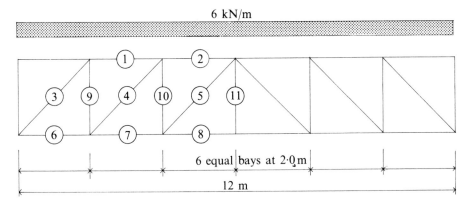

Fig. 21.30

the outer planes (1) and (5). The diagonals are twin members in planes (2) and (4) and the verticals are a single member in plane (3).

Load per panel = 2 × 6 = 12 kN = node point loading to top chord. Secondary bending moments occur in the top chord at and between the node points. Figure 27.10 gives coefficients for these on two-, three- and four-span continuous beams. Although this is a six-span continuous chord it is sufficiently accurate (see § 21.2) to take the values from Fig. 27.10 as $0.107WL_N$ at the node points and $0.036WL_N$ between node points, where $L_N$ is the distance between node points.

$$0.107WL_N = 0.107 \times 12 \times 2 = 2.6 \text{ kN.m}$$
$$0.036WL_N = 0.036 \times 12 \times 2 = 0.86 \text{ kN.m}$$

The axial loads are tabulated in Table 21.2 calculated from the coefficients given in Table 21.18.

**Table 21.2**

| Location | Member | Force (kN) |
|---|---|---|
| Top chord (compression) | 1 | 30 |
| | 2 | 48 |
| Diagonals (compression) | 3 | 42.4 |
| | 4 | 25.4 |
| | 5 | 8.6 |
| Bottom chord (tension) | 6 | 30 |
| | 7 | 48 |
| | 8 | 54 |
| Verticals (tension) | 9 | 18 |
| | 10 | 6 |
| | 11 | 0 |

### 21.4.2 Design of members

The decision on whether to determine member or joint requirement first is a matter of individual preference. Most designers first carry out the design of members, realising that some sizes may have to be increased or changed when the requirements of the joints are established. The authors emphasise that it is unwise for an engineer to finalise drawings, cost etc. on the basis of the design of members without first having established that the connections can be accommodated. To design tension members the net section applies, and it is necessary to assume connector sizes. In this design it is assumed that 64 mm diameter split rings are the preferred choice.

Chapter 13 details the general design recommendations for tension members. The bottom chord has a maximum tension of 54 kN, and being located in the outer planes, connectors occur only in one face of the member.

From Table 13.2 a twin 47 X 169 mm member with a single connector and bolt (i.e. $N_b = 1$ and $N_c = 1$) appears to have adequate medium term capacity (2 X 22.8 X 1.25 = 57 kN) on the net area.

The verticals are single tension members, and will have connectors on both faces. The maximum tension is 18 kN, and from Table 13.2 it can be seen that a 47 X 145 mm member with $N_b = 1$ and $N_c = 2$ appears adequate, having a medium term capacity based on the net area of 17.3 X 1.25 = 21.6 kN.

§ 16.6 gives general design recommendations for spaced compression members in triangulated frameworks. The top chord at member 2 has a secondary bending moment of 2.6 kN.m at the node positions and 0.86 kN.m at the centre of members with a maximum axial compression of 48 kN in member 2. With the principal axes as shown in Fig. 21.31, the effective lengths of the chords against compression are $L_{eX} = 0.85$ X 2 = 1.7 m and $L_{eY} = L_{eW} = 0.6$ m, the latter being set by the centres of joisting. The chord has to be designed on a trial-and-error basis.

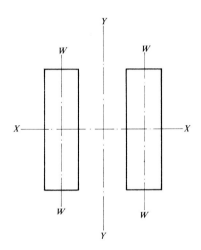

Fig. 21.31

*Top chord design* (see §§ 16.4, 16.5 and 16.6)
In certain cases it may be advantageous for the designer to consider two positions on member 2; one in the centre with $L_{eX} = 1.7$ m and the secondary moment of 0.86 kN.m; and the other at the node point with effective length equal to the distance between the points of contraflexure on each side of the node, or twice the distance to the point of contraflexure on one side, combined with the secondary moment of 2.6 kN.m. However, in this example, the moment of 2.6 kN.m will be combined with $L_{eX} = 1.7$ m.

The member may be considered to have adequate lateral restraint for bending on the *XX* axis to justify the full grade stress.

Try twin 47 X 219 mm GS grade Whitewood members spaced 141 mm apart (by three internal members each of 47 mm thickness).

$$\lambda_X = \frac{1700\sqrt{12}}{219} = 27$$

$$\lambda_W = \frac{600\sqrt{12}}{47} = 44$$

By inspection, $\lambda_Y < \lambda_W$.

Taking unmodified value of $E_{min}$ in determining $K_{12}$, and $\sigma_c$ as medium term:

$$\frac{E}{\sigma_c} = \frac{6000}{6.8 \times 1.25} = 706 \quad \text{and} \quad \lambda_{max} = 44$$

Therefore, from Table 14.2, $K_{12} = 0.756$.

$$\sigma_{c,adm} = 6.8 \times 1.25 \times 0.756 = 6.43 \, \text{N/mm}^2$$

$$\sigma_{c,a} = \frac{48\,000}{2 \times 47 \times 219} = 2.33 \, \text{N/mm}^2$$

$$\sigma_{m,adm} = 5.3 \times K_3 \times K_7 \times K_8$$
$$= 5.3 \times 1.25 \times 1.035 \times 1.10 = 7.54 \, \text{N/mm}^2$$

$$\sigma_{m,a} = \frac{2.6 \times 10^6}{2 \times 0.376 \times 10^6} = 3.46 \, \text{N/mm}^2$$

$$\sigma_e = \frac{\pi^2 E}{\lambda^2} = \frac{\pi^2 \times 6000}{44^2} = 30.6 \, \text{N/mm}^2$$

The interaction formula is:

$$\frac{3.46}{7.54 \left(1 - \dfrac{1.55 \times 2.33 \times 1.25}{30.6}\right)} + \frac{2.33}{6.43} = 0.90 < 1.0$$

which is satisfactory.

### Internals

The verticals are single members with connectors on each face. The maximum tension is 18 kN in member 9. From Table 13.2 use a 47 × 169 mm GS grade Whitewood member with capacity of 20.9 kN.

The maximum force in any diagonal is 42.4 kN compression, and the actual length is 2.8 m. Assume twin 47 mm thick members.

$L_{eX} = L_{eY} = 2.8$ m, but $L_{eW}$ may be less depending on whether or not packing pieces are fitted.

A check on twin 47 × 145 mm GS grade Whitewood members shows them to be inadequate for the maximum force. Therefore, for the maximum force, try twin 47 × 145 mm SS grade Whitewood (and consider dropping to GS grade for the less heavily loaded diagonals). Provide two packs at third points, fastened by connectors giving $L_{eW} = 0.933$ m. The diagonals will have a gap between them to accommodate the vertical internal members.

$$\lambda_X = \frac{2800\sqrt{12}}{145} = 66.9 < 180$$

From Table 16.1, with $c/b = 1$, $K_{13} = 1.8$.

$$\lambda_Y = \frac{2800 \times 1.8\sqrt{12}}{(2 \times 47) + \left(\frac{5 \times 47}{3}\right)} = 101 < 180$$

$$\lambda_W = \frac{933\sqrt{12}}{47} = 69 < 70$$

Taking unmodified value of $E_{min}$ in determining $K_{12}$; and $\sigma_c$ as medium term:

$$\frac{E}{\sigma_c} = \frac{6000}{6.8 \times 1.25} = 706$$

$$\lambda_{max} = 101$$

Therefore, from Table 14.2, $K_{12} = 0.339$.

$$\sigma_{c,adm} = 7.9 \times 1.25 \times 0.339 = 3.35 \text{ N/mm}^2$$

$$\sigma_{c,a} = \frac{42.4 \times 10^3}{2 \times 47 \times 145} = 3.11 \text{ N/mm}^2$$

Therefore the diagonal is satisfactory.

Because the section meets the requirements for the maximum case, it may be used for all the diagonals, or economies may be made by using smaller sections or GS grade sections nearer the centre of span. In practice, the tendency is to try to use the same size (and grade) for all diagonals as is the case with the vertical members. This simplifies material ordering, detailing and factory sub-assembly work. However, if many trusses are to be manufactured, the designer should consider using smaller sections where permissible.

From § 16.6.2, the shear force acting parallel to grain on one face of the intermediate packings

$$= \frac{0.65 \times 42.4 \times 145}{2 \times 94} = 21.2 \text{ kN}$$

This shear and the minimum length of pack will be considered when the connectors are designed (§ 21.4.7).

## 21.4.3 Heel joint

The vertical component of the load in member 3 can be isolated from the connector requirements by member 3 being extended below the horizontal member 6 (Fig. 21.32). Not only does this extension assist in providing a suitable end distance, but a vertical component of 30 kN is taken direct to the support point, bearing at an angle of 45° to the grain (note that 6 kN of total reaction travels down the last vertical, by-passing members 3 and 6), therefore the total reaction of 6 × 6 = 36 kN is transmitted to the bearing by 15 kN in each half of member 3, and 6 kN in the end vertical.

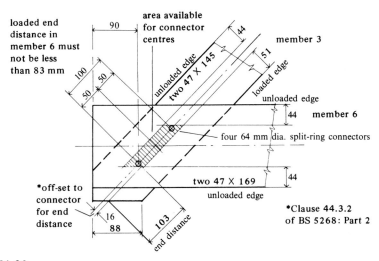

Fig. 21.32

Consider bearing of member 3 at 45° to grain, SS Whitewood, no wane permitted (see § 14.11). $K_8$ is taken in the calculations, despite the point made in § 4.4.1.

$$\sigma_{c,adm,par} = 7.9 \times K_3 \times K_8 = 6.8 \times 1.25 \times 1.1 = 10.9 \text{ N/mm}^2$$

$\sigma_{c,adm,tra}$ (SS grade, no wane)
$$= (2.1 \times 1.33) \times K_3 \times K_8 = 2.1 \times 1.33 \times 1.25 \times 1.1$$
$$= 3.84 \text{ N/mm}^2$$

$$\sigma_{c,adm,\alpha} = 10.9 - (10.9 - 3.84)0.7071 = 5.90 \text{ N/mm}^2$$

Therefore the bearing length required with two 47 mm thick members, at end of member 3:

$$= \frac{30 \times 10^3}{5.90 \times 2 \times 47} = 54 \text{ mm}$$

From Fig. 21.32 it can be seen that it is convenient to give 88 mm of bearing length.

The horizontal component in member 3 of 30 kN total is transferred at an angle of 45° into two faces of member 6. Because of the nominal end vertical member, member 3 may have connectors in both faces of each piece, but only the two faces abutting member 6 transfer the 30 kN. However, use the reduced capacity for one 64 mm split ring in two faces of 47 mm wide Whitewood from Table 18.43 (10.4 kN medium term). From Table 18.44, $K_\alpha = 0.824$ with $\alpha = 45°$.

$$\text{The number of connectors required} \ = \ \frac{30}{10.4 \times 0.824} \ = \ 3.50$$

There are two interfaces through which to transfer the load, so two connectors are used at each face on two bolts.

Four connectors give the percentage of allowable value used (PAV) as:

$$\frac{3.50}{4} \times 100\% \ = \ 87.5\% \qquad (\text{or a ratio of } 0.875)$$

In considering the actual end and edge distances and the spacing of connectors, BS 5268 : Part 2, Clause 45.5 calls for a factor $K_{64}$ to be taken as the lowest of values $K_S$, $K_C$ and $K_D$. However, in no case should the actual distances and spacing be less than the minimum tabulated values.

In this case, having chosen the connector size and number, and having calculated the PAV value in member 3 as 0.875, it can be seen that the connectors must be positioned such that end distance, edge distance and spacing provide, in each case, a value for $K_{64}$ (see Clause 45.5 of BS 5268 : Part 2) of at least 0.875. For this to happen, $K_S$ (spacing), $K_C$ (end distance) and $K_D$ (edge distance) must each have a value of 0.875 or more. Suitable spacings can be determined from Tables 78, 80 and 81 of BS 5268 : Part 2 or from the diagrams in Table 18.47 of this manual.

In this example the unloaded edge distance in members 3 and 6 must be at least 44 mm ($D_{min}$ in Table 18.47) which is independent of $K_S$, $K_C$ and $K_D$.

The loaded edge distance in member 3 (loaded at 45°) with a standard edge distance, $D_{st}$ of 70 mm, would give a value for $K_D$ of 1.0 and, with a minimum edge distance, $D_{min}$ of 44 mm, a value for $K_D$ of 0.875. For intermediate values of edge distance, $K_D$ is interpolated linearly. From Fig. 21.33 (reproduced from Table 18.47) with $K_D = 0.875$, it can be seen that the required edge distance must be (at least) 51 mm.

Consider the unloaded end distance to member 3 loaded at 45°. From Table 18.47, $C_{st} = 121$ mm and $C_{min} = 67$ mm. With $K_{67} = 0.875$, the required value for the loaded end distance $C$ is conveniently found from the diagram of $K_C$ values in Table 18.47, reproduced here as Fig. 21.34. Draw a line vertically from the $K_C$ value of 0.875 until it meets the diagonal line drawn from 121 mm ($C_{st}$). From the point of intersection draw a horizontal line to find the required minimum unloaded end distance. From the diagram it can be seen that 103 mm is required.

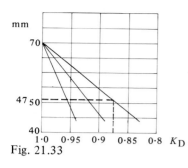

Fig. 21.33

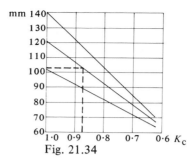

Fig. 21.34

The loaded end distance to member 6 requires a re-assessment of the percentage of allowable value used (PAV) as it applies to member 6, in which the connectors are loaded parallel to the grain on one face only, i.e. capacity = 11.0 kN (from Table 18.43).

The actual number of connectors required in member 6 is:

$$\frac{30}{11} = 2.73$$

so that PAV reduces (with four connectors shared to two faces) to $2.73/4 = 0.68$ related to $K_C$.

Consider the loaded end distance to member 6. From Table 18.47, $C_{st} = 140$ mm and $C_{min} = 70$ mm. With PAV = 0.68 the required value for loaded end distance $C$ is conveniently found from the diagram of $K_C$ values in Table 18.47, reproduced here as Fig. 21.35. Draw a line vertically from the $K_C$ value of 0.68 until it meets the diagonal line drawn from 140 mm ($C_{st}$). From the point of intersection draw a horizontal line to find the required loaded end distance. The diagram shows that this is 80 mm.

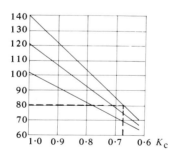

Fig. 21.35

Because two bolts are to be located, the spacing requirement must also be determined. Consider member 3 with loading at 45° and with $\theta$ provisionally taken as 0° (see Table 18.47). $S_{st} = 108$ mm and $S_{min} = 89$ mm. From § 18.6.3, the permissible spacing $S$ (with PAV = 0.875) is then calculated (from § 18.6.3) as:

$$= \frac{(0.875 - 0.75)(108 - 89)}{0.25} + 89 = 98.5 \text{ mm}$$

In member 6 with loading $\alpha = 0°$, $\theta = 45°$ and PAV $= 0.68$, a minimum spacing of 89 mm is permitted, therefore the spacing is set by the figure of 98.5 mm in member 3. (Alternatively, the value could be scaled from the lines for $K_D$ in Table 18.47, but the scale is rather small.)

With the above parameters fixed as calculated, the joint is set out to give a convenient arrangement for the connectors, having regard to the required connector positioning, convenience of detailing and ease of manufacture. The first step is to superimpose the permitted end and edge distances on the member arrangement to give an area within which the centres of the connectors can be placed. This is shown hatched in Fig. 21.32. In this case it is convenient to place the connectors on the centre line of the diagonal, 50 mm to each side of the intersection point of the axes.

### 21.4.4  Top chord joint

Consider the joint at the junction of members 1, 3 and 9. This joint is typical of the top chord joints, all of which are influenced by the applied vertical node point load, giving a direction of connector loading which in most cases does not coincide with the axes of the members.

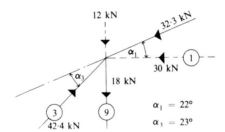

Fig. 21.36

The loads at the joint are summarised in Fig. 21.36. It should be noted that the 12 kN vertical load enters the joint as a shear on the two halves of member 1, and it is the vector sum of this shear and the axial load in member 1 which must be resisted by the connectors joining member 1 to member 3. The components of shear and axial load in member 1 are shown in Fig. 21.36 as dotted lines, and the vector and its angle as a full line.

Member 9 transfers 18 kN at 45° into member 3.

From Table 18.44, with $\alpha = 45°$, $K_\alpha = 0.824$. ∴ The number of 64 mm diameter split rings required:

$$= \frac{18.0}{10.4 \times 0.824} = 2.10$$

Using four connectors PAV $= 2.10/4 = 0.525$ which (as can be seen from Table 18.47) permits minimum loaded end, spacing and edge distances of $C_{min} = 70$ mm, $S_{min} = 89$ mm and $D_{min} = 44$ mm.

Member 1 transfers 32.3 kN into member 3 at 23°. $K_\alpha = 0.938$ (Table 18.44). ∴ The number of 64 mm diameter split rings required:

$$= \frac{32.3}{10.4 \times 0.938} = 3.31$$

Using four connectors, PAV $= 3.31/4 = 0.83$, which permits a minimum edge distance of 44 mm. The unloaded end distance to member 3 is $C_{st} = 121$ mm, $C_{min} = 67$ mm. With $K_C = 0.83$, from Table 18.47, $C$ is arrived at graphically as 96 mm. The actual spacing cannot be determined until the parallelogram of connector area is set out and $\theta$ determined, although it can be seen from the graph for $K_S$ in Table 18.47 that $S$ is between 89 and 108 mm.

The required edge distances may be set out on the member sizes as determined in § 21.4.2. The available parallelogram of area is large enough (see Fig. 21.37) for the connectors to be spaced 110 mm apart, which exceeds 108 mm ($S_{st}$) therefore no further detailed check is required.

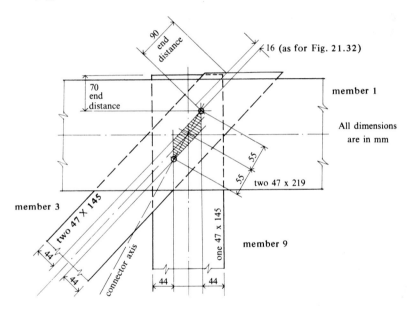

Fig. 21.37

### 21.4.5 Bottom chord joint

Consider the joint at the junction of members 9, 4, 6 and 7. This joint is typical of the bottom chord joints having no external node point load. All connector loads will align with the axes of the members.

The joints between members 9 and 4 have the same requirements as those between 9 and 3, which have already been established. The joint between members 4 and 6–7 transfers a horizontal load equal to the differential between 6 and 7 (i.e. 18 kN) which is at 45° to the grain of member 4. The requirements of the joint are the same as those required for the joint between 9 and 3. The joint is sketched in Fig. 21.38. The area available for the centres of connectors is a six-sided figure in this case. The connectors are placed on the longest diagonal to give a maximum spacing of 100 mm. PAV = 0.525 and $S_{min}$ of 89 mm would be acceptable.

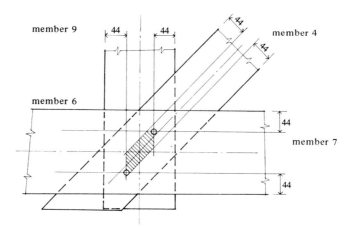

Fig. 21.38

## 21.4.6    Tabular method of joint analysis

In §§ 21.4.3–21.4.5, the three main joints were investigated and suitable connector arrangements derived. There may have been a number of alternative solutions which would have been equally acceptable. As a general rule, it will be found best to investigate the heavily loaded joints first. The remaining joints tend to be simpler to design and to show some repetition.

It is not usual to mix connectors of different sizes in a truss, but on occasions it may even be convenient to include different connector sizes on adjacent interfaces at one joint.

In practice, the design of joints is made easier if each joint is set out to a reasonably large scale. Various possible arrangements should be investigated to find the one most suited to detailing and the workshop. It is convenient to establish the number of connectors, the percentage of allowable value used and the spacing requirement in tabular form. A typical format for the truss under consideration is given in Table 21.3. Because the angle of connector axis to grain $\theta$ is not known prior to the detailing of the joint, spacing $S$ is tabulated assuming $\theta = 0°$. Then, if this spacing can be provided, no further calculation is required. If, however, the tabulated value of $S$ cannot be accommodated, further calculations with values of $\theta$ derived from the joint detail may indicate a possible reduction in spacing, otherwise at least one member width must be increased.

### 21.4.7 Packings for spaced compression member

Base the design on member 3.

The capacity of one 64 mm dia. split-ring connector (medium term load) is 10.4 kN in SC3 and SC4 material if a connector occurs on both faces (Table 18.43). Therefore, for the 21.2 kN shear acting on one face of the internal diagonals, use three connectors in a single row. $K_\alpha = 1.00$.

$$PAV = \frac{21.2}{10.4 \times 3} = 0.68$$

From Table 18.47 make:

end distance at least 70 mm
spacing at least 89 mm say 90 mm and
edge distance at least 44 mm.

Therefore length of pack required

$$= (2 \times 70) + (2 \times 90) = 320\,mm$$

which exceeds the minimum length of 230 mm (Clause 15.8.2 of BS 5268: Part 2).

## 21.5 BOWSTRING TRUSSES

### 21.5.1 Introduction

In bowstring trusses, the upper chord is usually circular (as distinct from the ideal parabolic shape), which gives a truss profile closely resembling the bending-moment diagram for a UDL. With balanced loading, the internal members take little load and are usually very small considering the span. Because of this, single internal members can be used on the vertical centre line of the truss plane (2) with the top and bottom chords each constructed in two halves and placed in planes (1) and (3). The resulting secondary moments on the chords caused by eccentricity of connections can usually be disregarded under balanced loading, but must be considered for unbalanced loading.

The truss is loaded either by purlins placed at node points or between the node points to cancel out part of the bending moment induced by the curvature. The bending stress in the top chord can vary significantly as a result of the placing of the purlins or secondary system.

The axial loading in members can be derived by the construction of a vector diagram, although in doing so a large scale must be used to obtain any degree of accuracy (particularly for the web members). The axial loading in the top chord is fairly consistent across the whole span of the truss and

**Table 21.3**

| Joint | Member to member | Conn. load (kN) | Member | α | $K_\alpha$ | Conns. one side or two | Medium term $\overline{F_0}^*$ (kN) | Number of conns. required | Use | PAV | End distance† C (mm) | Spacing S (mm) | Loaded edge distance D (mm) | Type of connector |
|---|---|---|---|---|---|---|---|---|---|---|---|---|---|---|
| 3 | 3 | 30 | 6 | 0° | 1.00 | 1 | 11.0 | 2.73 | 4 | 0.68 | 80 | 89 | – | 64 mm dia. |
| 6 | 6 |  | 3 | 45° | 0.824 | 2 | 10.4 | 3.50 |  | 0.875 | 103 | 98.5 | 51 | split rings |
| 1 | 9 | 18 | 9 | 0° | 1.00 | 2 | 10.4 | 1.73 | 4 | 0.43 | 70 | 89 | 44 | 64 mm dia. |
| 3 | 3 |  | 3 | 45° | 0.824 | 2 | 10.4 | 2.10 |  | 0.525 | 67 | 89 | 44 | split rings |
| 9 | 3 | 32.3 | 3 | 23° | 0.938 | 2 | 10.4 | 3.31 | 4 | 0.83 | 97 | 109 | 44 | 64 mm dia. |
|  | 1 |  | 1 | 22° | 0.943 | 1 | 11.0 | 3.11 |  | 0.78 | – | 97 | 44 | split rings |
| 9 | 9 | 18 | 9 | 0° | 1.00 | 2 | 10.4 | 1.73 | 4 | 0.43 | 70 | 89 | 44 | 64 mm dia. |
| 4 | 4 |  | 4 | 45° | 0.824 | 2 | 10.4 | 2.10 |  | 0.525 | 67 | 89 | 44 | split rings |
| 6/7 | 4 | 18 | 4 | 45° | 0.824 | 2 | 10.4 | 2.10 | 4 | 0.525 | 67 | 89 | 44 | 64 mm dia. |
|  | 6/7 |  | 6/7 | 0° | 1.00 | 1 | 11.0 | 1.64 |  | 0.41 | – | 89 | 44 | split rings |

| Member | | α† | | | | | | n | F₀* | | | | 64 mm dia. split rings |
|---|---|---|---|---|---|---|---|---|---|---|---|---|---|
| 10 | 10 | 0° | 1.0 | 2 | 10.4 | 0.58 | 2 | | 0.29 | 70 | – | 44 | 64 mm dia. split rings |
| 4 | 4 | 45° | 0.824 | 2 | 10.4 | 0.70 | | 6 | 0.35 | 67 | – | 44 | split rings |
| 1/2 | 4 | 11° | 0.984 | 2 | 10.4 | 2.11 | 4 | | 0.53 | 67 | 89 | 44 | 64 mm dia. split rings |
| | 1/2 | 34° | 0.882 | 1 | 11.0 | 2.23 | | 21.6 | 0.56 | – | 89 | 44 | split rings |
| 10 | 10 | 0° | 1.0 | 2 | 10.4 | 0.58 | 2 | | 0.29 | 70 | – | 44 | 64 mm dia. split rings |
| 5 | 5 | 45° | 0.824 | 2 | 10.4 | 0.70 | | 6 | 0.35 | 67 | – | 44 | split rings |
| 5 | 5 | 45° | 0.824 | 2 | 10.4 | 0.70 | 2 | | 0.35 | 67 | – | 44 | 64 mm dia. split rings |
| 7/8 | 7/8 | 0° | 1.0 | 1 | 11.0 | 0.55 | | 6 | 0.27 | – | – | 44 | split rings |
| 5 | 5 | 0° | 1.0 | 1 | 11.0 | 0.78 | 2 | | 0.39 | 64 | – | 44 | 64 mm dia. split rings |
| 2 | 2 | 45° | 0.824 | 1 | 11.0 | 0.95 | | 8.6 | 0.47 | – | – | 44 | split rings |

\* $F_0$ is the connector capacity from Table 18.43 at α = 0°.

† Loaded or unloaded end distance.

can be established mathematically. The curvature of the top chord results in a bending moment being induced into the member equal to the product of the axial load tangential at mid point between nodes and its eccentricity from node points. This moment must be added algebraically to any moment induced in the member by the local loading from purlins to give the net moment on the member.

$$\text{The moment in the top chord} = M = M_0 - F_N e$$

where $M_0$ = the mid panel moment due to localised purlin loading
$F_N$ = the axial load tangentially at mid node points
$e$ = the eccentricity to the line of action of $F_N$ from the node points.

The eccentricity is determined from the radius of the top chord and the distance between node points:

$$e = \frac{L_N^2}{8r}$$

where $L_N$ = the distance between node points
$r$ = the radius of the chord.

If the chord is made as a curved laminated beam (as is likely) of rectangular cross-section and $r/t$ is less than 240, the bending, tension and compression parallel to the grain stresses should be multiplied by a modification factor $K_{33}$. BS 5268 : Part 2 gives:

$$K_{33} = 0.76 + 0.001\,\frac{r}{t}$$

and states that $K_{33}$ should not be greater than 1.0.

The value of $t$ is taken as the thickness of one lamination if the member is laminated to the curvature, as is usual, or the full depth of the member if it is laminated straight and then bent to the curvature. The ratio $r/t$ should be greater than 125 for softwoods (or 100 for hardwoods). Even if the $r/t$ ratio permits the chord to be made straight and then bent, the engineer must anticipate the loads on the connectors due to the chord trying to straighten again. The method of bending a manufactured chord leads to difficulties and the method of laminating to the profile is preferred. Values of $K_{33}$ are given in Fig. 21.39.

In curved beams where the ratio of the minimum mean radius of curvature ($r_{mean}$) (assuming that there may be differing radii along the length), to the depth ($h$) is less than or equal to 15, the bending stress induced by a moment $M$ should be taken as:

(a) in the extreme fibre on the concave side

$$= K_{34} \cdot \frac{6M}{bh^2}$$

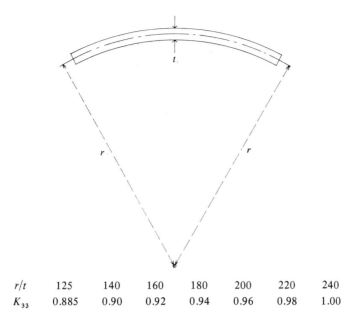

| $r/t$ | 125 | 140 | 160 | 180 | 200 | 220 | 240 |
|---|---|---|---|---|---|---|---|
| $K_{33}$ | 0.885 | 0.90 | 0.92 | 0.94 | 0.96 | 0.98 | 1.00 |

Fig. 21.39

where $K_{34} = 1 + \left(0.5 \dfrac{h}{r_{mean}}\right)$     for $r_{mean}/h \leqslant 10$

or $= 1.15 - \left(0.01 \dfrac{r_{mean}}{h}\right)$     for $r_{mean}/h$ greater than 10 but not greater than 15.

(b) in the extreme fibre on the convex face

$$= \frac{6M}{bh^2}$$

The bending moment in a curved member will produce a radial stress which should be calculated as:

$$\sigma_r = \frac{3M}{2bhr_{mean}}$$

where $\sigma_r$ = the radial stress perpendicular to grain

$M$ = the bending moment

$r_{mean}$ = the radius of curvature to the centre of the member (minimum $r_{mean}$ if $r_{mean}$ varies)

$b$ = the breadth of the member

$h$ = the depth of the member.

When the bending moment tends to increase the radius of curvature (i.e. to flatten the arc) $\sigma_r$ should not be greater than one-third the permissible shear parallel to the grain stress (Clause 21.2.2 of BS 5268: Part 2).

When the bending moment tends to reduce the radius of curvature, $\sigma_r$ should not be greater than 1.33 times the SS grade compression perpendicular to the grain stress for the species.

### 21.5.2   Preferred geometry for a bowstring truss

Some simplification of design is obtained with a bowstring truss if a preferred geometrical arrangement is adopted. The span-to-depth ratio should be approximately 8 to 1 to avoid large deflections on one hand and excessive height and curvature on the other. It is usually convenient to adopt a radius for the top chord equal to the span of the truss, so that the tangent angle at the support is 30° (Fig. 21.40). The rise of the truss is:

$$H = r - r \cos 30° = 0.134r$$

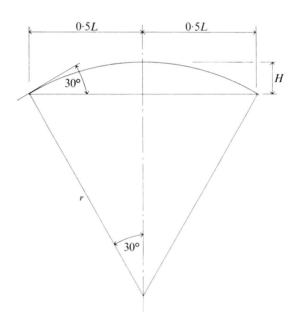

Fig. 21.40

By the provision of 0.009$L$ camber to the bottom chord, $0.134 - 0.009 = 0.125$ and the span-to-depth ratio becomes 8 to 1 (i.e. the distance between the centre lines of the upper and lower chords is $L/8$ at mid span). The length along the arc is:

$$\frac{60 \times 2\pi r}{360} = 1.05L$$

With a UDL of $F$ applied to the full span of the truss, the heel-joint axial loads are resolved as $N_A = F$, and $T_A = 0.866F$, and the mid span axial load is $N_C = T_C = F$ (Fig. 21.41).

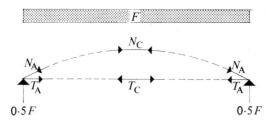

Fig. 21.41

### 21.5.3 Example of bowstring truss design

The special design requirements of a bowstring truss are illustrated by the following example.

*Example*
Consider 16 m span bowstring trusses at 6 m centres carrying 0.50 kN/m² dead load and 0.75 kN/m² imposed medium term load. Purlins are placed at the node points (Fig. 21.42) and the height is made 2 m.

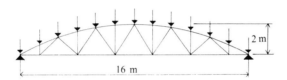

Fig. 21.42

The arc length = 16 × 1.05 = 16.8 m, which is divided into twelve equal lengths of 1.4 m. Note that the division is along the arc and not on true plan.
The total loading on the truss is

$$F_d = 16.8 \times 6 \times 0.5 = 50\,\text{kN}$$
$$F_i = 16 \times 6 \times 0.75 = 72$$
$$\overline{\phantom{xxxxxx}}$$
$$122\,\text{kN medium term}$$

$$\text{Eccentricity on panel length} = \frac{1.4^2}{8 \times 16} = 0.0153\,\text{m}$$

Axial load in the top chord = 122 kN (medium term)

Secondary moment (due to curvature) in the top chord

$$= 122 \times 0.0153 = 1.87\,\text{kN.m (medium term)}$$

For the top chord try two 65 × 180 mm LB grade Whitewood sections (four laminations in each) spaced 60 mm apart (Fig. 21.43).

$$i_Y = \frac{1}{\sqrt{12}}\left(2b + \frac{5c}{3}\right) = 66.4\,\text{mm} \qquad \text{(see § 16.4)}$$

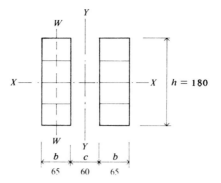

Fig. 21.43

With restraint about both the $XX$ and $YY$ axes at 1.4 m centres and with packs at mid panel length to restrain the $WW$ axis at 0.7 m centres, the various slenderness ratios are:

$$\lambda_X = \frac{1400 \times 0.85 \sqrt{12}}{180} = 23$$

$$\lambda_Y = \frac{1400}{66.4} = 21$$

$$\lambda_W = \frac{700 \sqrt{12}}{65} = 37$$

The $E$ value for entry into Table 14.2 to derive $K_{12}$ (see § 14.4)

$$= E_{\text{mean}} \text{ for SS grade Whitewood} \times K_{20}$$
$$= 10\,500 \times 0.90$$
$$= 9450 \text{ N/mm}^2$$

$$\sigma_c \text{ medium term} = \text{SS grade stress} \times K_{17} \times K_3$$
$$= 7.50 \times 1.04 \times 1.25$$
$$= 9.75 \text{ N/mm}^2$$

$E/\sigma_c = 969$. Therefore, with $\lambda_{\text{max}} = 37$, $K_{12} = 0.816$.

$$\therefore \sigma_{c,\text{adm}} = 7.90 \times K_{17} \times K_{12} \times K_3$$
$$= 7.90 \times 1.04 \times 0.816 \times 1.25 = 8.38 \text{ N/mm}^2$$

If it is desired to fabricate the top chord straight and then bend to the 16 m radius:

$$\frac{r}{h} = \frac{16}{0.18} = 89 < 125$$

This is not permitted, therefore the chord must be laminated to the curvature. Using 45 mm laminations:

$$\frac{r}{t} = \frac{16}{0.045} = 356 \quad \text{and hence} \quad K_{33} = 1.00 \text{ (Fig. 21.39)}$$

$$\sigma_{c,a} = \frac{122 \times 10^3}{2 \times 65 \times 180} \times K_{33} = 5.21 \times 1.00 = 5.21 \text{ N/mm}^2$$

$$r_{mean} = 16\,090 \text{ mm} \quad h = 180 \text{ mm} \quad \therefore \frac{r_{mean}}{h} = 89.4$$

Because this is greater than fifteen there is no need to increase the bending stress by $K_{34}$ (see § 21.5.1), or alternatively $K_{34}$ can be taken as 1.00.

$$\sigma_{m,adm} = 7.50 \times K_{15} \text{ (four laminates)} \times K_{33} \times K_3 \times K_7$$
$$= 7.50 \times 1.26 \times 1.00 \times 1.25 \times 1.058 = 12.50 \text{ N/mm}^2$$

$$\sigma_{m,a} = \frac{1.87 \times 10^6 \times 6}{2 \times 65 \times 180^2} = 2.66 \text{ N/mm}^2$$

$$\sigma_e = \frac{\pi^2 E}{\lambda^2} = \frac{\pi^2 \times 9450}{37^2} = 68.1 \text{ N/mm}^2$$

The interaction formula is:

$$\frac{5.21}{9.74 \left(1 - \dfrac{1.5 \times 2.66 \times 1.25}{68.1}\right)} + \frac{2.66}{12.50} = 0.577 + 0.213 = 0.79 < 1.0$$

which is acceptable.

$$\text{The radial stress} = \sigma_{r,a} = \frac{3M}{2bhr_{mean}} \quad (§ 21.5.1)$$

$$= \frac{3 \times 1.87 \times 10^6}{2 \times (65 \times 2) \times 180 \times 16\,090} = 0.0075 \text{ N/mm}^2$$

The moment induced into the top chord (assuming the force line is between node points) tends to reduce the radius of curvature therefore the radial stress (Clause 21.2.2 of BS 5268 : Part 2):

$$\sigma_{r,adm} = 1.33 \times \text{SS grade value of } \sigma_{c,tra} \text{ (medium term)}$$
$$= 1.33 \times 2.1 \times 1.25 = 3.49 \text{ N/mm}^2$$

which is acceptable.

It can be shown by a vector diagram that the forces in the internal members under balanced loading are extremely small for a truss of this size, whereas unbalanced loading on one half of the span can produce larger axial compression in the critical members near the centre (shown dotted in Fig. 21.44) of a value of approximately three times the imposed panel point loading, the exact value depending on the number of bays.

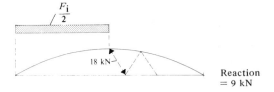

Fig. 21.44

In this case the maximum length of an internal member is $\sqrt{(2^2 + 1.4^2)} =$ 2.44 m, which must be made to carry an axial compression of $9 \times 2.44/2$ = 11 kN (medium term). Force from balanced dead loading in this member is zero.

A 60 $\times$ 145 mm SS grade Whitewood member would be satisfactory.

The designer can see how important it is to consider unbalanced loading. Had only balanced loading been considered, this longest of the web members would have been calculated as carrying no load at all.

## 21.6   DEFLECTION OF TRUSSES

### 21.6.1   Elastic deflection

The deflection of a triangulated framework results from a combination of axial shortening of the compression members, lengthening of tension members, and slip at the joints.

Deflection due to axial strain is conveniently calculated from the strain-energy formula:

$$\delta_e = \sum \frac{FUL}{AE}$$

where $\delta_e$ = the elastic deflection at a selected node point of the truss

$F$ = the load in each member of the truss caused by the applied loading (+ for compression, − for tension is the usual sign convention)

$U$ = the load in each member of the truss caused by a unit load placed at the node point for which the deflection is required and in the direction in which the deflection is to be calculated (+ for compression, − for tension is the usual sign convention)

$L$ = the actual length to node points of each member

$A$ = the area of each member (not the net area)

$E$ = the modulus of elasticity of each member.

Values of $F$ and $U$ are determined by normal structural analysis. Coefficients for various truss configurations are tabulated in § 21.7 and also in § 20.8 (although using $W$ instead of $F$).

Most trusses act as principal members (trussed rafters are an exception), therefore the $E$ value to use in calculations is $E_{min}$ for a single member, or

$E_{min} \times K_9$ or $K_{28}$ for values of $N$, or $E_{mean} \times K_{20}$ or $K_{26}$ for horizontally laminated members (see § 2.3 and Tables 7.1, 7.2 and 7.3).

Note that if the load in a member due to the unit load is of the same sign as the load in the member due to the applied loading, the deformation of this member adds to the external deflection being considered, whereas if the signs are opposite the deformation of the member reduces the external deflection.

The elastic deflection of the truss designed in § 21.4 is calculated in Table 21.4, in the form usual with strain-energy methods. The unit load is placed at the centre of the top chord, therefore member 11 need not be considered.

Table 21.4

| Member | $F$ (kN) | $U$ | $L$ (m) | $A$ (mm²) | $E$ (N/mm²) | $FUL/AE$ (m) |
|--------|----------|-----|---------|-----------|-------------|--------------|
| 1 | +30 | +0.5 | 2 | 20 600 | 6840 | 0.000 213 |
| 2 | +48 | +1.0 | 2 | 20 600 | 6840 | 0.000 681 |
| 3 | +42.4 | +0.71 | 2.83 | 13 630 | 7980 | 0.000 783 |
| 4 | +25.4 | +0.71 | 2.83 | 13 630 | 7980 | 0.000 469 |
| 5 | + 8.6 | +0.71 | 2.83 | 13 630 | 7980 | 0.000 145 |
| 6 | −30 | −0.71 | 2 | 15 880 | 6840 | 0.000 392 |
| 7 | −48 | −1.0 | 2 | 15 880 | 6840 | 0.000 883 |
| 8 | −54 | −1.5 | 2 | 15 880 | 6840 | 0.001 490 |
| 9 | −18 | −0.5 | 2 | 7 940 | 6000 | 0.000 378 |
| 10 | − 6 | −0.5 | 2 | 7 940 | 6000 | 0.000 126 |

For half truss: $\Sigma$ = 0.005 560

The product $FU$ is always positive in this case, and only half the truss is tabulated. The elastic deflection = 2 × 005 56 = 0.0111 m.

Note that members 1, 2, 6, 7 and 8 are twin GS Whitewood members therefore $E_N$ is taken for $N = 2$ and equals $6000 \times K_9$ (even for compression and tension members) $= 6000 \times 1.14 = 6840 \, N/mm^2$. For the twin SS Whitewood members 3, 4 and 5, $E_2 = 7000 \times 1.14 = 7980 \, N/mm^2$. $E_1$ for members 9 and 10 is 6000 N/mm².

## 21.6.2   Slip deflection

Slip occurs at the joints in triangulated frameworks (unless gluing is used) and this slip leads to deflection of the overall framework. For various reasons such as assembly tolerances, centring of bolts or connectors etc., it is not possible to predict exactly the amount of slip at each joint, and hence not possible to predict exactly the overall deflection of a truss, but experience has shown that the type of calculation detailed below gives answers borne out fairly well in practice. Certainly the theory is sound. Approximate values

**Table 21.5**

| Connector type | Slip (m) |
|---|---|
| Glue | nil |
| Bolt* | 0.0026 |
| Tooth plates* | 0.0026 |
| 64 mm diameter split ring | 0.0008 |
| 104 mm diameter split ring | 0.0010 |
| 67 mm diameter shear plate* | |
| timber-to-timber | 0.0026 |
| timber-to-steel | 0.0018 |
| 104 mm diameter shear plate* | |
| timber-to-timber | 0.0031 |
| timber-to-steel | 0.0026 |

\* An allowance of 0.0016 m is included for the usual bolt-hole tolerance.

for joint deformation are given in Table 21.5 which should give satisfactory accuracy for normal loading conditions. If the engineer can establish more accurate joint slip characteristics and control the assembly, including moisture content, the performance on site will match the calculated deflections more closely. Where the truly permanent load at a connector is a high percentage (say 70% or more) of the permissible long term loading, the authors are inclined to advise the designer to build-in an additional estimate for 'slip' to take account of indentation.

The contribution of each member towards the total deflection of the truss is the product of the load in the member due to unit load placed at the point where deflection is being calculated, and the net change in effective length of the member caused by joint slip at each end of the member plus joint slip in any intermediate joint along the member, due to the actual load $F$ in the member.

$$\text{Slip deflection} = \Sigma U \Delta_s$$

where $\Delta_s$ = the total change in member length due to joint slip.

(Note that this formula is derived from the strain-energy formula:

$$E = \frac{\text{stress}}{\text{strain}} = \frac{Fe}{AL}$$

where $e$ is extension or contraction, given by $e = FL/AE$, and $FUL/AE$ becomes $Ue$ or $U\Delta_s$.)

When the change in length between connectors has the same nature as that due to $F$ in the member, $\Delta_s$ is given the same sign as $U$ and the product $U\Delta_s$

for that member is positive; otherwise it is negative. For example, with a member taking a compression load of $F$, when the effect of bolt slip is to shorten the distance between connectors, the product $U\Delta_s$, is positive. It is only rarely that $U\Delta_s$ is negative, but it can occur.

Special thought is required when considering the chords continuous past a node point. Examine the bottom chord of the truss in Fig. 21.30, which is isolated here in Fig. 21.45.

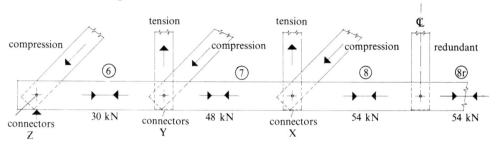

Fig. 21.45

The load in member 8 is the same as in member 8r, therefore there is no horizontal slip in the connection between them. The tension in member 8 is larger than in member 7, and therefore member 8 tends to pull past connector X, effectively increasing the length between connectors X and the centre line of the truss by the slip in connectors X. Therefore there is a change in length between the connectors (before and after loading) in member 8 which must be taken into account in calculating the deflection of the truss. Because the change in length is an increase in this tension member, it has the same type of effect on the distance between connectors as the tension force in the member, therefore the value tabulated for $U$ will have the same sign as the value for $F$, and the effect of this slip will be to increase the deflection of the truss. Providing the type of connector used at X is repeated at Y, the effect on member 7 of slip at both ends is cancelled out. If not, account must be taken of the difference.

The effect of the slip at the connectors Y in member 6 is to shorten the distance between connectors, but the effect at connectors Z is for the member to pull past and lengthen the distance between connectors. If connectors Y and Z are the same, the effect cancels out.

The mid span deflection of the truss in § 21.4 due to bolt slip is calculated in Table 21.6. 64 mm diameter split rings are used with an allowance for slip at one end of ±0.0008 m, or ±0.0016 m if slip occurs at both ends. Only half the truss is tabulated.

$$\text{The total slip deflection} = 2 \times 0.007\,008 = 0.014\,\text{m}$$
$$\text{The total deflection of the truss} = \text{elastic deflection} + \text{slip deflection}$$
$$= 0.0111 + 0.014$$
$$= 0.0251\,\text{m} = 25.1\,\text{mm (span/480)}$$

**Table 21.6**

| Member | $U$ | $\Delta_s(m)$ | $U\Delta_s$ |
|--------|------|--------|--------|
| 1 | +0.5 | 0 | 0 |
| 2 | +1.0 | +0.0008 | +0.000 8 |
| 3 | +0.71 | +0.0016 | +0.001 136 |
| 4 | +0.71 | +0.0016 | +0.001 136 |
| 5 | +0.71 | +0.0016 | +0.001 136 |
| 6 | −0.71 | 0 | 0 |
| 7 | −1.0 | 0 | 0 |
| 8 | −1.5 | −0.0008 | +0.001 2 |
| 9 | −0.5 | −0.0016 | +0.000 8 |
| 10 | −0.5 | −0.0016 | +0.000 8 |

For half truss: $\Sigma$ = +0.007 008

Note that the slip deflection is more than the elastic deflection in this example.

### 21.6.3   Standard deflection formulae

The American Institute of Timber Construction Standards quotes empirical values for the total deflection of flat, pitched and bowstring trusses which are purely empirical, being independent of loading and member size. These are reproduced in Table 21.7 in metric form.

**Table 21.7**

| Symmetrical flat and pitched trusses | Bowstring trusses |
|--------------------------------------|-------------------|
| $$\delta_t = \frac{L^2}{4290H}\left[\frac{L}{26.6} + 1\right]$$ | $$\delta_t = \frac{L^2}{19\,000H}\left[\frac{L}{6} + 1\right]$$ |

$\delta_t$ = the total deflection under total loading (m)
$L$ = the span of the truss (m)
$H$ = the height (m) of the truss at mid span.

Use the formula in Table 21.7 for flat trusses to check the calculated deflection of the example in § 21.6.2.

$$\text{Deflection} = \frac{12^2}{4290 \times 2}\left[\frac{12}{26.6} + 1\right] = 0.0244\,\text{m} = 24.4\,\text{mm}$$

The agreement is reasonable, particularly when one realises that the formula makes no allowance for the $E$ value used or the degree to which the truss is stressed.

## 21.7 COEFFICIENTS OF AXIAL LOADING

To save design-office time, Tables 21.8–21.25 give coefficients by which the rafter node point loading should be multiplied to arrive at the axial loading in the members. Coefficients are also tabulated for a unit load placed at the top chord of parallel-chord trusses, except in the case of Tables 21.23 and 21.25, where the unit load is placed at the centre of the bottom chord. In the case of Tables 21.11–21.25, coefficients are also given related to ceiling node point loading.

Although $F$ is used generally in BS 5268 and this manual to represent load, $W$ is retained in Tables 21.8–21.25 where there is no chance of confusion.

**Table 21.8**

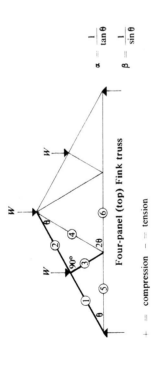

Four-panel (top) Fink truss

+ = compression   − = tension

$$\alpha = \frac{1}{\tan\theta} \qquad \beta = \frac{1}{\sin\theta}$$

| Member | Compression or tension | 15° W | 15° Unit | 17½° W | 17½° Unit | 20° W | 20° Unit | 22½° W | 22½° Unit | 25° W | 25° Unit | 27½° W | 27½° Unit | 30° W | 30° Unit | 32½° W | 32½° Unit | 35° W | 35° Unit | General formulae W | General formulae Unit |
|---|---|---|---|---|---|---|---|---|---|---|---|---|---|---|---|---|---|---|---|---|---|
| 1 | + | 5·80 | 1·87 | 5·00 | 1·59 | 4·39 | 1·37 | 3·92 | 1·21 | 3·55 | 1·07 | 3·25 | 0·96 | 3·00 | 0·87 | 2·79 | 0·78 | 2·61 | 0·71 | $1\cdot5\beta$ | $0\cdot5\alpha$ |
| 2 | + | 5·54 | 1·87 | 4·69 | 1·59 | 4·04 | 1·37 | 3·54 | 1·21 | 3·13 | 1·07 | 2·79 | 0·96 | 2·50 | 0·87 | 2·25 | 0·78 | 2·04 | 0·71 | $\dfrac{3\alpha^2+1}{2\beta}$ | $0\cdot5\alpha$ |
| 3 | + | 0·97 | 0 | 0·95 | 0 | 0·94 | 0 | 0·92 | 0 | 0·91 | 0 | 0·89 | 0 | 0·87 | 0 | 0·84 | 0 | 0·82 | 0 | $\alpha/\beta$ | 0 |
| 4 | − | 1·87 | 0 | 1·59 | 0 | 1·37 | 0 | 1·21 | 0 | 1·07 | 0 | 0·96 | 0 | 0·87 | 0 | 0·78 | 0 | 0·71 | 0 | $0\cdot5\alpha$ | 0 |
| 5 | − | 5·60 | 1·93 | 4·76 | 1·66 | 4·12 | 1·46 | 3·62 | 1·30 | 3·22 | 1·18 | 2·88 | 1·08 | 2·60 | 1·00 | 2·35 | 0·93 | 2·14 | 0·87 | $1\cdot5\alpha$ | $0\cdot5\beta$ |
| 6 | − | 3·73 | 1·93 | 3·17 | 1·66 | 2·75 | 1·46 | 2·41 | 1·30 | 2·14 | 1·18 | 1·92 | 1·08 | 1·73 | 1·00 | 1·57 | 0·93 | 1·43 | 0·87 | $\alpha$ | $0\cdot5\beta$ |
| α | | 3·73 | | 3·17 | | 2·74 | | 2·41 | | 2·14 | | 1·92 | | 1·73 | | 1·57 | | 1·43 | | | |
| β | | 3·86 | | 3·33 | | 2·92 | | 2·61 | | 2·36 | | 2·17 | | 2·00 | | 1·86 | | 1·74 | | | |

Table 21.9

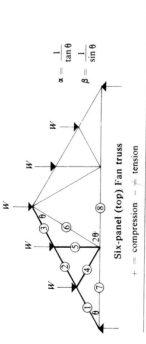

Six-panel (top) Fan truss

+ = compression   - = tension

$$\alpha = \frac{1}{\tan\theta}$$

$$\beta = \frac{1}{\sin\theta}$$

| Member | Compression or tension | 15° W | 15° Unit | 17½° W | 17½° Unit | 20° W | 20° Unit | 22½° W | 22½° Unit | 25° W | 25° Unit | 27½° W | 27½° Unit | 30° W | 30° Unit | 32½° W | 32½° Unit | 35° W | 35° Unit | General formulae W | General formulae Unit |
|---|---|---|---|---|---|---|---|---|---|---|---|---|---|---|---|---|---|---|---|---|---|
| 1 | + | 9·66 | 1·87 | 8·31 | 1·59 | 7·31 | 1·37 | 6·53 | 1·21 | 5·92 | 1·07 | 5·41 | 0·96 | 5·00 | 0·87 | 4·65 | 0·78 | 4·36 | 0·71 | $2\cdot5\beta$ | $0\cdot5\alpha$ |
| 2 | + | 8·19 | 1·87 | 7·01 | 1·59 | 6·10 | 1·37 | 5·40 | 1·21 | 4·83 | 1·07 | 4·39 | 0·96 | 4·00 | 0·87 | 3·67 | 0·78 | 3·39 | 0·71 | $\dfrac{13\beta^2-4}{6\beta}$ | $0\cdot5\alpha$ |
| 3 | + | 9·14 | 1·87 | 7·69 | 1·59 | 6·60 | 1·37 | 5·75 | 1·21 | 5·06 | 1·07 | 4·48 | 0·96 | 4·00 | 0·87 | 3·58 | 0·78 | 3·22 | 0·71 | $\dfrac{5\alpha^2+1}{2\beta}$ | $0\cdot5\alpha$ |
| 4 | + | 1·54 | 0 | 1·38 | 0 | 1·27 | 0 | 1·18 | 0 | 1·11 | 0 | 1·05 | 0 | 1·00 | 0 | 0·95 | 0 | 0·91 | 0 | $\dfrac{\alpha\sqrt{\alpha^2+9}}{3\beta}$ | 0 |
| 5 | + | 1·54 | 0 | 1·38 | 0 | 1·27 | 0 | 1·18 | 0 | 1·11 | 0 | 1·05 | 0 | 1·00 | 0 | 0·95 | 0 | 0·91 | 0 | $\dfrac{\alpha\sqrt{\alpha^2+9}}{3\beta}$ | 0 |
| 6 | − | 3·73 | 0 | 3·17 | 0 | 2·75 | 0 | 2·41 | 0 | 2·14 | 0 | 1·92 | 0 | 1·73 | 0 | 1·57 | 0 | 1·43 | 0 | $\alpha$ | 0 |
| 7 | − | 9·33 | 1·93 | 7·93 | 1·66 | 6·87 | 1·46 | 6·04 | 1·30 | 5·36 | 1·18 | 4·80 | 1·08 | 4·33 | 1·00 | 3·92 | 0·93 | 3·57 | 0·87 | $2\cdot5\alpha$ | $0\cdot5\beta$ |
| 8 | − | 5·60 | 1·93 | 4·76 | 1·66 | 4·12 | 1·46 | 3·62 | 1·30 | 3·22 | 1·18 | 2·88 | 1·08 | 2·60 | 1·00 | 2·35 | 0·93 | 2·14 | 0·87 | $1\cdot5\alpha$ | $0\cdot5\beta$ |
| α | | 3·73 | | 3·17 | | 2·74 | | 2·41 | | 2·14 | | 1·92 | | 1·73 | | 1·57 | | 1·43 | | | |
| β | | 3·86 | | 3·33 | | 2·92 | | 2·61 | | 2·34 | | 2·17 | | 2·00 | | 1·86 | | 1·74 | | | |

**Table 21.10**

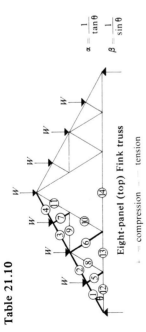

Eight-panel (top) Fink truss

+ = compression — = tension

$$\alpha = \frac{1}{\tan\theta}$$

$$\beta = \frac{1}{\sin\theta}$$

| Member | Compression or tension | 15° W | 15° Unit | 17½° W | 17½° Unit | 20° W | 20° Unit | 22½° W | 22½° Unit | 25° W | 25° Unit | 27½° W | 27½° Unit | 30° W | 30° Unit | 32½° W | 32½° Unit | 35° W | 35° Unit | General formulae W | General formulae Unit |
|---|---|---|---|---|---|---|---|---|---|---|---|---|---|---|---|---|---|---|---|---|---|
| 1 | + | 13·5 | 1·87 | 11·6 | 1·59 | 10·2 | 1·37 | 9·15 | 1·21 | 8·28 | 1·07 | 7·58 | 0·96 | 7·00 | 0·87 | 6·51 | 0·78 | 6·10 | 0·71 | $3\cdot5\beta$ | $0\cdot5\alpha$ |
| 2 | + | 13·2 | 1·87 | 11·3 | 1·59 | 9·88 | 1·37 | 8·75 | 1·21 | 7·84 | 1·07 | 7·13 | 0·96 | 6·50 | 0·87 | 5·97 | 0·78 | 5·51 | 0·71 | $\dfrac{3\cdot5\beta^2-1}{\beta}$ | $0\cdot5\alpha$ |
| 3 | + | 13·0 | 1·87 | 11·0 | 1·59 | 9·53 | 1·37 | 8·37 | 1·21 | 7·41 | 1·07 | 6·67 | 0·96 | 6·00 | 0·87 | 5·43 | 0·78 | 4·94 | 0·71 | $\dfrac{3\cdot5\beta^2-2}{\beta}$ | $0\cdot5\alpha$ |
| 4 | + | 12·7 | 1·87 | 10·7 | 1·59 | 9·19 | 1·37 | 7·98 | 1·21 | 6·99 | 1·07 | 6·21 | 0·96 | 5·50 | 0·87 | 4·90 | 0·78 | 4·36 | 0·71 | $\dfrac{3\cdot5\beta^2-3}{\beta}$ | $0\cdot5\alpha$ |
| 5 | + | 0·97 | 0 | 0·95 | 0 | 0·94 | 0 | 0·92 | 0 | 0·91 | 0 | 0·88 | 0 | 0·86 | 0 | 0·84 | 0 | 0·82 | 0 | $\alpha/\beta$ | 0 |
| 6 | + | 1·93 | 0 | 1·90 | 0 | 1·88 | 0 | 1·84 | 0 | 1·81 | 0 | 1·77 | 0 | 1·73 | 0 | 1·69 | 0 | 1·64 | 0 | $2\alpha/\beta$ | 0 |
| 7 | + | 0·97 | 0 | 0·95 | 0 | 0·94 | 0 | 0·92 | 0 | 0·91 | 0 | 0·88 | 0 | 0·86 | 0 | 0·84 | 0 | 0·82 | 0 | $\alpha/\beta$ | 0 |
| 8 | − | 1·87 | 0 | 1·59 | 0 | 1·37 | 0 | 1·21 | 0 | 1·07 | 0 | 0·96 | 0 | 0·87 | 0 | 0·78 | 0 | 0·71 | 0 | $0\cdot5\alpha$ | 0 |
| 9 | − | 1·87 | 0 | 1·59 | 0 | 1·37 | 0 | 1·21 | 0 | 1·07 | 0 | 0·96 | 0 | 0·87 | 0 | 0·78 | 0 | 0·71 | 0 | $0\cdot5\alpha$ | 0 |
| 10 | − | 3·73 | 0 | 3·17 | 0 | 2·75 | 0 | 2·41 | 0 | 2·14 | 0 | 1·92 | 0 | 1·73 | 0 | 1·57 | 0 | 1·43 | 0 | $1\cdot0\alpha$ | 0 |
| 11 | − | 5·6 | 0 | 4·76 | 0 | 4·12 | 0 | 3·62 | 0 | 3·22 | 0 | 2·88 | 0 | 2·60 | 0 | 2·35 | 0 | 2·14 | 0 | $1\cdot5\alpha$ | 0 |
| 12 | − | 13·1 | 1·93 | 11·1 | 1·66 | 9·62 | 1·46 | 8·45 | 1·30 | 7·51 | 1·18 | 6·72 | 1·08 | 6·06 | 1·00 | 5·49 | 0·93 | 5·00 | 0·87 | $3\cdot5\alpha$ | $0\cdot5\beta$ |
| 13 | − | 11·2 | 1·93 | 9·51 | 1·66 | 8·24 | 1·46 | 7·24 | 1·30 | 6·43 | 1·18 | 5·76 | 1·08 | 5·20 | 1·00 | 4·71 | 0·93 | 4·28 | 0·87 | $3\alpha$ | $0\cdot5\beta$ |
| 14 | − | 7·46 | 1·93 | 6·34 | 1·66 | 5·49 | 1·46 | 4·83 | 1·30 | 4·29 | 1·18 | 3·84 | 1·08 | 3·46 | 1·00 | 3·14 | 0·93 | 2·86 | 0·87 | $2\alpha$ | $0\cdot5\beta$ |
| α |  | 3·73 |  | 3·17 |  | 2·74 |  | 2·41 |  | 2·14 |  | 1·92 |  | 1·73 |  | 1·57 |  | 1·43 |  |  |  |
| β |  | 3·86 |  | 3·33 |  | 2·92 |  | 2·61 |  | 2·36 |  | 2·17 |  | 2·00 |  | 1·86 |  | 1·74 |  |  |  |

**Table 21.11**

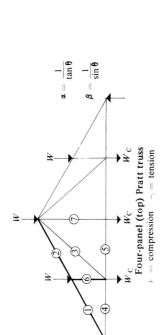

$$\alpha = \frac{1}{\tan \theta}$$

$$\beta = \frac{1}{\sin \theta}$$

**Four-panel (top) Pratt truss**

+ = compression   − = tension

**Coefficients for all slopes**

| Member | Compression or tension | $W$ | $W_c$ | Unit |
|---|---|---|---|---|
| 6 | + | 1 | 0 | 0 |
| 7 | − | 0 | 1 | 0 |

| Member | Compression or tension | $\theta$ 15° $W$ and $W_c$ | Unit | 17½° $W$ and $W_c$ | Unit | 20° $W$ and $W_c$ | Unit | 22½° $W$ and $W_c$ | Unit | 25° $W$ and $W_c$ | Unit | 27½° $W$ and $W_c$ | Unit | 30° $W$ and $W_c$ | Unit | 32½° $W$ and $W_c$ | Unit | 35° $W$ and $W_c$ | Unit | General formulae $W$ and $W_c$ | Unit |
|---|---|---|---|---|---|---|---|---|---|---|---|---|---|---|---|---|---|---|---|---|---|
| 1 | + | 5·80 | 1·87 | 5·00 | 1·59 | 4·39 | 1·37 | 3·92 | 1·21 | 3·55 | 1·07 | 3·25 | 0·96 | 3·00 | 0·87 | 2·79 | 0·78 | 2·61 | 0·71 | $1.5\beta$ | $0.5\alpha$ |
| 2 | + | 5·80 | 1·87 | 5·00 | 1·59 | 4·39 | 1·37 | 3·92 | 1·21 | 3·55 | 1·07 | 3·25 | 0·96 | 3·00 | 0·87 | 2·79 | 0 | 2·61 | 0 | $1.5\beta$ | $0.5\alpha$ |
| 3 | − | 2·12 | 0 | 1·87 | 0 | 1·70 | 0 | 1·57 | 0 | 1·47 | 0 | 1·39 | 0 | 1·32 | 0 | 1·27 | 0 | 1·23 | 0 | $0.5\sqrt{(\alpha^2+4)}$ | 0 |
| 4 | − | 5·60 | 1·93 | 4·76 | 1·66 | 4·12 | 1·46 | 3·62 | 1·30 | 3·22 | 1·18 | 2·88 | 1·08 | 2·60 | 1·00 | 2·35 | 0·93 | 2·14 | 0·87 | $1.5\alpha$ | $0.5\beta$ |
| 5 | − | 3·73 | 1·93 | 3·17 | 1·66 | 2·75 | 1·46 | 2·41 | 1·30 | 2·14 | 1·18 | 1·92 | 1·08 | 1·73 | 1·00 | 1·57 | 0·93 | 1·43 | 0·87 | $\alpha$ | $0.5\beta$ |
| $\alpha$ | | 3·73 | | 3·17 | | 2·74 | | 2·41 | | 2·14 | | 1·92 | | 1·73 | | 1·57 | | 1·43 | | | |
| $\beta$ | | 3·86 | | 3·33 | | 2·92 | | 2·61 | | 2·36 | | 2·17 | | 2·00 | | 1·86 | | 1·74 | | | |

**Table 21.12**

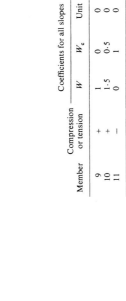

**Six-panel (top) Pratt truss**

+ = compression   − = tension

$$\alpha = \frac{1}{\tan\theta}$$

$$\beta = \frac{1}{\sin\theta}$$

θ

| Member | Compression or tension | 15° W and $W_c$ | 15° Unit | 17½° W and $W_c$ | 17½° Unit | 20° W and $W_c$ | 20° Unit | 22½° W and $W_c$ | 22½° Unit | 25° W and $W_c$ | 25° Unit | 27½° W and $W_c$ | 27½° Unit | 30° W and $W_c$ | 30° Unit | 32½° W and $W_c$ | 32½° Unit | 35° W and $W_c$ | 35° Unit |
|---|---|---|---|---|---|---|---|---|---|---|---|---|---|---|---|---|---|---|---|
| 1 | + | 9.66 | 1.87 | 8.31 | 1.59 | 7.31 | 1.37 | 6.53 | 1.21 | 5.92 | 1.07 | 5.41 | 0.96 | 5.00 | 0.87 | 4.65 | 0.78 | 4.36 | 0.71 |
| 2 | + | 9.66 | 1.87 | 8.31 | 1.59 | 7.31 | 1.37 | 6.53 | 1.21 | 5.92 | 1.07 | 5.41 | 0.96 | 5.00 | 0.87 | 4.65 | 0.78 | 4.36 | 0.71 |
| 3 | + | 7.73 | 1.87 | 6.65 | 1.59 | 5.85 | 1.37 | 5.23 | 1.21 | 4.73 | 1.07 | 4.33 | 0.96 | 4.00 | 0.87 | 3.72 | 0.78 | 3.49 | 0.71 |
| 4 | − | 2.12 | 0 | 1.87 | 0 | 1.70 | 0 | 1.57 | 0 | 1.47 | 0 | 1.39 | 0 | 1.32 | 0 | 1.27 | 0 | 1.23 | 0 |
| 5 | − | 2.39 | 0 | 2.18 | 0 | 2.03 | 0 | 1.92 | 0 | 1.84 | 0 | 1.78 | 0 | 1.73 | 0 | 1.69 | 0 | 1.66 | 0 |
| 6 | − | 9.33 | 1.93 | 7.93 | 1.66 | 6.87 | 1.46 | 6.04 | 1.30 | 5.36 | 1.18 | 4.80 | 1.08 | 4.33 | 1.00 | 3.92 | 0.93 | 3.57 | 0.87 |
| 7 | − | 7.46 | 1.93 | 6.34 | 1.66 | 5.49 | 1.46 | 4.83 | 1.30 | 4.29 | 1.18 | 3.84 | 1.08 | 3.46 | 1.00 | 3.14 | 0.93 | 2.86 | 0.87 |
| 8 | − | 5.6 | 1.93 | 4.76 | 1.66 | 4.12 | 1.46 | 3.62 | 1.30 | 3.22 | 1.18 | 2.88 | 1.08 | 2.60 | 1.00 | 2.35 | 0.93 | 2.14 | 0.87 |
| α | | 3.73 | | 3.17 | | 2.74 | | 2.41 | | 2.14 | | 1.92 | | 1.73 | | 1.57 | | 1.43 | |
| β | | 3.86 | | 3.33 | | 2.92 | | 2.61 | | 2.36 | | 2.17 | | 2.00 | | 1.86 | | 1.74 | |

Coefficients for all slopes

| Member | Compression or tension | W | $W_c$ | Unit |
|---|---|---|---|---|
| 9 | + | 1 | 0 | 0 |
| 10 | + | 1.5 | 0.5 | 0 |
| 11 | − | 0 | 1 | 0 |

General formulae

| Member | W and $W_c$ | Unit |
|---|---|---|
| 1 | $2.5\beta$ | $0.5\alpha$ |
| 2 | $2.5\beta$ | $0.5\alpha$ |
| 3 | $2\beta$ | $0.5\alpha$ |
| 4 | $0.5\sqrt{(\alpha^2+4)}$ | 0 |
| 5 | $0.5\sqrt{(\alpha^2+9)}$ | 0 |
| 6 | $2.5\alpha$ | $0.5\beta$ |
| 7 | $2\alpha$ | $0.5\beta$ |
| 8 | $1.5\alpha$ | $0.5\beta$ |

## Table 21.13

**Eight-panel (top) Pratt truss**

+ = compression    — = tension

$$\alpha = \frac{1}{\tan \theta}$$

$$\beta = \frac{1}{\sin \theta}$$

**Coefficients for all slopes**

| Member | W | $W_c$ | Unit |
|---|---|---|---|
| 12 | 1 | 0 | 0 |
| 13 | 1·5 | 0·5 | 0 |
| 14 | 2 | 1 | 0 |
| 15 | 0 | 1 | 0 |

$\theta$

| Member | Compression or tension | 15° W and $W_c$ | 15° Unit | 17½° W and $W_c$ | 17½° Unit | 20° W and $W_c$ | 20° Unit | 22½° W and $W_c$ | 22½° Unit | 25° W and $W_c$ | 25° Unit | 27½° W and $W_c$ | 27½° Unit | 30° W and $W_c$ | 30° Unit | 32½° W and $W_c$ | 32½° Unit | 35° W and $W_c$ | 35° Unit | General formulae W and $W_c$ | General formulae Unit |
|---|---|---|---|---|---|---|---|---|---|---|---|---|---|---|---|---|---|---|---|---|---|
| 1 | + | 13·5 | 1·87 | 11·6 | 1·59 | 10·2 | 1·37 | 9·15 | 1·21 | 8·28 | 1·07 | 7·58 | 0·96 | 7·00 | 0·87 | 6·51 | 0·78 | 6·10 | 0·71 | $3·5\beta$ | $0·5\alpha$ |
| 2 | + | 13·5 | 1·87 | 11·6 | 1·59 | 10·2 | 1·37 | 9·15 | 1·21 | 8·28 | 1·07 | 7·58 | 0·96 | 7·00 | 0·87 | 6·51 | 0·78 | 6·10 | 0·71 | $3·5\beta$ | $0·5\alpha$ |
| 3 | + | 11·6 | 1·87 | 9·98 | 1·59 | 8·77 | 1·37 | 7·84 | 1·21 | 7·10 | 1·07 | 6·50 | 0·96 | 6·00 | 0·87 | 5·58 | 0·78 | 5·23 | 0·71 | $3\beta$ | $0·5\alpha$ |
| 4 | + | 9·66 | 1·87 | 8·31 | 1·59 | 7·31 | 1·37 | 6·53 | 1·21 | 5·92 | 1·07 | 5·41 | 0·96 | 5·00 | 0·87 | 4·65 | 0·78 | 4·36 | 0·71 | $2·5\beta$ | $0·5\alpha$ |
| 5 | − | 2·12 | 0 | 1·87 | 0 | 1·70 | 0 | 1·57 | 0 | 1·47 | 0 | 1·39 | 0 | 1·32 | 0 | 1·27 | 0 | 1·23 | 0 | $0·5\sqrt{(\alpha^2 + 4)}$ | $0·5\alpha$ |
| 6 | − | 2·39 | 0 | 2·18 | 0 | 2·03 | 0 | 1·92 | 0 | 1·84 | 0 | 1·78 | 0 | 1·73 | 0 | 1·69 | 0 | 1·66 | 0 | $0·5\sqrt{(\alpha^2 + 9)}$ | 0 |
| 7 | − | 2·74 | 0 | 2·55 | 0 | 2·43 | 0 | 2·34 | 0 | 2·27 | 0 | 2·22 | 0 | 2·18 | 0 | 2·15 | 0 | 2·12 | 0 | 0 | 0 |
| 8 | − | 13·1 | 1·93 | 11·1 | 1·66 | 9·62 | 1·46 | 8·45 | 1·30 | 7·51 | 1·18 | 6·72 | 1·08 | 6·06 | 1·00 | 5·49 | 0·93 | 5·00 | 0·87 | $3·5\alpha$ | $0·5\beta$ |
| 9 | − | 11·2 | 1·93 | 9·51 | 1·66 | 8·24 | 1·46 | 7·24 | 1·30 | 6·43 | 1·18 | 5·76 | 1·08 | 5·20 | 1·00 | 4·71 | 0·93 | 4·28 | 0·87 | $3\alpha$ | $0·5\beta$ |
| 10 | − | 9·33 | 1·93 | 7·93 | 1·66 | 6·87 | 1·46 | 6·04 | 1·30 | 5·36 | 1·18 | 4·80 | 1·08 | 4·33 | 1·00 | 3·92 | 0·93 | 3·57 | 0·87 | $2·5\alpha$ | $0·5\beta$ |
| 11 | − | 7·46 | 1·93 | 6·34 | 1·66 | 5·49 | 1·46 | 4·83 | 1·30 | 4·29 | 1·18 | 3·84 | 1·08 | 3·46 | 1·00 | 3·14 | 0·93 | 2·86 | 0·87 | $2\alpha$ | $0·5\beta$ |
| $\alpha$ | | 3·73 | | 3·17 | | 2·74 | | 2·41 | | 2·14 | | 1·92 | | 1·73 | | 1·57 | | 1·43 | | | |
| $\beta$ | | 3·86 | | 3·33 | | 2·92 | | 2·61 | | 2·36 | | 2·17 | | 2·00 | | 1·86 | | 1·74 | | | |

**Table 21.14**

$$\alpha = \frac{1}{\tan\theta} \qquad \beta = \frac{1}{\sin\theta}$$

**Four-panel (top) Howe truss**

$+$ = compression   $-$ = tension

**Coefficients for all slopes**

| Member | Compression or tension | $W$ | $W_c$ | Unit |
|---|---|---|---|---|
| 6 | | 0 | 1·0 | 0 |
| 7 | | 1·0 | 2·0 | 0 |

$\theta$

| Member | Compression or tension | 15° $W$ and $W_c$ | 15° Unit | 17½° $W$ and $W_c$ | 17½° Unit | 20° $W$ and $W_c$ | 20° Unit | 22½° $W$ and $W_c$ | 22½° Unit | 25° $W$ and $W_c$ | 25° Unit | 27½° $W$ and $W_c$ | 27½° Unit | 30° $W$ and $W_c$ | 30° Unit | 32½° $W$ and $W_c$ | 32½° Unit | 35° $W$ and $W_c$ | 35° Unit | General formulae $W$ and $W_c$ | General formulae Unit |
|---|---|---|---|---|---|---|---|---|---|---|---|---|---|---|---|---|---|---|---|---|---|
| 1 | + | 5·80 | 1·87 | 5·00 | 1·59 | 4·39 | 1·37 | 3·92 | 1·21 | 3·55 | 1·07 | 3·25 | 0·96 | 3·00 | 0·87 | 2·79 | 0·78 | 2·61 | 0·71 | 1·5β | 0·5α |
| 2 | + | 3·86 | 1·87 | 3·33 | 1·59 | 2·92 | 1·37 | 2·61 | 1·21 | 2·37 | 1·07 | 2·16 | 0·96 | 2·00 | 0·87 | 1·86 | 0·78 | 1·74 | 0·71 | 1·0β | 0·5α |
| 3 | + | 1·93 | 0 | 1·66 | 0 | 1·46 | 0 | 1·30 | 0 | 1·18 | 0 | 1·08 | 0 | 1·00 | 0 | 0·93 | 0 | 0·87 | 0 | 0·5β | 0 |
| 4 | − | 5·60 | 1·93 | 4·76 | 1·66 | 4·12 | 1·46 | 3·62 | 1·30 | 3·22 | 1·18 | 2·88 | 1·08 | 2·60 | 1·00 | 2·35 | 0·93 | 2·14 | 0·87 | 1·5α | 0·5β |
| 5 | − | 5·60 | 1·93 | 4·76 | 1·66 | 4·12 | 1·46 | 3·62 | 1·30 | 3·22 | 1·18 | 2·88 | 1·08 | 2·60 | 1·00 | 2·35 | 0·93 | 2·14 | 0·87 | 1·5α | 0·5β |
| α | | 3·73 | | 3·17 | | 2·74 | | 2·41 | | 2·14 | | 1·92 | | 1·73 | | 1·57 | | 1·43 | | | |
| β | | 3·86 | | 3·33 | | 2·92 | | 2·61 | | 2·36 | | 2·17 | | 2·00 | | 1·86 | | 1·74 | | | |

# Table 21.15

**Six-panel (top) Howe truss**

+ = compression    − = tension

$$\alpha = \frac{1}{\tan\theta}$$

$$\beta = \frac{1}{\sin\theta}$$

**Coefficients for all slopes**

| Member | W | $W_c$ | Unit |
|---|---|---|---|
| 9 | 0 | 1·0 | 0 |
| 10 | 0·5 | 1·5 | 0 |
| 11 | 2 | 3 | 0 |

θ

| Member | Compression or tension | 15° W and $W_c$ | 15° Unit | 17½° W and $W_c$ | 17½° Unit | 20° W and $W_c$ | 20° Unit | 22½° W and $W_c$ | 22½° Unit | 25° W and $W_c$ | 25° Unit | 27½° W and $W_c$ | 27½° Unit | 30° W and $W_c$ | 30° Unit | 32½° W and $W_c$ | 32½° Unit | 35° W and $W_c$ | 35° Unit | General formulae W and $W_c$ | Unit |
|---|---|---|---|---|---|---|---|---|---|---|---|---|---|---|---|---|---|---|---|---|---|
| 1 | + | 9·66 | 1·87 | 8·31 | 1·59 | 7·31 | 1·37 | 6·53 | 1·21 | 5·92 | 1·07 | 5·41 | 0·96 | 5·00 | 0·87 | 4·65 | 0·78 | 4·36 | 0·71 | 2·5β | 0·5α |
| 2 | + | 7·73 | 1·87 | 6·65 | 1·59 | 5·85 | 1·37 | 5·23 | 1·21 | 4·73 | 1·07 | 4·33 | 0·96 | 4·00 | 0·87 | 3·72 | 0·78 | 3·49 | 0·71 | 2β | 0·5α |
| 3 | + | 5·80 | 1·87 | 5·00 | 1·59 | 4·39 | 1·37 | 3·92 | 1·21 | 3·55 | 1·07 | 3·25 | 0·96 | 3·00 | 0·87 | 2·79 | 0·78 | 2·61 | 0·71 | 1·5β | 0·5α |
| 4 | + | 1·93 | 0 | 1·66 | 0 | 1·46 | 0 | 1·30 | 0 | 1·18 | 0 | 1·08 | 0 | 1·00 | 0 | 0·93 | 0 | 0·87 | 0 | 0·5β | 0 |
| 5 | + | 2·12 | 0 | 1·87 | 0 | 1·70 | 0 | 1·57 | 0 | 1·47 | 0 | 1·39 | 0 | 1·32 | 0 | 1·27 | 0 | 1·23 | 0 | $0·5\sqrt{(\alpha^2 + 4)}$ | 0 |
| 6 | − | 9·33 | 1·93 | 7·93 | 1·66 | 6·87 | 1·46 | 6·04 | 1·30 | 5·36 | 1·18 | 4·80 | 1·08 | 4·33 | 1·00 | 3·92 | 0·93 | 3·57 | 0·87 | 2·5α | 0·5β |
| 7 | − | 9·33 | 1·93 | 7·93 | 1·66 | 6·87 | 1·46 | 6·04 | 1·30 | 5·36 | 1·18 | 4·80 | 1·08 | 4·33 | 1·00 | 3·92 | 0·93 | 3·57 | 0·87 | 2·5α | 0·5β |
| 8 | − | 7·46 | 1·93 | 6·34 | 1·66 | 5·49 | 1·46 | 4·83 | 1·30 | 4·29 | 1·18 | 3·84 | 1·08 | 3·46 | 1·00 | 3·14 | 0·93 | 2·86 | 0·87 | 2α | 0·5β |
| α | | 3·73 | | 3·17 | | 2·74 | | 2·41 | | 2·14 | | 1·92 | | 1·73 | | 1·57 | | 1·43 | | | |
| β | | 3·86 | | 3·33 | | 2·92 | | 2·61 | | 2·36 | | 2·17 | | 2·00 | | 1·86 | | 1·74 | | | |

**Table 21.16**

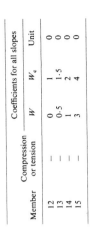

Eight-panel (top) Howe truss

+ = compression   − = tension

$$\alpha = \frac{1}{\tan\theta} \qquad \beta = \frac{1}{\sin\theta}$$

Coefficients for all slopes

| Member | Compression or tension | W | $W_c$ | Unit |
|---|---|---|---|---|
| 12 | — | 0 | 1 | 0 |
| 13 | — | 0.5 | 1.5 | 0 |
| 14 | — | 1 | 2 | 0 |
| 15 | — | 3 | 4 | 0 |

θ

| Member | Compression or tension | 15° W and $W_c$ | Unit | 17½° W and $W_c$ | Unit | 20° W and $W_c$ | Unit | 22½° W and $W_c$ | Unit | 25° W and $W_c$ | Unit | 27½° W and $W_c$ | Unit | 30° W and $W_c$ | Unit | 32½° W and $W_c$ | Unit | 35° W and $W_c$ | Unit | General formulae W and $W_c$ | Unit |
|---|---|---|---|---|---|---|---|---|---|---|---|---|---|---|---|---|---|---|---|---|---|
| 1 | + | 13.5 | 1.87 | 11.6 | 1.59 | 10.2 | 1.37 | 9.15 | 1.21 | 8.28 | 1.07 | 7.58 | 0.96 | 7.00 | 0.87 | 6.51 | 0.78 | 6.10 | 0.71 | 3.5β | 0.5α |
| 2 | + | 11.6 | 1.87 | 9.98 | 1.59 | 8.77 | 1.37 | 7.84 | 1.21 | 7.10 | 1.07 | 6.50 | 0.96 | 6.00 | 0.87 | 5.58 | 0.78 | 5.23 | 0.71 | 3.0β | 0.5α |
| 3 | + | 9.66 | 1.87 | 8.31 | 1.59 | 7.31 | 1.37 | 6.53 | 1.21 | 5.92 | 1.07 | 5.41 | 0.96 | 5.00 | 0.87 | 4.65 | 0.78 | 4.36 | 0.71 | 2.5β | 0.5α |
| 4 | + | 7.73 | 1.87 | 6.65 | 1.59 | 5.85 | 1.37 | 5.23 | 1.21 | 4.73 | 1.07 | 4.33 | 0.96 | 4.00 | 0.87 | 3.72 | 0.78 | 3.49 | 0.71 | 2.0β | 0.5α |
| 5 | + | 1.93 | 0 | 1.66 | 0 | 1.46 | 0 | 1.30 | 0 | 1.18 | 0 | 1.08 | 0 | 1.00 | 0 | 0.93 | 0 | 0.87 | 0 | 0.5β | 0 |
| 6 | + | 2.12 | 0 | 1.87 | 0 | 1.70 | 0 | 1.57 | 0 | 1.47 | 0 | 1.39 | 0 | 1.32 | 0 | 1.27 | 0 | 1.23 | 0 | $0.5\sqrt{(\alpha^2+4)}$ | 0 |
| 7 | + | 2.39 | 0 | 2.18 | 0 | 2.03 | 0 | 1.92 | 0 | 1.84 | 0 | 1.78 | 0 | 1.73 | 0 | 1.69 | 0 | 1.66 | 0 | $0.5\sqrt{(\alpha^2+9)}$ | 0 |
| 8 | − | 13.1 | 1.93 | 11.1 | 1.66 | 9.62 | 1.46 | 8.45 | 1.30 | 7.51 | 1.18 | 6.72 | 1.08 | 6.06 | 1.00 | 5.49 | 0.93 | 5.00 | 0.87 | 3.5α | 0.5β |
| 9 | − | 13.1 | 1.93 | 11.1 | 1.66 | 9.62 | 1.46 | 8.45 | 1.30 | 7.51 | 1.18 | 6.72 | 1.08 | 6.06 | 1.00 | 5.49 | 0.93 | 5.00 | 0.87 | 3.5α | 0.5β |
| 10 | − | 11.2 | 1.93 | 9.51 | 1.66 | 8.24 | 1.46 | 7.24 | 1.30 | 6.43 | 1.18 | 5.76 | 1.08 | 5.20 | 1.00 | 4.71 | 0.93 | 4.28 | 0.87 | 3.0α | 0.5β |
| 11 | − | 9.33 | 1.93 | 7.93 | 1.66 | 6.87 | 1.46 | 6.04 | 1.30 | 5.36 | 1.18 | 4.80 | 1.08 | 4.33 | 1.00 | 3.92 | 0.93 | 3.57 | 0.87 | 2.5α | 0.5β |
| α |  | 3.73 |  | 3.17 |  | 2.74 |  | 2.41 |  | 2.14 |  | 1.92 |  | 1.73 |  | 1.57 |  | 1.43 |  |  |  |
| β |  | 3.86 |  | 3.33 |  | 2.92 |  | 2.62 |  | 2.36 |  | 2.17 |  | 2.00 |  | 1.86 |  | 1.74 |  |  |  |

**Table 21.17**

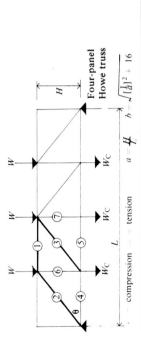

**Four-panel Howe truss**

$$a = \frac{H}{L}. \qquad b = \sqrt{\left[\tfrac{1}{a}\right]^2 + 16}$$

+ compression − tension

Constant coefficients

| Member | Compression or tension | W | $W_c$ | Unit |
|---|---|---|---|---|
| 6 | — | 0·5 | 1·5 | 0 |
| 7 | — | 0 | 1 | 0 |

Values of $a$

| Member | Compression or tension | 0·15 W and $W_c$ | Unit | 0·20 W and $W_c$ | Unit | 0·25 W and $W_c$ | Unit | 0·30 W and $W_c$ | Unit | 0·35 W and $W_c$ | Unit | 0·40 W and $W_c$ | Unit | General formulae W and $W_c$ | Unit |
|---|---|---|---|---|---|---|---|---|---|---|---|---|---|---|---|
| 1 | + | 2·50 | 0·83 | 1·87 | 0·62 | 1·50 | 0·50 | 1·25 | 0·42 | 1·07 | 0·36 | 0·94 | 0·31 | $\frac{3}{8a}$ | $\frac{1}{8a}$ |
| 2 | + | 2·91 | 0·97 | 2·40 | 0·80 | 2·12 | 0·71 | 1·95 | 0·65 | 1·84 | 0·61 | 1·77 | 0·59 | $\frac{3b}{8}$ | $\frac{b}{8}$ |
| 3 | + | 0·97 | 0·97 | 0·80 | 0·80 | 0·71 | 0·71 | 0·65 | 0·65 | 0·61 | 0·61 | 0·59 | 0·59 | $\frac{b}{8}$ | $\frac{b}{8}$ |
| 4 | − | 2·50 | 0·83 | 1·87 | 0·62 | 1·50 | 0·50 | 1·25 | 0·42 | 1·07 | 0·36 | 0·94 | 0·31 | $\frac{3}{8a}$ | $\frac{1}{8a}$ |
| 5 | − | 3·33 | 1·67 | 2·5 | 1·25 | 2·00 | 1·00 | 1·67 | 0·83 | 1·43 | 0·71 | 1·25 | 0·62 | $\frac{1}{2a}$ | $\frac{1}{4a}$ |
| Values of $b$ | | 7·77 | | 6·40 | | 5·66 | | 5·21 | | 4·92 | | 4·72 | | | |
| Values of $\theta$ | | 31° | | 39° | | 45° | | 50° | | 55° | | 58° | | | |

**Table 21.18**

Six-panel Howe truss

+ = compression    − = tension

$a = \dfrac{H}{L}$     $b = \sqrt{\left(\dfrac{1}{a}\right)^2 + 36}$

Legend: loads $W$ at top chord panel points, $W_C$ at bottom chord; span $L$, height $H$, angle $\theta$; members 1–11.

**Constant coefficients**

| Member | Compression or tension | $W$ | $W_c$ | Unit |
|---|---|---|---|---|
| 9 | — | 1·5 | 2·5 | 0·5 |
| 10 | — | 0·5 | 1·5 | 0·5 |
| 11 | — | 0 | 1 | 0 |

**Values of $a$**

| Member | Compression or tension | 0·10 $W$ and $W_c$ | 0·10 Unit | 0·15 $W$ and $W_c$ | 0·15 Unit | 0·167 $W$ and $W_c$ | 0·167 Unit | 0·20 $W$ and $W_c$ | 0·20 Unit | 0·25 $W$ and $W_c$ | 0·25 Unit | 0·30 $W$ and $W_c$ | 0·30 Unit | General $W$ and $W_c$ | General Unit |
|---|---|---|---|---|---|---|---|---|---|---|---|---|---|---|---|
| 1 | + | 4·17 | 0·83 | 2·78 | 0·56 | 2·50 | 0·50 | 2·08 | 0·42 | 1·67 | 0·33 | 1·39 | 0·28 | $\dfrac{5}{12a}$ | $\dfrac{1}{12a}$ |
| 2 | + | 6·67 | 1·67 | 4·44 | 1·11 | 4·00 | 1·00 | 3·33 | 0·83 | 2·67 | 0·67 | 2·22 | 0·56 | $\dfrac{2}{3a}$ | $\dfrac{1}{6a}$ |
| 3 | + | 4·86 | 0·97 | 3·74 | 0·75 | 3·53 | 0·71 | 3·25 | 0·65 | 3·00 | 0·60 | 2·86 | 0·57 | $\dfrac{5b}{12}$ | $\dfrac{b}{12}$ |
| 4 | + | 2·92 | 0·97 | 2·24 | 0·75 | 2·12 | 0·71 | 1·95 | 0·65 | 1·80 | 0·60 | 1·72 | 0·57 | $\dfrac{b}{4}$ | $\dfrac{b}{12}$ |
| 5 | + | 0·97 | 0·97 | 0·75 | 0·75 | 0·71 | 0·71 | 0·65 | 0·65 | 0·60 | 0·60 | 0·57 | 0·57 | $\dfrac{b}{12}$ | $\dfrac{b}{12}$ |
| 6 | − | 4·17 | 0·83 | 2·78 | 0·56 | 2·50 | 0·50 | 2·08 | 0·42 | 1·67 | 0·33 | 1·39 | 0·28 | $\dfrac{5}{12a}$ | $\dfrac{1}{12a}$ |
| 7 | − | 6·67 | 1·67 | 4·44 | 1·11 | 4·00 | 1·00 | 3·33 | 0·83 | 2·67 | 0·67 | 2·22 | 0·56 | $\dfrac{2}{3a}$ | $\dfrac{1}{6a}$ |
| 8 | − | 7·50 | 2·50 | 5·00 | 1·67 | 4·50 | 1·50 | 3·75 | 1·25 | 3·00 | 1·00 | 2·50 | 0·83 | $\dfrac{3}{4a}$ | $\dfrac{1}{4a}$ |
| Values of $b$ | | 11·7 | | 8·97 | | 8·48 | | 7·81 | | 7·21 | | 6·86 | | | |
| Values of $\theta$ | | 31° | | 42° | | 45° | | 50° | | 56° | | 61° | | | |

**Table 21.19**

Eight-panel Howe truss

$+$ = compression $\quad$ $-$ = tension $\quad$ $a = \dfrac{H}{L}$ $\quad$ $b = \sqrt{[\tfrac{L}{H}]^2 + 64}$

Constant coefficients

| Member | Compression or tension | $W$ | Unit | $W_c$ | Unit |
|---|---|---|---|---|---|
| 12 | — | 2.5 | 0.5 | 3.5 | 0.5 |
| 13 | — | 1.5 | 0.5 | 2.5 | 0.5 |
| 14 | — | 0.5 | 0.5 | 1.5 | 0.5 |
| 15 | — | 0 | 0 | 1.0 | 0 |

Values of $a$

| Member | Compression or tension | 0.075 $W$ and $W_c$ | Unit | 0.10 $W$ and $W_c$ | Unit | 0.125 $W$ and $W_c$ | Unit | 0.15 $W$ and $W_c$ | Unit | 0.175 $W$ and $W_c$ | Unit | 0.20 $W$ and $W_c$ | Unit | General formulae $W$ and $W_c$ | Unit |
|---|---|---|---|---|---|---|---|---|---|---|---|---|---|---|---|
| 1 | + | 5.83 | 0.83 | 4.37 | 0.62 | 3.50 | 0.50 | 2.92 | 0.42 | 2.50 | 0.36 | 2.19 | 0.31 | $\frac{7}{16a}$ | $\frac{1}{16a}$ |
| 2 | + | 10.0 | 1.67 | 7.50 | 1.25 | 6.00 | 1.00 | 5.00 | 0.83 | 4.28 | 0.71 | 3.75 | 0.62 | $\frac{3}{4a}$ | $\frac{1}{8a}$ |
| 3 | + | 12.5 | 2.22 | 9.37 | 1.67 | 7.50 | 1.33 | 6.25 | 1.11 | 5.36 | 0.95 | 4.69 | 0.83 | $\frac{15}{16a}$ | $\frac{1}{6a}$ |
| 4 | + | 6.80 | 0.97 | 5.60 | 0.80 | 4.95 | 0.71 | 4.55 | 0.65 | 4.30 | 0.61 | 4.13 | 0.59 | $\frac{7b}{16}$ | $\frac{b}{16}$ |
| 5 | + | 4.86 | 0.97 | 4.00 | 0.80 | 3.54 | 0.71 | 3.25 | 0.65 | 3.07 | 0.61 | 2.95 | 0.59 | $\frac{5b}{16}$ | $\frac{b}{16}$ |
| 6 | + | 2.92 | 0.97 | 2.40 | 0.80 | 2.12 | 0.71 | 1.95 | 0.65 | 1.84 | 0.61 | 1.77 | 0.59 | $\frac{3b}{16}$ | $\frac{b}{16}$ |
| 7 | + | 0.97 | 0.97 | 0.80 | 0.80 | 0.71 | 0.71 | 0.65 | 0.65 | 0.61 | 0.61 | 0.59 | 0.59 | $\frac{b}{16}$ | $\frac{b}{16}$ |
| 8 | − | 5.83 | 0.83 | 4.37 | 0.62 | 3.50 | 0.50 | 2.92 | 0.42 | 2.50 | 0.36 | 2.19 | 0.31 | $\frac{7}{16a}$ | $\frac{1}{16a}$ |
| 9 | − | 10.0 | 1.67 | 7.50 | 1.25 | 6.00 | 1.00 | 5.00 | 0.83 | 4.28 | 0.71 | 3.75 | 0.62 | $\frac{3}{4a}$ | $\frac{1}{8a}$ |
| 10 | − | 12.5 | 2.22 | 9.37 | 1.67 | 7.50 | 1.33 | 6.25 | 1.11 | 5.36 | 0.95 | 4.69 | 0.83 | $\frac{15}{16a}$ | $\frac{1}{6a}$ |
| 11 | − | 13.3 | 3.33 | 10.0 | 2.50 | 8.00 | 2.00 | 6.67 | 1.67 | 5.71 | 1.43 | 5.00 | 1.25 | $\frac{1}{a}$ | $\frac{1}{4a}$ |
| Values of $b$ | | 15.5 | | 12.8 | | 11.3 | | 10.4 | | 9.83 | | 9.43 | | | |
| Values of $\theta$ | | 31° | | 39° | | 45° | | 50° | | 55° | | 58° | | | |

## Table 21.20

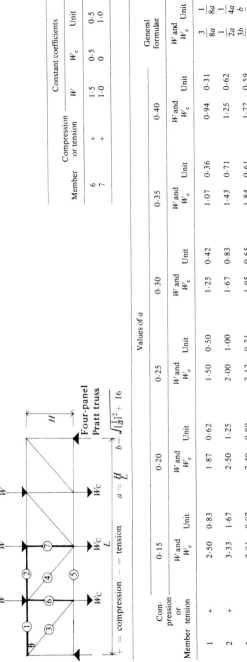

**Four-panel Pratt truss**

$+$ = compression    $-$ = tension    $a = \dfrac{H}{L}$    $b = \sqrt{\left(\dfrac{1}{a}\right)^2 + 16}$

| | Compression or tension | W | $W_c$ | Unit |
|---|---|---|---|---|
| Member | 6 | + | 1·5 | 0·5 | 0·5 |
| | 7 | + | 1·0 | 0 | 1·0 |

Constant coefficients

| | Compression or tension | \multicolumn{12}{c}{Values of $a$} | General formulae | |
|---|---|---|---|

| Member tension | Compression or tension | 0·15 W and $W_c$ | Unit | 0·20 W and $W_c$ | Unit | 0·25 W and $W_c$ | Unit | 0·30 W and $W_c$ | Unit | 0·35 W and $W_c$ | Unit | 0·40 W and $W_c$ | Unit | W and $W_c$ | Unit |
|---|---|---|---|---|---|---|---|---|---|---|---|---|---|---|---|
| 1 | + | 2·50 | 0·83 | 1·87 | 0·62 | 1·50 | 0·50 | 1·25 | 0·42 | 1·07 | 0·36 | 0·94 | 0·31 | $\dfrac{3}{8a}$ | $\dfrac{1}{8a}$ |
| 2 | + | 3·33 | 1·67 | 2·50 | 1·25 | 2·00 | 1·00 | 1·67 | 0·83 | 1·43 | 0·71 | 1·25 | 0·62 | $\dfrac{1}{2a}$ | $\dfrac{1}{4a}$ |
| 3 | − | 2·91 | 0·97 | 2·40 | 0·80 | 2·12 | 0·71 | 1·95 | 0·65 | 1·84 | 0·61 | 1·77 | 0·59 | $\dfrac{3b}{8}$ | $\dfrac{b}{8}$ |
| 4 | − | 0·97 | 0·97 | 0·80 | 0·80 | 0·71 | 0·71 | 0·65 | 0·65 | 0·61 | 0·61 | 0·59 | 0·59 | $\dfrac{b}{8}$ | $\dfrac{b}{8}$ |
| 5 | − | 2·50 | 0·83 | 1·87 | 0·62 | 1·50 | 0·50 | 1·25 | 0·42 | 1·07 | 0·36 | 0·94 | 0·31 | $\dfrac{3}{8a}$ | $\dfrac{1}{8a}$ |
| Values of $b$ | | 7·77 | | 6·40 | | 5·66 | | 5·21 | | 4·92 | | 4·72 | | | |
| Values of $\theta$ | | 31° | | 39° | | 45° | | 50° | | 55° | | 58° | | | |

**Table 21.21**

Six-panel Pratt truss

$a = \dfrac{H}{L}$  $b = \sqrt{\left[\dfrac{H}{L}\right]^2 + 36}$

+ = compression  − = tension

**Constant coefficients**

| Member | Compression or tension | W | $W_c$ | Unit |
|---|---|---|---|---|
| 10 | + | 2·5 | 1·5 | 0·5 |
| 11 | + | 1·5 | 0·5 | 0·5 |
| 12 | + | 1·0 | 0 | 1·0 |

| Member | Compression or tension | Values of $a$ | | | | | | | | | | | | General formulae | |
|---|---|---|---|---|---|---|---|---|---|---|---|---|---|---|---|
| | | 0·10 | | 0·15 | | 0·167 | | 0·20 | | 0·25 | | 0·30 | | | |
| | | *W* and $W_c$ | Unit | *W* and $W_c$ | Unit | *W* and $W_c$ | Unit | *W* and $W_c$ | Unit | *W* and $W_c$ | Unit | *W* and $W_c$ | Unit | *W* and $W_c$ | Unit |
| 1 | + | 4·17 | 0·83 | 2·78 | 0·56 | 2·50 | 0·50 | 2·08 | 0·42 | 1·67 | 0·33 | 1·39 | 0·28 | $\frac{5}{12a}$ | $\frac{1}{12a}$ |
| 2 | + | 6·67 | 1·67 | 4·44 | 1·11 | 4·00 | 1·00 | 3·33 | 0·83 | 2·67 | 0·67 | 2·22 | 0·56 | $\frac{2}{3a}$ | $\frac{1}{6a}$ |
| 3 | + | 7·5 | 2·5 | 5·00 | 1·67 | 4·50 | 1·50 | 3·75 | 1·25 | 3·00 | 1·00 | 2·50 | 0·83 | $\frac{3}{4a}$ | $\frac{1}{4a}$ |
| 4 | − | 4·86 | 0·97 | 3·74 | 0·75 | 3·53 | 0·71 | 3·25 | 0·65 | 3·00 | 0·60 | 2·86 | 0·57 | $\frac{5b}{12}$ | $\frac{b}{12}$ |
| 5 | − | 2·92 | 0·97 | 2·24 | 0·75 | 2·12 | 0·71 | 1·95 | 0·65 | 1·80 | 0·60 | 1·72 | 0·57 | $\frac{b}{4}$ | $\frac{b}{12}$ |
| 6 | − | 0·97 | 0·97 | 0·75 | 0·75 | 0·71 | 0·71 | 0·65 | 0·65 | 0·60 | 0·60 | 0·57 | 0·57 | $\frac{b}{12}$ | $\frac{b}{12}$ |
| 7 | − | 4·17 | 0·83 | 2·78 | 0·56 | 2·50 | 0·50 | 2·08 | 0·42 | 1·67 | 0·33 | 1·39 | 0·28 | $\frac{5}{12a}$ | $\frac{1}{12a}$ |
| 8 | − | 6·67 | 1·67 | 4·44 | 1·11 | 4·00 | 1·00 | 3·33 | 0·83 | 2·67 | 0·67 | 2·22 | 0·56 | $\frac{2}{3a}$ | $\frac{1}{6a}$ |
| Values of $b$ | | 11·7 | | 8·97 | | 8·48 | | 7·81 | | 7·21 | | 6·86 | | | |
| Values of $\theta$ | | 31° | | 42° | | 45° | | 50° | | 56° | | 61° | | | |

**Table 21.22**

Eight-panel Pratt truss

$+ = \text{compression}$   $- = \text{tension}$   $a = \dfrac{H}{L}$   $b = \sqrt{\left(\dfrac{1}{a}\right)^2 + 64}$

Values of $a$

| Member | Compression or tension | 0·075 $W$ and $W_c$ | Unit | 0·10 $W$ and $W_c$ | Unit | 0·125 $W$ and $W_c$ | Unit | 0·15 $W$ and $W_c$ | Unit | 0·175 $W$ and $W_c$ | Unit | 0·20 $W$ and $W_c$ | Unit |
|---|---|---|---|---|---|---|---|---|---|---|---|---|---|
| 1 | + | 5·83 | 0·83 | 4·37 | 0·62 | 3·50 | 0·50 | 2·92 | 0·42 | 2·50 | 0·36 | 2·19 | 0·31 |
| 2 | + | 10·0 | 1·67 | 7·50 | 1·25 | 6·00 | 1·00 | 5·00 | 0·83 | 4·28 | 0·71 | 3·75 | 0·62 |
| 3 | + | 12·5 | 2·22 | 9·37 | 1·67 | 7·50 | 1·33 | 6·25 | 1·11 | 5·36 | 0·95 | 4·69 | 0·83 |
| 4 | + | 13·3 | 3·33 | 10·0 | 2·50 | 8·00 | 2·00 | 6·67 | 1·67 | 5·71 | 1·43 | 5·00 | 1·25 |
| 5 | − | 6·80 | 0·97 | 5·60 | 0·80 | 4·95 | 0·71 | 4·55 | 0·65 | 4·30 | 0·61 | 4·13 | 0·59 |
| 6 | − | 4·86 | 0·97 | 4·00 | 0·80 | 3·54 | 0·71 | 3·25 | 0·65 | 3·07 | 0·61 | 2·95 | 0·59 |
| 7 | − | 2·92 | 0·97 | 2·40 | 0·80 | 2·12 | 0·71 | 1·95 | 0·65 | 1·84 | 0·61 | 1·77 | 0·59 |
| 8 | − | 0·97 | 0·97 | 0·80 | 0·80 | 0·71 | 0·71 | 0·65 | 0·65 | 0·61 | 0·61 | 0·89 | 0·59 |
| 9 | − | 5·83 | 0·83 | 4·37 | 0·62 | 3·50 | 0·50 | 2·92 | 0·42 | 2·50 | 0·36 | 2·19 | 0·31 |
| 10 | − | 10·0 | 1·67 | 7·50 | 1·25 | 6·00 | 1·00 | 5·00 | 0·83 | 4·28 | 0·71 | 3·75 | 0·62 |
| 11 | − | 12·5 | 2·22 | 9·37 | 1·67 | 7·50 | 1·33 | 6·25 | 1·11 | 5·36 | 0·95 | 4·69 | 0·83 |
| Values of $b$ | | 15·5 | | 12·8 | | 11·3 | | 10·4 | | 9·83 | | 9·43 | |
| Values of $\theta$ | | 31° | | 39° | | 45° | | 50° | | 55° | | 58° | |

Constant coefficients

| Member | Compression or tension | $W$ | $W_c$ | Unit |
|---|---|---|---|---|
| 12 | + | 3·5 | 2·5 | 0·5 |
| 13 | + | 2·5 | 1·5 | 0·5 |
| 14 | + | 1·5 | 0·5 | 0·5 |
| 15 | + | 1·0 | 0 | 1·0 |

General formulae

| Member | $W$ and $W_c$ | Unit |
|---|---|---|
| 1 | $\dfrac{7}{16a}$ | $\dfrac{1}{16a}$ |
| 2 | $\dfrac{3}{4a}$ | $\dfrac{1}{8a}$ |
| 3 | $\dfrac{15}{16a}$ | $\dfrac{1}{6a}$ |
| 4 | $\dfrac{1}{a}$ | $\dfrac{1}{4a}$ |
| 5 | $\dfrac{7b}{16}$ | $\dfrac{b}{16}$ |
| 6 | $\dfrac{5b}{16}$ | $\dfrac{b}{16}$ |
| 7 | $\dfrac{3b}{16}$ | $\dfrac{b}{16}$ |
| 8 | $\dfrac{b}{16}$ | $\dfrac{b}{16}$ |
| 9 | $\dfrac{7}{16a}$ | $\dfrac{1}{16a}$ |
| 10 | $\dfrac{3}{4a}$ | $\dfrac{1}{8a}$ |
| 11 | $\dfrac{15}{16a}$ | $\dfrac{1}{6a}$ |

**Table 21.23**

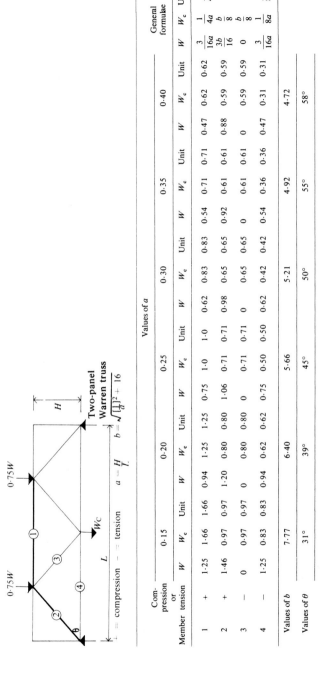

Two-panel Warren truss

$+$ = compression   $-$ = tension   $a = \dfrac{H}{L}$   $b = \sqrt{\left[\dfrac{1}{a}\right]^2 + 16}$

Values of $a$

| Member | Compression or tension | 0.15 W | 0.15 $W_c$ | 0.15 Unit | 0.20 W | 0.20 $W_c$ | 0.20 Unit | 0.25 W | 0.25 $W_c$ | 0.25 Unit | 0.30 W | 0.30 $W_c$ | 0.30 Unit | 0.35 W | 0.35 $W_c$ | 0.35 Unit | 0.40 W | 0.40 $W_c$ | 0.40 Unit | General W | General $W_c$ | General Unit |
|---|---|---|---|---|---|---|---|---|---|---|---|---|---|---|---|---|---|---|---|---|---|---|
| 1 | + | 1·25 | 1·66 | 1·66 | 0·94 | 1·25 | 1·25 | 0·75 | 1·0 | 1·0 | 0·62 | 0·83 | 0·83 | 0·54 | 0·71 | 0·71 | 0·47 | 0·62 | 0·62 | $\dfrac{3}{16a}$ | $\dfrac{1}{4a}$ | $\dfrac{1}{4a}$ |
| 2 | + | 1·46 | 0·97 | 0·97 | 1·20 | 0·80 | 0·80 | 1·06 | 0·71 | 0·71 | 0·92 | 0·65 | 0·65 | 0·88 | 0·61 | 0·61 | 0·88 | 0·59 | 0·59 | $\dfrac{3b}{16}$ | $\dfrac{b}{8}$ | $\dfrac{b}{8}$ |
| 3 | − | 0 | 0·97 | 0·97 | 0 | 0·80 | 0·80 | 0 | 0·71 | 0·71 | 0 | 0·65 | 0·65 | 0 | 0·61 | 0·61 | 0 | 0·59 | 0·59 | 0 | $\dfrac{b}{8}$ | $\dfrac{b}{8}$ |
| 4 | − | 1·25 | 0·83 | 0·83 | 0·94 | 0·62 | 0·62 | 0·75 | 0·50 | 0·50 | 0·62 | 0·42 | 0·42 | 0·54 | 0·36 | 0·36 | 0·47 | 0·31 | 0·31 | $\dfrac{3}{16a}$ | $\dfrac{1}{8a}$ | $\dfrac{1}{8a}$ |
| Values of $b$ | | 7·77 | | | 6·40 | | | 5·66 | | | 5·21 | | | 4·92 | | | 4·72 | | | | | |
| Values of $\theta$ | | 31° | | | 39° | | | 45° | | | 50° | | | 55° | | | 58° | | | | | |

**Table 21.24**

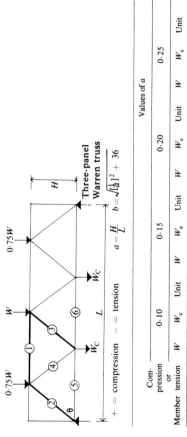

Three-panel Warren truss

$+$ = compression  $-$ = tension   $a = \dfrac{H}{L}$   $b = \sqrt{[4a]^2 + 36}$

| Member | Com-pression or tension | 0.10 W | 0.10 Wc | 0.10 Unit | 0.15 W | 0.15 Wc | 0.15 Unit | 0.20 W | 0.20 Wc | 0.20 Unit | 0.25 W | 0.25 Wc | 0.25 Unit | 0.30 W | 0.30 Wc | 0.30 Unit | General W | General Wc | General Unit |
|---|---|---|---|---|---|---|---|---|---|---|---|---|---|---|---|---|---|---|---|
| 1 | + | 2·92 | 3·33 | 1·67 | 1·94 | 2·22 | 1·11 | 1·46 | 1·67 | 0·83 | 1·17 | 1·33 | 0·67 | 0·97 | 1·11 | 0·56 | $\frac{7}{24a}$ | $\frac{1}{3a}$ | $\frac{1}{6a}$ |
| 2 | + | 2·43 | 1·94 | 0·97 | 1·87 | 1·49 | 0·75 | 1·63 | 1·30 | 0·65 | 1·50 | 1·20 | 0·60 | 1·43 | 1·14 | 0·57 | $\frac{5b}{24}$ | $\frac{b}{6}$ | $\frac{b}{12}$ |
| 3 | + | 0·97 | 0 | 0·97 | 0·75 | 0 | 0·75 | 0·65 | 0 | 0·65 | 0·60 | 0 | 0·60 | 0·57 | 0 | 0·57 | $\frac{b}{12}$ | 0 | $\frac{b}{12}$ |
| 4 | − | 0·97 | 1·94 | 0·97 | 0·75 | 1·49 | 0·75 | 0·65 | 1·30 | 0·65 | 0·60 | 1·20 | 0·60 | 0·57 | 1·14 | 0·57 | $\frac{b}{12}$ | $\frac{b}{6}$ | $\frac{b}{12}$ |
| 5 | − | 2·08 | 1·67 | 0·83 | 1·39 | 1·11 | 0·56 | 1·04 | 0·83 | 0·41 | 0·83 | 0·67 | 0·33 | 0·69 | 0·56 | 0·28 | $\frac{5}{24a}$ | $\frac{1}{6a}$ | $\frac{1}{12a}$ |
| 6 | − | 3·75 | 3·33 | 2·50 | 2·50 | 2·22 | 1·67 | 1·87 | 1·67 | 1·25 | 1·50 | 1·33 | 1·00 | 1·25 | 1·11 | 0·83 | $\frac{3}{8a}$ | $\frac{1}{3a}$ | $\frac{1}{4a}$ |
| Values of *b* | | 11·7 | | | 8·97 | | | 7·81 | | | 7·21 | | | 6·86 | | | | | |
| Values of *θ* | | 31° | | | 42° | | | 50° | | | 56° | | | 61° | | | | | |

**Table 21.25**

0·75W   0·75W   W   W   W

Four-panel Warren truss

$\dashv$ = compression   $- -$ = tension   $a = \dfrac{H}{L}$   $b = \sqrt{\left(\dfrac{1}{a}\right)^2 + 64}$

Values of $a$

| Member | Compression or tension | 0·075 | | | 0·10 | | | 0·125 | | | 0·15 | | | 0·175 | | | 0·20 | | | General formulae | | |
|---|---|---|---|---|---|---|---|---|---|---|---|---|---|---|---|---|---|---|---|---|---|---|---|---|
| | | $W$ | $W_c$ | Unit | $W$ | $W_c$ | Unit | $W$ | $W_c$ | Unit | $W$ | $W_c$ | Unit | $W$ | $W_c$ | Unit | $W$ | $W_c$ | Unit | $W$ | $W_c$ | Unit |
| 1 | + | 4·58 | 5·00 | 1·67 | 3·44 | 3·75 | 1·25 | 2·75 | 3·00 | 1·00 | 2·29 | 2·50 | 0·83 | 1·96 | 2·14 | 0·71 | 1·72 | 1·87 | 0·62 | $\frac{11}{32a}$ | $\frac{3}{8a}$ | $\frac{1}{8a}$ |
| 2 | + | 6·25 | 6·67 | 3·33 | 4·69 | 5·00 | 2·50 | 3·75 | 4·00 | 2·00 | 3·12 | 3·33 | 1·67 | 2·70 | 2·86 | 1·43 | 2·34 | 2·50 | 1·25 | $\frac{15}{32a}$ | $\frac{1}{2a}$ | $\frac{1}{4a}$ |
| 3 | + | 3·39 | 2·91 | 0·97 | 2·80 | 2·40 | 0·80 | 2·47 | 2·12 | 0·71 | 2·27 | 1·95 | 0·65 | 2·15 | 1·84 | 0·61 | 2·06 | 1·77 | 0·59 | $\frac{7b}{32}$ | $\frac{3b}{16}$ | $\frac{b}{16}$ |
| 4 | + | 1·94 | 0·97 | 0·97 | 1·60 | 0·80 | 0·80 | 1·41 | 0·71 | 0·71 | 1·30 | 0·65 | 0·65 | 1·23 | 0·61 | 0·61 | 1·18 | 0·59 | 0·59 | $\frac{b}{8}$ | $\frac{b}{16}$ | $\frac{b}{16}$ |
| 5 | − | 1·94 | 2·91 | 0·97 | 1·60 | 2·40 | 0·80 | 1·41 | 2·12 | 0·71 | 1·30 | 1·95 | 0·65 | 1·23 | 1·84 | 0·61 | 1·18 | 1·77 | 0·59 | $\frac{b}{8}$ | $\frac{3b}{16}$ | $\frac{b}{16}$ |
| 6 | − | 0 | 0·97 | 0·97 | 0 | 0·80 | 0·80 | 0 | 0·71 | 0·71 | 0 | 0·65 | 0·65 | 0 | 0·61 | 0·61 | 0 | 0·59 | 0·59 | 0 | $\frac{b}{16}$ | $\frac{b}{16}$ |
| 7 | − | 2·92 | 2·50 | 0·83 | 2·19 | 1·87 | 0·62 | 1·75 | 1·50 | 0·50 | 1·46 | 1·25 | 0·41 | 1·25 | 1·07 | 0·36 | 1·09 | 0·94 | 0·31 | $\frac{7}{32a}$ | $\frac{3}{16a}$ | $\frac{1}{16a}$ |
| 8 | − | 6·25 | 5·83 | 2·22 | 4·69 | 4·37 | 1·67 | 3·75 | 3·50 | 1·33 | 3·12 | 2·92 | 1·11 | 2·70 | 2·50 | 0·95 | 2·34 | 2·19 | 0·83 | $\frac{15}{32a}$ | $\frac{7}{16a}$ | $\frac{1}{6a}$ |
| Values of $b$ | | 15·5 | | | 12·8 | | | 11·3 | | | 10·4 | | | 9·83 | | | 9·43 | | | | | |
| Values of $\theta$ | | 31° | | | 39° | | | 45° | | | 50° | | | 55° | | | 58° | | | | | |

# Chapter Twenty-two
# Structural Design for Fire Resistance

## 22.1 INTRODUCTION

Research into the effects of fire on timber has reached a stage at which the performance of timber can be predicted with sufficient accuracy for a design method to be evolved to calculate the period of fire resistance of even completely exposed load-bearing solid timber or glulam sections. This is presented in BS 5268:Part 4:Section 4.1:1978, 'Method of calculating fire resistance of timber members'.

Also, considerable testing has been carried out on many combinations of standard wall components (load-bearing and non-load-bearing), and floor components of timber joists with various top and bottom skins. Fire resistance periods for the 'stability', 'integrity' and 'insulation' performance levels of these can be found in Building Regulations and various trade association and company publications. It is the intention that a design method to predict the stability, integrity and insulation performance levels of combinations without the necessity of a fire test will be presented in BS 5268: Part 4:Section 4.2 (at present under preparation).

'Stability' refers to the ability of an 'element' to resist collapse. 'Integrity' relates to resistance to the development of gaps, fissures or holes which allow flames or hot gases to breach the integrity of an element of construction. 'Insulation' relates to the resistance to the passage of heat which would cause the unheated face of an element of construction to rise to an unacceptable level. Not all elements are required to meet all three criteria. For example, an exposed beam or column would be required to meet only the 'stability' criterion, whereas a wall or floor designed to contain an actual fire would usually be required to meet all three criteria. A component (or requirement) rated as 30/30/15 indicates 'stability' for 30 minutes, 'integrity' for 30 minutes, 'insulation' for 15 minutes.

The intention of fire regulations in the UK is to save life, not property. The required fire resistance period quoted in regulations for a component or element represents a length of time during which it is considered that people can escape, and is related to the position in a structure, availability of fire appliances, access etc. Although the regulations are not designed to save property, it will be obvious from the figures given below that, if a fire on a clad or exposed timber component or element is extinguished quickly, there is a chance that the structure will continue to be serviceable and may be able to be repaired *in situ.*

The fire resistance 'stability' period for an element is basically the time during which it is required to support the *design* load whilst subjected to fire without *failure* taking place. In the case of beams, it is sensible for a deflection limit to be set (as well as an ultimate stress limitation) and BS 5268 : Part 4 : Section 4.1 gives this as $L/30$.

## 22.2 PROPERTIES OF TIMBER IN FIRE

As defined by BS 476 : Part 4, timber is 'combustible' but, as defined by BS 476 : Part 5, timber is 'not easily ignitable' in the sizes normally used for building purposes.

Timber having a density in excess of 400 kg/m³ (i.e. most structural softwoods and hardwoods) falls naturally into the Class 3 'surface spread of flame' category of BS 476 : Part 7 and can be upgraded by specialist surface or pressure treatments to Class 1 or Class 0.

Without a source of flame or prolonged pre-heating, timber will not ignite spontaneously until a temperature of around 450–500°C is reached. It is worth realising that this is around the temperature at which steel and aluminium lose most of their useful strength, hence the photographs one sees of timber members still standing after a fire and supporting tangled steel secondary members.

Once timber has ignited, as it burns it builds-up a layer of charcoal on the surface and this, being a good thermal insulant, protects the timber immediately beneath the charred layer. Figure 22.1 is reproduced by permission of the Swedish Finnish Timber Council and shows a typical temperature plot through the section of a timber member which has been exposed in a furnace to a temperature of some 930°C. The temperature immediately under the charcoal is less than 200°C and the temperature in the centre of the member is less than 90°C. Timber has a very low coefficient of thermal expansion and the strength properties of the uncharred part of the member are virtually unaffected, even during the fire period. After the fire is extinguished the properties of the residual section can be considered not to have been affected by the fire. Obviously this gives the possibility of developing a design method based on the strength of the residual section.

Another important positive property of the performance of timber in fire is that the charring rate when exposed to a standard fire test is sufficiently constant and predictable for values to be quoted. For the softwoods quoted in BS 5268 : Part 2 (except Western Red Cedar) the notional charring rates are taken as 20 mm in 30 minutes and 40 mm in 60 minutes, with linear interpolation and extrapolation permitted for times between 15 and 90 minutes.

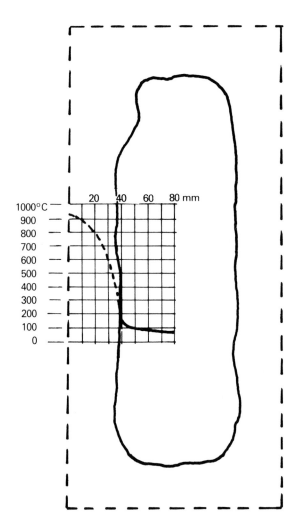

Fig. 22.1

The 30 minute value for Western Red Cedar is 25 mm, and for oak, utile, keruing, teak, greenheart and jarrah is 15 minutes.

The charring rates for glulam may be taken as the same as solid timber providing the adhesive used in assembly is resorcinol-formaldehyde, phenol-formaldehyde, phenol/resorcinol-formaldehyde, urea-formaldehyde, or urea/melamine-formaldehyde.

A column exposed to the fire on all four faces (including one which abuts or forms part of a wall which does not have the required fire resistance) should be assumed to char on all four faces at 1.25 times the rates given above. However, when a column abuts a construction which has a fire resistance at least as high as the required fire resistance of the column, it may be assumed that this prevents charring of the member at that point or surface (see Fig. 22.2).

The corners of a rectangular section become rounded when subjected to

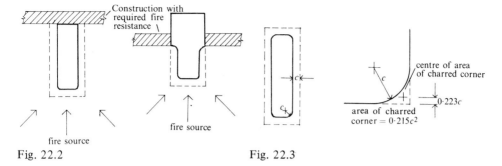

Fig. 22.2                                    Fig. 22.3

fire. The radius has been found to approximate to the depth of charring (see Fig. 22.3) but, for exposure periods of 30 minutes or less, if the least dimension of a rectangular *residual* section is not less than 50 mm, the rounding is considered insignificant and can be disregarded.

Metal fasteners within the charring line will conduct heat into the timber. They must either be fully protected, or charring of each individual piece of timber must be considered to act on all faces, even if two timber surfaces are butting.

## 22.3   DESIGN METHOD FOR THE RESIDUAL SECTION

Knowledge of the charring rate and the fire stresses enables a design method to be evolved to check whether the chosen section will withstand the design load for the period of required fire resistance. In the case of exposed sections, this is often referred to as the 'sacrificial timber' design method or the 'residual section' design method. The designer is referred to BS 5268: Part 4 : Section 4.1 for detailed rules for this method.

In this method the load-bearing capacity of a flexural member should be calculated using the residual section and permissible stresses equal to the permissible long term dry stresses (as given in the code) $\times$ 2.25 when the minimum initial breadth (i.e. width not thickness) is 70 mm or more, and $\times$ 2.00 (instead of 2.25) when less than 70 mm.

The designer must judge the degree of load sharing which will remain at the end of the charring period in choosing values of $K_8$, $E$ etc. to use in calculations. The deflection should be limited to $L/30$. The notional charring rate is as detailed in § 22.2.

In checking a column, no restraint in direction (as against positional restraint) should be assumed in determining the effective length, unless this can be assured at the end of the charring period. $L_e/i$ can be up to 250. The permissible compression stress parallel to the grain can be taken as 2.00 $\times$ the grade long term dry stress modified for slenderness. The charring rate is as detailed in § 22.2.

The residual section of a tension member should be determined using a charring rate 1.25 times the notional rates given in §22.2, and the permissible stress may be taken as 2.00 X the permissible long term dry stress (including $K_{14}$ based on the reduced greater dimension).

Where there is combined bending and compression (or combined bending and tension) one presumes that the size of the residual section should be calculated using a charring rate of 1.25 times the notional rate. The permissible stresses are calculated as above and combined by the usual interaction formula.

## 22.4   STRESS GRADE

As the outside of the section chars away, the knot area ratio changes, and this affects the stress grade to a certain extent. However, in the context of the accuracy of the residual section method, it is suggested that the designer assumes that the stress grade remains the same. Where there has been a shake, charring will tend to increase the stress grade. Charring will also usually reduce knot area ratios.

## 22.5   PLY WEB BEAMS

The residual section method is generally not appropriate in checking ply web beams when a fire resistance is required. When using ply web components for which a fire resistance is required, it is normal to protect them by a ceiling or cladding. In a special case a fire test might be used to validate a beam design which is to be exposed.

## 22.6   CONNECTIONS

The design method outlined in §22.3 is for solid timber or glulam sections, and special consideration must be given to the resistance of connections, particularly if metal is included in the connection, and particularly if the connection is required to develop a fixing moment. The effect of fire on a frame may be to induce failure by changing moment connections into hinges.

It may be possible to protect connections locally, either by adding material or ensuring that all the connection unit is behind the charring line. If even a part of a steel connection is within the charring area, it will tend to conduct heat into the section. The engineer may justify a particular connection by design on the net section or by test under conditions of fire.

BS 5268:Part 2:Section 4.1 gives certain 'deemed to satisfy' clauses for metal hangers protected by plasterboard.

## 22.7 TESTING FOR FIRE RESISTANCE

When the residual section method is not appropriate, or when desired, one can test for fire resistance. The test procedure for the UK is detailed in BS 476:Part 8. It is worth realising that the fire test described in BS 476:Part 8 is arbitrary and does not necessarily reflect exactly what will occur in a real fire, although it has proved a useful yardstick for many years.

## 22.8 PROPRIETARY TREATMENTS FOR SURFACE SPREAD OF FLAME

There are several proprietary chemical fluid methods of treating timber to up-grade the surface spread of flame classification (see BS 476:Part 7) from Class 3 to Class 1 (even to Class 0). Current formulations do not increase the fire resistance to any measurable extent.

An up to date list of manufacturers of this type of product can be obtained from the British Wood Preserving Association.

## 22.9 CHECK ON THE FIRE RESISTANCE OF A GLULAM BEAM

Check whether an exposed 90 X 180 mm glulam beam (Fig. 22.4) of European Whitewood subject to a design bending moment of 3.0 kN.m (4 kN UDL over 6 m), has a fire resistance of 30 minutes. The beam is not part of a load-sharing system. The timber grade is LB for all laminations. Full lateral stability can be assumed to be given by the decking, which also has a fire resistance of at least 30 minutes. Beam cambered 10 mm.

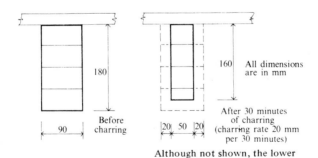

Before charring

90

180

After 30 minutes of charring (charring rate 20 mm per 30 minutes)

160   All dimensions are in mm

20  50  20

Although not shown, the lower corners will have rounded.

Fig. 22.4

After 30 minutes, the residual width is 50 mm therefore rounding can be disregarded (§ 22.2).

$$Z \text{ of depleted section} = 50 \times 160^2/6 = 0.213 \times 10^6 \text{ mm}^3$$
$$\text{Fire bending stress} = 2.25 \times \sigma_{m,g} \times K_{15}$$
$$= 2.25 \times 7.50 \times 1.26 = 21.26 \text{ N/mm}^2$$

(still assuming four laminates even though one is badly charred).

$$\text{Moment capacity} = 21.6 \times 0.213 \times 10^6 = 4.60 \times 10^6 \, \text{N.mm}$$
$$= 4.60 \, \text{kN.m}$$

which is greater than the design moment.

Check the deflection $\not> L/30 = 6000/30 = 200 \, \text{mm}$

$$I = 50 \times 160^3/12 = 17.07 \times 10^6 \, \text{mm}^4$$
$$E = (E_{\text{mean}} \text{ for SS}) \times K_{20} = 10\,500 \times 0.90$$
$$= 9450 \, \text{N/mm}^2$$
$$EI = 161 \, \text{kN.m}^2$$

$$\text{Bending deflection} = \frac{5FL^3}{384EI} = \frac{5 \times 4 \times 6^3}{384 \times 161} = 0.070 \, \text{m} = 70 \, \text{mm}$$

$L/h = 6000/160 = 37.5$, therefore, from Fig. 4.26 it can be seen that shear deflection is less than $2\frac{1}{2}\%$ of bending deflection. Say total deflection = 72 mm, which is satisfactory even without taking account of the 10 mm camber, therefore the section has 30 minute fire resistance. Actually, the phrase in BS 5268 : Part 4 : Section 4.1 which relates to the deflection limit is 'ability to resist deflection during the fire test to span/30' therefore it is doubtful if the designer could justify deducting camber from calculations of deflection. One feels that the limit is intended to be a (generous) limit on the displacement/deformation of the member, rather than the normal interpretation of deflection.

## 22.10    CHECK ON THE FIRE RESISTANCE OF A GLULAM COLUMN

Check whether a glulam column 115 $\times$ 180 mm (Fig. 22.5) of European Whitewood exposed on all four sides has a fire resistance of 30 minutes. Grade is LB throughout.

The charring rate is 1.25 times 20 mm in 30 minutes. The size of the residual section is shown in Fig. 22.5. Because the lesser dimension of the residual section is more than 50 mm, rounding of corners can be disregarded.

The residual section is checked as a normal column but with the permissible (fire) compression stress increased by 2.00, and with the effective length determined assuming no directional restraint at ends.

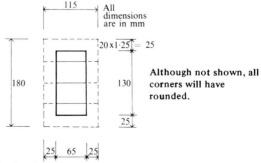

Fig. 22.5: *Section as depleted after 30 minutes.*

# Chapter Twenty-three
# Considerations of Overall Stability

## 23.1   GENERAL DISCUSSION

It is not sufficient to design individual components for what one may regard as their isolated loading conditions without considering how they are affected by, and how they affect the overall stability of the building or frame in which they occur. For example, a horizontal beam which is also part of the horizontal bracing of a building, may have to carry axial compression or tension due to side wind on the building; a column which is part of a vertical bracing frame may have to carry tension or additional compression due to wind on the end of the building; the bottom tie of a truss, or the beam of a portal frame may have to take compression or additional tension from the action of side wind.

Strictly speaking, the stability calculations, calculations of forces in bracing systems etc. should be carried out before individual components are designed. In a major or tall structure it is essential for this to be done, but in many small buildings it is quite normal to design individual components first for the localised loading, and subsequently to check those which are called upon to carry additional loads from stability considerations. The stability loads are usually of short or very short term duration from wind, and consequently the component as designed for localised loading is usually adequate to carry the additional stability loading, if restrained laterally.

It is dangerous to forget considerations of stability. The classic error in disregarding stability calculations is the 'four-pin frame'. This can occur if a simply supported beam is supported at each end by vertical columns which have no fixity at either end, in a building where the roof is neither provided with horizontal bracing nor acts as a horizontal diaphragm (Fig. 23.1). On an

Fig. 23.1

architect's drawing this looks sound, but failure occurs as sketched on the right of Fig. 23.1 unless the roof is braced or is a horizontal membrane, or unless fixity is provided at least at one joint.

Before proceeding with the calculations of any building, the designer must decide whether the design is to be based on the building being permitted to sway or being restrained from swaying. This question is absolutely fundamental to any building design, yet so often is not stated (although often it is 'understood').

## 23.2   NO SWAY

When a designer has a choice it is normally more economical to design a building not to sway. This requires that the roof plane is either braced or acts as a horizontal diaphragm or membrane, and that the ends of the bracing or diaphragm are connected through stiffened end walls and/or cross-walls to the ground (Fig. 23.2).

With the type of construction sketched in Fig. 23.2, the columns can be hinged top and bottom because the overall stability comes from the bracing. If the columns are rather tall, the decision may be taken to fix the base to reduce deflection (Fig. 23.3).

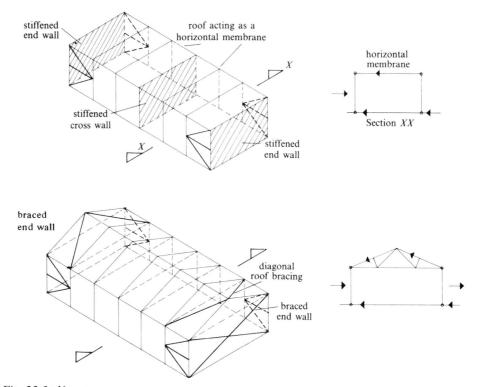

Fig. 23.2: *No cross sway.*

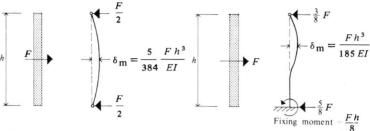

Fig. 23.3: *No sway.*

Whether or not a building is designed to permit local sway, it is usually restrained from sway in the longitudinal direction by roof and side-wall bracing, diaphragms, or stiffened panels.

In deciding the basic design of a building, it is necessary to distinguish between 'sway' and 'no sway', but the designer should realise that even when a building can be designed for 'no sway' due to there being adequate bracing or membrane effect, the bracing or membrane deforms in itself. A bracing frame deforms like a lattice beam, and a membrane deforms rather like a ply web beam. However the proportion of span to effective depth, or of height to width, is usually rather small, which generally leads to small deformations. In small buildings this can usually be disregarded. In a major structure, however, the effect should be calculated and checked for acceptability.


## 23.3   WITH SWAY

With several types of building, the designer has no choice but to permit the building to sway and to design accordingly, by providing the necessary fixing moments and sufficient stiffness to limit sway. For example, a tower or mast can not be prevented from swaying (other than by guy ropes). A long narrow building without cross-walls, or split by expansion bays, must be designed for sway.

When designing a building for sway, it is quite usual to provide two-pin or fully fixed portal frames. The types of portal shown in Fig. 23.4 can be

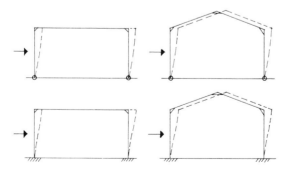

Fig. 23.4

analysed by the formulae of Kleinlogel and the *Steel Designers' Manual*. If, however, roof bracing is provided to prevent sway at the eaves, these formulae do not apply and the portal can be designed using simple moment distribution without sinking supports.

A three-pin frame, as sketched in Fig. 23.5, is designed as a three-pin arch (see § 17.6).

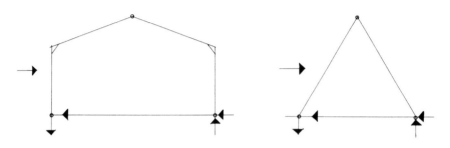

Fig. 23.5

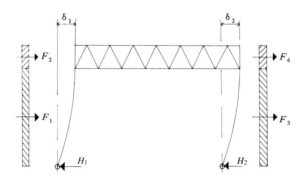

Fig. 23.6

The frame sketched in Fig. 23.6 will sway as indicated under the action of side loading. Whether or not the columns have equal stiffness and whether or not the loading on each side is equal, the columns will sway to the same degree at the eaves. This sway is the algebraic sum of the deflection of a cantilever affected by the side loading and by the horizontal thrust at the base. A point load $H$ at the end of a cantilever causes a bending deflection at the end of $HL^3/3EI$. Similarly, a UDL of total $F$ on a cantilever causes a deflection of $FL^3/8EI$.

$$H_2 = F_1 + F_2 + F_3 + F_4 - H_1$$

$$\delta_1 = \frac{H_1 h^3}{3EI} - \frac{F_1 h^3}{8EI} = \delta_2 = \frac{H_2 h^3}{3EI} - \frac{F_3 h^3}{8EI}$$

where $h$ = the height to the underside of the cross beam.

The force in the lower connection to the beam is found by taking moments about the top where the moment is zero. The force in the top connection is found by equating horizontal forces.

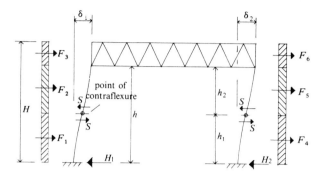

Fig. 23.7

The frame sketched in Fig. 23.7 will sway as indicated. Even though the horizontal loadings on each side may differ, the points of contraflexure may be assumed to occur at the same level. The horizontal shear at the points of contraflexure will be the sum of the horizontal loadings above the level of the points of contraflexure shared to the columns in proportion to their stiffness (i.e. shared equally if columns are of equal stiffness).

$$h_1 = \frac{h}{2}\left(\frac{h + 2H}{2h + H}\right)$$

For columns of equal stiffness:

$$S = \frac{F_2 + F_3 + F_5 + F_6}{2}$$

$$\delta_1 = \frac{1}{EI}\left(\frac{Sh_2^3}{3} - \frac{F_2 h_2^3}{8} + \frac{Sh_1^3}{3} + \frac{F_1 h_1^3}{8}\right)$$

$$\delta_2 = \frac{1}{EI}\left(\frac{Sh_2^3}{3} - \frac{F_5 h_2^3}{8} + \frac{Sh_1^3}{3} + \frac{F_4 h_1^3}{8}\right)$$

$$\delta_1 = \delta_2 \qquad H_1 = S - F_1 \qquad H_2 = S + F_4$$

The frame sketched in Fig. 23.8 will sway as indicated. $h_1$ is calculated as for Fig. 23.7.

$$\delta_1 = \frac{H_1 h^3}{3EI} - \frac{F_1 h^3}{8EI}$$

$$\delta_2 = \frac{Sh_2^3}{3EI} - \frac{F_4 h_2^3}{8EI} + \frac{Sh_1^3}{3EI} + \frac{F_3 h_1^3}{8EI}$$

$$\delta_1 = \delta_2 \qquad H_2 = S + F_3$$
$$H_1 + H_2 = F_1 + F_2 + F_3 + F_4 + F_5$$

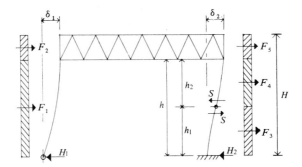

Fig. 23.8

## 23.4   BRICK WALLS

With single-storey buildings with brick walls, a designer very rarely considers sway to occur, and providing that the walls are designed to CP 121 : Part 1 : 1973 and that piers or cross-walls occur at regular intervals, and that there are end walls, it is usually fair to assume no sway. However, when the distance between cross-walls is quite long and there are no piers, the roof should be designed as a diaphragm or bracing system. In doing so, the diaphragm is usually considered to take half the wind loading on the side walls, although there is a case for treating the wall as a propped cantilever, thus transferring three-eighths of the side wind to the roof, providing that the bending moment induced at the base is not sufficient to overcome the vertical downward loading and cause excessive tension along either face of the wall at the level of the damp-proof course.

## 23.5   HORIZONTAL DIAPHRAGMS

### 23.5.1   General discussion

With timber construction, a designer often uses a horizontal diaphragm as a means of bracing a building. Such a diaphragm takes the form of plywood nailed to cross noggings and beams or joists at fairly close centres, or a similar stiffened construction such as solid decking. Providing that the building is single or double storey, and that the ratio of length between cross bracing walls and the span of the beams is fairly small (say not exceeding 3) it is quite normal for a designer to rely on past experience and not produce detailed calculations to prove the adequacy of the diaphragm. However, a designer who is not fully familiar with diaphragm action should carry out a check, particularly of fixing details. The example given in § 23.5.2 may be considered as one method of doing so. In the example, the stiffened plywood is treated as a deep thin beam spanning between cross-walls, the construction hardly seeming to justify a more complex design method. If a roof contains a large number of holes for roof lights, the designer should take these into consideration, perhaps providing localised additional bracing.

The Timber Research and Development Association also give guide rules for diaphragm effect in *Timber Frame Housing–Structural Recommendations.*

### 23.5.2   Example of checking a horizontal diaphragm

Check the adequacy of the roof diaphragm sketched in Fig. 23.9. The diaphragm is 12.5 mm Canadian softwood plywood nailed to noggings which are at 0.4 m centres, and beams at 1.2 m centres. Beams 6 m span. Building 21.6 m long, no cross-walls. Horizontal loading at roof level from wind is a total of 0.7 kN per metre run of building.

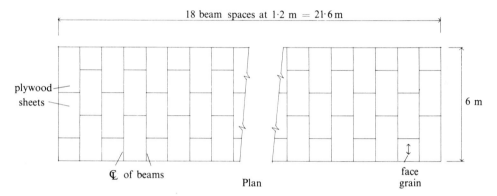

Fig. 23.9

$$\text{Total load } F = 0.7 \times 21.6 = 15.12 \text{ kN}$$
$$\text{Reactions} = 7.56 \text{ kN}$$

$$M = \frac{15.12 \times 21.6}{8} = 40.8 \text{ kN.m}$$

$$\text{Minimum thickness of plywood} = 12.0 \text{ mm}$$

$$Z \text{ of plywood 'beam'} = \frac{12.0 \times 6000^2}{6} = 72 \times 10^6 \text{ mm}^3$$

$$\text{Bending stress} = \frac{40.8 \times 10^6}{72.0 \times 10^6} = 0.57 \text{ N/mm}^2$$

The permissible plywood stress for bending in the plane of the roof is the permissible tension or compression stress (whichever is the lesser), face grain perpendicular to the direction of loading (see Table 42 of BS 5268 : Part 2 for $\sigma_{t,adm,tra}$).

$$1.98 \times K_3 = 1.98 \times 1.75 = 3.47 \text{ N/mm}^2$$

which is well above the actual stress.

$$I \text{ of plywood 'beam'} = \frac{12.0 \times 6000^3}{12} = 216\,000 \times 10^6\,\text{mm}^4$$

Modulus of elasticity in tension or compression, face grain perpendicular to span, is $3350\,\text{N/mm}^2$ (from Table 42 of BS 5268:Part 2).

$$\therefore EI = 723\,600\,\text{kN.m}^2$$

$$\delta_m = \frac{5 \times 15.12 \times 21.6^3}{384 \times 723\,600} = 0.0027\,\text{m}$$

This value $0.0027\,\text{m} = 0.000\,125$ of the span.

Even accepting that the joints are not glued and that slip can take place, and that shear deflection has not been calculated, this value seems acceptable, particularly bearing in mind that the wind gust is only for 5 seconds, and that there will be a certain amount of composite action.

$$\text{Panel shear} = \frac{3 \times 7560}{2 \times 12.0 \times 6000} = 0.16\,\text{N/mm}^2$$

$$\text{Permissible} = 0.93 \times K_3 = 0.93 \times 1.75 = 1.63\,\text{N/mm}^2$$

which is acceptable.

With the layout of plywood sheets shown in Fig. 23.9, there is a continuous joint every 1.2 m. The nailing must be adequate.

When considering the plywood as a horizontal diaphragm spanning between end walls it is simpler if one can obtain sufficient strength without having to provide a 'flange' along each edge. However, the absence of a flange means that the nail connections from the plywood into the main beams (or framing on top of the beams) are called upon to take a bending effect as well as a shearing effect. The bending effect is maximum at the centre of the building. Shear is a maximum at the ends where the load from wind must be transmitted 'down' into the end walls or framing.

$$\text{The maximum shear at ends} = 7560\,\text{N}$$
$$\text{The maximum moment} = 40.8\,\text{kN.m}$$

Assume forty nails along each line (each side of the joint between the plywood sheets).

Shear per nail $= 7560/40 = 189\,\text{N}$ parallel to face grain in the plywood and parallel to grain in the sub-framing (or beam).

From Table 18.5 it can be seen that, for very short term loading with $K_{48} = 1.25$ for the pointside loading, a 2.65 mm diameter nail with a pointside penetration of 25 mm would be adequate ($180 \times 1.25 = 225\,\text{N}$) if only shear has to be considered.

From Table 18.7 it can be seen that, for very short term loading with $K_{48} = 1.25$ for the headside loading in 12 mm thick Canadian softwood plywood, a 2.65 mm diameter nail would be adequate providing it has a length of 40 mm ($145 \times 1.25 = 181\,\text{N}$).

At mid span, moment $= 40.8$ kN.m (shear $= 0$).

For a single row of $N$ fasteners at pitch $p$ resisting a moment $M$, the lateral load on the extreme nail is:

$$\frac{6M}{pN^2}$$

Although forty nails are satisfactory for the shear, try sixty nails (at 100 mm centres).

Load on the extreme nail from bending

$$= \frac{6 \times 40.8 \times 10^6}{100 \times 60^2} = 680 \text{ N}$$

From the calculations for shear above it can be seen that it would be difficult to develop this load per nail. If 120 nails are used the load reduces to 170 N which is more manageable. At quarter span, $M = 0.75 \times 40.8 = 30.6$ kN.m and shear $= 3780$ N. This condition may be more critical and should be checked.

The alternative to dense nailing is to utilise 'flanges' to the diaphragm, placed along each of the longer sides of the building, providing joints in the 21.6 m length. The section required to take the bending moment should have an area $A$:

$$= \frac{M}{\sigma_{adm} \times d_e}$$

where $d_e =$ the distance between centres of flanges (say $= 5940$ mm)
and $\sigma_{adm}$ (based on tension for SC3 timber)
$\qquad = 4.50 \times K_3 = 4.50 \times 1.75 = 7.875$ N/mm² (disregarding $K_{14}$ for simplicity)

$$\therefore A = \frac{40.8 \times 10^6}{7.875 \times 5940} = 870 \text{ mm}^2$$

A $60 \times 47$ mm member would give ample area.

With a 'flange' to the diaphragm the nails at the joint between plywood sheets are called upon only to carry shear, and the deflection calculated above as 0.0027 m would be reduced.

To calculate the load on the nails between the plywood and the edge flanges, refer to § 8.8.1 and modify equation (8.7), by deleting $h_f$, to arrive at a load per unit length rather than a stress. The nail fixing is on one plane therefore $n$ in the formula becomes 1.0.

$$\tau_r \text{ per unit length} = \frac{F_v S_{Xf}}{I_X \times 1.0}$$

where $F_v$ = the external shear at the section

$S_{Xf}$ = the first moment of area of one flange about the $XX$ axis (of the roof on plan)

$I_X$ = the second moment of area of the section about the $XX$ axis.

Take the flange as shown in Fig. 23.10 as GS Whitewood.

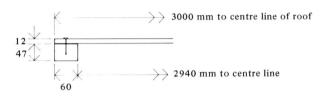

Fig. 23.10

$$S_{Xf} = 47 \times 60 \times 2970 = 8.375 \times 10^6 \, \text{mm}^3$$

$I_X$ (disregarding self $I$ of flange)

$$= \frac{12 \times 6000^3}{12} + 47 \times 60 \times 2970^2$$

$$= 216 \times 10^9 + 25 \times 10^9 = 241 \times 10^9 \, \text{mm}^4$$

$$\tau = \frac{7560 \times 8.375 \times 10^6}{241 \times 10^9} = 0.26 \, \text{N per mm run}$$

$$= 260 \, \text{N per metre run}$$

From the nail values used above it can be seen that nails at 100 or 150 mm spacing would be more than adequate to carry this shear.

A joint would have to be provided at any discontinuity in the 'flange'.

## 23.6    VERTICAL DIAPHRAGMS

### 23.6.1    General discussion on design method

With timber construction it is very common to brace a wall by cladding vertical studs placed at close centres (say 400–600 mm) with plywood sheets to act as a vertical diaphragm. Medium density fibreboard is also being used as a bracing membrane, or even plasterboard, chipboard and bitumen-impregnated fibreboard when the racking loads are not large. Work on establishing a design method which will allow the racking performance of sheathed walls to be predicted is proceeding within the CSB/32 (BS 5268) committee. In the meantime, designers are using several simplified methods of design or carrying out prototype testing. Simplified design methods are described in *Timber Frame Housing–Structural Recommendations* by TRADA and *Design and Detail of Timber Stud Walls* by the Swedish Finnish Timber Council, and information on racking is available from the Princes Risborough Laboratory, TRADA and the Fibre Building Board Development Organisation Ltd.

What is already known is that the normal magnitude of racking forces which board materials will be called upon to resist is unlikely to deform the shape of the board to any significant extent. The racking of the panel is affected much more by localised indentation of the fixings. The normal type of fixing of plywood sheathing to the wall studs is by nails at 75, 100 or 150 mm centres along the perimeter of the board, with wider spacing along any intermediate stud or cross noggings. If staples are used as an alternative to nails, they should be placed at closer centres. If the racking loads are high and maximum resistance is required, the designer may find it advantageous to glue the sheathing to the studs.

In addition to the racking and overturning effects discussed below, there is horizontal shear due to wind on the face of the panel which must be transmitted to the base (and top) usually by nails or ballistic nail fixings.

In considering horizontal loads acting at the base of a stud wall the designer may consider the restraining effect of friction between the stud wall/damp-proof course/base (providing of course that there is no residual uplift for the design case being considered). Tests have shown the coefficient of friction to be as high as 0.4 and a figure of 0.2 may be considered to be well on the safe side. However, even if the designer decides to take advantage of friction, sufficient locating fixings will be used.

Narrow panels with large openings may have to be disregarded in considering racking strength. A certain amount of engineering judgement is called for.

### 23.6.2  Simple design methods

Designers of timber-framed construction have evolved several relatively simple methods of checking the adequacy of sheathed timber stud walls as discussed below. In addition, based on accumulated tests, the Timber Research and Development Association have published permissible horizontal racking loads parallel to panels which are reproduced in Table 23.1 by permission of TRADA. The figures are for permissible racking loads in kN per metre of plan length for plywood diaphragms without openings, fixed to dry softwood of Group S2 species of CP 112:Part 2 which, related to BS 5268: Part 2 are:

  imported European Whitewood and Redwood
  imported North American Spruce-Pine-Fir
  imported North American Hem-Fir
  British-grown Scots Pine.

**Table 23.1**

| Nominal plywood thickness (mm) | Nail size | Permissible shear load in kN per metre run for nail spacing on perimeter members of: | | |
|---|---|---|---|---|
| | | 150 mm | 100 mm | 75 mm |
| 8 | 3.00 mm dia × 50 mm | 2.44 | 3.65 | 4.09 |
| 9.5 | 3.35 mm dia × 65 mm | 3.50 | 5.15 | 5.84 |
| 12.5 | 3.75 mm dia × 75 mm | 4.13 | 6.23 | 7.11 |

For the figures to apply, noggings must occur at all edges of plywood not fixed to a stud or plate, and the nailing is to be as shown in Table 23.1. The centres of nails to members intermediate in each sheet of plywood is not to exceed 300 mm. Studs are to be at centres not exceeding 600 mm, and panels are to be adequately fixed down against any residual overturning moment.

Whether using a design method or Table 23.1 values the designer must distinguish between racking effects and any residual overturning effect (which is not covered by Table 23.1). By reference to Fig. 23.11 it can be seen that the racking force acts on the sheathing, the fixings between the sheathing and the studs and other framing, and on the fixings into the base. On the other hand, any residual overturning effect (causing rotation about Point O in Fig. 23.11 rather than horizontal shear above Point O) by-passes the sheathing; other than perhaps the nailing between the sheathing and the bottom plate (see Fig. 23.17) if holding-down fixings are not used between the studs and the foundations.

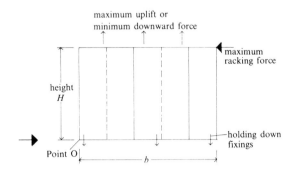

Fig. 23.11

### 23.6.3    Consideration of the racking effect

If the designer decides to develop a design method to consider racking it is necessary to decide how the racking force will be shared to the portions of a stud wall.

If a wall is made in one long panel with no significant openings, there is little doubt that the racking force is shared equally per metre run of wall.

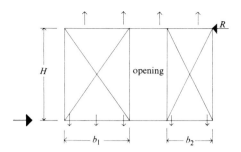

Fig. 23.12

If two individually assembled panels are joined only at the top by a horizontal member (as indicated in Fig. 23.12) with a complete opening between the panels, the racking force $R$ will be shared to each panel. The designer may decide that force $R$ should be shared in proportion to the width of each panel. Although it is probably more theoretically correct to share in proportion to the relative stiffness of each separate panel, slip in the connections between the sheathing and the framing would tend to invalidate this proportion, and tests tend to confirm that it is sufficiently accurate to share racking force in proportion to width.

If a wall is constructed of two or more individually assembled panels without full height gaps between them, and adjacent panels are not fixed together through their perimeter studs, the racking force will tend to deform the panels as shown by the firm lines in Fig. 23.13, and the force $R$ will be shared to each panel in relation to its width or stiffness as discussed above.

The dotted lines indicate lines taken up
by unstiffened panels

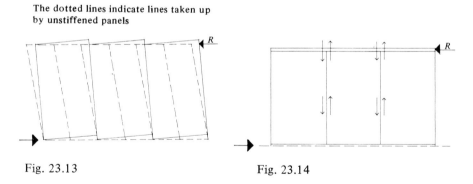

Fig. 23.13                                      Fig. 23.14

However, it is virtually certain that adjacent panels will be fixed together through perimeter studs and will have a continuous head runner as indicated in Fig. 23.14. In such a case it is reasonable to assume that racking force $R$ is shared equally per metre run of panel (certainly where there are no large openings).

Where a designer is using a design method (rather than simply referring to the figures in Table 23.1) it is necessary to check the strength of the fixings between the sheathing and the framing. To do this, the designer must share the racking force $R$ to each sheet of plywood (or other sheathing) which can move separately from other sheets.

If a racking force $R_s$ acts on one sheet it will cause a panel shear stress in the panel of:

$$\frac{3R_s}{2bt}$$

where $t$ = the thickness of the plywood
and $b$ = the width of the sheet.

Also, the racking action causes a moment on the nail group fixing the sheathing to the studs and the horizontal members. This moment is resisted by the polar modulus of the nail group. The nails in the extreme corners

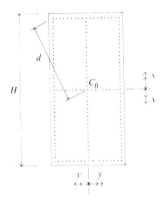

Fig. 23.15

(Fig. 23.15) are those most heavily loaded. The maximum lateral load

$$= \frac{R_s H d}{2(\Sigma x^2 + \Sigma y^2)}$$

where $x$ = the dimension from the horizontal axis through $C_0$ to each nail
$y$ = the dimension from the vertical axis through $C_0$ to each nail
$d$ = the dimension from $C_0$ to the furthest nail.

With this method, the designer can ensure that there is no over-stress, but the horizontal deflection or sway of the panel cannot be calculated unless the slip characteristic of the fasteners is known and evaluated as a group. To make the apparent conservative assumption that the $EI$ value of the plywood only could be taken in a sway calculation does not necessarily give a conservative answer, because the rotation rather than the deformation of the plywood panel is what contributes mostly to sway. The designer is referred to the values quoted in Table 23.1 and to information of a similar nature available from board manufacturers' associations. These contain figures per unit width of panel which lead to acceptable deformations. Otherwise, the designer is advised to carry out prototype testing. One alternative is to assemble selected panels with adhesive and assume that all racking forces are taken by these panels as simple cantilevers, having a construction similar to a ply web beam.

### 23.6.4 Consideration of residual overturning effect

In considering whether or not it is necessary to provide positive holding-down fixings into the foundation (or structure) below the stud panel or stud wall, it is necessary to assume the worst combination of minimum vertical residual downward loading (from dead loading, imposed loading and wind loading, which may of course lead to a residual vertical uplift) and maximum horizontal loading. In calculating the load on holding-down connections it is usually good practice to increase the overturning moment by a factor. Perhaps surprisingly, values for the factor are not given in the Codes, but it is relatively common to use a value of 1.33 as below.

Basically, three design conditions can occur which are illustrated in Fig. 23.16. For ease of explanation, they are shown for panels having only three studs.

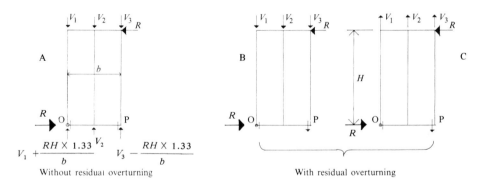

Fig. 23.16

In Case A of Fig. 23.16 the vertical downward loading is of sufficient magnitude to overcome the overturning moment $RH \times 1.33$ without any residual uplift occurring at P. The vertical loads are considered to be spread at the top to the stud positions and to pass down the studs to the base, without causing stress in the sheathing or in the fasteners between the sheathing and the studs, or any racking of the sheathing. The only load taken by the fixings between the bottom plate and the base is a shear from load $R$ (i.e. if there are two fixings, the horizontal shear to each is $0.5R$ and there is no tension in either fixing).

In Case B of Fig. 23.16 the residual vertical loading is still downwards as in Case A, however, it is not of sufficient magnitude to overcome the overturning moment $RH \times 1.33$, therefore there is a tendency for the panel to lift at P, assuming rotation about Point O.

There are two methods of preventing the panel from lifting. One is to provide sufficient fixings between the sheathing and the lower plate of the panel (see Fig. 23.17) to take the overturning effect (in addition to the load indicated in Fig. 23.17 which is a maximum for nail N), and to provide

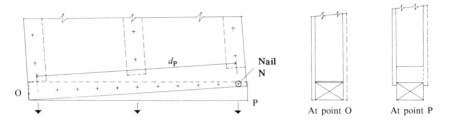

Fig. 23.17

sufficient fixings through the lower plate into the base. If the designer has decided to consider the wall as consisting of a few narrow panels acting separately it is unlikely that it will be easy to resist the overturning effect with these fixings. However, if the designer is able to design or assume one wide panel (as in Fig. 23.14) it is possible that the fixings will be adequate, particularly if account is also taken of the influence of panels at right angles to resist overturning.

The second method of preventing the panel from lifting is to provide fasteners at a few stud positions which take the overturning effect without the fixings between the sheathing and the lower plate taking any load from overturning. A possible bracket connection is illustrated in Fig. 23.18 although it is more usual to use a galvanised steel strap anchored in the base and taken up the line of the stud to a sufficient height to accommodate the necessary fixings.

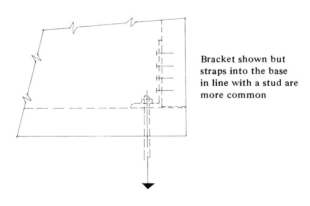

Bracket shown but straps into the base in line with a stud are more common

Fig. 23.18

Referring to Case B in Fig. 23.16, if there are two such fixings at P and O between the panel and the base, the tension in the fixing at P:

$$= \frac{(RH \times 1.33) - V_3 \times b - V_2 \times b/2}{b}$$

If there are more than two fixings along a stud wall, they should be placed at stud positions and the tension in each may be assumed to be proportional to its distance from O.

Case C of Fig. 23.16 is generally as Case B and, in addition, the designer must check the connections at the top of the panel against uplift. Because $V_1$ and $V_2$ are residual uplift forces and hence increase the overturning moment, the moment caused by them about Point O should be increased by a factor 1.33 when designing the holding down fixings. Therefore the tension in the fixing at P:

$$= \frac{1.33}{b} (RH + V_3 \times b + V_2 \times b/2)$$

# Chapter Twenty-four
# Preservation. Durability. Moisture Content

## 24.1   INTRODUCTION, PRESERVATION

The subject of preservation is covered extensively in many publications, and it is not intended that this chapter or manual should attempt to cover the subject in depth. The intention is to guide a designer in deciding when preservation is necessary, when there is a choice, or when preservation is unnecessary, and also to put forward several points of which a designer should be aware.

The notes are related to UK conditions, with the more commonly available softwoods in mind, and related to structural components rather than joinery.

## 24.2   DURABILITY

Each species of timber has a certain natural durability which is usually greater for the heartwood than the sapwood. Durability in this instance is a measure of the natural resistance to fungal decay (not insect attack). The Princes Risborough Laboratory of the Building Research Establishment evolved a classification for durability which is shown in Table 24.1. It is important to realise, however, that it is possible to refer to durability only in relative and indicative terms, and that the classifications refer to 50.8 $\times$ 50.8 mm (2 $\times$ 2 inch) pieces of heartwood in the ground. In general, larger pieces would have longer life.

**Table 24.1**

| Grade of durability | Approximate life in contact with the ground (years) |
|---|---|
| Very durable | More than 25 |
| Durable | 15–25 |
| Moderately durable | 10–15 |
| Non-durable | 5–10 |
| Perishable | Less than 5 |

The life stated refers to 50.8 $\times$ 50.8 mm sections. Larger sections would have longer life.

It is also important to realise that even species with low natural durability will not decay if the moisture content in service is kept below 20–22%. Even occasional short periods when the moisture content is raised above 20–22% will not lead to decay, although discolouration (e.g. blue stain) may occur in the sapwood of some species. Even a piece of timber given the emotive description of 'perishable' (Table 24.1) will not decay if the service conditions lead to a moisture content of 20–22% or less. On the other hand, it is as well to realise that if a species with high natural durability (or properly preserved timber) is used where moisture is trapped (e.g. by a poor building detail or use of an unsuitable paint), although the timber will not decay, the trapped moisture may cause trouble in another way (e.g. paint may peel off). Therefore, the use of a durable timber or preserved timber should never be considered as an alternative to good detailing.

The sapwood of almost all timbers (not just softwoods) is either perishable or non-durable. Engineers should not fall into the trap of assuming that all hardwoods are durable. Some of the cheaper hardwoods currently available (usually used for joinery rather than structural use) are less durable than several commonly available softwoods. Table 24.2 lists the durability rating of the heartwood and sapwood of commonly available softwoods used for structural purposes, mainly extracted from a *Handbook of Softwoods* by the Building Research Establishment.

**Table 24.2**

| Timber in ground contact | Grade of durability of: | |
|---|---|---|
| | unpreserved heartwood | unpreserved sapwood |
| European Redwood | Non-durable | Not known to the same extent as heartwood but usually close to or better than the boundary between non-durable and perishable |
| European Whitewood | Non-durable | |
| Hem-Fir | Non-durable | |
| Spruce-Pine-Fir | Non-durable | |
| Douglas-Fir-Larch | Moderately durable | |
| UK-grown Sitka Spruce | Non-durable | |

## 24.3 AMENABILITY TO PRESERVATIVE TREATMENT

The ease with which timber can be impregnated or treated with preservative can be an important consideration where it is decided to increase the natural durability. When correctly treated, non-durable timber can be made at least as durable as naturally durable timbers.

The UK Princes Risborough Laboratory evolved a classification system (before the use of organic-solvent preservatives became common) which gives a guide to the extent to which timber can be impregnated with preservative. The classifications are:

*Permeable.* These timbers can be penetrated completely under pressure without difficulty, and can usually be readily impregnated by the open-tank process.

*Moderately resistant.* These timbers are fairly easy to treat and it is usually possible to obtain a lateral penetration of the order of 6–19 mm in 2–3 hours under pressure.

*Resistant.* These timbers are difficult to impregnate under pressure and require a long period of treatment. It is often very difficult to penetrate them laterally more than about 3–6 mm. Incising can be used to obtain a better treatment (although not common in the UK).

*Extremely resistant.* These timbers absorb only a small amount of preservative even under long pressure treatments. They cannot be penetrated to an appreciable depth laterally, and only to a very small extent longitudinally.

The classification of the heartwood and sapwood of a few softwoods is given in Table 24.3. These classifications are related to pressure impregnation.

**Table 24.3**

|  | Heartwood | Sapwood |
|---|---|---|
| European Whitewood | Resistant | Resistant |
| European Redwood | Moderately resistant | Permeable |
| Douglas Fir | Resistant | Resistant |
| Spruce-Pine-Fir* | Resistant* | Resistant* |
| Hem-Fir† | Resistant† | Resistant† |
| UK Sitka Spruce | Resistant | Resistant |

* Amenability based on the Spruce.
† Amenability based on the Hemlock.

Since the classification described above was evolved, double-vacuum organic-solvent treatments have been developed by which 'resistant' species such as Whitewood can be treated commercially to acceptable levels, although still not as easily as less resistant species.

## 24.4  RISK AND AVOIDANCE

### 24.4.1  Risk and avoidance of decay

As stated in § 24.2, Codes and Standards are in agreement that there is little risk of decay, even in sapwood, if the timber is maintained at a moisture content in service of 20–22% or less. Therefore, in most cases of timber used inside a building there is rarely a 100% case for the use of preservation unless there is a local hazard. Also, by good practice and detailing, any risk of decay can be minimised with reasonable certainty for many other cases. However, over recent years it has become quite common to preserve many components where, strictly speaking, the risk of decay hardly warrants it. Where this happens the preservation can be regarded as an extra insurance against mischance.

In deciding on whether or not to preserve, and how to preserve, the specifier is faced with several Codes, Standards and regulations to consult. Some of these are written in general terms whilst others are very specific. Those likely to be most pertinent to a UK engineer are discussed in § 24.7.

### 24.4.2   Risk and avoidance of insect attack

Reference is made below to the four most common forms of insect attack encountered in the UK in softwoods. In dry timber, it is a fair assumption that only the House longhorn beetle, if present, could be significant regarding the risk of structural failure and, in the areas of England where it occurs, the Building Regulations call for preservation in roofs.

*Furniture beetles* (Anobium punctatum *De Geer*)
The furniture beetle is the most common form of insect attack in the UK. These insects can attack dried sapwood, and some heartwood.

*Ambrosia beetles* (Pinhole borer)
Although attack is less common in softwoods than hardwoods, the standing tree or recently felled logs of several softwoods can be attacked by Pinhole borer. Attack ceases when the timber is dry. Attack is more common in the sapwood than the heartwood and takes the form of circular holes or short tunnels of 0.5–3 mm diameter. The holes are dark-stained and contain no dust.

*Wood wasps* (Siricidae)
These insects attack the standing tree and logs. The attack dies when the timber is dried. The tunnels are circular and filled with tightly packed bore dust.

*House longhorn beetles* (Hylotrupes bajulus *L.*)
There is one area in the Home Counties of England (delineated in B3 of the Building Regulations for England and Wales) where House longhorn beetles are a risk to softwood in roofs. They can attack dried softwood. Although the Building Regulations call for all softwoods in roofs in this area to be preserved, only four of the species listed in Table 24.4 are recorded as having been attacked (perhaps because the other two were not used before preservation became mandatory in the risk area).

**Table 24.4**

| Species | Furniture beetles | House longhorn beetles | Pinhole borer | Wood wasp |
|---|:---:|:---:|:---:|:---:|
| European Whitewood | √ | √ | √ | √ |
| European Redwood | √ | √ | √ | √ |
| Douglas Fir | ** | √ | √ | ** |
| Hem-Fir | √ | ** | √ | √ |
| Spruce-Pine-Fir | √ | √ | √ | ** |
| UK Sitka Spruce | √ | ** | ** | √ |

√   Species recorded as having been attacked by the insect indicated in the column.
**   Not immune although not recorded as having been attacked.

Termites are not a hazard in the UK but must be considered if an engineer is producing a design for many other areas (e.g. the Middle East). There are many types. Although not all will climb, the designer must realise that timber can be attacked when in store on the ground. Useful reading on preserving against termite attack is contained in AMD 3916 to BS 5589.

In the area of the Home Counties where the House longhorn beetles are a risk, the Building Regulations require softwood to be treated with a suitable preservative. Two 'deemed-to-satisfy' provisions are water-borne CCA in accordance with BS 4072 (to a retention level of 5.3 kg/m³), or immersion for not less than 10 minutes in an organic-solvent solution. (Although this is said to apply to all softwoods there must be some doubt about the value of a 10 minute immersion for 'resistant' timbers in this situation. A double-vacuum process would give more protection.)

If it is desired to treat against insect attack (as well as decay) by an organic-solvent preservative, the specifier should ensure that the preservative contains an insecticide. Some of the preservatives used for preserving external joinery contain no insecticides because insect attack is not usually a problem. CCA is effective against the insects encountered in the UK.

## 24.5   TYPES OF PRESERVATIVE

The two principal types of preservatives used in the UK for building components are:

(a) The water-borne salt types of which by far the most common in the UK are the formulations based on solutions of copper sulphate, sodium dichromate and arsenic pentoxide covered by BS 4072 (usually referred to as copper/chrome/arsenic or simple CCA). They are applied by pressure or vacuum/pressure.

(b) Organic-solvent types. There are several different formulations, generally covered by BS 5707:Part 1. The solutions consist of one or more organic fungicides in an organic solvent such as a white spirit.

Common fungicides are pentachlorophenol, tributyltin oxide, zinc naphthenate and pentachlorophenyl laurate. There are others. In addition, an insecticide may be added if resistance to insect attack is required. Two listed in BS 5707:Part 1 are lindane and dieldrin. Organic solvents can be applied by a double-vacuum process. Some species (such as European Redwood, see Table 24.3) are relatively easy to preserve and can be treated for certain applications by dipping/immersion in an organic-solvent solution.

A tar oil type of preservative (e.g. creosote) can be used but such preservatives have a strong smell which makes them unsuitable for most components inside a building. (The relevant British Standards are BS 144 and BS 3051.)

The diffusion process (e.g. the boron process) must be carried out on 'green' timber, usually at a sawmill. Most mills have discontinued this type of treatment.

There are several preservatives formulated and designed for remedial work. They are not covered in this chapter.

### 24.5.1 General uses

If a water-borne process (to BS 4072) is used the timber should have been dried to about 25% moisture content or less, but it is extremely important for a specifier to realise that the use of a water-borne process increases the moisture content of the timber very considerably, causes an increase in the cross-section and will raise grain. Therefore timber for building usually has to be re-dried before being used. (See §7.7 for preservation of glulam members.) With some components (e.g. small battens) it is usually acceptable to air dry the timber after preservation and before use, but for certain uses it will be necessary to kiln the timber before use. The salts take about 7 days to 'fix' in the timber and the timber should not be used before this period is over. The preservative salts are chemically 'fixed' in the timber after treatment, however soluble salts (e.g. sodium sulphate) which form as a by-product of the reactions between CCA and timber may appear as a white deposit on the surface of the timber.

If a CCA treatment is used on a section which has already been planed or moulded there is a risk of distortion and therefore the final machining is usually carried out after preservation and re-drying (although, for example, cladding boards are often preserved with CCA after moulding). Processing after preserving removes part of the preserved timber. In the UK, organic solvents have largely taken over for preservation of processed construction and joinery sections, and where drying time and lack of distortion are important.

Organic-solvent processes require timber to be at a moisture content of about 22% or less (preferably less) at time of treatment. They do not increase the moisture content of the timber, nor do they affect the dimensions/profile of the sections or raise the grain. Where an organic solvent is used to treat exterior timber it may be advantageous to use one containing a water repellent.

If timber treated with an organic solvent is to be painted it is essential to limit the amount of free solvent left in the timber to be compatible with the paint, and to leave sufficient time for sufficient solvent to evaporate. This point is particularly important if the timber is to be factory finished.

As far as possible all cutting, notching etc., must be carried out before the timber is preserved. If some cutting after treatment is unavoidable the cut surfaces should be given a thorough application ('swabbing') of a suitable preservative.

## 24.6    ADDITIONAL NOTES ON PRESERVATION

### 24.6.1    Effect of preservatives on strength

Normally, preservatives do not affect the strength of timber to any measurable extent, but there are certain fire-retardant treatments which are said to reduce the strength. When using a specific fire-retardant the designer is advised to check on this point with the manufacturer (and on the compatibility with adhesives).

### 24.6.2    Compatibility of glues and preservatives

It is usually relatively easy to glue preserved timber or preserve glued timber, providing certain precautions are taken. This may take the form of gluing shortly after machining, brushing off salts, drying before gluing etc. Many points are covered in Chapter 19, but if in doubt the designer should check with the manufacturers of the glue and the preservative. With certain fire-retardant solutions, particularly those including ammonia or inorganic salts, certain precautions must be taken. It may not be possible to glue at all on timber already treated with certain of these solutions and, if preservation is carried out after gluing, usually at least 7 days must elapse between gluing and preservation. See BRE Current Paper No. 54 (Laidlaw and Paxton, 1974). If the preservative contains any additive it is as well to check the compatibility with the adhesive.

Dried timber treated with CCA is generally not corrosive to metals in normal building applications but, if the timber is wet or becomes wet, the preservation may increase the rate of corrosion of some metals. Preservatives containing copper should not be used in direct contact with aluminium.

Currently there is disagreement on whether or not timber pre-treated at the sawmill by the boron diffusion process can be glued. The treatment process is not practised to any extent. Few mills offer it but, if the designer is called upon to use timber treated in this way, it would be prudent to obtain the latest information on whether or not it can be glued.

### 24.6.3   Compatibility of paint and preservatives

It is usually possible to paint successfully on the surface of preserved timber, although with certain preservatives containing water-repellent waxes, there may be some difficulties, or some precautions which have to be taken. Once more the advice of the manufacturers should be sought.

### 24.6.4   Flame retardant

Special formulations are available which will raise the surface spread of flame classification of timber to Class 1 when tested according to BS 476 : Part 7, or achieve a Building Regulation rating of Class 0, which requires a fire propagation index ($I$) not exceeding 12 and a sub-index ($i_1$) not exceeding 6, when tested in accordance with BS 476 : Part 6. They do not improve the fire resistance (time) of timber to any noticeable extent. Galvanised fasteners should not be used in timber which has been treated with a flame retardant unless the formulation is a non-salt type.

### 24.6.5   Blue stain/anti-stain

The sapwood of certain species (e.g. European Redwood) is susceptible to discolouration of the sapwood (e.g. blue stain). Blue stain is not a structural defect, nor is it a sign of incipient decay, and it will not spread whilst the timber is at less than about 20–22% moisture content. Some sawmills treat the sawn timber with anti-stain chemicals. It is mentioned here because it is quite common for a designer to be questioned about blue stain. It is not a structural defect, and is permitted by BS 4978 in stress-graded timber.

### 24.6.6   Size of preserving plant

If the engineer wishes to call for pressure impregnation or immersion of components in their final form, a check should be made that a treatment plant is available large enough to take the component. The major suppliers of preservatives will know the size and disposition of plants.

### 24.6.7   Moisture content readings

The readings of a moisture meter are affected by CCA preservatives and by fire retardants, therefore the moisture content of such treated timber can be measured only by destructive methods unless the manufacturer of the meter can give a correction factor. Organic-solvent treatments do not affect moisture meter readings to any significant extent (with the possible exception of those containing copper naphthanate).

### 24.6.8   Treating mixed species

When treating mixed species the rule has generally been to use a treatment schedule designed for the species most resistant to preservation. However, this has led to over-absorption in certain cases (e.g. Spruce-Pine-Fir studs) with subsequent trouble sometimes from preservative exuding at a later date, and the NHBC/BRE have agreed modified treatment schedules for the BWPA Code C9 which is specified in NHBC Practice Note 5 for studs in external walls of timber framed housing. (Also see BWPA Code C122.)

### 24.6.9 Exterior wood stains

Finishes given the title 'preservative stains' should not be confused with preservatives. The stain gives virtually no added preservation to the timber, and the term 'exterior wood stain' is a better description even where it contains a fungicide. The fungicide is mainly to give protection to the finish not the timber.

### 24.6.10 Surface degradation

All species of preserved timber left exposed externally will be subject to surface degradation ('greying'). Although there is no firm evidence that preservative treatment slows down the process there are indications that some treatments may do so to some degree.

### 24.6.11 Safety

Preservatives for use in the UK must be cleared through the government's Pesticides Safety Precaution Scheme (PSPS).

Obviously preservatives are toxic and must be treated with care during storage and application. Some are flammable until fixed in the timber. Once 'fixed' in the timber they do not constitute a health hazard and the flammability of the timber is not increased. Care must be taken not to taint foodstuffs.

If moulding, cutting etc. are carried out after preservation, great care should be taken to ensure safe disposal of the waste. It should not be used, for example, for animal litter.

## 24.7 PUBLICATIONS GIVING GUIDANCE OR RULES ON WHEN TO PRESERVE

### 24.7.1 BS 5268:Part 5

BS 5268:Part 5:1977, 'Preservative treatments for constructional timber', is part of the Code of Practice for the structural use of timber. Clause 5.2 details four hazard categories A, B, C and D (not to be confused with the performance categories A and B of BS 5589:1978, 'Preservation of timber'). These are:

A   Where preservative treatment of timbers, even those with low inherent resistance to biological degradation, is unnecessary. This is because the conditions of use involve negligible risk, render the consequences acceptable, or make the cost of preservative treatment generally unfavourable.

B   Where there is a low risk of decay or insect attack or where remedial action or replacement is simple. In such situations preservation, if adopted, may be regarded as an insurance against the cost of subsequent repairs.

C   Where experience has shown that there is an unacceptable risk of decay, whether due to the nature of the design or the standard of workmanship and maintenance, or where there is a substantial risk of decay or insect attack which, if it occurs, would be difficult and expensive to remedy.

D   Where timber is exposed to a continually hazardous environment and cannot be protected by design or where there is a high risk of decay or insect attack in structures the collapse of which would constitute a serious danger to persons or property.

This wording does permit the specifier an element of discretion, but leaves an element of doubt. However, Table 3 (and 4) of BS 5268 : Part 5 does give more precise guidance. The way of stating the extent to which various species should be preserved in various situations is being changed. The method now favoured is to quote a treatment schedule for the plant operator to use. However, BS 5268 : Part 5 : 1977 is written in a different way. For organic-solvent methods it simply states 'yes' or 'no' leaving the specifier/applier to refer to schedules given by the preservative manufacturer or the BWPA (British Wood Preserving Association). For CCA methods retentions in kg/m$^3$ are quoted. It is the intention to amend BS 5268 : Part 5 in the future to give preservation treatment schedules (perhaps with retention levels also being given for CCA).

### 24.7.2   BS 5589

BS 5589 : 1978, 'Preservation of timber', is complementary to BS 5268 : Part 5 in that it aims to cover non-structural external timber in buildings and some special uses such as fencing. Although printed only one year later than BS 5268 : Part 5 it is written in different terms. It gives tables of use-conditions (e.g. external timber in buildings and out of contact with the ground), leaves it to the specifier or user to decide on a desired service life (e.g. 60, 50, 30, 20, 15 years), and then gives suitable treatment schedules (or even immersion periods) for various species. Thus if a specifier quotes BS 5589 : 1978 it is necessary also to quote the desired service life.

Because the use of Whitewood is beginning to be more popular for external cladding (to which Table 2 of BS 5589 only gives a 30 year treatment schedule for Whitewood, it is worth noting that schedules are given in BWPA Code C6 for European Whitewood and Canadian Hemlock (and European Redwood) for a 60 year service life.

### 24.7.3   Codes of the British Wood Preserving Association

Several useful publications are available from the BWPA. Amongst these there are ten 'codes' for various applications as listed below:

C1   (1975)   Timber for use as packing in cooling towers
C2   (1975)   Timber for use permanently or intermittently in contact with sea or fresh water
C3   (1975)   Fencing
C4   (1975)   Agricultural and horticultural timbers
C5   (1975)   External joinery and external fittings not in ground contact
C6   (1975)   External cladding
C7   (1979)   Prefabricated timber buildings for use in termite infested areas
C8   (1979)   Constructional timbers
C9   (1982)   Timber-framed housing (The constructional frame of external walls.)
C10 (1984)   Treatment of hardwood exterior joinery with organic-solvent preservatives by double vacuum.

BWPA Codes are offered to BSI for consideration as British Standards, therefore, if the date of the relevant BS is later than the date of the BWPA Code, it is probably better to specify to the BS. NHBC refer to C9 in their Practice Note 5 on timber-framed housing. One of the advantages of Code C9 is that it gives the latest 'agreed' treatment schedules for various components (e.g. for exterior stud walls), providing that equipment able to withstand 2 bar pressure is available.

### 24.7.4   The Building Regulations

The Building Regulations for England and Wales 1976 require softwood timber in certain defined areas of the Home Counties of England in the construction of a roof or fixed within a roof to be treated with a suitable preservative to prevent infestation by the House longhorn beetle (Regulations B3 and B4).

Regulation A16(1) and Schedule 5 combine to require external softwood cladding of Douglas Fir, Hemlock, Larch, European Redwood, Scots Pine, Sitka Spruce and Whitewood or European Spruce to be preserved in accordance with Table 5 of Schedule 5. The Building Regulations (Northern Ireland) 1973 have similar requirements.

The Building Standards (Scotland) Regulations 1981 have a deemed-to-satisfy provision which is relevant to the preservation of weatherboarding fixed direct to studs. Most softwoods can be used if (with the exception of Western Red Cedar which does not need to be preserved), they are preserved in accordance with BS 5589. (See Schedule 13.G9 (6) and Schedule 14. Part I.8.) Where weatherboarding is fixed in a position where it is readily accessible for inspection and maintenance or renewal it can be claimed that Regulation B2(ii) makes preservation unnecessary, particularly if the weatherboarding is fixed to battens (although probably it makes sense to preserve).

### 24.7.5   The National House-building Council Manual

In their handbook and in Practice Note 5 for timber-framed housing the NHBC call for preservation as required by the Building Regulations and, in addition for preservation of the following items:

|  | CCA | Double-vacuum organic solvent |
|---|---|---|
| Lintels in brick or blockwork external walls | ✓ | |
| Battens as fixings for claddings | ✓ | ✓ |
| Any embedded timber | ✓ | |
| Joists in flat roofs | ✓ | ✓ |
| Joists with ends built into solid (non-cavity) walls | ✓ | ✓ |
| Door frames (for external doors) | ✓ | ✓ |
| Windows | ✓ | ✓ |
| Surrounds to metal windows | ✓ | ✓ |
| External doors other than flush doors | ✓ | ✓ |
| External timber features other than fencing | ✓ | ✓ |

NHBC also refer to BWPA Code C9 for the preservation of the constructional timber frame of external walls of timber-framed housing.

### 24.7.6   Further reading

BS 1282:1975, 'Guide to the choice, use and application of wood preservatives'.

BS 4072:1974, 'Specification for wood preservation by means of waterborne copper/chrome/arsenic compositions'.

BS 5707:Part 1:1979, 'Specification for solutions for general purpose applications, including timber that is to be painted'.

## 24.8   MOISTURE CONTENT

### 24.8.1   Equilibrium moisture content in service

As far as joinery is concerned, BS 1186:Part 1:1984 gives guidance on the likely service moisture contents. These are quoted in Table 24.5.

**Table 24.5**

| Category | Requirement or likely average moisture content (%) |
|---|---|
| External joinery* | 16 ± 3 |
| Internal joinery | |
|   Buildings with intermittent heat | 15 ± 2 |
|   Buildings with continuous heating providing room temperatures in the range 12–19°C | 12 ± 2 |
|   Buildings with continuous heating providing room temperatures in the range 20–24°C | 10 ± 2 |

\* Except hardwood sills at floor or ground level or hardwood thresholds for which a figure of 19% ± 3% is given.

Table 24.6 gives guidance on moisture content for end-use categories as given in BS 5268:Part 2 for structural components.

Ideally, all timber should have a moisture content at manufacture and installation close to the moisture content it will have in service. If timber has been correctly dried once, if it is subjected to rain for relatively short periods during delivery or erection it is most unlikely that anything other than the

**Table 24.6**

| Position of timber in building | Average moisture content attained in service (%) | Moisture content which should not be exceeded at time of erection |
|---|---|---|
| External uses fully exposed | 18 or more | Not defined |
| Covered and generally unheated areas | 18 | 24 |
| Covered and generally heated areas | 16 | 21 |
| Internal in continuously heated building | 14 | 19 |

outer 2–3 mm will be affected. A normal progressive or chamber kiln takes several days to extract moisture from timber and this gives a measure of how little effect rain will have. The designer should not be misled by high surface readings into assuming that all the timber is equally wet throughout the thickness.

### 24.8.2 Moisture content at time of manufacture

The moisture content at the time of manufacture should be within a few per cent of the service conditions. In addition, the type of connection used may set the upper limit. If the timber is being joined by simple nailing which is only lightly loaded, a moisture content of up to 22% at time of manufacture is probably satisfactory. With most mechanical connectors (Chapter 18) an upper limit of 20% is likely to be satisfactory. With a glued joint (Chapter 19) a slightly lower limit is more appropriate, but depends on the type of glue and the method of curing. Some WBP glues have been used successfully at timber moisture contents of 22% whilst radio-frequency curing requires moisture contents of around 12%.

In the manufacture of a component from timber which has been delivered 'green' to the factory and is still in the process of drying 'naturally', much more care is required in checking the moisture content than for timber which has been kiln dried at the sawmill (usually to 18% + 4% − 2%) or air dried at the sawmill (usually to 22% or less) some weeks or months before. Softwoods 'move' less during any subsequent changes in moisture content than they do during initial drying from the green condition.

### 24.8.3 Moisture content at time of erection

Generally speaking the moisture content at time of erection can be as high as 19–24% providing that there is natural air circulation which will permit the timber to continue to dry. The notes in § 4.16 on creep deflection are particularly relevant, as are those in § 24.6.5 on 'blue stain'.

If timber is erected at 22% moisture content or over, or the moisture content after erection becomes higher than around 18–20% due to site conditions, the sections should not be completely enclosed until the moisture content has dropped to below 20%.

### 24.8.4 Measuring moisture content

In the range of moisture contents of interest to structural engineers (up to 22%) small portable moisture meters have sufficient accuracy. If the engineer has reason to believe that the initial moisture content is likely to be higher than the surface reading, deep probes (20–25 mm long) can be used. With these, the 'moisture gradient' can be traced through a piece. Also see § 24.6.7.

Most Codes and textbooks on timber technology detail a way of measuring the moisture content of timber by weighing a sample piece as presented, drying it in an oven to constant weight, then expressing the weight of water removed as a percentage of the oven-dry weight. However, this method obviously destroys the piece and it is only of use to a structural engineer if wishing to check the factory quality control before or during manufacture of components, or in extreme cases of site 'trouble-shooting'.

Often one hears the oven-drying method referred to as being more accurate than the use of a moisture meter. This is not so. They have different uses. A properly calibrated meter gives an accurate method of measuring the

**Table 24.7   Relative humidity**

| Dry-bulb temperature ($^\circ$C) | Difference between dry-bulb temperature and wet-bulb depression ($^\circ$C) | | | | | | | | | | | |
|---|---|---|---|---|---|---|---|---|---|---|---|---|
| | 1 | 2 | 3 | 4 | 5 | 6 | 7 | 8 | 9 | 10 | 11 | 12 |
| 1 | 83 | 66 | 49 | 33 | 17 | | | | | | | |
| 2 | 84 | 68 | 52 | 37 | 22 | 7 | | | | | | |
| 3 | 84 | 70 | 55 | 40 | 26 | 12 | | | | | | |
| 4 | 85 | 71 | 57 | 43 | 30 | 16 | | | | | | |
| 5 | 86 | 72 | 58 | 45 | 33 | 20 | 7 | | | | | |
| 6 | 86 | 73 | 60 | 48 | 35 | 24 | 11 | | | | | |
| 7 | 87 | 74 | 62 | 50 | 38 | 26 | 15 | | | | | |
| 8 | 87 | 75 | 63 | 51 | 40 | 29 | 19 | 8 | | | | |
| 9 | 88 | 76 | 64 | 53 | 42 | 32 | 22 | 12 | | | | |
| 10 | 88 | 77 | 66 | 55 | 44 | 34 | 24 | 15 | 6 | | | |
| 11 | 89 | 78 | 67 | 56 | 46 | 36 | 27 | 18 | 9 | | | |
| 12 | 89 | 78 | 68 | 58 | 48 | 39 | 29 | 21 | 12 | | | |
| 13 | 89 | 79 | 69 | 59 | 50 | 41 | 32 | 23 | 15 | 7 | | |
| 14 | 90 | 79 | 70 | 60 | 51 | 42 | 34 | 26 | 18 | 10 | | |
| 15 | 90 | 80 | 71 | 61 | 53 | 44 | 36 | 27 | 20 | 13 | 6 | |
| 16 | 90 | 81 | 71 | 63 | 54 | 46 | 38 | 30 | 23 | 15 | 8 | |
| 17 | 90 | 81 | 72 | 64 | 55 | 47 | 40 | 32 | 25 | 18 | 11 | |
| 18 | 91 | 82 | 73 | 65 | 57 | 49 | 41 | 34 | 27 | 20 | 14 | 7 |
| 19 | 91 | 82 | 74 | 65 | 58 | 50 | 43 | 36 | 29 | 22 | 16 | 10 |
| 20 | 91 | 83 | 74 | 66 | 59 | 51 | 44 | 37 | 31 | 24 | 18 | 13 |
| 21 | 91 | 83 | 75 | 67 | 60 | 53 | 46 | 39 | 32 | 26 | 20 | 14 |
| 22 | 92 | 83 | 76 | 68 | 61 | 54 | 47 | 40 | 34 | 28 | 22 | 17 |
| 23 | 92 | 84 | 76 | 68 | 62 | 55 | 48 | 42 | 36 | 30 | 24 | 19 |
| 24 | 92 | 84 | 77 | 69 | 62 | 56 | 49 | 43 | 37 | 31 | 26 | 20 |
| 25 | 92 | 84 | 77 | 70 | 63 | 57 | 50 | 44 | 39 | 33 | 28 | 22 |
| 26 | 92 | 85 | 78 | 71 | 64 | 58 | 51 | 46 | 40 | 34 | 29 | 24 |
| 27 | 92 | 85 | 78 | 71 | 65 | 58 | 52 | 47 | 41 | 36 | 31 | 26 |
| 28 | 93 | 85 | 78 | 72 | 65 | 59 | 53 | 48 | 42 | 37 | 32 | 27 |
| 29 | 93 | 86 | 79 | 72 | 66 | 60 | 54 | 49 | 43 | 38 | 33 | 28 |
| 30 | 93 | 86 | 79 | 73 | 67 | 61 | 55 | 50 | 44 | 39 | 35 | 30 |

The values given are for a 2 metres per second movement of air past the hygrometer.

moisture content at a point and can therefore be useful in tracing the source of moisture, or the pattern of moisture in a piece of timber. The oven-drying method is a (destructive) method of giving an average moisture content. Hence the sample being dried must be kept small and the result can be influenced by the presence of resin etc. in the sample.

### 24.8.5 Measuring relative humidity

Relative humidity is the amount of water in the air compared to the amount of moisture which the air would contain at the same temperature if fully saturated. The ratio is expressed as a percentage. The pressure which moisture exerts is closely related to the amount of moisture in the air, therefore the use of a dry- and wet-bulb thermometer gives a convenient method of establishing the relative humidity of a specific air condition. Table 24.7 gives values of relative humidity for differences in degrees Celsius between the dry- and wet-bulb thermometers, based on a 2 metre per second movement of air past the hygrometer. The values are derived from Swedish sources.

Lightweight meters which measure relative humidity are available and, in view of the importance of humidity rather than temperature in setting the moisture content of timber (see Tables 24.7 and 24.8), it is rather surprising that one sees them (and humidifiers) used so infrequently.

### 24.8.6 Relative humidity/temperature/moisture content

The moisture content of timber depends on conditions of relative humidity and air temperature. With constant relative humidity and temperature, the timber will assume an equilibrium moisture content. Changes in humidity and temperature do not lead to an instantaneous measurable change in the moisture content of the timber. Moisture content in buildings is more dependent on humidity than temperature, therefore the service moisture

**Table 24.8   Equilibrium moisture content**

| Relative humidity (%) | Temperature (°C) | | | | | | | |
|---|---|---|---|---|---|---|---|---|
| | 16 | 18 | 20 | 22 | 24 | 26 | 28 | 30 |
| 85 | 18 | 18 | 18 | 18 | 18 | 18 | 18 | 18 |
| 80 | 17 | 16 | 16 | 16 | 16 | 16 | 16 | 16 |
| 75 | 15 | 15 | 15 | 14 | 14 | 14 | 14 | 14 |
| 70 | 13 | 13 | 13 | 13 | 13 | 13 | 13 | 13 |
| 65 | 12 | 12 | 12 | 12 | 12 | 12 | 12 | 11 |
| 60 | 11 | 11 | 11 | 11 | 11 | 11 | 11 | 10 |
| 55 | 10 | 10 | 10 | 10 | 10 | 10 | 10 | 10 |
| 50 | 9 | 9 | 9 | 9 | 9 | 9 | 9 | 9 |
| 45 | 9 | 8 | 8 | 8 | 8 | 8 | 8 | 8 |

content of timber can be established with reasonable accuracy for a building in which the relative humidity is fairly constant, even though the temperature varies. In textbooks it is usual to present the relationship in the form of curves (approximate) of moisture content, but for convenience moisture content figures to the nearest degree are given in Table 24.8 for the limited temperature range likely to be of most interest to engineers.

# Chapter Twenty-five
# Considerations for the Structural Use of Hardwood

## 25.1 INTRODUCTION

In Chapter 1 it was made clear that the contents of this manual are based on softwoods because these are the timbers used mainly in timber engineering. However, hardwoods are used and do fulfil a useful role. They can have the advantage of greater strength and durability than most softwoods, although the designer must not assume that all hardwoods are either more durable or stronger than softwoods. As well as possible or actual advantages, there can be disadvantages. By and large, a designer should not specify a hardwood without having investigated the properties of interest, and the availability of sizes and lengths. Table 25.1 may assist with guidance notes on properties.

## 25.2 SPECIES

The species listed in BS 5268:Part 2 and allocated to a Strength Class (for the purpose of allocating strength and permissible fastener loads etc.) are given below with a note of the Strength Class to which they are allocated:

| | | | |
|---|---|---|---|
| Iroko | SC5 | Balau | SC8 |
| Jarrah | SC5 | Ekki | SC8 |
| Teak | SC5 | Kapur | SC8 |
| Merbau | SC6 | Kempas | SC8 |
| Opepe | SC6 | Greenheart | SC9 |
| Karri | SC7 | | |
| Keruing | SC7 | | |

## 25.3 STRESS GRADE

BS 5756, 'Tropical hardwoods graded for structural use', was first printed in 1980. It describes one stress grade called 'Hardwood Structural', even for glulam, and each piece purporting to match this grading must be marked with HS and BS 5756 (unless, as with softwood, the marking is unacceptable

**Table 25.1**

|  | Iroko | Jarrah | Teak | Merbau | Opepe |
|---|---|---|---|---|---|
| Weight (kg/m³) at 12% m.c. | Approx. 640 | 690–1040 | 610–690 | 740–900 | Average 740 |
| Durability of heartwood | Very durable | Very durable | Very durable | Durable | Very durable |
| Amenability to preservation: heartwood sapwood | Extremely resistant<br>Permeable | Extremely resistant<br>Permeable | Extremely resistant | Extremely resistant | Moderately resistant<br>Permeable |
| Drying | Dries well with little splitting | Care must be taken to limit distortion | Dries well but slowly | Dries well | Variable. There can be serious splitting and distortion |
| Movement characteristics | Small | Medium | Small | Small | Small |
| Machinability | Satisfactory with experience | Satisfactory with experience | Relatively easy | Reported as variable | Satisfactory with experience |
| Nailability (also see § 25.7) | Satisfactory | Difficult | Pre-boring recommended | Pre-boring advisable | Pre-boring necessary |
| Gluability | Good | Good | Good |  | Good |
| Resinous (or gum) nature | No trouble reported | May contain gum streaks or pockets | No trouble reported | Unconfirmed reports of resinous nature | No trouble reported |
| Sapwood identification | Distinct from heartwood | Pale | Light/ pale | Pale yellow | Whitish/ pale |
| Other comment | Has been used as a substitute for teak. Not recommended for heavy duty flooring | High resistance to wear but inclined to splinter under heavy wear | A valuable timber, now expensive | Liable to stain in contact with iron in wet conditions |  |

| Karri | Keruing | Balau | Ekki | Kapur | Kempas | Greenheart |
|---|---|---|---|---|---|---|
| Average 880 | 720–800 | Usually less than 800 | 950–1100 | 720–800 | 770–1000 | Approx. 1030 |
| Durable | Moderately durable | Very durable to mod. dur. | Very durable | Very durable | Durable | Very durable |
| Extremely resistant | Mod./resist. to resistant | Extremely resistant | Extremely resistant | Extremely resistant | Resistant | |
| Permeable | Mod. resist. | Permeable | | Permeable | | |
| Pronounced tendency to check in thick pieces and to distort in thick pieces | Dries slowly and distortion may occur | Care necessary to prevent splitting and distortion | Dries slowly with splitting and some distortion likely | Dries slowly but well | Normally dries well | Dries slowly with degrade (splits) |
| Large | Medium to large | Small to medium | Medium | Medium | Stable when dry | Medium |
| Satisfactory with special care | Satisfactory with experience and care | | Difficult | Rather difficult | Somewhat difficult | Can be difficult due to high density |
| Difficult | Satisfactory | Pre-boring necessary | Pre-boring necessary | Satisfactory | Pre-boring advisable | Unsuitable |
| Good | Variable | | Variable | | | Variable to fairly good |
| No trouble reported | Known to exude resin | Resin canals present | No trouble reported | Non-resinous | No trouble reported | No trouble reported |
| | Grey | Paler than the heartwood | Paler than the heartwood | Pale | White or pale yellow | Pale yellow or green |
| | | It is a Shorea and can be variable. Possible confusion with Red Balau | | Acidic, can stain fabrics and corrode some metals | Slightly acidic | Noted for its strength |

for the appearance of the final component, in which case special arrangements with the client are necessary). The company or grader responsible for
the grading, and the standard name of the timber (or an abbreviation) must
also appear. Generally speaking, sawn sections of hardwood are relatively
free of knots, but slope of grain and fissures may need to be checked against
BS 5756.

The Strength Class to which the various species are allocated is shown in
§ 25.2. Before using dry stresses the designer should ensure that the timber
will be dried. For thicknesses of 100 mm and over, it will normally be
necessary to use wet exposure stresses.

## 25.4   PROPERTIES/CHARACTERISTICS

Most sawn sections of hardwood are relatively free from knots and wane,
and are fairly straight grained, however there is often a tendency to distort
and split, particularly if sections are resawn and particularly as the $h/b$ ratio
increases.

Table 25.1 has been prepared as a starting point to assist a designer to
choose a suitable species, or check if a suggested species is likely to be
suitable for an intended use. The comments have been extracted from
reference books on hardwoods such as the *Handbook of Hardwoods* by the
Building Research Establishment. However, the authors emphasise that it can
be considered to be only indicative of properties. Descriptions such as
'satisfactory', 'pale', etc. are not precise. Before using a species for the first
time, a designer should discuss with a hardwood specialist, preferably someone who has used the species and is also aware of availability and cost.

However, even with these reservations, Table 25.1 can be a useful starting
point. For example, if a designer is looking for a species to use for glulam
members (which requires to have stresses allocated by BS 5268, and needs to
be dried, machined, relatively free of distortion, and to be glued) it looks as
though Iroko is a possibility, whereas Ekki and Greenheart are unlikely
starters despite their advantages in heavy marine uses.

A gap in Table 25.1 indicates that nothing documented has been found
about the property by the authors.

## 25.5   MOISTURE CONTENT

If the designer wishes to use 'dry' stresses, a check is necessary that the
timber to be supplied will be dried to around 20% or less before erection, as
hardwoods can take rather long to dry out *in situ.* For thicknesses of
100 mm and over, it will normally be necessary to design using wet exposure
stresses.

## 25.6   DURABILITY/PRESERVATION

The heartwood of all the twelve hardwoods named in BS 5268:Part 2 is at least 'moderately durable' (see Table 24.1). Although the heartwood is 'moderately to extremely resistant' (see § 24.3) to preservation this is not usually a disadvantage due to the natural durability.

If present, the sapwood of these hardwoods is likely to be 'non-durable' or less durable. If, therefore, preservation is necessary, most sapwood is 'permeable' (see § 24.3 and Table 25.1).

If using a hardwood not listed in BS 5268 the designer is advised to check on the durability (and of course the Strength Class to use in designs).

## 25.7   CONNECTIONS

Many hardwoods can be glued with normal structural glues, but rather more care is necessary than with softwoods. In addition to all the quality control requirements detailed in Chapter 19, the surface must be reasonably free from resin. Also, particularly if the fastenings which are used to hold the pieces in place during curing are to be removed after curing, the manufacturer must ensure that individual pieces can be pulled into place easily for gluing. If excessive force is necessary, it is quite possible that the piece will burst open the glue line when subsequently it tries to regain its previous shape. In glulam or similar, it is prudent to consider using thinner laminations to prevent this action. Information on the gluability of nineteen hardwoods is contained in TRADA Research Report C/RR/22 (1965). Also see Table 25.1 for guidance.

When considering nailing a hardwood it may be necessary to pre-drill (see Table 25.1 for guidance). BS 5268:Part 2 advises that all hardwoods in Strength Classes 6–9 will probably require pre-drilling.

BS 5268:Part 2 advises that toothed-plate connectors can not normally be used in a hardwood of Strength Classes 6–9 due to the difficulty (or inability) of embedding the teeth.

In BS 5268:Part 2 the strength of mechanical connections used in hardwoods which fall into Strength Classes 6–9 are allocated the same values immaterial of species or Strength Class.

# Chapter Twenty-six
# Prototype Testing

## 26.1 GENERAL

Testing of prototypes or parts of construction can be useful to a designer in many ways, some of which are listed below. Where a test or tests is being carried out to satisfy BS 5268:Part 2 then Section 8 of that Code should be studied before testing is commenced.

Testing may be used to study the performance and arrive at the failure load and the load-deflection curve of a component. The test may be carried out to check a design which has already been prepared or as an alternative to design, for example when a component is a redundant framework. Usually prototype testing is carried out by or for an individual company or designer, but in the case of trussed rafters the information from several hundred tests carried out at a few testing stations for many companies has been pooled. Although satisfactory designs can be carried out mathematically for trussed rafters, the accumulated test information has justified increasing the span of standard trusses.

The permissible stresses in BS 5268 are largely set at the lower 5 in 100 exclusion limit, and if a designer wishes to justify a higher strength for a particular component, it is usually possible to do so by testing, particularly if two or more members occur in cross-section. It is a matter of balancing the benefits with the cost of testing.

With a composite section constructed without glued joints, with the joints taking horizontal shear, or any framework constructed with mechanical joints, although it may well be possible to calculate the strength it is almost certain that the deflection will have to be found by test, either of the whole component, or of joints to arrive at the load-slip characteristics.

A designer may use testing to check that a roof or floor construction is capable of lateral distribution of loading.

Codes of Practice are not always exactly applicable to all components or constructions, and the designer may wish to use a test to prove that a particular clause may be amended or disregarded in a particular application.

There are still cases in which the basic information required to produce a design is not available or is available only in a very conservative form (e.g.

information to calculate sway of a wall panel braced with a different board material on each side). In these cases the designer may have to revert to testing.

In certain constructions (e.g. thin plywood decking), it is the 'feel' rather than the strength which is the limiting criterion. In such cases testing can be preferable to a mathematical design.

Prototype testing can be expensive, and great care should be taken in planning the test to ensure that it represents the actual conditions, restraint and loading as far as practical and to ensure that the correct readings are taken. Before the test is started, it is advisable for the supervising engineer to calculate how the component can be anticipated to deflect, perform and fail, and to check against a load-deflection curve as the test proceeds to see that nothing untoward is happening. A record of the test as it proceeds must be made.

Any testing of full-size components can be dangerous. A supervising engineer must be appointed and must ensure that the method of applying the load, the strength limit of the test rig etc. will not place the testing staff at risk.

Part of the object of a prototype test is to find where and describe how failure occurs. Once failure has occurred or has started to occur at one point, this can lead to failure at other points, particularly if a triangulated framework is being tested. To prevent any secondary failures occurring which might make it impossible to see where the first failure started, the test rig should be designed so that the component is supported once it begins to fail.

## 26.2   TEST FACTOR OF ACCEPTANCE

From the notes in Chapter 2 on variability, it will be obvious that the timber from which the prototype component is made should contain defects as close to the grade (lower) limit as possible. Even then it is not certain that the chosen material will be the weakest which could be built into production components. Therefore in analysing a prototype test to failure, before accepting the component as satisfactory, a fairly high factor against failure is required. BS 5268:Part 2 requires a factor of 2.50 is one component is tested, decreasing to 2.00 minimum if five or more components are tested. The required value of the factor is quoted in BS 5268 as $K_{72}$ (Table 26.1). The moisture content must be appropriate to the service conditions. See BS 5268:Part 2 for the method of applying $K_{72}$ and Part 3 as one example of doing so.

## 26.3   TEST PROCEDURE

The test should be divided into three separate parts.

First, loading should be applied for a short period of time and then released, load-deflection readings being recorded. The purpose of this first

**Table 26.1    Value of $K_{72}$ in BS 5268:Part 2**

| Number of similar components tested | Value of $K_{72}$ (minimum) |
|---|---|
| 2 | 2.30 |
| 3 | 2.15 |
| 4 | 2.05 |
| 5 or more | 2.00 |

test is to take up any slack at supports and any initial slip at connections. By measuring the height of the component at supports, any vertical movement at the supports can be eliminated from analysis of the load-deflection curve over the span. BS 5268:Part 2 requires that for this 'pre-load' test the design dead loading should be applied for 30 minutes and then released.

Secondly, and straight away, the dead load should be applied again, maintained for a short period (15 minutes in BS 5268:Part 2) and then the loading should be increased to the maximum design load. BS 5268 requires this further loading to be applied over a period of 30–45 minutes. This load is left in place for a day and then released. The deflection readings and condition of the component should approximate to the required performance, but BS 5268:Part 2 stipulates that the deflection after 24 hours should be only 80% of the calculated deflection and that the rate of deflection with time over the 24 hours should decrease and certainly not increase at any time during this period. Deflection readings are taken immediately before and after load release.

The final stage of the test is to reload up to the maximum design load, checking deflection then, in the case of BS 5268, to load until 2.5 times the design load is applied (unless failure occurs before this). If the component is still intact, loading can be continued until failure occurs. If failure occurs at a factor of safety between 2 and 2.5, the engineer has the option of strengthening the component and retesting, or testing a sufficient number of components to satisfy the $K_{72}$ factor (Table 26.1).

The testing method described above does not apply to simple joints or individual members, nor to quality control testing, for which a simpler procedure is necessary, which should comply with a national standard if one exists.

# Chapter Twenty-seven

# Miscellaneous Tables

## 27.1 WEIGHTS OF BUILDING MATERIALS

Selected weights of building materials are tabulated below. When the weight of a manufactured item is known, that value should be used in preference to the weight given in this section. 'Schedules of weights of building materials' are given in BS 648 : 1964.

Although the unit of mass is the kg, because an engineer is interested in loads on structures the weights tabulated below are given in kN. To convert kilonewtons to kilograms (force) multiply by 101.97.

| | Thickness (mm) | Loading (kN/m²) |
|---|---|---|
| Asbestos cement sheeting  76 mm pitch | 5.5 | 0.15 |
| 146 mm pitch | 6.4 | 0.16 |
| Asphalt | 20 | 0.46 |
| Felt underlay for asphalt | 2 | 0.02 |
| Chipboard | 25 | 0.19 |
| Chippings | | 0.20* |
| 3 layers of bitumen felt | 6 | 0.11 |
| Fibreboard insulation | 12 | 0.04 |
| | 18 | 0.06 |
| Glassfibre | 80 | 0.016 |
| Mineral wool | 80 | 0.02 |
| Plaster (gypsum lime) | 12 | 0.30 |
| Plasterboard (no skim coat) | 9.5 | 0.10 |
| | 12.7 | 0.13 |
| Plaster (skim coat) | – | 0.05 |
| Plywood | 6.5 | 0.03 |
| | 9 | 0.05 |
| | 12 | 0.07 |
| Screed. Sand/Cement | 25 | 0.58 |
| Vermiculite | 25 | 0.12 |

|  | Thickness (mm) | Loading (kN/m²) |
|---|---|---|
| Timber boarding | 12 | 0.07 |
| Water | 25 | 0.25 |
| Wood wool slab. Standard | 38 | 0.21 |
|  | 50 | 0.25 |
|  | 63 | 0.30 |
|  | 76 | 0.33 |
| Extra density | 50 | 0.32 |
|  | 63 | 0.37 |
|  | 76 | 0.42 |
| Wood wool slab. Channel reinforced | 50 | 0.30 |
|  | 76 | 0.38 |
| Interlocking | 50 | 0.36 |
|  | 76 | 0.47 |

* Minimum. It is important to check the actual weight which will occur.

## Tile Weights (as Laid Up Slope)

|  | Lap or gauge (mm) | Loading (kN/m²) |
|---|---|---|
| *Broughton Moor* |  |  |
| Best |  | 0.48 |
| Seconds |  | 0.56 |
| Thirds |  | 0.64 |
| Special Peggies |  | 0.52 |
| Second Peggies |  | 0.59 |
| *Hardrow* |  |  |
| 457 mm × 305 mm | 76 lap | 0.79 |
| 457 mm × 457 mm | 102 lap | 0.79 |
| 711 mm × 457 mm | 127 lap | 0.96 |
| *Marley* |  |  |
| Anglia | 75 lap | 0.47 |
|  | 100 lap | 0.51 |
| Bold Roll | 75 lap | 0.47 |
|  | 100 lap | 0.51 |
| Ludlow major | 75 lap | 0.45 |
|  | 100 lap | 0.49 |
| Ludlow plus | 75 lap | 0.47 |
|  | 100 lap | 0.51 |
| Mendip | 75 lap | 0.47 |
|  | 100 lap | 0.51 |

| | Lap or gauge (mm) | Loading (kN/m²) |
|---|---|---|
| Modern | 75 lap | 0.54 |
| | 100 lap | 0.58 |
| Plain | 100 gauge | 0.73 |
| | 90 gauge | 0.80 |
| Wessex | 75 lap | 0.54 |
| | 100 lap | 0.58 |
| Yeoman | 75 lap | 0.50 |
| | 100 lap | 0.54 |
| *Redland* | | |
| Delta | 345 gauge | 0.59 |
| Double Pantile | 345 gauge | 0.50 |
| Double Roman | 345 gauge | 0.46 |
| Interlocking | 280 gauge | 0.50 |
| Plain | 100 gauge | 0.80 |
| Regent | 345 gauge | 0.46 |
| Renown | 345 gauge | 0.46 |
| Stonewold | 355 gauge | 0.54 |
| *Speakers* | | |
| Eternit | | 0.21 |
| Duchess | | 0.21 |
| Countess | | 0.21 |
| Ladies | | 0.23 |
| *Others* | | |
| Asbestos slate | | 0.23 |
| Bangor slate.  Best | | 0.27 |
| Seconds | | 0.35 |
| Thirds | | 0.49 |
| Westmorland.  Best | | 0.44 |
| Seconds | | 0.55 |
| Thirds | | 0.70 |
| Western Red Cedar Shingles | | 0.07– 0.12 |

## 27.2    BENDING AND DEFLECTION FORMULAE

The groups of bending and deflection formulae presented in Figs 27.1–27.11, for simply supported beams and continuous beams, are extracted from *Steel Designers' Manual* by permission of Granada Publishing.

For simple span beams the additional deflection attributable to shear should be considered.

For continuous beams moment and reaction coefficients should be adjusted to allow for shear deflection. As quoted they apply to designs where only bending deflection is considered.

In this manual, total load is usually given the symbol '*F*', however '*W*' or '*P*' is retained in Figs. 27.1–27.11 from the *Steel Designers' Manual*. Likewise '*d*' is retained as the symbol for deflection.

In the two left hand examples shown in Fig. 27.7 the designer must establish if the deflections at the end of the cantilevers and the deflections in the centre of the span are upwards or downwards. This is not clear from the diagrams/formulae. (See § 12.6.9.)

In calculating intermediate values of bending moment the designer should recall the proportions of a parabola as sketched below:

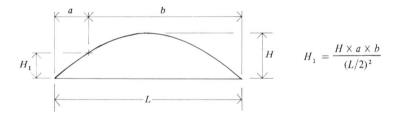

$$H_1 = \frac{H \times a \times b}{(L/2)^2}$$

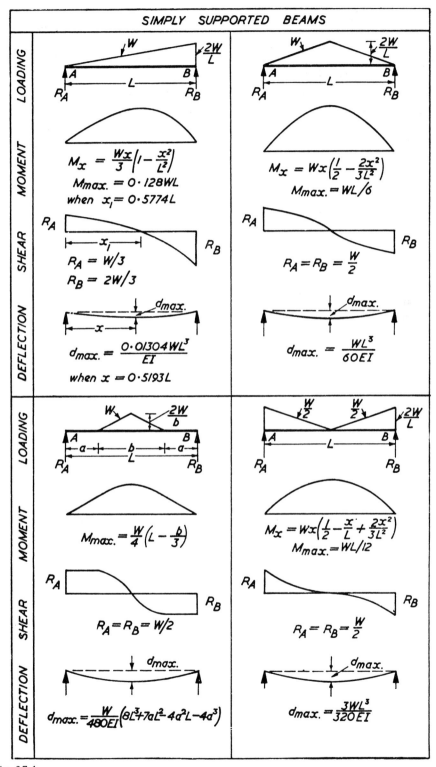

Fig. 27.1

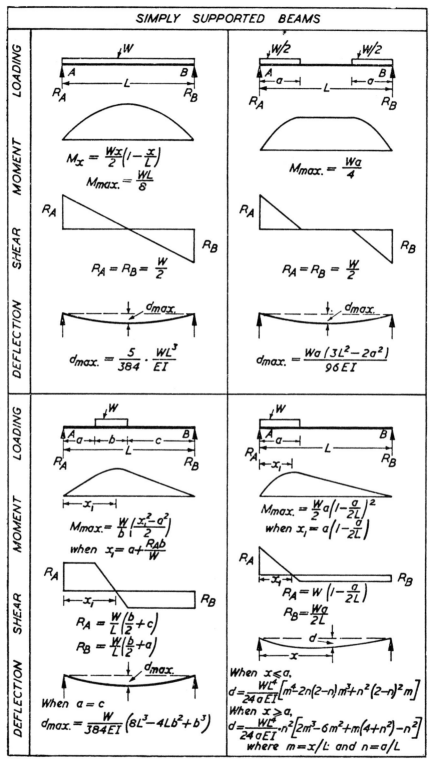

Fig. 27.2

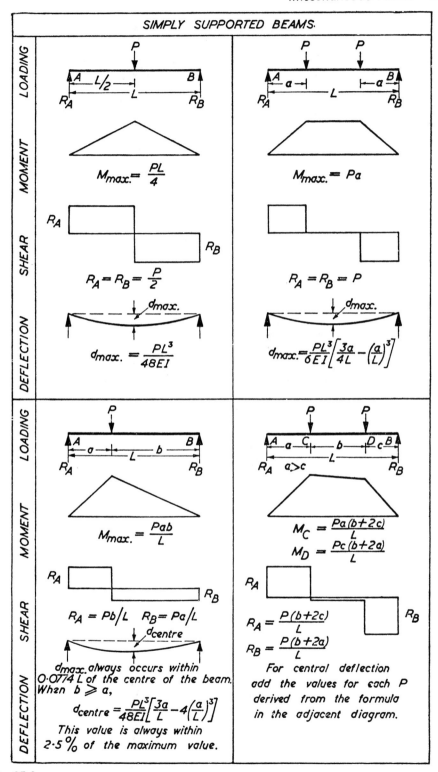

Fig. 27.3

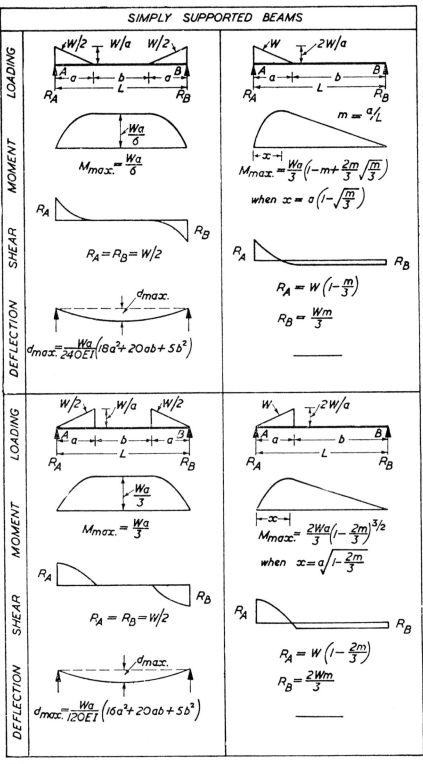

Fig. 27.4

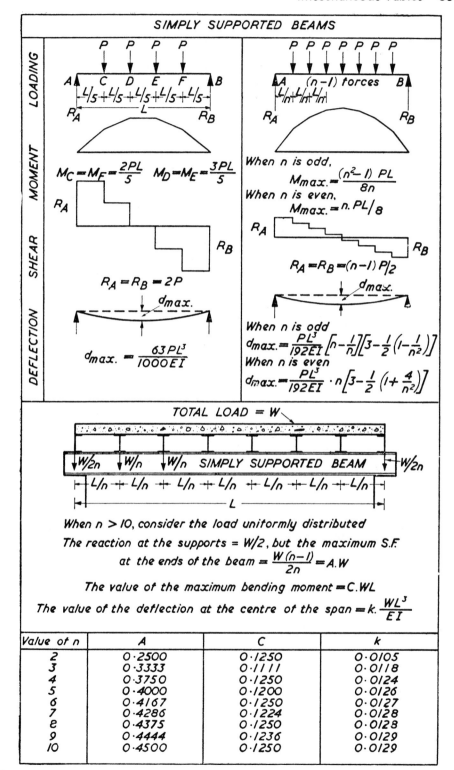

Fig. 27.5

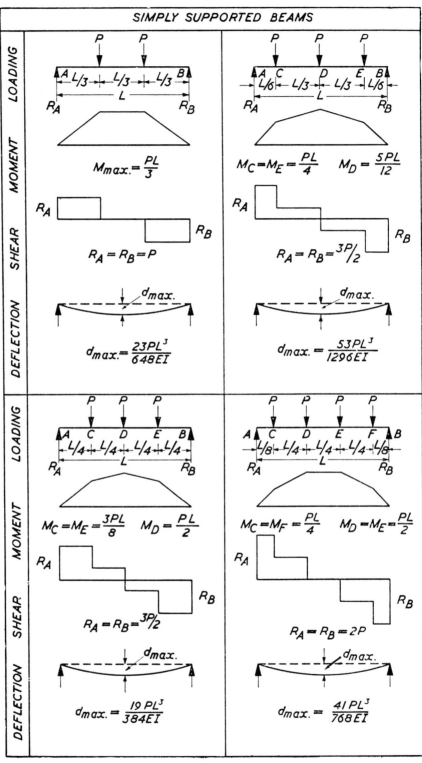

Fig. 27.6

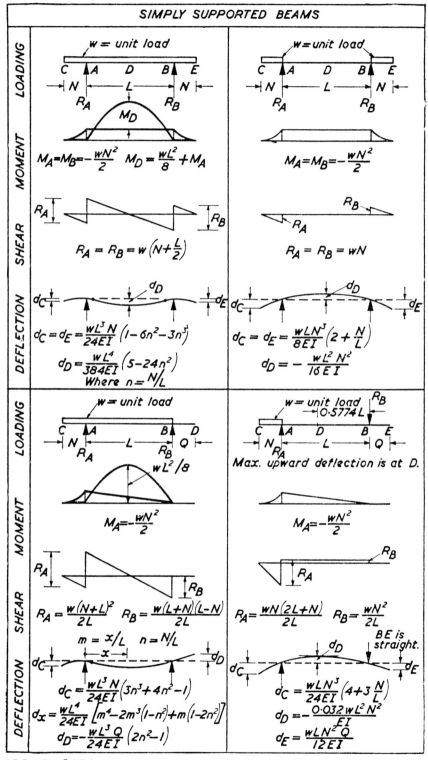

## SIMPLY SUPPORTED BEAMS

**LOADING** — $w$ = unit load

$C \quad A \quad D \quad B \quad E$

$M_A = M_B = -\dfrac{wN^2}{2} \quad M_D = \dfrac{wL^2}{8} + M_A$

$R_A = R_B = w\left(N + \dfrac{L}{2}\right)$

$d_C = d_E = \dfrac{wL^3 N}{24EI}\left(1 - 6n^2 - 3n^3\right)$

$d_D = \dfrac{wL^4}{384EI}\left(5 - 24n^2\right)$

Where $n = N/L$

---

**LOADING** — $w$ = unit load

$M_A = M_B = -\dfrac{wN^2}{2}$

$R_A = R_B = wN$

$d_C = d_E = \dfrac{wLN^3}{8EI}\left(2 + \dfrac{N}{L}\right)$

$d_D = -\dfrac{wL^2 N^2}{16EI}$

---

**LOADING** — $w$ = unit load

$wL^2/8$

$M_A = -\dfrac{wN^2}{2}$

$R_A = \dfrac{w(N+L)^2}{2L} \quad R_B = \dfrac{w(L+N)(L-N)}{2L}$

$m = x/L \quad n = N/L$

$d_C = \dfrac{wL^3 N}{24EI}\left(3n^3 + 4n^2 - 1\right)$

$d_x = \dfrac{wL^4}{24EI}\left[m^4 - 2m^3(1-n^2) + m(1-2n^2)\right]$

$d_D = -\dfrac{wL^3 Q}{24EI}\left(2n^2 - 1\right)$

---

**LOADING** — $w$ = unit load, $R_B$

$\leftarrow 0.5774L \rightarrow$

$C \quad A \quad D \quad B \quad E$

Max. upward deflection is at $D$.

$M_A = -\dfrac{wN^2}{2}$

$R_A = \dfrac{wN(2L+N)}{2L} \quad R_B = \dfrac{wN^2}{2L}$

$BE$ is straight.

$d_C = \dfrac{wLN^3}{24EI}\left(4 + 3\dfrac{N}{L}\right)$

$d_D = -\dfrac{0.032 \, wL^2 N^2}{EI}$

$d_E = \dfrac{wLN^2 Q}{12EI}$

Fig. 27.7  *See* §27.2.

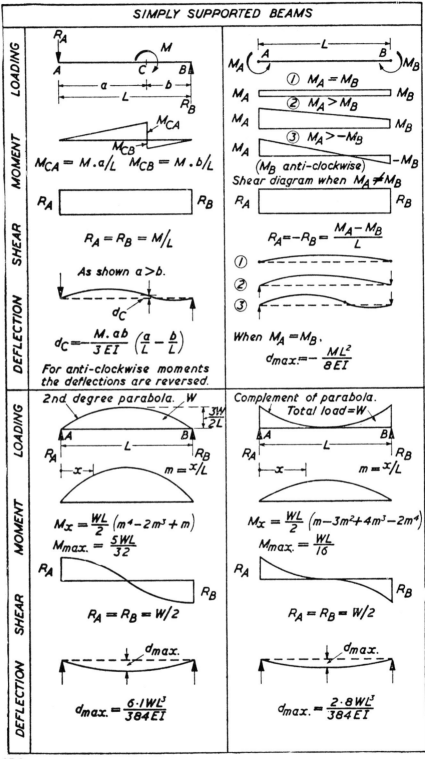

Fig. 27.8

## EQUAL SPAN CONTINUOUS BEAMS
## CENTRAL POINT LOADS

Moment = coefficient x W x L
Reaction = coefficient x W
where W is the Load on one span only and L is one span

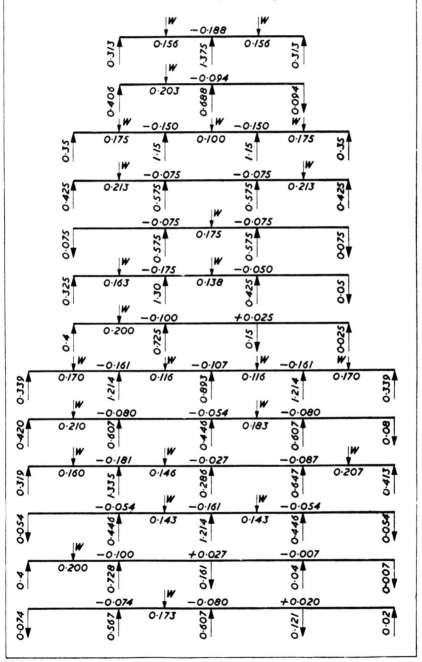

Fig. 27.9

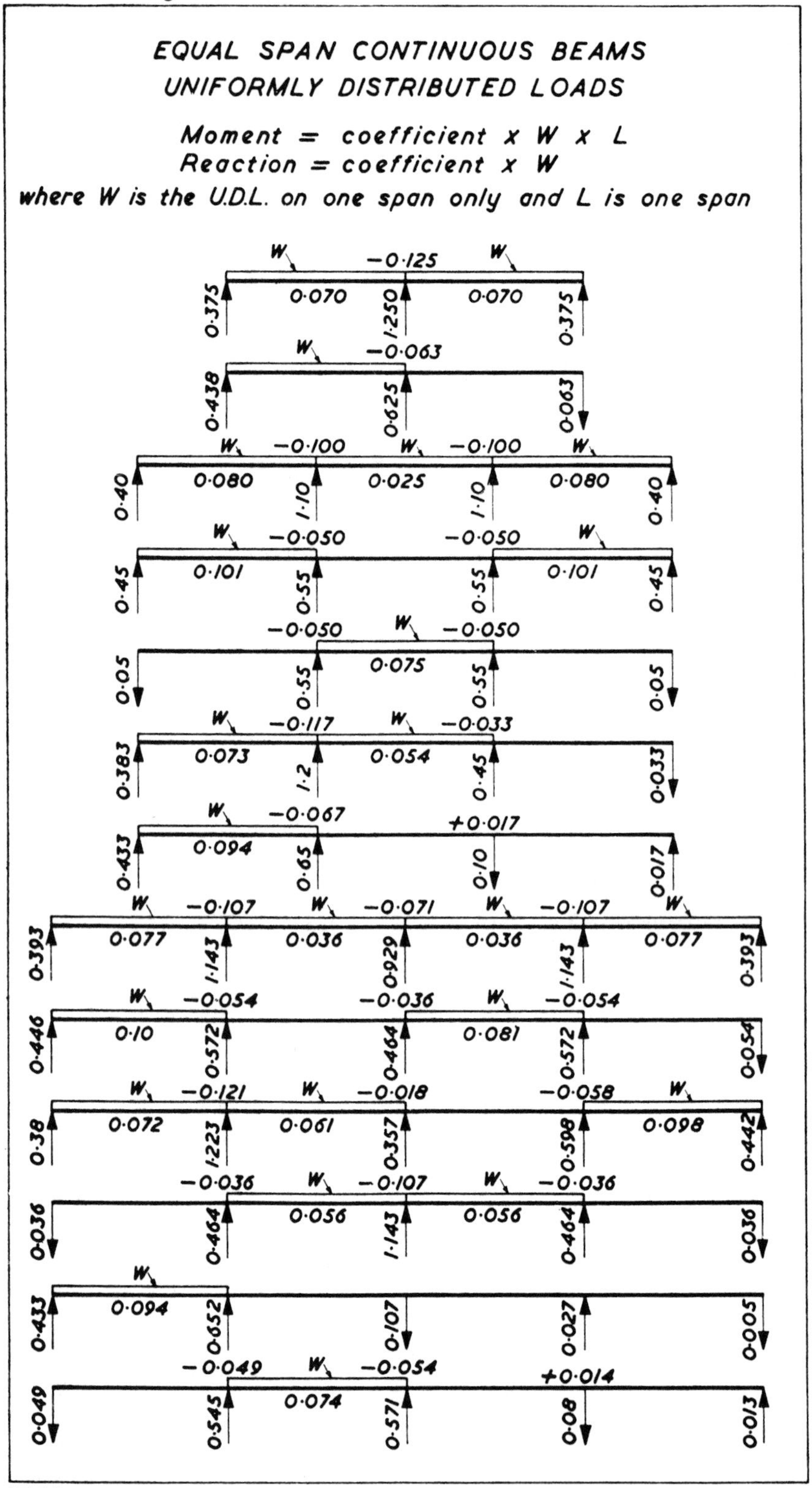

## EQUAL SPAN CONTINUOUS BEAMS
## UNIFORMLY DISTRIBUTED LOADS

Moment = coefficient x W x L
Reaction = coefficient x W

where W is the U.D.L. on one span only and L is one span

Fig. 27.10

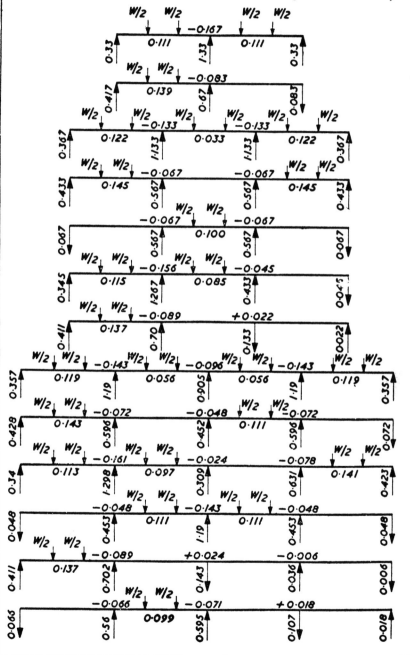

**EQUAL SPAN CONTINUOUS BEAMS**
**POINT LOADS AT THIRD POINTS OF SPANS**

Moment = coefficient x W x L
Reaction = coefficient x W
where W is the _total_ load on one span only & L is one span

Fig. 27.11

## 27.3 PERMISSIBLE LORRY OVERHANGS

Although the transporting of components is not usually the actual responsibility of a design engineer, it is reasonable for the transport manager or road haulage company to expect that the engineer has given some thought to delivery when setting the maximum size of components. Guidance is given in this section on the sizes which can be transported without special requirements. There are additional requirements, including requirements for Central London.

If the site to which the components are to be delivered is some distance from main roads, the engineer would be well advised to check at the design stage with the highway authorities through whose areas the vehicle will pass, for details of weight or height restriction of any bridges, or details of narrow sections of road and bends which might restrict the height, width or length of the acceptable load.

Where it is necessary to give notification to the police in each police area through which a vehicle will pass, at least two clear day's notice is required (excluding Saturdays, Sundays and bank holidays). Police area headquarters are a source of information on route guidance. Occasionally a special event (e.g. a County Show) will mean that a vehicle is re-routed or delayed.

*Weight*
Compared to some loads (e.g. steelwork) timber is not particularly heavy, and the type of vehicle used to transport timber components will usually be able to have a total laden weight of up to 38 000 kg.

*Width* (Fig. 27.12)
Providing the overhang on either side does not exceed 305 mm and the width of the load does not exceed 2.9 m there are no special requirements.

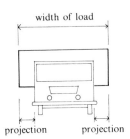

Fig. 27.12

If the projection on either side of an 'indivisible' load exceeds 305 mm or the width of the load exceeds 2.9 m police notification is required. If a load in this width category can be divided in width, it must be divided. If the width is in excess of 4.3 m, a written note must be given to the Secretary of State at the Department of Transport.

*Length* (Fig. 27.13)

Providing the overall length of a standard articulated vehicle does not exceed 15.5 m, or 17.33 m over the vehicle and load, there are no special requirements. If this length is exceeded then police notification is required and the driver must be accompanied by an attendant.

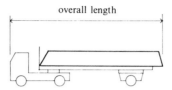

overall length

Fig. 27.13

In the case of an articulated vehicle specially constructed for abnormally long loads, providing the length overall the load and trailer (but not the drawing vehicle) does not exceed 18.3 m there are no special requirements. If this length is exceeded, then police notification is required and the driver must be accompanied by an attendant.

In the case of a combination of vehicles and load, providing the overall length including projections does not exceed 25.9 m there are no special requirements. If this length is exceeded, then police notification is required and the driver must be accompanied by an attendant.

*Rear overhang* (Fig. 27.14)

Providing the overhang does not exceed 1.07 m there are no special requirements.

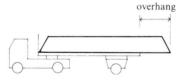

overhang

Fig. 27.14

If the overhang exceeds 1.07 m the overhang must be made 'clearly visible'.

If the overhang exceeds 1.83 m standard end marker boards or a reflective marker are required.

If the overhang is over 3.05 m standard end marker boards are required, the police must be notified and the driver must be accompanied by an attendant.

If the overhang is over 5.1 m additional side marker boards are required within 3.6 m of the normal marker boards.

Marker boards must be illuminated after lighting-up time.

*Front overhang* (Fig. 27.15)
Providing the overhang does not exceed 1.83 m there are no special requirements.

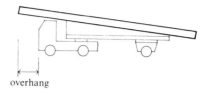

overhang

Fig. 27.15

If the overhang exceeds 1.83 m standard end and side marker boards are required and the driver must be accompanied by an attendant.

If the overhang exceeds 3.05 m standard end and side marker boards are required, the police must be notified and the driver must be accompanied by an attendant.

If the overhang is over 4.5 m additional side marker boards are required within 2.4 m of the normal marker boards.

*Height*
The minimum clearance height under motorway bridges is 5.03 m. On minor roads the clearance is likely to be much less and, if a loaded lorry requires a clearance approaching 3.6 m and will be travelling on minor roads, it is advisable to contact the highway authorities or police forces through which the load will pass for details of clearance under all over-bridges. Very occassionally a bridge with much less clearance is encountered.

## 27.4   CONVERSION FACTORS

**Table 27.1**

| To convert | to | multiply by |
|---|---|---|
| inches | millimetres | 25.4 |
| feet | metres | 0.304 8 |
| millimetres | inches | 0.039 37 |
| metres | feet | 3.281 |
| square inches | square millimetres | 645.16 |
| square feet | square metres | 0.092 903 |
| square millimetres | square inches | 0.001 55 |
| square metres | square feet | 10.763 9 |
| cubic inches | cubic millimetres | 16 387.064 |
| cubic feet | cubic metres | 0.028 316 8 |
| cubic millimetres | cubic inches | 0.000 061 |
| cubic metres | cubic feet | 35.31 |
| pounds (mass) | kilogrammes | 0.453 592 4 |
| tons (long) | tonnes | 1.016 05 |
| kilogrammes | pounds (mass) | 2.204 62 |
| tonnes | tons (long) | 0.984 21 |
| pounds (mass) per linear foot | kilogrammes per linear metre | 1.488 2 |
| pounds (mass) per square inch | kilogrammes per square millimetre | 0.000 703 |
| pounds (mass) per square foot | kilogrammes per square metre | 4.882 4 |
| pounds (mass) per cubic foot | kilogrammes per cubic metre | 16.018 5 |
| kilogrammes per linear metre | pounds (mass) per linear foot | 0.672 |
| kilogrammes per square millimetre | pounds (mass) per square inch | 1 422.34 |
| kilogrammes per square metre | pounds (mass) per square foot | 0.204 8 |
| kilogrammes per cubic metre | pounds (mass) per cubic foot | 0.062 4 |
| pounds (force) | newtons | 4.448 22 |
| kilogrammes (force) | newtons | 9.806 65 |
| newtons | pounds (force) | 0.244 8 |
| newtons | kilogrammes (force) | 0.101 97 |
| pounds (force) per square inch | newtons per square millimetre | 0.006 894 76 |
| pounds (force) per square inch | kiloponds (force) per square centimetre | 0.070 31 |
| pounds (force) per square foot | newtons per square metre | 47.88 |
| newtons per square millimetre | pounds (force) per square inch | 145.039 |
| newtons per square metre | pounds (force) per square foot | 0.020 89 |
| pounds (force) per linear foot | newtons per metre run | 14.594 6 |
| newtons per metre run | pounds (force) per linear foot | 0.068 52 |
| kiloponds (force) per square centimetre | pounds (force) per square inch | 14.223 |
| kiloponds (force) per square centimetre | newtons per square millimetre | 0.098 066 |
| kiloponds (force) | newtons | 9.806 65 |
| degrees Celsius | degrees Fahrenheit | multiply by $\frac{9}{5}$ then add 32 |
| degrees Fahrenheit | degrees Celsius | deduct 32 then multiply by $\frac{5}{9}$ |
| British thermal units per hour | watts | 0.293 1 |
| British thermal units per square foot per hour per degree Fahrenheit | watts per square metre per degree Celsius | 5.678 |
| British thermal units inch per square foot per hour per degree Fahrenheit | watts per metre per degree Celsius | 0.144 2 |
| Square foot hour degree Fahrenheit per British thermal unit | square metre degree Celsius per watt | 0.176 1 |
| Square foot hour degree Fahrenheit per British thermal unit per inch | metre degree Celsius per watt | 6.933 5 |
| Petrograd 'Standard' (timber measure) | cubic metres | 4.671 |

# Bibliography and References

## BOOKS

Bleich, F. *Buckling Strength of Metal Structures.* Engineering Societies Monograph. McGraw-Hill, New York, 1952.

Booth, L. G. and Reece, P. O. *The Structural Use of Timber. A Commentary on CP 112.* Spon, London, 1967.

Hanson, H. J. *Timber Engineers' Handbook.* Wiley, New York, 1948.

Kleinlogel, A. *Rigid Frame Formulas.* Crosby Lockwood, London, 1952.

Roark, R. J. *Formulas for Stress and Strain*, 4th edn. McGraw-Hill, New York, 1956.

*Steel Designers' Manual* 4th edn. (Revised.) Prepared by the Constructional Steel Research and Development Organisation. Crosby Lockwood Staples, London, 1982.

Timber Engineering Company, *Timber Design and Construction Handbook.* F. W. Dodge Corporation, New York, 1956.

Tuma, J. J. and Munshi, R. K. *Advanced Structural Analysis.* McGraw-Hill, New York, 1971.

## PUBLICATIONS OF THE SWEDISH FINNISH TIMBER COUNCIL
–about Swedish and Finnish, Redwood and Whitewood.

*The Sawn Timber and Products.* Notts. 1982.

*Stress Graded to BS 4978.* Notts. 1984.

*Finger Jointed.* Notts. 1983.

*Design and Detail of Timber Stud Walls.* Notts. 1984.

*The Properties* (excluding strength properties). Notts. 1979.

*Performance in Fire.* Notts. 1981.

*Various Span Tables.* Notts. 1984.

*Principles of Timber Framed Construction.* Notts. 1983.

## PUBLICATIONS OF THE COUNCIL OF FOREST INDUSTRIES OF BRITISH COLUMBIA

*Fir Plywood Folded Plate Design (PMBC).* Vancouver, 1967.
*Fir Plywood Diaphragms.* Vancouver, 1973.
*Fir Plywood Web Beam Design.* Vancouver, 1970.
*Fir Plywood Stressed Skin Panels.* Vancouver, 1971.
*Canadian Fir Plywood Data for Designers.* London (undated).
*Plywood Design Fundamentals.* Vancouver, 1978.
*Plywood Construction Manual.* Vancouver, 1976.
*Canadian Fir Plywood Stiffened Panel Design (PMBC).* Vancouver, 1967.
*Plywood Barrel Vaults (PMBC).* Vancouver, 1964.
*Plywood Panel Arches (PMBC).* Vancouver, 1965.
*Canadian COFI Exterior Plywood: Guide to Use.* London, 1981.
*Nailed Plywood Beams.* London, 1977.
*Stress Graded Canadian Timber.* London, 1981.
*CLS Hem-Fir.* London, 1981.
*CLS Kiln-dried Spruce-Pine-Fir.* London, 1982.
(Several COFI publications will probably be updated in 1984.)


## PUBLICATIONS OF THE AMERICAN PLYWOOD ASSOCIATION

*Guide to Plywood Grades.* Tacoma (undated).
*Plywood Technical Data.* Tacoma. (Binder of several publications.)
*American Plywood for Floors, Walls and Roofs.* London, 1981.


## PUBLICATIONS OF THE BRITISH WOOD PRESERVING ASSOCIATION

### BWPA Codes on preservation

C1 (1975) *Timber for use as packing in cooling towers.*
C2 (1975) *Timber for use permanently or intermittently in contact with sea or fresh water.*
C3 (1975) *Fencing.*
C4 (1975) *Agricultural and horticultural timbers.*
C5 (1975) *External joinery and external fittings not in ground contact.*
C6 (1975) *External cladding.*
C7 (1979) *Prefabricated timber buildings for use in termite infested areas.*
C8 (1979) *Constructional timbers.*
C9 (1982) *Timber frame housing. (The constructional frame of external walls.)*
C10 (1984) *Treatment of hardwood external joinery with organic solvent preservatives by double vacuum.*
C122 (1982) *The application of organic-solvent wood preservatives by double vacuum.*

## PUBLICATIONS OF THE FIBRE BUILDING BOARD ASSOCIATION

Product Data Sheets
  PD/2 *Tempered hardboard.*
  PD/3 *Medium board.*
  PD/5 *Bitumen-impregnated insulating board.*
Design Data Sheets
  DD/1 *Building Regulations.*
  DD/2 *Fire test performance.*
Applications Data Sheets
  AD/3 *Partitions and internal linings.*
Technical Bulletins
  TB/001 *Bitumen-impregnated insulating board sheathing for timber framed houses.*
  TB/002 *Medium board sheathing in timber framed buildings.*
Sitework recommendations
  SR/6 *Recommended fasteners.*
*Strength properties and structural use of tempered hardboard.* London, 1979.
*Design guide: The structural use of tempered hardboard.* London, 1980.

## PUBLICATIONS OF THE BUILDING RESEARCH ESTABLISHMENT

### Published by H.M. Stationery Office, London

Bulletin No. 50. *The Strength Properties of Timber*, 1969.
Bulletin No. 53. *Grade Stresses for Structural Laminated Timber.* 1970.
Bulletin No. 54. *The Resistance of Timber to Impregnation with Creosote.* 1971.
*Handbook of Softwoods.* 1957.
*Handbook of Hardwoods.* 1972.
Brazier, J. D. and Laidlaw, R. A. The implications of using inorganic salt flame-retardant treatments with timber. *BRE Information*, December, 1974.
Laidlaw, R. A. Influence of preservatives and wax on assembly gluing of dipped softwoods. *BRE Information*, March, 1974.
Laidlaw, R. A. and Paxton, B. H. *The Effect of Moisture Contents and Wood Preservatives on the Assembly Gluing of Timber.* Current Paper No. 54. 1974.
Choice of Glues for Wood. *BRE Digest* No. 175. 1975.
*Timber Drying Manual*, 1974

### Obtainable from the Princes Risborough Laboratory

Curry, W. T. *Mechanical Stress Grading of Timber.* Timberlab Paper No. 18. 1969.
*Duration of Load Factors in Timber Design.* Information Sheet No. 4. 1975.

IP1/83 *Depth factor adjustments in the determination of characteristic bending stresses for visually stress graded timber.* A. R. Fewell and W. T. Curry.

IP28/80 *Relations between the moduli of elasticity of structural timber in bending.* A. R. Fewell.

IP 16/79 *The effect of dimensional tolerances on machine stress graded timber.* M. F. Mouldsworth. The determination of softwood strength properties for grades, strength classes and laminated timber for BS 5268: Part 2.

Technical Note No. 31. *Gluing Preservative-Treated Wood.* 1975.

Technical Note No. 40. *The Natural Durability Classification of Timber.* 1969.

## PUBLICATIONS OF THE TIMBER RESEARCH AND DEVELOPMENT ASSOCIATION, HIGH WYCOMBE

Ashton, L. A. *Fire and Timber in Modern Building Design.* 1970. (Revised 1977.)

Chugg, W. A. and James, P. E. *The Gluability of Hardwoods for Structural Purposes.* TRADA Research Report C/RR/22. 1965.

Kay, J. A. and Sabatini, M. *Racking Tests on Roofing Panels.* Test Record E/TR/27. 1966.

*Timber Frame Housing–Structural Recommendations.*

Various span charts and span tables based around BS 5268:Part 2

Wood Information Sheets on:
   *Timber bridges*
   *Structural use of hardwoods*
   *Calculation for the racking resistance of timber framed walls*
   *Timber frame separating walls*
   *Flame retardant treatments for timber*
   *Mechanical fasteners for structural timberwork.*

## PUBLICATIONS OF THE BRITISH STANDARDS INSTITUTION, LONDON

### General series

BS 144:1973. *Coal tar creosote for the preservation of timber.*

BS 373:1957. *Testing small clear specimens of timber.*

BS 449:Part 2:1969. *The use of structural steel in building.*

BS 476:Part 4:1970. *Non-combustibility test for materials.*

BS 476:Part 5:1979. *Method of test for ignitability.*

BS 476:Part 6:1981. *Method of test for fire propagation for products.*

BS 476 : Part 7 : 1971. *Surface spread of flame test for materials.*
BS 476 : Part 8 : 1972. *Test methods and criteria for the fire resistance of elements of building construction.*
BS 565 : 1972. *Glossary of terms relating to timber and woodwork* (being rewritten as BS 6100 : Part 4).
BS 648 : 1964. *Schedule of weights of building materials.*
BS 913 : 1973. *Wood preservation by means of pressure creosoting.*
BS 1142 : Part 1 : 1971. *Fibre building boards. Methods of test.*
BS 1142 : Part 2 : 1971. *Fibre building boards. Medium board and hardboard.*
BS 1142 : Part 3 : 1972. *Fibre building boards. Insulating board (softboard).*
BS 1186 : Part 1 : 1984. *Quality of timber in manufactured joinery.* (There is a possibility of delay.)
BS 1186 : Part 2 : 1971. *Quality of workmanship in joinery* (under review).
BS 1202 : Part 1 : 1974. *Steel nails.*
BS 1203 : 1979. *Synthetic resin adhesives for plywood.*
BS 1204 : Part 1 : 1979. *Synthetic resin adhesives for wood. Gap filling.*
BS 1204 : Part 2 : 1979 *Synthetic resin adhesives for wood. Close-contact adhesives.*
BS 1210 : 1963. *Wood screws.*
BS 1282 : 1975. *Classification of wood preservatives and their methods of application.*
BS 1444 : 1970. *Cold-setting casein adhesive powders for wood.*
BS 1579 : 1960. *Connectors for timber.*
BS 2482 : 1981. *Timber scaffold boards.*
BS 2989 : 1982. *Continuously hot-dip zinc coated and iron-zinc alloy coated steel.*
BS 3051 : 1972. *Coal tar creosotes for wood preservation (other than creosotes to BS 144).*
BS 3452 : 1962. *Copper/chrome water-borne wood preservatives and their application.*
BS 3842 : 1965. *Treatment of plywood with preservatives.*
BS 4072 : 1974. *Wood preservation by means of water-borne copper/chrome/ arsenic compositions.*
BS 4169 : 1970. *Glued-laminated timber structural members* (incl. AMD 768 and AMD 3453).
BS 4190 : 1967. *ISO metric black hexagon bolts, screws and nuts.*
BS 4261 : 1968. *Glossary of terms relating to timber preservation.*
BS 4471 : Part 1 : 1978. *Dimensions for softwood. Basic sections.*
BS 4471 : Part 2 : 1971. *Dimensions for softwood. Small resawn sections.*
BS 4978 : 1973. *Timber grades for structural use* (being rewritten).
BS 5268 : Part 2 : 1984. *Structural use of timber. Code of practice. Permissible stress design, materials and workmanship* (formerly CP 112 : Part 2).
BS 5268 : Part 3. *Trussed rafter roofs* (under preparation to replace CP 112 : Part 3).

BS 5268 : Part 4 : Section 4.1 : 1978. *Fire resistance of timber structures.*
BS 5268 : Part 5 : 1977. *Preservative treatments for constructional timber* (under review).
BS 5291 : 1984. *Finger joints in structural softwood.* (There is a possibility of delay.)
BS 5400 : Part 3 : 1982. *Code of practice for design of steel bridges.*
BS 5450 : 1977. *Sizes of hardwoods and methods of measurement.*
BS 5502 : Part 1 : Section 1.2 : 1980. *Code of practice for the design, construction and loading of buildings and structures for agriculture.*
BS 5534 : Part 1 : 1978. *Code of practice for slating and tiling* (incl. AMD 2734 and AMD 3554).
BS 5589 : 1978. *Code of practice for preservation of timber.*
BS 5669 : 1979. *Wood chipboard and methods of test for particle board.*
BS 5707 : Parts 1, 2 and 3. *Solutions of wood preservatives in organic solvents.*
BS 5756 : 1980. *Tropical hardwoods graded for structural use.*
BS 6178 : Part 1 : 1982. *Joist hangers for building into masonry walls of domestic dwellings.*
BS 6446 : 1984. *The manufacture of glued structural components of timber and wood based panel products.* (There is a possibility of delay.)

**Codes of Practice** (still with CP number)

CP 3 : Chapter V : Part 1 : 1967. *Loading. Dead and imposed loads* (incl. AMD 141, AMD 587, AMD 1024). (Being rewritten as BS 6399 : Part 1.)
CP 3 : Chapter V : Part 2 : 1972. *Loading. Wind loads.*
CP 112 : Part 1 : 1967. *The structural use of timber. Imperial units* (now out of print).
CP 112 : Part 2 : 1971. *The structural use of timber. Metric units* (replaced by BS 5268 : Part 2).
CP 112 : Part 3 : 1973. *Trussed rafters for roofs of dwellings* (to be replaced by BS 5268 : Part 3).

**OTHER PUBLICATIONS**

*The Building Regulations 1976* including the First, Second and Third Amendments. HMSO, London.
*The Building Standards (Scotland) Regulations 1981.* HMSO, London.
*Building Regulations (Northern Ireland) 1977.* HMSO, London.
Chipboard Promotion Association. *Technical Manual.* Surrey, 1981.
*Corrosion of Metals by Wood.* National Physical Laboratory. Department of Industry and Central Office of Information. 1979.
Curry, W. T. and Brown, B. Can 19 mm floorboards be used with joists at 600 mm centres? *Timber Trades Journal*, 15 March, 1969.

*Fire and Structural Use of Timber in Buildings.* Ministry of Technology and Fire Officers' Committee, Joint Fire Research Organisation. Symposium No. 3. HMSO, London, 1970.

Foschi, R. O. *Stress Distribution in Plywood Stressed Skin Panels with Longitudinal Stiffeners.* Canadian Forest Service Publication No. 1261. Department of Fisheries and Forestry, Ottawa, 1969.

*Handbook of the National House-Building Council.* NHBC, London. Also, Practice Note 5 (1982), *Timber Framed Dwellings.* (Revised 1983.)

*Instructions for Marking and Sorting T-Timber.* T-Timber Association, Stockholm, 1966 (English translation available from Swedish Finnish Timber Council).

*Standard Grading Rules for Canadian Lumber.* Canadian National Lumber Grades Authority, Ottawa, 1980.

*Export R List Grading and Dressing Rules.* Pacific Lumber Inspection Bureau. Seattle. 1951 (Revised 1971).

Structural finger joints. Chapter 9 in *NKB13. Joints in Timber Structures.* Nordic Committee for Building Regulations, Stockholm, 1970 (English translation available from Swedish Finnish Timber Council).

*Wind Loading Handbook.* BRE. HMSO. 1976.

# Index